AN INTRODUCTION TO CLIFFORD ALGEBRAS AND SPINORS

An Introduction to Clifford Algebras and Spinors

Jayme Vaz, Jr.
IMECC, Universidade Estadual de Campinas, Campinas, SP, Brazil

Roldão da Rocha, Jr.
CMCC – Universidade Federal do ABC, Santo André, SP, Brazil

Great Clarendon Street, Oxford, OX2 6DP,
United Kingdom

Oxford University Press is a department of the University of Oxford.
It furthers the University's objective of excellence in research, scholarship,
and education by publishing worldwide. Oxford is a registered trade mark of
Oxford University Press in the UK and in certain other countries

First published 2016
First published in paperback 2019

Published in the United States of America by Oxford University Press
198 Madison Avenue, New York, NY 10016, United States of America

British Library Cataloguing in Publication Data
Data available

Library of Congress Cataloging in Publication Data
Data available

ISBN 978-0-19-878292-6 (Hbk.)
ISBN 978-0-19-883628-5 (Pbk.)

To Maria Clara and Liliane
J.

To my Family
R.

Preface

In 1878 William Kingdom Clifford published an article in the *American Journal of Mathematics* entitled 'Applications of Grassmann's Extensive Algebra' (Clifford, 1878). Clifford presented a new mathematical structure which he called 'Geometric Algebra', but which is now usually known as Clifford Algebra. Some years earlier, in 1873, Clifford had published in *Proceedings of the London Mathematical Society* an article called 'Preliminary Sketch of Biquaternions', where he presented the first rudiments of his ideas. These later evolved in 1876 in two unpublished manuscripts – called 'Further note on so-called biquaternions' and 'On the classification of geometric algebra', respectively (published posthumously as a single article entitled 'On the Classification of Geometric Algebras' (Clifford, 1882)) – culminating in the ideas presented in the 1978 article. Briefly, what Clifford accomplished was the synthesis of two apparently dissociated mathematical structures: the quaternions of Sir William Rowan Hamilton, and the algebra of extensions (*Ausdehnungslehre*) of Hermann Grassmann. We can trace the origin of the Clifford algebras to the efforts to geometrically represent a complex number.

The idea of unifying geometric and algebraic operations was first advocated by Gottfried Leibniz in 1679, in a letter to Christiaan Huygens (this letter was later published in 1833). This idea was manifested for complex numbers in terms of the Argand, or Argand–Gauss plane, which has been given the name tribute to the results published on this topic in 1806 by Jean-Robert Argand and then in 1831 by Carl Friedrich Gauss. Meanwhile, the first to succeed in representing the complex numbers in a plane was the Norwegian Caspar Wessel, who presented his results in 1797 to the Royal Danish Academy of Sciences and Letters, in his work '*Om Directionens analytiske Betegning*'. This work was published in 1799 but remained in obscurity until it was translated into French in 1897. The success of the geometrical representation of complex numbers led Hamilton to search for a generalisation of this approach to three-dimensional space. After some years of failed attempts, Hamilton succeeded in 1843 with the discovery of quaternions.[1] Afterwards, Hamilton (and his followers) dedicated years to developing the theory and applications of quaternions. Meanwhile, in 1844 Grassmann published his outstanding work, *Die lineale Ausdehnungslehre*, (Grassmann, 1844, 1894). In this work – influenced by ideas from his father, Justus Grassmann – he introduced an algebraic system based on a geometric product that, because of its innovative and abstract character, proved to be difficult to understand at the time. In an attempt to facilitate the knowledge of his work, Grassmann presented in 1862 a new version of *Die lineale Ausdehnungslehre*, which he called *Die Ausdehnungslehre: Vollstandig und in strenger Form bearbeitet* (Grassmann, 1862). Nevertheless, Grassmann's work

[1]The present-day formalism of vector algebra was extracted from the quaternion product of two vectors, by Josiah Willard Gibbs in 1901.

remained obscure for but a few like Clifford. Interestingly, in 1844, after the publication in 1833 of Leibniz's 1679 letter to Huygens, the *Jablonowskischen Gesellschaft der Wissenschaft* offered an award to those who developed Leibniz's ideas. The prize was granted to Grassmann in 1846 for taking Leibniz's ideas into the realm of the algebraic system presented in *Die lineale Ausdehnungslehre* and was later published as a separate work (Grassmann, 1847). Yet, this award was not enough to draw the attention of the majority of the scientific community to Grassmann's work.

More information on the work of Hamilton, Grassmann and other important names in the history of the vector analysis can be found in, for example, the book by Crown (1994). Clifford's major achievement in 1878 was essentially introducing a quaternion framework into Grassmann's algebra of extensions, thus obtaining a system naturally adapted to the orthogonal geometry of an arbitrary space: in this system, quaternions represent merely a very particular case of a Clifford algebra. Furthermore, one of the many (at least ten) products defined by Grassmann is the so-called 'stereometrisches Produkt', which implements quaternion multiplication for three generators, and is thus itself a Clifford product. Unfortunately, Clifford was unable to continue his work as he died in the following year, at the age of 33. However, it was not long before the Clifford algebras were reinvented. In 1880, Lipschitz studied representations of rotations by complex numbers and quaternions, and their generalisations to high dimensions led to the Clifford geometric algebra and to the group Spin.

Many scientists have contributed to the development of the Clifford algebras. One of the most important is certainly Élie Cartan. As well as making other contributions, he described Clifford algebras as algebras of matrices and found the periodicity 8 of these algebras (the periodicity theorem). Cartan introduced in 1913 the concept of the spinor, and in 1938 the concept of the pure spinor. Although the term 'spinor' had been coined by Paul Ehrenfest in the 1920s, the intrinsic concept of a spinor is much older than that, as a spinor is a special linear structure, which had been studied (and had been used in civil engineering for calculating statics) long before Ehrenfest used it in quantum theory.

It was through the concept of the spinor that Clifford algebras had a decisive presence in science. It was prominently Wolfgang Pauli who introduced the concept of spin to physics, first to explain, for example, the Zeeman effect, and later to formulate the exclusion principle (his Nobel-prize winning work). Paul Dirac showed in 1928 that the main equation in relativistic quantum mechanics is written in terms of a Clifford algebra and that the electron is described by a spinor, or rather, by a spinor field (Dirac, 1928). Cornelius Lanczos rewrote Dirac's equation in terms of quaternions (Gsponer and Hurni, 1994) in 1929, and Gustave Juvet and Fritz Sauter in 1930 replaced column spinors by square matrix spinors, in minimal left ideals, where only the first column contained non-zero elements (Lounesto, 2001*a*). Marcel Riesz in 1947 was the first one to consider spinors as elements of a minimal left ideal in a Clifford algebra. Thereafter, Feza Gürsey (1956, 1958) expressed the Dirac equation in terms of 2×2 quaternionic matrices. After Kustaanheimo presented the spinor regularisation of Kepler motion in 1964, proposing the so-called Kustaanheimo–Stiefel transformation, David Hestenes (1966, 1967) reformulated the Dirac theory in this context, providing new aspects and insights. Since then, a variety of other applications of Clifford algebras have been found, not only in physical sciences, but also in engineering and computation (Doran and Lasenby, 2003; Baylis, 1996; Dorst, Fontijne, and Mann, 2007; Perwass, 2008).

This book is divided into six chapters. In the first one, a review of the main concepts and results from linear algebra is presented, as these are essential in order to develop the subsequent ideas in this book. Then, tensor algebra is introduced, as it is required for the definitions of exterior algebras and Clifford algebras. The second chapter is dedicated to the presentation of exterior algebras and Grassmann algebras. Although some authors use these terms as synonyms, that will not be the case here. The exterior algebra is understood in this book as a structure devoid of a metric, whereas Grassmann algebras are structures constructed on a vector space endowed with a metric. In this chapter, some fundamental concepts, such as the exterior product, the contraction, and the quasi-Hodge isomorphism, and the Hodge isomorphism are introduced. Clifford algebras are then defined in the third chapter, together with their properties. Three different definitions of real Clifford algebras are introduced, emphasising the number of excellent features possessed by Clifford algebras. In the fourth chapter, theorems about the structure of Clifford algebras are presented; these theorems are later used for the classification and representation of the Clifford algebras in terms of matrix algebras. Moreover, we introduce procedures for explicitly building these matrix representations, as these are important when it comes to computations in physics and other applications. The groups associated with the Clifford algebras are the objects of discussion in the fifth chapter. The groups Pin and Spin are comprehensively examined. In addition, the Lie algebras associated with those groups are discussed and some of their applications provided. In particular, conformal transformations and twistors are introduced. The sixth chapter is dedicated to the study of spinors. Three different definitions of spinors are presented, as well as the properties associated with each definition. In addition, the relations between the different definitions are examined. The so-called pure spinors are also introduced and the triality principle and the Penrose flagpole are discussed. A detailed study of the so-called Weyl spinors, which form the basis of the Penrose and Rindler formalism (Penrose and Rindler, 1984), concludes this chapter. Finally, in an appendix, the Lorentz transformations and the classical counterparts of dotted/undotted Weyl spinors, the Penrose flagpole and supersymmetry algebra are presented, in the context of the Van der Waerden framework.

This book is based on lecture notes written by Jayme Vaz Jr for two courses on Clifford algebras in 1999 and 2005 at the Institute of Mathematics, Statistics and Scientific Computation (IMECC) at the University of Campinas (Unicamp). From 2008 to 2015 Roldão da Rocha Jr used and improved the lecture notes, in particular for the courses 'Clifford Algebras and Spinors', 'Spinors in Hilbert Spaces', and 'Quantum Field Theory' for graduate programs at the Center of Mathematics, Computation and Cognition and at the Center of Natural and Human Sciences at ABC Federal University, Santo André, Brazil, in the process of completing and ameliorating the manuscript, as well as adding topics. Those who are familiar with the Clifford algebras and their applications certainly will realise that an important topic, the Dirac operators, which play a significant role in many areas of modern mathematics and physics, is missing. The reason for its absence is that, although a discussion of the Dirac operators would have been welcome in the course, it was beyond the scope of these lecture notes, which aimed to provide material for a 30-hour course. For a study of Dirac operators, we recommend Gilbert and Murray (1991) book, which contains a complete introduction to Clifford algebras as well as to representations of the Spin group. Moreover, it defines Dirac-type operators in analysis and geometry, as well as providing their applications. Examples of Dirac-type operators include, for instance, the Hodge–Dirac operator on a Riemannian manifold, the Laplacian in Euclidean spaces, and the Atiyah–Singer–Dirac operator on a spin manifold. Therefore, the local Atiyah–Singer index theorem can be demonstrated. Moreover, Gilbert and Murray (1991) book offers a systematic and comprehensive presentation of spinor fields, Dirac operators, and the Atiyah–Singer index theorem.

Acknowledgements

We would like to thank all colleagues and friends who in some way or other helped us in the preparation of this work. In particular, we wish to thank R. T. Cavalcanti, R. A. Mosna, E. C. de Oliveira, W. A. Rodrigues Jr. and M. A. Traesel, A. Y. Martinho and R. Lopes, for many important discussions and suggestions. We also thank S. Adlung from OUP for all support in the publishing process, and the referees for the invaluable advices and suggestions. Finally, we acknowledge the support of CNPq, Capes and Fapesp research agencies over the years through which this work was accomplished.

Contents

1 Preliminaries

In this chapter, we briefly review essential concepts regarding vector spaces and their associated dual vector spaces. Moreover, the tensor algebra, including its universal character, is introduced, with some computational examples as well. Moreover, we fix the notation to be used throughout the text. Some demonstrations can be also found in the standard literature – for a good reference, see for example the work by Hoffman and Kunze (1971), Birkhoff and MacLane (1997), or Kostrykin and Manin (1997).

1.1 Vectors and Covectors

The concept of *vector space* is fundamental throughout the text, and so deserves to be discussed in detail. A vector space V over a field $\mathbb{K}$ (from here on, elements of a field are called *scalars*) is a set of elements called vectors, endowed with an additive operation $+: V \times V \to V$ and also equipped with a product by scalars $.\ : \mathbb{K} \times V \to V$, satisfying the following properties:

(1) With each pair of vectors $\mathbf{u}, \mathbf{v} \in V$, an element $\mathbf{u} + \mathbf{v} \in V$ is associated, denominated the sum of $\mathbf{u}$ and $\mathbf{v}$, with the properties:
(a) commutativity: $\mathbf{u} + \mathbf{v} = \mathbf{v} + \mathbf{u}$, for all $\mathbf{u}, \mathbf{v} \in V$;
(b) associativity: $\mathbf{u} + (\mathbf{v} + \mathbf{w}) = (\mathbf{u} + \mathbf{v}) + \mathbf{w}$, for all $\mathbf{u}, \mathbf{v}, \mathbf{w} \in V$;
(c) there exists a vector denoted by 0, named the null vector, such that $\mathbf{u} + 0 = \mathbf{u}$ for all $\mathbf{u} \in V$;
(d) for each $\mathbf{u} \in V$, there exists a unique vector denoted by $(-\mathbf{u})$ such that $\mathbf{u}+(-\mathbf{u}) = 0$.

(2) For each pair $(a, \mathbf{u}) \in \mathbb{K} \times V$, there exists a vector denoted by $a.\mathbf{u} \in V$, satisfying the properties:
(a) associativity: $a.(b.\mathbf{u}) = (ab).\mathbf{u}$, for all $a, b \in \mathbb{K}$ and $\mathbf{u} \in V$, where ab is the product of a and b in $\mathbb{K}$;
(b) $1.\mathbf{u} = \mathbf{u}$ for all $\mathbf{u} \in V$, where 1 denotes the unity of $\mathbb{K}$;
(c) distributivity: $a.(\mathbf{u} + \mathbf{v}) = a.\mathbf{u} + a.\mathbf{v}$, for all $a \in \mathbb{K}$, and all $\mathbf{u}, \mathbf{v} \in V$;
(d) distributivity: $(a + b).\mathbf{u} = a.\mathbf{u} + b.\mathbf{u}$, for all $a, b \in \mathbb{K}$, and $\mathbf{u} \in V$.

This text concerns vector spaces V over the field $\mathbb{R}$ of real numbers and over the field $\mathbb{C}$ of complex numbers as well and, in some specific cases, we consider modules over the quaternions $\mathbb{H}$.[1] Some attention is needed when $\mathbb{H}$ is regarded, to circumvent

[1]The quaternions $\mathbb{H}$ are defined by the set of elements $\{a + b\boldsymbol{i} + c\boldsymbol{j} + d\boldsymbol{k} \mid a, b, c, d \in \mathbb{R}\}$, with the associative product defined as $\boldsymbol{ij} = -\boldsymbol{ji} = \boldsymbol{k}$; $\boldsymbol{jk} = -\boldsymbol{kj} = \boldsymbol{i}$; $\boldsymbol{ki} = -\boldsymbol{ik} = \boldsymbol{j}$; and $\boldsymbol{i}^2 = \boldsymbol{j}^2 = \boldsymbol{k}^2 = -1$ (Hamilton, 1866).

An Introduction to Clifford Algebras and Spinors. First Edition. Jayme Vaz, Jr. and Roldão da Rocha, Jr.

some subtleties, since multiplication by scalars (in this case, elements of $\mathbb{H}$) is not commutative. Consequently, both the *left* and the *right* multiplications must be evaluated. Such details, however, will not be discussed here, being postponed to section 6.10. We restrict our discussion to vector spaces over $\mathbb{R}$ for most of our purposes, taking into account that $\mathbb{C}$ is a straightforward generalisation that is sometimes simpler in the context of Clifford algebras than in other contexts. In addition, finite-dimensional vector spaces, where we denote $\dim(V) = n$, will be examined. It is worth emphasising that various demonstrations throughout the book are general and also hold for infinite-dimensional spaces. From now on, we denote the product $a.\mathbf{v}$ by the juxtaposition $a\mathbf{v}$. Moreover, unless otherwise stated, we use the *Einstein summation convention*. For instance, if $\mathfrak{B} = \{\mathbf{e}_1, \ldots, \mathbf{e}_n\}$ is a basis for V, then

$$\mathbf{v} = \sum_{i=1}^{n} v^i \mathbf{e}_i = v^i \mathbf{e}_i.$$

Not so much discussed as the vector space, although as important and inseparable from it, is the *dual vector space*. A mapping $\alpha : V \to \mathbb{K}$ is said to be a *linear functional* if it is a homomorphism of vector spaces,[2] namely if it satisfies

$$\alpha(a\mathbf{u} + \mathbf{v}) = a\alpha(\mathbf{u}) + \alpha(\mathbf{v}),$$

for all $a \in \mathbb{K}$ and for all $\mathbf{u}, \mathbf{v} \in V$. A linear functional is also known as a *covector*, and the latter term will be preferentially used throughout the text. When the sum of covectors

$$(\alpha + \beta)(\mathbf{v}) = \alpha(\mathbf{v}) + \beta(\mathbf{v})$$

and the multiplication by scalars

$$(a\alpha)(\mathbf{v}) = a(\alpha(\mathbf{v}))$$

are defined, the space of covectors is endowed with a vector space structure. Such a covector space is called the *dual vector space* associated with the vector space V, and it is denoted by V^*.

The evaluation of the covector α on a vector $\mathbf{v} = v^i \mathbf{e}_i$ is performed by

$$\alpha(\mathbf{v}) = \alpha(v^i \mathbf{e}_i) = v^i \alpha(\mathbf{e}_i) = v^i \alpha_i,$$

where $\alpha_i = \alpha(\mathbf{e}_i)$. This expression reveals that a covector α is completely defined by its components α_i with respect to a basis $\mathfrak{B} = \{\mathbf{e}_i\}$ of the vector space V.

[2]A homomorphism is a mapping of a structure X into a structure Y such that the structural properties in the domain are preserved in the image. More explicitly, if $*$ is an operation in X and $\bullet$ is another operation in Y, then $\varphi : X \to Y$ is a homomorphism when $\varphi(a * b) = \varphi(a) \bullet \varphi(b)$. If $Y = X$, then this homomorphism is called an endomorphism.

Consider now the set of covectors $\{\mathbf{e}^i\}$ $(i = 1, \dots, n)$ defined by

$$\boxed{\mathbf{e}^i(\mathbf{e}_j) = \delta^i_j = \begin{cases} 1 & \text{if } i = j \\ 0 & \text{if } i \neq j \end{cases}}. \tag{1.1}$$

The set $\{\mathbf{e}^i\}$ in eqn (1.1) is a basis for the dual vector space V^*. The basis $\mathfrak{B}^* = \{\mathbf{e}^i\}$ is said to be the *dual basis* associated with the basis $\mathfrak{B} = \{\mathbf{e}_i\}$. As an immediate consequence, it follows that

$$\dim(V^*) = \dim(V). \tag{1.2}$$

Covariant Transformations and Contravariant Transformations

The intrinsic difference between vectors and covectors can be explored by considering, for instance, the effect of a change of basis. Let us consider a change of basis $\mathfrak{B} \mapsto \mathfrak{B}'$ as described by $\mathbf{e}'_j = B^i{}_j\mathbf{e}_i$. A vector $\mathbf{v} \in V$ has components $\{v^i\}$ with respect to the basis $\mathfrak{B}$, and components $\{v'^i\}$ with respect to the basis $\mathfrak{B}'$, namely $\mathbf{v} = v^i\mathbf{e}_i = v'^i\mathbf{e}'_i$. The vector components are related by $v^j = B^j{}_iv'^i$.

Now, let $\mathfrak{B}^* = \{\mathbf{e}^i\}$, and $\mathfrak{B}'^* = \{\mathbf{e}'^i\}$, be the dual bases respectively associated with the bases $\mathfrak{B} = \{\mathbf{e}_i\}$, and $\mathfrak{B}' = \{\mathbf{e}'_i\}$. By definition, we have $\mathbf{e}^i(\mathbf{e}_j) = \mathbf{e}'^i(\mathbf{e}'_j) = \delta^i_j$. The components of $\alpha \in V^*$ in the bases $\mathfrak{B}^*$ and $\mathfrak{B}'^*$ are given by the evaluation of α on the bases $\mathfrak{B}$ and $\mathfrak{B}'$, respectively. Hence, $\alpha = \alpha_i\mathbf{e}^i = \alpha'_i\mathbf{e}'^i$, where $\alpha_i = \alpha(\mathbf{e}_i)$, and $\alpha'_i = \alpha(\mathbf{e}'_i)$. Since $\mathbf{e}'_j = B^i{}_j\mathbf{e}_i$, the components of α transform as $\alpha'_j = B^i{}_j\alpha_i$, and consequently the dual bases are related by $\mathbf{e}^j = B^j{}_i\mathbf{e}'^i$.

To summarise, with a change of basis, the components of a covector transform as the basis vectors

$$\mathbf{e}'_j = B^i{}_j\mathbf{e}_i, \qquad \alpha'_j = B^i{}_j\alpha_i,$$

whereas the components of a vector transform as the basis covectors

$$\mathbf{e}^j = B^j{}_i\mathbf{e}'^i, \qquad v^j = B^j{}_iv'^i.$$

The inverse transformations are given by

$$\mathbf{e}_j = (B^{-1})^i{}_j\mathbf{e}'_i, \qquad \alpha_j = (B^{-1})^i{}_j\alpha'_i,$$

and

$$\mathbf{e}'^j = (B^{-1})^j{}_i\mathbf{e}^i, \qquad v'^j = (B^{-1})^j{}_iv^i,$$

accordingly.

The transformations regarding these basis vectors are usually called *covariant* transformations, and the transformations of the vector components are called *contravariant* transformations. Such a denomination asserts that the dual basis vectors transform in a contravariant way, and the covectors components transform in a covariant way.

Example 1.1 Let $\mathfrak{B} = \{\mathbf{e}_i\}$ be the standard basis of $\mathbb{R}^3$, namely $\mathbf{e}_1 = (1,0,0)^\intercal$; $\mathbf{e}_2 = (0,1,0)^\intercal$; $\mathbf{e}_3 = (0,0,1)^\intercal$; and let $\mathfrak{B}' = \{\mathbf{e}'_i\}$ be another basis such that $\mathbf{e}'_1 = (-1,-1,1)^\intercal$; $\mathbf{e}'_2 = (-1,0,1)^\intercal$; and

$\mathbf{e}'_3 = (2, 1, -1)^{\mathsf{T}}$. Given a vector $\mathbf{v} = v^i \mathbf{e}_i = v'^i \mathbf{e}'_i$, the components $\{v^i\}$ and $\{v'^i\}$ with respect to these bases are expressed as

$$\begin{pmatrix} v^1 \\ v^2 \\ v^3 \end{pmatrix} = \begin{pmatrix} -1 & -1 & 2 \\ -1 & 0 & 1 \\ 1 & 1 & -1 \end{pmatrix} \begin{pmatrix} v'^1 \\ v'^2 \\ v'^3 \end{pmatrix}, \qquad \begin{pmatrix} v'^1 \\ v'^2 \\ v'^3 \end{pmatrix} = \begin{pmatrix} 1 & -1 & 1 \\ 0 & 1 & 1 \\ 1 & 0 & 1 \end{pmatrix} \begin{pmatrix} v^1 \\ v^2 \\ v^3 \end{pmatrix},$$

and the bases vectors are related by

$$\begin{aligned} \mathbf{e}'_1 &= -\mathbf{e}_1 - \mathbf{e}_2 + \mathbf{e}_3, & \mathbf{e}_1 &= \mathbf{e}'_1 + \mathbf{e}'_3, \\ \mathbf{e}'_2 &= -\mathbf{e}_1 + \mathbf{e}_3, & \mathbf{e}_2 &= -\mathbf{e}'_1 + \mathbf{e}'_2, \\ \mathbf{e}'_3 &= 2\mathbf{e}_1 + \mathbf{e}_2 - \mathbf{e}_3, & \mathbf{e}_3 &= \mathbf{e}'_1 + \mathbf{e}'_2 + \mathbf{e}'_3. \end{aligned}$$

Let $\mathfrak{B}^* = \{\mathbf{e}^i\}$ be the dual basis related to $\mathfrak{B}$. By writing the vectors $\{\mathbf{e}_i\}$ as

$$\mathbf{e}_1 = \begin{pmatrix} 1 \\ 0 \\ 0 \end{pmatrix}, \qquad \mathbf{e}_2 = \begin{pmatrix} 0 \\ 1 \\ 0 \end{pmatrix}, \qquad \mathbf{e}_3 = \begin{pmatrix} 0 \\ 0 \\ 1 \end{pmatrix},$$

the dual basis reads

$$\mathbf{e}^1 = (1\ 0\ 0), \qquad \mathbf{e}^2 = (0\ 1\ 0), \qquad \mathbf{e}^3 = (0\ 0\ 1).$$

Let now $\mathfrak{B}'^* = \{\mathbf{e}'^i\}$ be the dual basis associated with $\mathfrak{B}'$; therefore, $\mathbf{e}'^i(\mathbf{e}'_j) = \delta^i_j$, and in particular $\mathbf{e}'^1(\mathbf{e}'_1) = 1$; $\mathbf{e}'^1(\mathbf{e}'_2) = \mathbf{e}'^1(\mathbf{e}'_3) = 0$; and so on. In order to find a relationship among the dual basis vectors, it is possible to evaluate the action of the covector $\mathbf{e}'^i$ on the vectors $\mathbf{e}_j$, expressed in terms of $\mathfrak{B}'$. For instance,

$$\begin{aligned} \mathbf{e}'^1(\mathbf{e}_1) &= \mathbf{e}'^1(\mathbf{e}'_1 + \mathbf{e}'_3) = \mathbf{e}'^1(\mathbf{e}'_1) + \mathbf{e}'^1(\mathbf{e}'_3) = 1, \\ \mathbf{e}'^1(\mathbf{e}_2) &= \mathbf{e}'^1(-\mathbf{e}'_1 + \mathbf{e}'_2) = -\mathbf{e}'^1(\mathbf{e}'_1) + \mathbf{e}'^1(\mathbf{e}'_2) = -1, \\ \mathbf{e}'^1(\mathbf{e}_3) &= \mathbf{e}'^1(\mathbf{e}'_1 + \mathbf{e}'_2 + \mathbf{e}'_3) = \mathbf{e}'^1(\mathbf{e}'_1) + \mathbf{e}'^1(\mathbf{e}'_2) + \mathbf{e}'^1(\mathbf{e}'_3) = 1, \end{aligned}$$

where we can conclude that $\mathbf{e}'^1 = \mathbf{e}^1 - \mathbf{e}^2 + \mathbf{e}^3$. An analogous procedure can be used to express the set $\{\mathbf{e}^i\}$ with respect to the basis $\mathfrak{B}'^*$. The result yields

$$\begin{aligned} \mathbf{e}'^1 &= \mathbf{e}^1 - \mathbf{e}^2 + \mathbf{e}^3, & \mathbf{e}^1 &= -\mathbf{e}'^1 - \mathbf{e}'^2 + 2\mathbf{e}'^3, \\ \mathbf{e}'^2 &= \mathbf{e}^2 + \mathbf{e}^3, & \mathbf{e}^2 &= -\mathbf{e}'^1 + \mathbf{e}'^3, \\ \mathbf{e}'^3 &= \mathbf{e}^1 + \mathbf{e}^3, & \mathbf{e}^3 &= \mathbf{e}'^1 + \mathbf{e}'^2 - \mathbf{e}'^3. \end{aligned}$$

Consider now the linear functional α acting on a vector $\mathbf{v}$:

$$\alpha(\mathbf{v}) = \alpha_1 v^1 + \alpha_2 v^2 + \alpha_3 v^3,$$

where the set $\{\alpha_i\}$ contains the components of α with respect to the basis $\mathfrak{B}^*$. If the components of $\mathbf{v}$ are placed in a row matrix, then the components of the linear functional α can be written as the column matrix

$$[\alpha]_{\mathfrak{B}^*} = (\alpha_1\ \alpha_2\ \alpha_3);$$

therefore,

$$\alpha(\mathbf{v}) = (\alpha_1\ \alpha_2\ \alpha_3) \begin{pmatrix} v^1 \\ v^2 \\ v^3 \end{pmatrix} = \alpha_1 v^1 + \alpha_2 v^2 + \alpha_3 v^3.$$

As usual, by denoting by α'_i the components of α with respect to the basis $\mathfrak{B}'^*$ ($\alpha = \alpha_i \mathbf{e}^i = \alpha'_i \mathbf{e}'^i$), the respective components are related by:

$$(\alpha_1\ \alpha_2\ \alpha_3) = (\alpha'_1\ \alpha'_2\ \alpha'_3) \begin{pmatrix} 1 & -1 & 1 \\ 0 & 1 & 1 \\ 1 & 0 & 1 \end{pmatrix},$$

$$(\alpha'_1\ \alpha'_2\ \alpha'_3) = (\alpha_1\ \alpha_2\ \alpha_3) \begin{pmatrix} -1 & -1 & 2 \\ -1 & 0 & 1 \\ 1 & 1 & -1 \end{pmatrix}.$$

When a matrix is multiplied from the right by a column vector, we relate the components of a vector in a basis A as components with respect to a basis B. Moreover, when a matrix is multiplied from the left by a row vector, the components of a covector in a basis B are related to the components with respect to a basis A.

Example 1.2 Let $\mathcal{F}$ be the space of continuous functions $f : \mathbb{R} \to \mathbb{R}$. The integral

$$L(f) = \int_{x_0}^{x_1} f(x)\,\mathrm{d}x$$

defines a linear functional L over $\mathcal{F}$. Consider the subset $\mathcal{P}_2$ of $\mathcal{F}$ and which consists of the polynomial functions P of degree less than or equal to 2 and is expressed as $P(x) = a + bx + cx^2$, where $a, b, c \in \mathbb{R}$. The canonical basis for $\mathcal{P}_2$ is $\mathfrak{B} = \{1, x, x^2\}$, and we denote

$$\mathbf{e}_1 = 1, \qquad \mathbf{e}_2 = x, \qquad \mathbf{e}_3 = x^2.$$

Let us define the following linear functionals on $\mathcal{P}_2$:

$$L^1(P) = \int_0^1 p(x)\,\mathrm{d}x, \qquad L^2(P) = \int_0^2 p(x)\,\mathrm{d}x, \qquad L^3(P) = \int_0^3 p(x)\,\mathrm{d}x.$$

Explicitly, these integrals read

$$L^1(P) = a + \frac{1}{2}b + \frac{1}{3}c, \qquad L^2(P) = 2a + 2b + \frac{8}{3}c, \qquad L^3(P) = 3a + \frac{9}{2}b + 9c.$$

If $\{\mathbf{e}^i\}$ is the dual basis of $\{\mathbf{e}_i\}$, then these equations yield

$$L^1 = \mathbf{e}^1 + \frac{1}{2}\mathbf{e}^2 + \frac{1}{3}\mathbf{e}^3, \qquad L^2 = 2\mathbf{e}^1 + 2\mathbf{e}^2 + \frac{8}{3}\mathbf{e}^3, \qquad L^3 = 3\mathbf{e}^1 + \frac{9}{2}\mathbf{e}^2 + 9\mathbf{e}^3.$$

Let now $\{L_i\}$ be the basis related to the dual basis $\{L^i\}$, so that $L^i(L_j) = \delta^i_j$. Since $L^i = \mathbf{e}^j L^i(\mathbf{e}_j)$, and $\mathbf{e}_j = L^i(\mathbf{e}_j)L_j$, the expressions for $\{\mathbf{e}_i\}$ in terms of $\{L_i\}$ are thus given by

$$1 = \mathbf{e}_1 = L_1 + 2L_2 + 3L_3, \quad x = \mathbf{e}_2 = \frac{1}{2}L_1 + 2L_2 + \frac{9}{2}L_3, \quad x^2 = \mathbf{e}_3 = \frac{1}{3}L_1 + \frac{8}{3}L_2 + 9L_3.$$

The inverse relation can be obtained and derived by usual manipulation, which results in

$$L_1 = 3 - 5x + \frac{3}{2}x^2, \qquad L_2 = -\frac{3}{2} + 4x - \frac{3}{2}x^2, \qquad L_3 = \frac{1}{3} - x + \frac{1}{2}x^2.$$

Since $L_i = \mathbf{e}^j(L_i)\mathbf{e}_j$, and $\mathbf{e}^j = \mathbf{e}^j(L_i)L^i$, from these expressions, $\{\mathbf{e}^i\}$ can be expressed from $\{L^i\}$ as

$$\mathbf{e}^1 = 3L^1 - \frac{3}{2}L^2 + \frac{1}{3}L^3, \qquad \mathbf{e}^2 = -5L^1 + 4L^2 - L^3, \qquad \mathbf{e}^3 = \frac{3}{2}L^1 - \frac{3}{2}L^2 + \frac{1}{2}L^3.$$

Finally, an arbitrary functional L,

$$L(P) = \int_{x_0}^{x_1} p(x)\,\mathrm{d}x,$$

can be, for instance, written in the form

$$L = \lambda_1\mathbf{e}^1 + \lambda_2\mathbf{e}^2 + \lambda_3\mathbf{e}^3 = l_1L^1 + l_2L^2 + l_3L^3,$$

where

$$\lambda_1 = (x_1 - x_0), \qquad \lambda_2 = \frac{x_1^2 - x_0^2}{2}, \qquad \lambda_3 = \frac{x_1^3 - x_0^3}{3},$$

and the components in the set $\{l_i\}$ are given by

$$(l_1\ l_2\ l_3) = (\lambda_1\ \lambda_2\ \lambda_3)\begin{pmatrix} 3 & -3/2 & 1/3 \\ -5 & 4 & -1 \\ 3/2 & -3/2 & 1/2 \end{pmatrix}.$$

The Bidual

Since V^* is a vector space, one can argue whether it is also possible to define a dual space associated with the dual vector space V^*. Define the linear functionals $\xi_{\mathbf{v}} : V^* \to \mathbb{R}$ as

$$\xi_{\mathbf{v}}(\alpha) = \alpha(\mathbf{v}). \tag{1.3}$$

The sum of such linear functionals is given by $\xi_{\mathbf{v}} + \xi_{\mathbf{u}} = \xi_{(\mathbf{v}+\mathbf{u})}$, and the multiplication by a scalar by $a\xi_{\mathbf{v}} = \xi_{(a\mathbf{v})}$. Such space is denoted by $(V^*)^*$ and $\dim((V^*)^*) = \dim(V)$.

The fundamental result here is that the spaces V and $(V^*)^*$ are *isomorphic*. Although it is well known that all vector spaces having the same dimension are isomorphic, in this context, the assertion is stronger than usual: there exists a *natural* (namely, a canonical identification) isomorphism between V and $(V^*)^*$, given by ξ and which is the mapping $\xi : V \to (V^*)^*$, and $\xi : \mathbf{v} \mapsto \xi_{\mathbf{v}}$. By 'natural' we mean that such an isomorphism does not depend upon the choice of a basis. In fact, eqn (1.3) defines the natural isomorphism, since $\alpha(\mathbf{v})$ is a scalar, as being invariant under basis change.

However, *there is no natural isomorphism* (in this sense) between a vector space V and its respective dual V^*. An additional structure is needed to define the isomorphism between V and V^*! This additional structure is called a *correlation*. In other words, an isomorphism between V and V^* must be chosen.

The Kernel and the Image of a Linear Transformation

Given two sets X and Y, a mapping $f : X \to Y$ is said to be a surjective (or onto) mapping if, for each $y \in Y$, there exists an element $x \in X$ such that $f(x) = y$. However, the mapping is said to be an injective (or one-to-one) mapping if, for each $y \in Y$, there exists *at most* a unique element $x \in X$ such that $f(x) = y$.[3] A mapping that is onto and one-to-one is said to be bijective.

Consider now the two vector spaces V and W and let f be a linear mapping $f : V \to W$. The *rank* of a linear mapping f, denoted by $\operatorname{rank} f$, is the dimension of the image of the mapping (the image is itself a vector subspace $f(V) \subseteq W$). The *kernel* of f (denoted by $\ker f$) consists of the set $\{\mathbf{v} \in V \mid f(\mathbf{v}) = 0\}$. The dimension of the kernel of f is also called the *nullity* of f. For finite-dimensional vector spaces, the expression

$$\operatorname{rank} f + \dim \ker f = \dim V$$

holds. It is straightforward to show that a linear mapping f is injective if and only if $\ker f = \{0\}$. Moreover, when V and W are vector spaces of the same finite dimension, any one-to-one surjective linear mapping $f : V \to W$ is an *isomorphism* – and, in this case, $\ker f = \{0\}$. For spaces of infinite dimension, we must also impose the mapping f to be onto (surjective), for the definition of isomorphism.

Correlations and Bilinear Functionals

We can introduce the linear mapping $\tau : V \to V^*$, the so-called correlation, which naturally defines a bilinear functional $B : V \times V \to \mathbb{R}$ by the equation

$$B(\mathbf{v}, \mathbf{u}) = \tau(\mathbf{v})(\mathbf{u}), \quad \forall\, \mathbf{u}, \mathbf{v} \in V. \tag{1.4}$$

Bilinearity means linearity in each one of the entries, namely $B(a\mathbf{v}+\mathbf{u}, \mathbf{w}) = aB(\mathbf{v}, \mathbf{w}) + B(\mathbf{u}, \mathbf{w})$, and $B(\mathbf{v}, a\mathbf{u} + \mathbf{w}) = aB(\mathbf{v}, \mathbf{u}) + B(\mathbf{v}, \mathbf{w})$, $\forall\, \mathbf{u}, \mathbf{v}, \mathbf{w} \in V$, and $a, b \in \mathbb{R}$.

If $\ker \tau = \{0\}$, the correlation is said to be *non-degenerate*. In this case, the vector space V and the bilinear functional associated with τ are also said to be non-degenerate. It is possible to show that the bilinear functional B is non-degenerate if and only if, for each vector $\mathbf{v} \neq 0$, there exists a vector $\mathbf{u} \neq 0$ such that $B(\mathbf{v}, \mathbf{u}) \neq 0$.

[3]There can be a situation wherein there does not exist any element.

Since $\dim V = \dim V^*$, if $\ker \tau = \{0\}$, then τ is an *isomorphism*; thus, the non-degenerate correlation provides an isomorphism between a vector space and its respective dual space.

Quadratic Spaces and Symplectic Spaces

A bilinear functional, in this case denoted by g, is said to be *symmetric* if

$$g(\mathbf{v}, \mathbf{u}) = g(\mathbf{u}, \mathbf{v}), \qquad \forall\, \mathbf{u}, \mathbf{v} \in V. \tag{1.5}$$

The correlation τ associated with g and defined by $\tau(\mathbf{v})(\mathbf{u}) = g(\mathbf{v}, \mathbf{u})$ satisfies $\tau(\mathbf{v})(\mathbf{u}) = \tau(\mathbf{u})(\mathbf{v})$. A vector space endowed with a non-degenerate symmetric bilinear functional is said to be a *quadratic space*. A symmetric bilinear functional is completely determined by the *quadratic form* $Q(\mathbf{v}) = g(\mathbf{v}, \mathbf{v})$ via the so-called polarisation procedure. Indeed, by using the bilinearity property to calculate $Q(\mathbf{v} + \mathbf{u}) = g(\mathbf{v} + \mathbf{u}, \mathbf{v} + \mathbf{u})$, we can write

$$g(\mathbf{v}, \mathbf{u}) = \frac{1}{2}\left(Q(\mathbf{v} + \mathbf{u}) - Q(\mathbf{v}) - Q(\mathbf{u})\right). \tag{1.6}$$

A bilinear functional, in this case denoted by σ, is said to be *antisymmetric* when it satisfies the property

$$\sigma(\mathbf{v}, \mathbf{u}) = -\sigma(\mathbf{u}, \mathbf{v}). \tag{1.7}$$

The correlation τ in this case satisfies $\tau(\mathbf{v})(\mathbf{u}) = -\tau(\mathbf{u})(\mathbf{v})$. A vector space endowed with a non-degenerate antisymmetric bilinear functional is said to be a *symplectic space*.

If $g(\mathbf{v}, \mathbf{u}) = 0$, the vectors $\mathbf{v}$ and $\mathbf{u}$ are said to be *orthogonal* with respect to g, when g is non-degenerate. A non-trivial vector $\mathbf{v}$ can be orthogonal to itself, namely $g(\mathbf{v}, \mathbf{v}) = 0$, and such vectors are called *isotropic*. It is immediately obvious that, in a symplectic space, all the vectors are isotropic.

Musical Isomorphisms

In this section, only quadratic spaces are considered with respect to the symmetric correlations $\tau : V \to V^*$, and $\tau^{-1} : V^* \to V$. Here, we shall use a notation which is much more convenient than the one we have used so far: we shall denote those correlations respectively by

$$\flat : V \to V^*, \quad \sharp : V^* \to V, \tag{1.8}$$

where $\flat = \sharp^{-1}$, and $\sharp = \flat^{-1}$. Such isomorphisms are called *musical isomorphisms*. In general, the expression

$$\mathbf{v}_\flat = \flat(\mathbf{v}), \quad \alpha^\sharp = \sharp(\alpha) \tag{1.9}$$

will be used. By definition, it follows that

$$\mathbf{v}_\flat(\mathbf{u}) = g(\mathbf{v}, \mathbf{u}), \quad g(\alpha^\sharp, \mathbf{v}) = \alpha(\mathbf{v}). \tag{1.10}$$

For the vectors $\mathbf{v} = v^i \mathbf{e}_i$, and $\mathbf{u} = u^i \mathbf{e}_i$, we can write $g(\mathbf{v}, \mathbf{u}) = g_{ij} v^i u^j$, where $g_{ij} = g(\mathbf{e}_i, \mathbf{e}_j) = g_{ji}$. As $\mathbf{v}_\flat(\mathbf{u}) = v_{\flat i} u^i$, where $v_{\flat i}$ represents the covector $\mathbf{v}_\flat$ components in the basis $\{\mathbf{e}^i\}$, namely $\mathbf{v}_\flat = v_{\flat i} \mathbf{e}^i$, it follows that

$$v_{\flat i} = g_{ij}v^j. \tag{1.11}$$

Similarly,

$$\mathbf{e}_{i\flat} = g_{ij}\mathbf{e}^j. \tag{1.12}$$

On the other hand, by writing $\alpha^\sharp = \alpha_i \mathbf{e}^{i\sharp}$, it follows from $g(\alpha^\sharp, \mathbf{v}) = \alpha(\mathbf{v})$ that $\alpha_k g(\mathbf{e}^{k\sharp}, \mathbf{e}_i)v^i = \alpha_i v^i$; thus,

$$\mathbf{e}^{i\sharp} = g^{ij}\mathbf{e}_j, \tag{1.13}$$

where $g^{ij} = g^{ji}$ is defined as

$$g^{ik}g_{kj} = \delta^i_j. \tag{1.14}$$

This 'numerical inverse' is not an inverse mapping, but an adjoint. It is also possible to write

$$\alpha^{\sharp i} = g^{ij}\alpha_j, \tag{1.15}$$

where $\alpha^{\sharp i}$ is the ith component of the vector $\alpha^\sharp$ in the basis $\{\mathbf{e}_i\}$.

Observation ☞ In some texts, these equations read

$$v_i = g_{ij}v^j, \qquad \mathbf{e}_i = g_{ij}\mathbf{e}^j, \tag{1.16}$$

and

$$\alpha^i = g^{ij}\alpha_j, \qquad \mathbf{e}^i = g^{ij}\mathbf{e}_j. \tag{1.17}$$

In such texts, it is common to assert that such equations are responsible for the indices *raising* and *lowering*, respectively. However, rigorously speaking, these equations make no sense, since g^{ij} is not the *inverse* (as a mapping) of g_{ij} (although it is a numerical inverse with respect to both a chosen basis and its dual basis); for instance, there is a vector at the left-hand side of the equation $\mathbf{e}_i = g_{ij}\mathbf{e}^j$ but a linear combination of covectors at the right-hand side. Like the mapping $g_{ij} : \mathbb{R} \to \mathbb{R}$, these functions could be defined properly, for example by using a closed structure on the category of finite-dimensional vector spaces over $\mathbb{R}$; however this approach outside the scope of this book.

However, as $\sharp$ and $\flat$ are isomorphisms and sometimes isomorphism symbols can be concealed in order to avoid a heavy notation, we shall use the notation outlined in this section. Indeed, the explicit use of the indices in these equations should virtually eliminate any confusion there might otherwise be. However, when there is no reference to any particular basis, there is no way to avoid using notation using $\sharp$ and $\flat$.

Example 1.3 Let $\mathfrak{B} = \{\mathbf{e}_i\}$ $(i = 1, 2, 3)$ be the standard basis of $\mathbb{R}^3$ and let $\mathfrak{B}^* = \{\mathbf{e}^i\}$ be its respective dual basis. Let us define a correlation $\flat$ as

$$\mathbf{e}_{1\flat} = 3\mathbf{e}^1 + \mathbf{e}^2, \qquad \mathbf{e}_{2\flat} = \mathbf{e}^1 + 3\mathbf{e}^2, \qquad \mathbf{e}_{3\flat} = 2\mathbf{e}^3.$$

If we represent a vector $\mathbf{v} = v^i\mathbf{e}_i$ by a column matrix, this correlation reads

$$\flat\begin{pmatrix} v^1 \\ v^2 \\ v^3 \end{pmatrix} = \left((3v^1 + v^2)\ (3v^2 + v^1)\ (2v^3)\right).$$

This correlation is non-degenerate, in view of $\ker\flat = \{0\}$. Indeed, if $\flat(\mathbf{v}) = 0$, then $3v^1 + v^2 = 0$; $3v^2 + v^1 = 0$; and $2v^3 = 0$; thus $v^1 = v^2 = v^3 = 0$. Moreover, the correlation $\flat$ is symmetric, since

$$\mathbf{v}_\flat(\mathbf{u}) = \left((3v^1 + v^2)\ (3v^2 + v^1)\ (2v^3)\right)\begin{pmatrix} u^1 \\ u^2 \\ u^3 \end{pmatrix} = 3v^1u^1 + v^2u^1 + 3v^2u^2 + v^1u^2 + 2v^3u^3$$

$$= \left((3u^1 + u^2)\ (3u^2 + u^1)\ (2u^3)\right)\begin{pmatrix} v^1 \\ v^2 \\ v^3 \end{pmatrix} = \mathbf{u}_\flat(\mathbf{v}).$$

Therefore, a symmetric bilinear form g can be defined as $g(\mathbf{v}, \mathbf{u}) = \mathbf{v}_\flat(\mathbf{u})$. Since

$$\mathbf{v}_\flat = \left(v^1\ v^2\ v^3\right)\begin{pmatrix} 3 & 1 & 0 \\ 1 & 3 & 0 \\ 0 & 0 & 2 \end{pmatrix},$$

it follows that g can be represented in this basis by the matrix

$$\begin{pmatrix} 3 & 1 & 0 \\ 1 & 3 & 0 \\ 0 & 0 & 2 \end{pmatrix}.$$

On the other hand, the inverse correlation $\sharp$ is given by

$$\mathbf{e}^{1\sharp} = \frac{1}{8}(3\mathbf{e}_1 - \mathbf{e}_2), \qquad \mathbf{e}^{2\sharp} = \frac{1}{8}(3\mathbf{e}_2 - \mathbf{e}_1), \qquad \mathbf{e}^{3\sharp} = \frac{1}{2}\mathbf{e}_3,$$

which can be represented by

$$\sharp\left(\alpha_1\ \alpha_2\ \alpha_3\right) = \begin{pmatrix} (3\alpha_1 - \alpha_2)/8 \\ (3\alpha_2 - \alpha_1)/8 \\ \alpha_3/8 \end{pmatrix}.$$

Since

$$\begin{pmatrix} (3\alpha_1 - \alpha_2)/8 \\ (3\alpha_2 - \alpha_1)/8 \\ \alpha_3/2 \end{pmatrix} = \begin{pmatrix} 3/8 & -1/8 & 0 \\ -1/8 & 3/8 & 0 \\ 0 & 0 & 1/2 \end{pmatrix}\begin{pmatrix} \alpha_1 \\ \alpha_2 \\ \alpha_3 \end{pmatrix},$$

it follows that, in this basis the symmetric bilinear form g^{-1} (here it must be clear that this inverse is taken in the group of automorphisms $\mathrm{Aut}(\mathbb{R}^3)$) such that $g^{-1}(\alpha, \beta) = \alpha(\beta^\sharp) = \beta(\alpha^\sharp) = g^{-1}(\beta, \alpha)$ can be represented by

$$\begin{pmatrix} 3/8 & -1/8 & 0 \\ -1/8 & 3/8 & 0 \\ 0 & 0 & 1/2 \end{pmatrix}.$$

As expected, the matrices g and g^{-1} are inverse with respect to each other, as can be clearly seen. Indeed, the elements g_{ij} which are entries of the matrix representation of g, and the elements g^{ij} of the matrix which represents g^{-1}, must satisfy eqn (1.14); therefore, in this case, the matrices are the inverse of each other.

In addition, the basis $\{\mathbf{e}_i\}$ is not orthogonal with respect to g, inasmuch as $g(\mathbf{e}_1, \mathbf{e}_2) = g(\mathbf{e}_2, \mathbf{e}_1) = 1 \neq 0$. Frequently, it is convenient to work with an orthogonal basis, as in this example. In order to obtain such an orthogonal basis, the eigenvalues of g are calculated when the characteristic equation

$$\begin{vmatrix} 3-r & 1 & 0 \\ 1 & 3-r & 0 \\ 0 & 0 & 2-r \end{vmatrix} = 0,$$

is solved, having $r_1 = 2$; $r_2 = 2$; and $r_3 = 4$ as solutions. The corresponding eigenvectors are

$$\boldsymbol{\xi}_1 = \begin{pmatrix} 1 \\ -1 \\ 0 \end{pmatrix}, \qquad \boldsymbol{\xi}_2 = \begin{pmatrix} 0 \\ 0 \\ 1 \end{pmatrix}, \qquad \boldsymbol{\xi}_3 = \begin{pmatrix} 1 \\ 1 \\ 0 \end{pmatrix},$$

so the matrix T that diagonalises g, and its inverse T^{-1}, are respectively represented by

$$T = \begin{pmatrix} 1 & 0 & 1 \\ -1 & 0 & 1 \\ 0 & 1 & 0 \end{pmatrix}, \qquad T^{-1} = \begin{pmatrix} 1/2 & -1/2 & 0 \\ 0 & 0 & 1 \\ 1/2 & 1/2 & 0 \end{pmatrix}.$$

The set of vectors $\{\boldsymbol{\xi}_i\}$ is an orthogonal basis. In addition, an orthonormal basis can be constructed as

$$g(\boldsymbol{\xi}_1,\boldsymbol{\xi}_1)=4, \qquad g(\boldsymbol{\xi}_2,\boldsymbol{\xi}_2)=2, \qquad g(\boldsymbol{\xi}_3,\boldsymbol{\xi}_3)=8,$$

and the orthonormal basis $\mathfrak{B}'=\{\mathbf{e}_1,\mathbf{e}_2,\mathbf{e}_3\}$ is forthwith elicited by

$$\mathbf{e}'_1=\frac{1}{2}\begin{pmatrix}1\\-1\\0\end{pmatrix}=\frac{1}{2}(\mathbf{e}_1-\mathbf{e}_2), \quad \mathbf{e}'_2=\frac{1}{\sqrt{2}}\begin{pmatrix}0\\0\\1\end{pmatrix}=\frac{1}{\sqrt{2}}\mathbf{e}_2, \quad \mathbf{e}'_3=\frac{1}{2\sqrt{2}}\begin{pmatrix}1\\1\\0\end{pmatrix}=\frac{1}{2\sqrt{2}}(\mathbf{e}_1+\mathbf{e}_2).$$

1.2 The Tensor Product

Let us consider the vector spaces U, V, and W. $\mathrm{Lin}(V,W)$ denotes the space of the linear mappings from V to W, and $\mathrm{Lin}_{(2)}(V,W;U)$ denotes the space of the bilinear mappings from $V\times W$ to U. The tensor product of the two vector spaces (over the same field) V and W can be defined from the bilinear transformations $\phi: V\times W\to U$. One of these transformations is *universal*, in the sense that it describes all others. A bilinear mapping $\phi: V\otimes W\to U$ is said to be universal if the following linear mapping is bijective for every vector space S:

$$\mathrm{Lin}(U,S)\to\mathrm{Lin}_{(2)}(V,W;S), \qquad f\mapsto f\circ\phi\,.$$

It is straightforward to apply this definition to vector spaces of finite dimensions.

Theorem 1.1 ▶ *If $\{\mathbf{e}_1,\ldots,\mathbf{e}_n\}$ is a basis of V and $\{\mathbf{e}'_1,\ldots,\mathbf{e}'_p\}$ is a basis of W, the following two assertions are equivalent:*

(a) the bilinear mapping $\phi: V\otimes W\to U$ is universal;

(b) the set of np elements $\{\phi(\mathbf{e}_j,\mathbf{e}'_k)\}$ constitutes a basis of U.

Proof: Remember that a family $\{\epsilon_1,\ldots,\epsilon_m\}$ of elements of U is a basis if and only if the following linear mapping is bijective for every vector space S:

$$\mathrm{Lin}(U,S)\to\underbrace{S\times S\times\cdots\times S}_{m\text{ times}}=S^m, \qquad f\mapsto\{f(\epsilon_1),\ldots,f(\epsilon_m)\}\,.$$

Now, let us consider this sequence of two linear mappings:

$$\mathrm{Lin}(U,S)\to\mathrm{Lin}_{(2)}(V,W;S)\to S^{np}\,,$$

where the first arrow is defined as above, and the second arrow maps every F to the family of all $F(\mathbf{e}_j,\mathbf{e}'_k)$. The second arrow is well known to be bijective. Therefore, the bijectiveness of the first arrow, namely the universality of ϕ, is equivalent to the bijectiveness of the mapping $\mathrm{Lin}(U,S)\to S^{np}$, which maps every f to the family of all $f(\phi(\mathbf{e}_j,\mathbf{e}'_k))$. The bijectiveness of $\mathrm{Lin}(U,S)\to S^{np}$ means that the set of elements $\{\phi(\mathbf{e}_j,\mathbf{e}'_k)\}$ constitutes a basis of U. ✓

An immediate corollary of this theorem states that property (b) is true for every basis of V and every basis of W as soon as it is true for a particular basis of V and a particular basis of W.

In particular, given a bilinear mapping $\phi : V \times W \to U$, there exists a unique linear transformation $\psi : V \otimes W \to U$ such that

$$\phi(\mathbf{v}, \mathbf{w}) = \psi(\mathbf{v} \otimes \mathbf{w}), \qquad \mathbf{v} \in V,\ \mathbf{w} \in W. \tag{1.18}$$

We can also denote the tensor product by $V \otimes_{\mathbb{K}} W$ when we explicitly want to emphasise on which field the tensor product is performed. However, unless otherwise stated, we shall not use this notation; instead, from here on, $V \otimes W$ will be used to denote the tensor product when $\mathbb{K} = \mathbb{R}$.

The tensor product between covectors can be defined mutatis mutandis, and the tensor product defines a bilinear functional acting on the Cartesian product $V \times V$ by

$$(\alpha \otimes \beta)(\mathbf{v}, \mathbf{u}) = \alpha(\mathbf{v})\beta(\mathbf{u}). \tag{1.19}$$

The quantity $\alpha \otimes \beta$ is denominated the *tensor product* between α and β. Consequently, from this definition, the tensor product is not commutative, in general:

$$\alpha \otimes \beta \neq \beta \otimes \alpha.$$

It is also possible to define a tensor product between vectors acting on covectors. In this case, the tensor product between vectors must be a quantity that acts on the Cartesian product $V^* \times V^*$ and results in a scalar. Indeed, we can use the natural isomorphism between V and the bidual V^{**} and, since $\xi_{\mathbf{v}}(\alpha) = \alpha(\mathbf{v})$, the tensor product $\mathbf{v} \otimes \mathbf{u}$ can be defined as

$$(\mathbf{v} \otimes \mathbf{u})(\alpha, \beta) = \alpha(\mathbf{v})\beta(\mathbf{u}). \tag{1.20}$$

The expression $(\mathbf{v} \otimes \mathbf{u})(\alpha, \beta)$ formally means that $(\xi_{\mathbf{v}} \otimes \xi_{\mathbf{u}})(\alpha, \beta) = \xi_{\mathbf{v}}(\alpha)\xi_{\mathbf{u}}(\beta) = \alpha(\mathbf{v})\beta(\mathbf{u})$, since V^{**} and V are canonically isomorphic.

The space defined by the tensor product between covectors is itself a vector space, denoted by $\mathsf{T}^2(V) = V^* \otimes V^*$. Similarly, the space defined by the tensor product between vectors is also a vector space, denoted by $\mathsf{T}_2(V) = V \otimes V$. It follows immediately that $\mathsf{T}_2(V) \simeq \mathsf{T}^2(V^*)$; this result implies that, if $\dim V = n$, then $\dim \mathsf{T}^2(V) = \dim \mathsf{T}_2(V) = n^2$. The spaces $\mathsf{T}_2(V)$ and $\mathsf{T}^2(V)$ can be straightforwardly related by the correlation.

Example 1.4 Let $\mathbf{v}, \mathbf{u} \in \mathbb{R}^2$ be the vectors $\mathbf{v} = 2\mathbf{e}_1 - \mathbf{e}_2$ and $\mathbf{u} = \mathbf{e}_1 + 3\mathbf{e}_2$. Then, the following tensor products can be evaluated as

$$\mathbf{v} \otimes \mathbf{u} = 2\mathbf{e}_1 \otimes \mathbf{e}_1 + 6\mathbf{e}_1 \otimes \mathbf{e}_2 - \mathbf{e}_2 \otimes \mathbf{e}_1 - 3\mathbf{e}_2 \otimes \mathbf{e}_2,$$
$$\mathbf{u} \otimes \mathbf{v} = 2\mathbf{e}_1 \otimes \mathbf{e}_1 - \mathbf{e}_1 \otimes \mathbf{e}_2 + 6\mathbf{e}_2 \otimes \mathbf{e}_1 - 3\mathbf{e}_2 \otimes \mathbf{e}_2.$$

This result explicitly shows that the tensor product is, in general, not commutative: $\mathbf{v} \otimes \mathbf{u} \neq \mathbf{u} \otimes \mathbf{v}$. Indeed,

$$\mathbf{v} \otimes \mathbf{u} - \mathbf{u} \otimes \mathbf{v} = 7(\mathbf{e}_1 \otimes \mathbf{e}_2 - \mathbf{e}_2 \otimes \mathbf{e}_1).$$

Moreover, $\mathbf{v} \otimes \mathbf{u} = \mathbf{u} \otimes \mathbf{v}$ if and only if $\mathbf{u}$ and $\mathbf{v}$ are collinear. Indeed, if $\mathbf{u}$ and $\mathbf{v}$ are not collinear, there is a basis $\{\mathbf{e}_1, \ldots, \mathbf{e}_n\}$ such that $\mathbf{e}_1 = \mathbf{u}$. We can then write $\mathbf{v} = \sum_{i=1}^n v^i \mathbf{e}_i$ and, if $\mathbf{u} \otimes \mathbf{v} = \mathbf{v} \otimes \mathbf{u}$, it follows that

$$0 = (v^1 - v^1)\mathbf{e}_1 \otimes \mathbf{e}_1 + \sum_{i=2}^n v^i \mathbf{e}_1 \otimes \mathbf{e}_i - \sum_{i=2}^n v^i \mathbf{e}_i \otimes \mathbf{e}_1 .$$

This result implies that $v^i = 0$ for $i = 2, 3, \ldots, n$ and that v^1 can be arbitrary. Hence, $\mathbf{v}$ and $\mathbf{u}$ are collinear. Conversely, if $\mathbf{u}$ and $\mathbf{v}$ are collinear, there exists $\lambda \in \mathbb{K}$ such that $\mathbf{u} = \lambda \mathbf{v}$, and therefore

$$\mathbf{u} \otimes \mathbf{v} = \lambda \mathbf{v} \otimes \mathbf{v} = \mathbf{v} \otimes (\lambda \mathbf{v}) = \mathbf{v} \otimes \mathbf{u}.$$

Bases

Let $\mathfrak{B} = \{\mathbf{e}_i\}$ be a basis for V and let $\mathfrak{B}^* = \{\mathbf{e}^i\}$ be its dual basis. We know that $\alpha(\mathbf{v}) = \alpha_i v^i$, and $\beta(\mathbf{u}) = \beta_i u^i$, where $\alpha_i = \alpha(\mathbf{e}_i)$; $\beta_i = \beta(\mathbf{e}_i)$; $v^i = \mathbf{e}^i(\mathbf{v})$; and $u^i = \mathbf{e}^i(\mathbf{u})$. In this way,

$$(\alpha \otimes \beta)(\mathbf{v}, \mathbf{u}) = \alpha_i v^i \beta_j u^j.$$

On the other hand, we have

$$(\mathbf{e}^i \otimes \mathbf{e}^j)(\mathbf{v}, \mathbf{u}) = v^i u^j.$$

Comparing these equations, we can write

$$\alpha \otimes \beta = \alpha_i \beta_j \mathbf{e}^i \otimes \mathbf{e}^j.$$

The set of bilinear functionals $\{\mathbf{e}^i \otimes \mathbf{e}^j\}$ $(i, j = 1, \ldots, n)$ is a basis for the space $\mathsf{T}^2(V)$. If B is an arbitrary bilinear functional, it follows that

$$B = b_{ij} \mathbf{e}^i \otimes \mathbf{e}^j, \tag{1.21}$$

where the scalars b_{ij}, which are the components of B in this basis $\mathfrak{B}$, are given by

$$b_{ij} = B(\mathbf{e}_i, \mathbf{e}_j).$$

Consequently,

$$B(\mathbf{v}, \mathbf{u}) = b_{ij} v^i u^j.$$

Analogously, the set $\{\mathbf{e}_i \otimes \mathbf{e}_j\}$ $(i, j = 1, \ldots, n)$ forms a basis for the space $\mathsf{T}_2(V)$, namely $A \in \mathsf{T}_2(V)$, which can be written as

$$A = a^{ij} \mathbf{e}_i \otimes \mathbf{e}_j, \tag{1.22}$$

where $a^{ij} = A(\mathbf{e}^i, \mathbf{e}^j)$ are components of A in this basis.

The bilinear functionals in $\mathsf{T}^2(V)$ and $\mathsf{T}_2(V)$ can be decomposed into the sum of symmetric and antisymmetric bilinear functionals. Given $B \in \mathsf{T}^2(V)$, a symmetric bilinear functional B_{sym} can be written as

$$B_{\text{sym}}(\mathbf{v}, \mathbf{u}) = \frac{B(\mathbf{v}, \mathbf{u}) + B(\mathbf{u}, \mathbf{v})}{2},$$

and an antisymmetric bilinear functional B_{alt} can be expressed as

$$B_{\text{alt}}(\mathbf{v}, \mathbf{u}) = \frac{B(\mathbf{v}, \mathbf{u}) - B(\mathbf{u}, \mathbf{v})}{2}.$$

Indeed, it is always possible to decompose $B = B_{\text{sym}} + B_{\text{alt}}$.

Still considering the space of the symmetric bilinear functionals $\mathsf{T}^2_{\text{sym}}(V)$ and the space of the antisymmetric bilinear functionals $\mathsf{T}^2_{\text{alt}}(V)$, it is clear that the sets

$$\{\mathbf{e}^i \otimes \mathbf{e}^j + \mathbf{e}^j \otimes \mathbf{e}^i\} \qquad \text{and} \qquad \{\mathbf{e}^i \otimes \mathbf{e}^j - \mathbf{e}^j \otimes \mathbf{e}^i\}$$

are, respectively, bases for $\mathsf{T}^2_{\text{sym}}(V)$ and $\mathsf{T}^2_{\text{alt}}(V)$.

Mixed Bilinear Functionals

It is also possible to define a tensor product between a vector and a covector; this product is called a mixed bilinear functional. We denote this vector space by $\mathsf{T}^1{}_1(V) = V^* \otimes V$. Obviously, $\dim \mathsf{T}^1{}_1(V) = n^2$. A basis for $\mathsf{T}^1{}_1(V)$ is given by the tensor products $\{\mathbf{e}^i \otimes \mathbf{e}_j\}$. Thus, a tensor $C \in \mathsf{T}^1{}_1(V)$ can be written as

$$C = c_i{}^j \mathbf{e}^i \otimes \mathbf{e}_j, \tag{1.23}$$

where $c_i{}^j = C(\mathbf{e}_i, \mathbf{e}^j)$.

Similarly, we can define the tensor product $V \otimes V^*$, denoting it by $\mathsf{T}_1{}^1(V)$. An element $D \in \mathsf{T}_1{}^1(V)$ therefore can be written as

$$D = d^i{}_j \mathbf{e}_i \otimes \mathbf{e}^j, \tag{1.24}$$

where $d^i{}_j = D(\mathbf{e}^i, \mathbf{e}_j)$.

There exists an isomorphism between the spaces $\mathsf{T}^1{}_1(V)$ and $\mathsf{T}_1{}^1(V)$. However, correct positioning of the indices is necessary in order to avoid ambiguity about which space we are considering when there is a correlation between lowering and raising indices (see Example 1.5). By convention, we do not take into account the position of the indices, and the notation $\mathsf{T}_1^1(V)$ refers to the space $\mathsf{T}^1{}_1(V)$.

Example 1.5 The isomorphism between the spaces $\mathsf{T}^1{}_1(V)$ and $\mathsf{T}_1{}^1(V)$ is defined by $\mu_{V^*,V} : \mathsf{T}^1{}_1(V) \to \mathsf{T}_1{}^1(V)$ given by

$$\mu_{V^*,V}(\alpha \otimes \mathbf{v}) = \mathbf{v} \otimes \alpha.$$

With respect to a basis of $\mathsf{T}^1{}_1(V)$, it follows that $\mu_{V^*,V}(\mathbf{e}^i \otimes \mathbf{e}_j) = \mathbf{e}_j \otimes \mathbf{e}^i$. It is straightforward to see that this definition *does not* depend upon the chosen basis. Indeed, given another basis $\{\mathbf{e}'_i\}$ for V, and $\{\mathbf{e}'^i\}$ for V^*, this equation reads $\mu_{V^*,V}(\mathbf{e}'^i \otimes \mathbf{e}'_j) = \mathbf{e}'_j \otimes \mathbf{e}'^i$. Despite the isomorphism, care is demanded with respect to the notation in order to distinguish the spaces $\mathsf{T}^1{}_1(V)$ and $\mathsf{T}_1{}^1(V)$. Suppose that, given a symmetric bilinear form g defined with respect to a basis $\{\mathbf{e}_1, \mathbf{e}_2\}$ of $\mathbb{R}^2$ by

$$g_{11} = 1, \quad g_{12} = 1, \quad g_{21} = 1, \quad g_{22} = -1,$$

where $g_{ij} = g(\mathbf{e}_i, \mathbf{e}_j)$, it is then possible to raise and lower indices, respectively, via

$$\mathbf{e}_1 = \mathbf{e}^1 + \mathbf{e}^2, \quad \mathbf{e}_2 = \mathbf{e}^1 - \mathbf{e}^2,$$

and

$$\mathbf{e}^1 = \frac{1}{2}(\mathbf{e}_1 + \mathbf{e}_2), \quad \mathbf{e}^2 = \frac{1}{2}(\mathbf{e}_1 - \mathbf{e}_2).$$

Let us consider $B \in \mathsf{T}^1{}_1(V)$ given by

$$B = \mathbf{e}^1 \otimes \mathbf{e}_2.$$

In this case,

$$B_1{}^1 = 0, \quad B_1{}^2 = 1, \quad B_2{}^1 = 0, \quad B_2{}^2 = 0.$$

By raising and lowering indices via these formulæ, we can write

$$B = \frac{1}{2}\mathbf{e}_1 \otimes \mathbf{e}^1 - \frac{1}{2}\mathbf{e}_1 \otimes \mathbf{e}^2 + \frac{1}{2}\mathbf{e}_2 \otimes \mathbf{e}^1 - \frac{1}{2}\mathbf{e}_2 \otimes \mathbf{e}^2,$$

so that

$$B^1{}_1 = \frac{1}{2}, \quad B^1{}_2 = -\frac{1}{2}, \quad B^2{}_1 = \frac{1}{2}, \quad B^2{}_2 = -\frac{1}{2}.$$

It follows that $B_i{}^j \neq B^j{}_i$. Hence, it is necessary to distinguish the spaces $\mathsf{T}^1{}_1(V)$ and $\mathsf{T}_1{}^1(V)$ and thus to adopt the convention of writing B_i^j for $\mathsf{T}^1{}_1(V)$. Thus, it is then legitimate to write

$$B_1^1 = 0, \quad B_1^2 = 1, \quad B_2^1 = 0, \quad B_2^2 = 0.$$

This example is one where using the notation $\sharp$ and $\flat$ for the correlations associated with g was very useful for avoiding possible misunderstandings. Indeed, what we accomplished here was defining another tensor

B' given by $B' = (\sharp \otimes \flat)B$, where $\sharp \otimes \flat : \mathsf{T}^1{}_1(V) \to \mathsf{T}_1{}^1(V)$ denotes the tensor product between the mappings $\sharp$ and $\flat$ and which is defined by $(\sharp \otimes \flat)(\alpha \otimes \mathbf{v}) = \sharp(\alpha) \otimes \flat(\mathbf{v})$. As seen in this example, we thus obtain $\sharp \otimes \flat \neq \mu_{V^*,V}$.

Tensors of Type (p, q)

The above definitions can be straightforwardly generalised for the tensor product of an arbitrary number of covectors and vectors, defining in such a way the spaces $\mathsf{T}^p(V) = (V^*)^{\otimes^p}$; $\mathsf{T}_q(V) = V^{\otimes^q}$; $\mathsf{T}^p{}_q(V) = (V^*)^{\otimes^p} \otimes V^{\otimes^q}$; $\mathsf{T}_q{}^p(V) = V^{\otimes^q} \otimes (V^*)^{\otimes^p}$; $\mathsf{T}^1{}_q{}^p(V) = V^* \otimes V^{\otimes^q} \otimes (V^*)^{\otimes^p}$; and so on. We adopt the convention that $\mathsf{T}^p_q(V)$ refers to the space $\mathsf{T}^p{}_q(V)$.

Consider the space $\mathsf{T}^p_q(V)$. A basis for this space is given by the set of tensor products

$$\{\mathbf{e}^{\mu_1} \otimes \mathbf{e}^{\mu_2} \otimes \cdots \otimes \mathbf{e}^{\mu_p} \otimes \mathbf{e}_{\nu_1} \otimes \mathbf{e}_{\nu_2} \otimes \cdots \otimes \mathbf{e}_{\nu_q}\},$$

where $\{\mu_1, \ldots, \mu_p, \nu_1, \ldots, \nu_q\} = \{1, \ldots, n\}$. An arbitrary element $T \in \mathsf{T}^p_q(V)$ can be written as

$$T = T^{\nu_1\nu_2\cdots\nu_q}_{\mu_1\mu_2\cdots\mu_p} \mathbf{e}^{\mu_1} \otimes \mathbf{e}^{\mu_2} \otimes \cdots \otimes \mathbf{e}^{\mu_p} \otimes \mathbf{e}_{\nu_1} \otimes \mathbf{e}_{\nu_2} \otimes \cdots \otimes \mathbf{e}_{\nu_q}, \tag{1.25}$$

where

$$T^{\nu_1\nu_2\cdots\nu_q}_{\mu_1\mu_2\cdots\mu_p} = T(\mathbf{e}_{\mu_1}, \mathbf{e}_{\mu_2}, \ldots, \mathbf{e}_{\mu_p}, \mathbf{e}^{\nu_1}, \mathbf{e}^{\nu_2}, \ldots, \mathbf{e}^{\nu_q}).$$

The multilinear functional $T \in \mathsf{T}^p_q(V)$ is called a *tensor of type* (p, q). The quantities $T^{\nu_1\nu_2\cdots\nu_q}_{\mu_1\mu_2\cdots\mu_p}$ are the components of the tensor T in the given basis.

Tensors of type $(p, 0)$ are sometimes called *covariant tensors*, and tensors of type $(0, q)$ are called *contravariant tensors*. A covector is therefore a tensor of type (1,0) – an example of covariant tensor – whereas a vector is a tensor of type (0,1) and thus a contravariant tensor.

Transformations

Under a change of basis $\mathfrak{B} \mapsto \mathfrak{B}'$, described by $\mathbf{e}'_j = B^i_j \mathbf{e}_i$, the dual basis transforms according to $\mathbf{e}'^j = B^j_i \mathbf{e}'^i$, and the components of a vector $\mathbf{v}$ and a covector α respectively transform according to $v'^j = B^j_i v^i$, and $\alpha'_j = B^i_j \alpha_i$. The basis vectors $\mathfrak{B}$ and the components of a covector transform in a covariant way, and the dual basis vectors $\mathfrak{B}^*$ and the vector components transform in a contravariant way. A generalisation of those results for a tensor of type (p, q) is straightforward.

Indeed, considering a tensor T of type (p, q), the expression for this tensor in the bases $\mathfrak{B}$ and $\mathfrak{B}'$ is given by

$$\begin{aligned} T &= T^{\nu_1\nu_2\cdots\nu_q}_{\mu_1\mu_2\cdots\mu_p} \mathbf{e}^{\mu_1} \otimes \mathbf{e}^{\mu_2} \otimes \cdots \otimes \mathbf{e}^{\mu_p} \otimes \mathbf{e}_{\nu_1} \otimes \mathbf{e}_{\nu_2} \otimes \cdots \otimes \mathbf{e}_{\nu_q} \\ &= (T')^{\nu_1\nu_2\cdots\nu_q}_{\mu_1\mu_2\cdots\mu_p} \mathbf{e}'^{\mu_1} \otimes \mathbf{e}'^{\mu_2} \otimes \cdots \otimes \mathbf{e}'^{\mu_p} \otimes \mathbf{e}'_{\nu_1} \otimes \mathbf{e}'_{\nu_2} \otimes \cdots \otimes \mathbf{e}'_{\nu_q}. \end{aligned} \tag{1.26}$$

Now, when we substitute the basis change $\mathbf{e}'_{\mu_i} = B^{\nu_i}_{\mu_i} \mathbf{e}_{\nu_i}$ and the corresponding transformation $\mathbf{e}'^{\mu_i} = (B^{-1})^{\mu_i}_{\nu_i} \mathbf{e}'^{\nu_i}$ in this expression, it then reads

$$T^{\rho_1\rho_2\cdots\rho_q}_{\sigma_1\sigma_2\cdots\sigma_p} = (T')^{\nu_1\nu_2\cdots\nu_q}_{\mu_1\mu_2\cdots\mu_p}(B^{-1})^{\mu_1}_{\sigma_1}\cdots(B^{-1})^{\mu_p}_{\sigma_p}(B^{-1})^{\rho_1}_{\nu_1}\cdots(B^{-1})^{\rho_q}_{\nu_q}. \qquad (1.27)$$

Hence, the covariant components (the ones with index p) transform in the same way as covector components, that is, e. g., in a covariant way. On the other hand, the *contravariant* components (that ones with index q) transform in the same way as the vector components do, that is, in a contravariant way.

Example 1.6 Let g be the symmetric bilinear form shown in Example 1.3. For this case,

$$\begin{aligned} g(\mathbf{v},\mathbf{u}) &= 3v^1u^1 + v^2u^1 + 3v^2u^2 + v^1u^2 + 2v^3u^3 \\ &= 3(\mathbf{e}^1\otimes\mathbf{e}^1)(\mathbf{v},\mathbf{u}) + (\mathbf{e}^2\otimes\mathbf{e}^1)(\mathbf{v},\mathbf{u}) + 3(\mathbf{e}^2\otimes\mathbf{e}^2)(\mathbf{v},\mathbf{u}) \\ &+ (\mathbf{e}^1\otimes\mathbf{e}^2)(\mathbf{v},\mathbf{u}) + 2(\mathbf{e}^3\otimes\mathbf{e}^3)(\mathbf{v},\mathbf{u}), \end{aligned}$$

and, therefore,

$$g = 3\mathbf{e}^1\otimes\mathbf{e}^1 + \mathbf{e}^2\otimes\mathbf{e}^1 + 3\mathbf{e}^2\otimes\mathbf{e}^2 + \mathbf{e}^1\otimes\mathbf{e}^2 + 2\mathbf{e}^3\otimes\mathbf{e}^3,$$

characterising a covariant type $(2,0)$ tensor. Now, g can be expressed in terms of another basis. In Example 1.3, the basis $\mathfrak{B}' = \{\mathbf{e}'_i\}$,

$$\mathbf{e}'_1 = \frac{1}{2}(\mathbf{e}_1 - \mathbf{e}_2), \qquad \mathbf{e}'_2 = \frac{1}{\sqrt{2}}\mathbf{e}_3, \qquad \mathbf{e}'_3 = \frac{1}{2\sqrt{2}}(\mathbf{e}_1 + \mathbf{e}_2),$$

is orthonormal with respect to g. From these expressions, we can straightforwardly express $\{\mathbf{e}^i\}$ in terms of $\{\mathbf{e}'^i\}$ by merely calculating $\mathbf{e}^i(\mathbf{e}'_j)$, since $\mathbf{e}^i = \mathbf{e}^i(\mathbf{e}'_j)\mathbf{e}'^j$. This calculation yields

$$\mathbf{e}^1 = \frac{1}{2}\mathbf{e}'^1 + \frac{1}{2\sqrt{2}}\mathbf{e}'^3, \qquad \mathbf{e}^2 = -\frac{1}{2}\mathbf{e}'^1 + \frac{1}{2\sqrt{2}}\mathbf{e}'^3, \qquad \mathbf{e}^3 = \frac{1}{\sqrt{2}}\mathbf{e}'^2.$$

Substituting these expressions into the one for g, we obtain

$$g = \mathbf{e}'^1\otimes\mathbf{e}'^1 + \mathbf{e}'^2\otimes\mathbf{e}'^2 + \mathbf{e}'^3\otimes\mathbf{e}'^3,$$

thus showing that the basis $\mathfrak{B}'$ is orthonormal with respect to g. On the other hand, g^{-1} is a type $(0,2)$ contravariant tensor, given by

$$g^{-1} = \frac{3}{8}\mathbf{e}_1\otimes\mathbf{e}_1 - \frac{1}{8}\mathbf{e}_1\otimes\mathbf{e}_2 - \frac{1}{8}\mathbf{e}_2\otimes\mathbf{e}_1 + \frac{3}{8}\mathbf{e}_2\otimes\mathbf{e}_2 + \frac{1}{2}\mathbf{e}_3\otimes\mathbf{e}_3.$$

In order to express g^{-1} in terms of the basis $\{\mathbf{e}'_i\}$ we use that

$$\mathbf{e}_1 = \mathbf{e}'_1 + \sqrt{2}\mathbf{e}'_3, \qquad \mathbf{e}_2 = -\mathbf{e}'_1 + \sqrt{2}\mathbf{e}'_3, \qquad \mathbf{e}_3 = \sqrt{2}\mathbf{e}'_2,$$

which implies, as expected, that

$$g^{-1} = \mathbf{e}'_1\otimes\mathbf{e}'_1 + \mathbf{e}'_2\otimes\mathbf{e}'_2 + \mathbf{e}'_3\otimes\mathbf{e}'_3.$$

Direct Sum of Vector Spaces

Let W_i be a subspace of a vector space V and let $\mathbf{v}_i \in W_i$ $(i = 1, 2, \ldots, n)$. The sum $W_1 + W_2 + \cdots + W_n$ is defined by the set of all the sums $\mathbf{v}_1 + \mathbf{v}_2 + \cdots + \mathbf{v}_n$. This sum is a subspace of V. If $W_i \cap W_j = \{0\}$ for $i \neq j$, and $V = W_1 + W_2 + \cdots + W_n$ (i.e. for each $\mathbf{v} \in V$, there are $\mathbf{w}_i \in W_i$ with $\mathbf{v} = \mathbf{v}_1 + \cdots + \mathbf{v}_n$) (Rotman, 2000), then V is said to be the *direct sum* of the W_i subspaces, denoted by

$$V = W_1 \oplus W_2 \oplus \cdots \oplus W_n = \bigoplus_{i=1}^{n} W_i.$$

Let now $\mathcal{P}$ be a linear operator in V. This operator is said to be a projection operator, or a *projection*, if $\mathcal{P}^2 = \mathcal{P}$.

Consider a vector space V defined as the direct sum $V = W_1 \oplus W_2 \oplus \cdots \oplus W_k$, in other words, that $\mathbf{v} = \mathbf{v}_1 + \mathbf{v}_2 + \cdots + \mathbf{v}_k \in V$, where $\mathbf{v}_i \in W_i$ $(i = 1, \ldots, k)$. Let $\mathcal{P}_j$ $(j = 1, \ldots, k)$ be a linear mapping defined by

$$\mathcal{P}_j(\mathbf{v}_i) = \delta_{ji}\mathbf{v}_i.$$

It follows that

$$\mathcal{P}_j(\mathbf{v}) = \mathcal{P}_j(\mathbf{v}_1 + \mathbf{v}_2 + \cdots + \mathbf{v}_k) = \mathbf{v}_j.$$

Thus, the operator $\mathcal{P}_j$ is a projection.

If $V = \bigoplus_{i=1}^k W_i$, there exists k operators $\mathcal{P}_1, \ldots, \mathcal{P}_k$ in V such that (i) $\mathcal{P}_j$ is a projector $(\mathcal{P}_j^2 = \mathcal{P}_j)$, $i = 1, \ldots, k$; (ii) $\mathcal{P}_i\mathcal{P}_j = \mathbf{0}$ for $i \neq j$; (iii) $\mathcal{P}_1 + \mathcal{P}_2 + \cdots + \mathcal{P}_k = \mathbf{1}$, where $\mathbf{1}$ denotes the identity operator $(\mathbf{1}(\mathbf{v}) = \mathbf{v})$; and (iv) $\mathcal{P}_j(V) = W_j$, that is, the image of $\mathcal{P}_j$ is W_j. Conversely, every family $(\mathcal{P}_1, \ldots, \mathcal{P}_k)$ such that $\mathcal{P}_i^2 = \mathcal{P}_i$ for $i = 1, \ldots, k$ and $\mathcal{P}_i\mathcal{P}_j = \mathbf{0}$, if $i \neq j$, determines a decomposition $V = W_1 \oplus \cdots \oplus W_k$.

Example 1.7 Let us consider the space $\mathsf{T}^2(V)$ of the covariant tensors of order 2. We characterise the spaces $\mathsf{T}^2_{\mathsf{alt}}(V)$ and $\mathsf{T}^2_{\mathsf{sym}}(V)$ by the so-called alternator and symmetriser operators, respectively, also adducing the concepts of direct sum and of projection operators. Let P be the permutation operator

$$P(\alpha \otimes \beta) = \beta \otimes \alpha,$$

and Alt and Sym the operators defined in $\mathsf{T}^2(V)$ by

$$\mathrm{Alt} = \frac{1}{2}(\mathrm{id} - P), \qquad \mathrm{Sym} = \frac{1}{2}(\mathrm{id} + P),$$

where id is the identity operator $\mathrm{id}(\alpha \otimes \beta) = \alpha \otimes \beta$. Then, the operators Alt and Sym are projection operators: $\mathrm{Alt}^2 = \mathrm{Alt} \circ \mathrm{Alt} = \mathrm{Alt}$, and $\mathrm{Sym}^2 = \mathrm{Sym} \circ \mathrm{Sym} = \mathrm{Sym}$. In order to prove this assertion, we can use the property $P^2 = P \circ P = \mathrm{id}$; for instance,

$$\begin{aligned} \mathrm{Alt}^2 &= \frac{1}{4}(\mathrm{id} - P) \circ (\mathrm{id} - P) = \frac{1}{4}(\mathrm{id} \circ \mathrm{id} - \mathrm{id} \circ P - P \circ \mathrm{id} + P \circ P) \\ &= \frac{1}{4}(\mathrm{id} - P - P + \mathrm{id}) = \frac{1}{2}(\mathrm{id} - P) = \mathrm{Alt}\,. \end{aligned}$$

In addition,

$$\mathrm{Alt} \circ \mathrm{Sym} = \mathrm{Sym} \circ \mathrm{Alt} = 0, \qquad \mathrm{Alt} + \mathrm{Sym} = \mathrm{id},$$

which implies that $\mathsf{T}^2(V) = \mathsf{T}^2_{\mathsf{alt}}(V) \oplus \mathsf{T}^2_{\mathsf{sym}}(V)$, where $\mathrm{Alt} : \mathsf{T}^2(V) \to \mathsf{T}^2_{\mathsf{alt}}(V)$, and $\mathrm{Sym} : \mathsf{T}^2(V) \to \mathsf{T}^2_{\mathsf{sym}}(V)$. Furthermore, $\ker \mathrm{Sym} = \mathsf{T}^2_{\mathsf{alt}}(V)$, and $\ker \mathrm{Alt} = \mathsf{T}^2_{\mathsf{sym}}(V)$.

1.3 Tensor Algebra

Preliminaries

An *algebra* $\mathcal{A}$ over a field $\mathbb{K}$ consists of a vector space over $\mathbb{K}$ (here, $\mathbb{K} = \mathbb{R}, \mathbb{C}$) and endowed with a bilinear product $m_{\mathcal{A}} : \mathcal{A} \times \mathcal{A} \to \mathcal{A}$.

Let G be an abelian group. An algebra $\mathcal{A}$ is said to be *G-graded* if there exists a subspace $\mathcal{A}_k$ $(k \in \mathrm{G})$ such that $\mathcal{A} = \bigoplus_k \mathcal{A}_k$ and, if given $x_k \in \mathcal{A}_k$, $y_l \in \mathcal{A}_l$, it follows that $x_k y_l \in \mathcal{A}_{k+l}$. The elements of $\mathcal{A}_k$ are said to be *homogeneous of degree* k. In general, we use the notation

$$k = \deg(x_k), \quad x_k \in \mathcal{A}_k. \tag{1.28}$$

Since G is abelian,

$$\deg(x_k y_l) = \deg(x_k) + \deg(y_l). \tag{1.29}$$

Note that, for the scalar $a \in \mathbb{K} = \mathbb{R}, \mathbb{C}$, it is assumed that $\deg(a) = 0$ and that the null vector must be considered homogeneous for all degrees, since every subspace $\mathcal{A}_k$ contains it. If the unique element which is negative graded is the null vector, the algebra is said to be positive graded.

The Tensor Algebra

Given two tensors T and S of type (p, q), it is possible to define their sum as the tensor $T + S$ of type (p, q) in terms of their components by

$$(T+S)^{\nu_1\nu_2\cdots\nu_q}_{\mu_1\mu_2\cdots\mu_p} = T^{\nu_1\nu_2\cdots\nu_q}_{\mu_1\mu_2\cdots\mu_p} + S^{\nu_1\nu_2\cdots\nu_q}_{\mu_1\mu_2\cdots\mu_p}. \tag{1.30}$$

If T is a tensor of type (p, q), and S is a tensor of type (r, s), we can define a tensor product $T \otimes S$ which is a tensor of type $(p+r, q+s)$. In terms of components, it follows that

$$(T \otimes S)^{\nu_1\nu_2\cdots\nu_q\rho_1\rho_2\cdots\rho_s}_{\mu_1\mu_2\cdots\mu_p\sigma_1\sigma_2\cdots\sigma_r} = T^{\nu_1\nu_2\cdots\nu_q}_{\mu_1\mu_2\cdots\mu_p} S^{\rho_1\rho_2\cdots\rho_s}_{\sigma_1\sigma_2\cdots\sigma_r}. \tag{1.31}$$

The tensor product is distributive with respect to the sum, namely $(T+S)\otimes R = T\otimes R + S\otimes R$ and $T \otimes (S+R) = T\otimes S + T\otimes R$; in addition, it is associative: $(T\otimes S)\otimes R = T\otimes(S\otimes R)$. As already seen, the tensor product is not commutative: in general, $T\otimes S \neq S\otimes T$. The direct sum of all the vector spaces $\mathsf{T}^p_q(V)$ endowed with the operations of sum and of tensor product is called the *tensor algebra* associated with the vector space V. The tensor algebra is a graded algebra. In the general case, the grading is given by $\mathrm{G} = \mathbb{Z}\times\mathbb{Z}$, and it is positive. Two cases are particularly important: the algebra of the covariant tensors and that of the contravariant tensors. The algebra of the covariant tensors – of type $(p, 0)$ – is denoted by $\mathsf{T}^*(V) = \bigoplus_{p=0}^{\infty} \mathsf{T}^p(V)$. We denote $\mathsf{T}^0(V) = \mathbb{R}$, and $\mathsf{T}^1(V) = V^*$. The algebra of the contravariant tensors – of type $(0, q)$ – is denoted by $\mathsf{T}(V) = \bigoplus_{q=0}^{\infty} \mathsf{T}_q(V)$, where $\mathsf{T}_0(V) = \mathbb{R}$, and $\mathsf{T}_1(V) = V$. The algebra of the covariant tensors, and the algebra of the contravariant tensors, are $\mathbb{Z}$-graded algebras.

Let us consider the algebra of the covariant tensors $\mathsf{T}^*(V)$. Since it is $\mathbb{Z}$-graded, it allows us to define a mapping called the *grade involution* as

$$\#(T_p) = (-1)^{\deg T_p} T_p = (-1)^p T_p, \tag{1.32}$$

where $T_p \in \mathsf{T}^p(V) \subset \mathsf{T}^*(V)$. Another notation that can be used for the grade involution is

$$\widehat{T_p} = \#(T_p). \tag{1.33}$$

Observation ☞ There are reasons for using both notations for the grade involution. The second notation (the 'hat') is much more convenient than the first, for its usefulness in formulæ involving the action of this mapping on an element. Nevertheless, when we solely want to denote the mapping itself, the first notation is much more appropriate than the second. As it will be seen, there are situations where one notation is more suitable than the other. In addition, there is no possibility of confusing the symbol $\#$ for the grade involution with the correlation $\sharp$. Besides the noticeable typographic difference, in general the correlation is written $\alpha^\sharp$ instead of $\sharp(\alpha)$, and in this text we never use $\#$ as an index, as in $x^\#$.

The mapping $\#$ is an automorphism. Indeed,

$$\begin{aligned}\#(T_p \otimes S_q) &= (-1)^{\deg(T_p \otimes S_q)}(T_p \otimes S_q) = (-1)^{(\deg(T_p)+\deg(S_q))}(T_p \otimes S_q) \\ &= (-1)^{\deg(T_p)}T_p \otimes (-1)^{\deg(S_q)}S_q = (\#T_p) \otimes (\#S_q).\end{aligned} \tag{1.34}$$

Besides, it also satisfies $\#^2 = 1$, where here 1 denotes the identity mapping, which justifies the denomination grade 'involution'.

Since $\#^2 = 1$, there is a refinement in the grading of $\mathsf{T}^*(V)$. An element $T_p \in \mathsf{T}_p(V)$ is said to be *even* or *odd* if $(-1)^p$ is respectively even or odd. In this way, the operators $\Pi_\pm$ can be defined as

$$\Pi_\pm = \frac{1}{2}(1 \pm \#). \tag{1.35}$$

These operators $\Pi_\pm$ are projectors, as can be straightforwardly verified. The subspace $\mathsf{T}^*_+(V) = \Pi_+(\mathsf{T}^*(V))$ consists of the even elements in $\mathsf{T}^*(V)$, and the subspace $\mathsf{T}^*_-(V) = \Pi_-(\mathsf{T}^*(V))$ consists of the odd elements. It is then possible to write $\mathsf{T}^*(V) = \mathsf{T}^*_+(V) \oplus \mathsf{T}^*_-(V)$, and

$$\begin{aligned}&\mathsf{T}^*_+(V) \otimes \mathsf{T}^*_+(V) \subset \mathsf{T}^*_+(V), \quad \mathsf{T}^*_+(V) \otimes \mathsf{T}^*_-(V) \subset \mathsf{T}^*_-(V), \\ &\mathsf{T}^*_-(V) \otimes \mathsf{T}^*_+(V) \subset \mathsf{T}^*_-(V), \quad \mathsf{T}^*_-(V) \otimes \mathsf{T}^*_-(V) \subset \mathsf{T}^*_+(V).\end{aligned} \tag{1.36}$$

The grade involution endows both the algebras of covariant and contravariant tensors with a $\mathbb{Z}_2$-grading.

Finally, another prominent and very useful mapping is called *reversion*, denoted by a tilde and defined by

$$\widetilde{(T_p \otimes S_q)} = \widetilde{S_q} \otimes \widetilde{T_p}, \tag{1.37}$$

for all $T_p \in \mathsf{T}_p(V)$, and $S_q \in \mathsf{T}_q(V)$, where

$$\begin{aligned}\tilde{a} &= a, \quad \forall a \in \mathbb{R}, \\ \tilde{\alpha} &= \alpha, \quad \forall \alpha \in \mathsf{T}_1(V) = V^*.\end{aligned} \tag{1.38}$$

This definition implies that

$$(\alpha \otimes \widetilde{\beta \otimes \cdots} \otimes \omega) = \omega \otimes \cdots \otimes \beta \otimes \alpha, \tag{1.39}$$

for $\alpha, \beta, \dots, \omega \in V^*$, which justifies the name reversion.

The composition of the grade involution and the reversion is called *conjugation* and is denoted by a bar:

$$\bar{T}_p = \widetilde{\widehat{T_p}} = \widehat{\widetilde{T_p}}. \tag{1.40}$$

Now, the tensor product of linear mappings can be defined. Let $f_i : V_i \to W_i$ $(i = 1, 2)$ be homomorphisms. Because of the universality of $V_1 \times V_2 \to V_1 \otimes V_2$, there is a linear bijection from $\mathrm{Lin}(V_1 \otimes V_2, W_1 \otimes W_2)$ onto $\mathrm{Lin}_{(2)}(V_1, V_2; W_1 \otimes W_2)$. The mapping $f_1 \otimes f_2$ is the $\mathrm{Lin}(V_1 \otimes V_2, W_1 \otimes W_2)$ element that corresponds to the $\mathrm{Lin}_{(2)}(V_1, V_2; W_1 \otimes W_2)$ element which maps every $(\mathbf{v}_1 \otimes \mathbf{v}_2)$ to $f_1(\mathbf{v}_1) \otimes f_2(\mathbf{v}_2)$. This definition can be straightforwardly extended for an arbitrary number of factors.

1.4 Exercises

(1) Let $\{\mathbf{e}_1, \mathbf{e}_2, \mathbf{e}_3\}$ be a basis of $V = \mathbb{R}^3$, where $\mathbf{e}_1 = (1, 0, 1)^\intercal$; $\mathbf{e}_2 = (1, 1, -1)^\intercal$; and $\mathbf{e}_3 = (0, 1, 2)^\intercal$. Let also α be the covector given by $\alpha(\mathbf{e}_1) = 4$; $\alpha(\mathbf{e}_2) = 1$; and $\alpha(\mathbf{e}_3) = 1$. Calculate $\alpha(\mathbf{v})$ for the vector $\mathbf{v} = (a, b, c)$ and express α in terms of the dual basis associated to the standard basis of $\mathbb{R}^3$.

(2) Let $\{\mathbf{e}_1, \mathbf{e}_2\}$ be a basis of $\mathbb{R}^2$, where $\mathbf{e}_1 = (1, -1)^\intercal$, and $\mathbf{e}_2 = (2, -1)^\intercal$, and let g be the non-degenerate symmetric bilinear form given by $g = 2\mathbf{e}^1 \otimes \mathbf{e}^1 + \mathbf{e}^1 \otimes \mathbf{e}^2 + \mathbf{e}^2 \otimes \mathbf{e}^1 + 2\mathbf{e}^2 \otimes \mathbf{e}^2$. Compute the correlation τ associated with g in terms of $\mathbf{v} = (a, b)$ and find a basis $\{\mathbf{e}'_1, \mathbf{e}'_2\}$ in terms of which g can be written in the form $g = \mathbf{e}'^1 \otimes \mathbf{e}'^1 + \mathbf{e}'^2 \otimes \mathbf{e}'^2$.

(3) Let $\mathcal{M}(n, \mathbb{R})$ be the vector space of the real matrices $n \times n$. Given $A = \{A_{ij}\} \in \mathcal{M}(n, \mathbb{R})$, define the *trace* function $\mathrm{Tr} : \mathcal{M}(n, \mathbb{R}) \to \mathbb{R}$ as $\mathrm{Tr}(A) = \sum_i A_{ii}$. (a) Show that Tr is a linear function over $\mathcal{M}(n, \mathbb{R})$. (b) Show that $\mathrm{Tr}(AB) = \mathrm{Tr}(BA)$ for all $A, B \in \mathcal{M}(n, \mathbb{R})$. (c) Show that there do not exist matrices A and B such that $AB - BA = I$, where $I \in \mathcal{M}(n, \mathbb{R})$ is the identity matrix of order n. (d) Consider now an infinite-dimensional space W. Exhibit operators $A, B \in$ in the space $\mathrm{End}(W)$ of endomorphisms of W such that $AB - BA = \mathrm{Id}_W$. (e) Suppose that $AB - BA = \mathrm{Id}_W$. Show that $A^m B - BA^m = mA^{m-1}, \quad m \in \mathbb{N}$.

(4) Let $\mathcal{M}(n, \mathbb{R})$ be the space of the real matrices $n \times n$. Consider the set of the n^2 matrices E_{ij} $(i, j = 1, \ldots, n)$ defined in the following way: all the matrix entries E_{ij} equal 0 except the entry that corresponds to the ith row and j^{th} column, as it equals 1. In other words, if the pairs kl denote the entries corresponding to the kth row and the lth column, then the matrix E_{ij} has the form $(E_{ij})_{kl} = \delta_{ik}\delta_{jl}$, where δ_{ik} equals 1 if $i = k$ and equals 0 if $i \neq k$. (a) Show that the set of the matrices $\mathfrak{B} = \{E_{ij}\}$ $(i, j = 1, \ldots, n)$ is a basis for $\mathcal{M}(n, \mathbb{R})$. (b) Show that $E_{ij}E_{mn} = \delta_{jm}E_{in}$. (c) Define the dual basis $\mathfrak{B}^* = \{E^{ij}\}$ such that $E^{ij}(E_{kl}) = \delta^i_k \delta^j_l$. Show that the components of the trace function Tr in this basis are given by $\mathrm{Tr}_{ij} = \delta_{ij}$, so that $\mathrm{Tr} = \sum_i E^{ii}$.

(5) Let $\mathrm{Lin}(V, V)$ be the set of the linear mappings of a vector space V in itself V. Consider the mapping $\phi_V : V \otimes V^* \to \mathrm{Lin}(V, V)$, defined as $(\phi_V(\mathbf{v} \otimes \alpha))(\mathbf{u}) = \mathbf{v}\alpha(\mathbf{u})$, for any $\mathbf{v}, \mathbf{u} \in V$. (a) Show that any linear transformation $T : V \to V$ can be written as $T = \phi_V(T^i_j \mathbf{e}_i \otimes \mathbf{e}^j)$ (here the summation convention is assumed!), where $\mathfrak{B} = \{\mathbf{e}_i\}$ is an arbitrary basis of V, $\mathfrak{B} = \{\mathbf{e}^i\}$ is its corresponding dual basis, and the scalar T^i_j is given by $T(\mathbf{e}_i) = T^j_i \mathbf{e}_j$. (b) Show that $\ker \phi_V = \{0\}$, in such a way that, when it is taken together with the result in item (a), we can conclude that ϕ_V is an isomorphism, namely that the spaces $\mathrm{Lin}(V, V)$ and $V \otimes V^*$ are isomorphic. (c) Show that $\phi_V(\mathbf{e}_i \otimes \mathbf{e}^i) = \mathrm{id}_V$. (d) Consider the mappings $\mathrm{ev}_V : V^* \otimes V \to \mathbb{R}$, and $\mu_{V,V^*} : V \otimes V^* \to V^* \otimes V$, defined by $\mathrm{ev}_V(\alpha \otimes \mathbf{v}) = \alpha(\mathbf{v})$, and $\mu_{V,V^*}(\mathbf{v} \otimes \alpha) = \alpha \otimes \mathbf{v}$, for all $\mathbf{v} \in V$ and $\alpha \in V^*$. Show that the trace function can be defined by the following composition of the mappings: $\mathrm{Tr} = \mathrm{ev}_V \circ \mu_{V,V^*} \circ \phi_V^{-1}$.

(6) Let $V = \mathcal{M}(n, \mathbb{K})$. Construct explicitly the isomorphism $\text{End}(V) \simeq (\text{End}(V))^*$ (Hint: show that, for any $\alpha \in V^*$, there exists a unique matrix $A \in V$ with the property $\alpha(X) = \text{Tr}(AX)$, $\forall X \in V$).

(7) Although the tensor product between two vector spaces V and W is not commutative, it is possible to establish the isomorphism

$$\mu_{V,W} : V \otimes W \to W \otimes V, \\ \mathbf{v} \otimes \mathbf{w} \mapsto \mathbf{w} \otimes \mathbf{v} \tag{1.41}$$

when the basis $\{\mathbf{e}_i \otimes \mathbf{f}_j\}$ of $V \otimes W$ is changed to the basis $\{\mathbf{f}_j \otimes \mathbf{e}_i\}$ of $W \otimes V$.

(a) Show that $\mu_{V,W} \circ \mu_{W,V} = \text{id}_{W \otimes V}$ and that $\mu_{W,V} \circ \mu_{V,W} = \text{id}_{V \otimes W}$.

(b) Given another vector space U, considering the vector space $U \otimes V \otimes W$, show that $(\mu_{V,W} \otimes \text{id}_U) \circ (\text{id}_V \otimes \mu_{U,W}) \circ (\mu_{U,V} \otimes \text{id}_W) = (\text{id}_W \otimes \mu_{U,V}) \circ (\mu_{U,W} \otimes \text{id}_V) \circ (\text{id}_U \otimes \mu_{V,W})$. This expression is called the Yang–Baxter equation and it defines the braid group.

(8) (a) Find the value of the tensor $\phi \otimes \psi - \psi \otimes \phi \in T^5(V)$ applied to $(\mathbf{v}_1, \mathbf{v}_2, \ldots, \mathbf{v}_5) \in V \times V \times V \times V \times V$, where $\phi = \mathbf{e}^1 \otimes \mathbf{e}^2 + \mathbf{e}^2 \otimes \mathbf{e}^3 + \mathbf{e}^2 \otimes \mathbf{e}^2 \in T^2(V)$; $\psi = \mathbf{e}^1 \otimes \mathbf{e}^1 \otimes (\mathbf{e}^1 - \mathbf{e}^3) \in T^3(V)$; $\mathbf{v}_1 = \mathbf{e}_1$; $\mathbf{v}_2 = \mathbf{e}_1 + \mathbf{e}_2$; $\mathbf{v}_3 = \mathbf{e}_2 + \mathbf{e}_3$; and $\mathbf{v}_4 = \mathbf{v}_5 = \mathbf{e}_2$. (b) Find the components $\tilde{T}^{12}_{123}$ of a tensor in ${T_2}^3(V)$, if all its components in the basis $\{\mathbf{e}_i\}$ are equal to 2, and the bases $\{\tilde{\mathbf{e}}_i\}$ and $\{\mathbf{e}_i\}$ are related by

$$(\tilde{\mathbf{e}}_1, \tilde{\mathbf{e}}_2, \tilde{\mathbf{e}}_3) = (\mathbf{e}_1, \mathbf{e}_2, \mathbf{e}_3) \begin{pmatrix} 1 & 2 & 3 \\ 0 & 1 & 2 \\ 0 & 0 & 1 \end{pmatrix}.$$

(9) Let $A = [a_{ij}]$ be a matrix associated with a linear operator $A : V \to V$ in the basis $\{\mathbf{e}_1, \ldots, \mathbf{e}_n\}$ of V and let $B = [b_{kl}]$ be a matrix associated with $B : V \to V$ in the basis $\{\mathbf{f}_1, \ldots, \mathbf{f}_m\}$ of W. The matrix associated with $A \otimes B$ in the basis $\{\mathbf{e}_1 \otimes \mathbf{f}_1, \mathbf{e}_1 \otimes \mathbf{f}_2, \ldots, \mathbf{e}_1 \otimes \mathbf{f}_m, \mathbf{e}_1 \otimes \mathbf{f}_1, \mathbf{e}_2 \otimes \mathbf{f}_2, \ldots, \mathbf{e}_2 \otimes \mathbf{f}_m, \ldots, \mathbf{e}_n \otimes \mathbf{f}_m\}$ of $V \otimes W$ is given by

$$A \otimes B = \begin{pmatrix} a_{11}B & a_{12}B & \cdots & a_{1n}B \\ a_{21}B & a_{22}B & \cdots & a_{2n}B \\ \vdots & \vdots & \ddots & \vdots \\ a_{n1}B & a_{n2}B & \cdots & a_{nn}B \end{pmatrix}.$$

(a) Show that $\text{Tr}(A \otimes B) = \text{Tr}\,A\,.\,\text{Tr}\,B$. Calculate $\text{Tr}(A \otimes A \otimes \cdots \otimes A)$. (b) Given $I = \sigma_0 = \begin{pmatrix} 1 & 0 \\ 0 & 1 \end{pmatrix}$, and the Pauli matrices $\sigma_1 = \begin{pmatrix} 0 & 1 \\ 1 & 0 \end{pmatrix}$, $\sigma_2 = \begin{pmatrix} 0 & -i \\ i & 0 \end{pmatrix}$, and $\sigma_3 = \begin{pmatrix} 1 & 0 \\ 0 & -1 \end{pmatrix}$ in $\mathcal{M}(2, \mathbb{C})$, compute $\sigma_i \otimes \sigma_j$ $(i, j = 0, 1, 2, 3)$.

(c) Show that, if $A : V \to V$ is diagonalisable, then $A \otimes A \otimes \cdots \otimes A$ is also diagonalisable. If $\{\lambda_i\}$ denotes the spectrum of A, what are the eigenvalues associated with the operator $A \otimes A \otimes \cdots \otimes A$?

2
Exterior Algebra and Grassmann Algebra

In this chapter, exterior algebras and Grassmann algebras are discussed. In some texts, the terms exterior algebra and Grassmann algebra are considered to be *synonyms*; however, this is **not** the case here! We aim to present the differences between the exterior algebra and the Grassmann algebra as clearly as possible, in order to avoid future confusion. The quasi-Hodge isomorphism is thus presented and studied. Basing our approach on the *Ausdehnungslehre* of 1862 (Grassmann, 1862), where Grassmann used the Hodge star operator to define the regressive product, here we will employ the quasi-Hodge operator to accomplish the quasi-Hodge isomorphism. We will end the chapter by examining the Hodge isomorphisms.

2.1 Permutations and the Alternator

A *permutation* of the set of p elements $\{1, 2, \ldots, p\}$ is a bijection $\sigma : \{1, 2, \ldots, p\} \to \{1, 2, \ldots, p\}$, represented by the cycle

$$\begin{pmatrix} 1 & 2 & \cdots & p \\ \sigma(1) & \sigma(2) & \cdots & \sigma(p) \end{pmatrix}.$$

The composition of two permutations is obviously another permutation, and the set of all permutations is a group called the *symmetric group*, denoted by S_p. The number of elements in S_p is $p!$.

A permutation σ of the set $\{1, 2, \ldots, p\}$ such that $\sigma(k) = k$ for all $k \neq i$, and $k \neq j$, and moreover $\sigma(i) = j$, and $\sigma(j) = i$, is called a *transposition*. A permutation of n elements is *even* or *odd* if the permutation is obtained respectively by an even or odd number of transpositions. The *sign* $\varepsilon(\sigma)$ of the permutation σ is defined to be $\varepsilon(\sigma) = +1$ if the permutation is *even*, and $\varepsilon(\sigma) = -1$ if the permutation is *odd*.

Let us consider now a tensor that is either contravariant or covariant of the form $X_1 \otimes X_2 \otimes \cdots \otimes X_p$, where X denotes respectively either a vector or a covector, and the indices enumerate such elements. The operator Alt called *alternator* is defined in the following way:

$$\mathrm{Alt}(X_1 \otimes X_2 \otimes \cdots \otimes X_p) = \frac{1}{p!} \sum_{\sigma \in S_p} \varepsilon(\sigma) X_{\sigma(1)} \otimes X_{\sigma(2)} \otimes \cdots \otimes X_{\sigma(p)}. \tag{2.1}$$

An Introduction to Clifford Algebras and Spinors. First Edition. Jayme Vaz, Jr. and Roldão da Rocha, Jr.

For other cases, this definition is generalised by linearity. The operator Alt is a projection operator ($\mathrm{Alt}^2 = \mathrm{Alt}$) and is a straightforward generalisation of the operator defined in example 1.7.

Example 2.1 Permutations of three elements form the symmetric group S_3. These permutations can be represented by

$$\begin{pmatrix}1\,2\,3\\1\,2\,3\end{pmatrix}, \quad \begin{pmatrix}1\,2\,3\\2\,3\,1\end{pmatrix}, \quad \begin{pmatrix}1\,2\,3\\3\,1\,2\end{pmatrix},$$
$$\begin{pmatrix}1\,2\,3\\3\,2\,1\end{pmatrix}, \quad \begin{pmatrix}1\,2\,3\\2\,1\,3\end{pmatrix}, \quad \begin{pmatrix}1\,2\,3\\1\,3\,2\end{pmatrix}.$$

The permutations represented by the matrices in the first row are all even permutations, whereas the ones in the second row are all odd permutations.

For a tensor $X_1 \otimes X_2 \otimes X_3$, which is either covariant or contravariant, this alternator is therefore written as

$$\begin{aligned}\mathrm{Alt}(X_1 \otimes X_2 \otimes X_3) =& \frac{1}{6}(X_1 \otimes X_2 \otimes X_3 + X_2 \otimes X_3 \otimes X_1 + X_3 \otimes X_1 \otimes X_2 \\ &- X_3 \otimes X_2 \otimes X_1 - X_2 \otimes X_1 \otimes X_3 - X_1 \otimes X_3 \otimes X_2).\end{aligned}$$

There is another way to express the action of the alternator. Let us, for instance, suppose that covariant tensors are taken into account. These objects are multilinear functionals whose arguments are vectors. Without loss of generality, let us consider a covariant tensor of the form $\alpha_1 \otimes \alpha_2 \otimes \cdots \otimes \alpha_p$. Therefore,

$$(\alpha_1 \otimes \alpha_2 \otimes \cdots \otimes \alpha_p)(\mathbf{v}_1, \mathbf{v}_2, \ldots, \mathbf{v}_p) = \alpha_1(\mathbf{v}_1)\alpha_2(\mathbf{v}_2)\cdots\alpha_p(\mathbf{v}_p).$$

The action of Alt on a contravariant tensor is then defined by writing the resulting tensor acting on vectors as

$$\mathrm{Alt}(\alpha_1 \otimes \cdots \otimes \alpha_p)(\mathbf{v}_1, \ldots, \mathbf{v}_k) = \frac{1}{p!}\begin{vmatrix}\alpha_1(\mathbf{v}_1) & \alpha_1(\mathbf{v}_2) & \ldots & \alpha_1(\mathbf{v}_p)\\ \alpha_2(\mathbf{v}_1) & \alpha_2(\mathbf{v}_2) & \ldots & \alpha_2(\mathbf{v}_p)\\ \vdots & \vdots & \ddots & \vdots\\ \alpha_p(\mathbf{v}_1) & \alpha_p(\mathbf{v}_2) & \ldots & \alpha_p(\mathbf{v}_p)\end{vmatrix}. \tag{2.2}$$

At the right-hand side of this equation, we denote the determinant of the associated matrix. This result follows immediately from the definition of the determinant, namely, if A is the matrix of order p with entries A_{ij}, then the determinant $\det A$ reads as follows (Hoffman and Kunze, 1971):

$$\det A = \sum_{\sigma \in S_p} \varepsilon(\sigma) A_{1\sigma(1)} A_{2\sigma(2)} \cdots A_{p\sigma(p)}.$$

2.2 p-Vectors and p-Covectors

A *p-vector* is an alternating contravariant tensor of order p. A p-vector is denoted by $A_{[p]}$ and characterised by

$$A_{[p]} = \mathrm{Alt}(A_{[p]}). \tag{2.3}$$

The brackets here are used to indicate the alternation of the p indices set. Thus, given $A_p \in \mathsf{T}_p(V)$, $\mathrm{Alt}(A_p)$ is a p-vector, since $\mathrm{Alt}(\mathrm{Alt}(A_p)) = \mathrm{Alt}(A_p)$.

Similarly, a *p-covector* is an alternating covariant tensor of order p. Denoted by $\Psi^{[p]}$, it is intrinsically alternating as well:

$$\Psi^{[p]} = \mathrm{Alt}(\Psi^{[p]}). \tag{2.4}$$

The symbols $\bigwedge_p(V)$ and $\bigwedge^p(V)$ respectively denote the space of the p-vectors and p-covectors; 1-vector is a synonym for vector, and 1-covector is a synonym for covector. Both 0-vectors or 0-covectors are scalars. Using this notation for all p, we thus consider

$$\bigwedge{}^0(V) = \bigwedge{}_0(V) = \mathbb{R}, \quad \bigwedge{}^1(V) = V^*, \quad \bigwedge{}_1(V) = V. \tag{2.5}$$

Observation ☞ It is common also to use the term bivector for a 2-vector, trivector for a 3-vector, and so on (analogously for the p-covectors).

Example 2.2 Let $\Psi = \Psi_{ij}\mathbf{e}^i \otimes \mathbf{e}^j \in \mathsf{T}^2(V)$. A 2-covector $\Psi^{[2]} \in \bigwedge^2(V)$ is alternating by definition:

$$\begin{aligned}\Psi^{[2]} &= \frac{1}{2}\Psi_{ij}(\mathbf{e}^i \otimes \mathbf{e}^j - \mathbf{e}^j \otimes \mathbf{e}^i)\\ &= \frac{1}{2}(\Psi_{ij} - \Psi_{ji})\mathbf{e}^i \otimes \mathbf{e}^j.\end{aligned}$$

Notice that, in this case, the inequality $\dim V \geq 2$ must hold since, if $\dim V = 1$, it follows that $\Psi^{[2]} = (1/2)\Psi_{11}(\mathbf{e}^1 \otimes \mathbf{e}^1 - \mathbf{e}^1 \otimes \mathbf{e}^1) = 0$.

Let now $A = A^{ijk}\mathbf{e}_i \otimes \mathbf{e}_j \otimes \mathbf{e}_k \in \mathsf{T}_3(V)$ be a 3-tensor. A 3-vector $A_{[3]} \in \bigwedge_3(V)$ can be written as an alternator $\mathrm{Alt}(A)$:

$$\begin{aligned}A_{[3]} &= \frac{1}{6}A^{ijk}(\mathbf{e}_i \otimes \mathbf{e}_j \otimes \mathbf{e}_k + \mathbf{e}_j \otimes \mathbf{e}_k \otimes \mathbf{e}_i + \mathbf{e}_k \otimes \mathbf{e}_i \otimes \mathbf{e}_j\\ &\qquad - \mathbf{e}_k \otimes \mathbf{e}_j \otimes \mathbf{e}_i - \mathbf{e}_j \otimes \mathbf{e}_i \otimes \mathbf{e}_k - \mathbf{e}_i \otimes \mathbf{e}_k \otimes \mathbf{e}_j)\\ &= \frac{1}{6}(A^{ijk} + A^{jki} + A^{kij} - A^{kji} - A^{jik} - A^{ikj})\mathbf{e}_i \otimes \mathbf{e}_j \otimes \mathbf{e}_k.\end{aligned}$$

In this case, $\dim V \geq 3$; otherwise, $A_{[3]} = 0$.

The Dimensions of $\bigwedge^p(V)$ and $\bigwedge_p(V)$

Obviously, $\dim \bigwedge^p(V) = \dim \bigwedge_p(V)$, but what are the dimensions of such spaces? Let us consider a basis for the space $\mathsf{T}^p(V)$ of the covariant tensors of order p:

$$\mathfrak{B}^p = \{\mathbf{e}^{i_1} \otimes \mathbf{e}^{i_2} \otimes \cdots \otimes \mathbf{e}^{i_p}\},$$

where the indices i_k $(k = 1, 2, \ldots, p)$ take the values $i_k = 1, 2, \ldots, n$, and where $n = \dim V$. A p-covector $\Psi^{[p]} \in \bigwedge^p(V)$ can be written as $\Psi^{[p]} = \mathrm{Alt}(\Psi^p)$, where $\Psi^p \in \mathsf{T}^p(V)$. In order to calculate the dimension of $\bigwedge^p(V)$, it suffices to realise how many elements in the basis of $\mathsf{T}^p(V)$ contribute to the space $\bigwedge^p(V)$.

Among the elements of $\mathfrak{B}^p$, we must select all the ones which are annihilated by the alternator. Hence, for the index i_1, we have n possible choices; for the index i_2, there are $n-1$ possible choices; and so on, until the pth index i_p, for which we have $n-p+1$ choices. Consequently, the number of $\mathfrak{B}^p$ elements that contributes is reduced to $n(n-1)\cdots(n-p+1)$.

Meanwhile, not all the $n(n-1)\cdots(n-p+1)$ elements of $\mathfrak{B}^p$ present distinct contributions! Any one of them can be obtained from another one by a permutation and contributes in the same way, because of the action of the alternator via permutations. For p elements, the alternator has $p!$ distinct permutations, in such a way that the quantity $n(n-1)\cdots(n-p+1)$ must be divided by $p!$. The number of elements of $\mathfrak{B}^p$ that contribute in a different way to the calculations for the dimension of $\bigwedge^p(V)$ is therefore

$$\frac{n(n-1)\cdots(n-p+1)}{p!} = \frac{n!}{(n-p)!p!} = \binom{n}{p},$$

which is the well-known number of p-combinations of n elements. Thus, we conclude that

$$\dim \bigwedge{}^p(V) = \dim \bigwedge{}_p(V) = \binom{n}{p}, \qquad (p = 0, 1, \ldots, n). \tag{2.6}$$

There can be no $(n+1)$-vectors or $(n+1)$-covectors if $\dim V = n$. This result is obvious since, in such a case, at least two indices of an alternating tensor of order $(n+1)$ would be equal, thus annihilating the alternating tensor.

Remembering that $\binom{n}{p} = \binom{n}{n-p}$, we can then obtain the following important result:

$$\boxed{\dim \bigwedge{}^p(V) = \dim \bigwedge{}^{n-p}(V)}. \tag{2.7}$$

Although the spaces $\bigwedge^p(V)$ and $\bigwedge^{n-p}(V)$ are isomorphic, *there is no* natural isomorphism between them. One isomorphism which is extremely useful and which requires the consideration of additional structures on the vector space V is called the *Hodge isomorphism*, and will be examined in section 2.10.

Observation ☞ It would be at the very least irritating (to the reader, and to the authors as well) if we were to repeat every construction involving p-vectors for p-covectors. Once the construction for one case has been accomplished, the same for the other case can be regarded mutatis mutandis. Hence, from this point on, *only the case involving p-vectors* shall be considered.

2.3 The Exterior Product

Let $A_{[p]}$ be a p-vector, and $B_{[q]}$ a q-vector. Since such quantities are alternating covariant tensors, it is natural to take the tensor product between these quantities, namely $A_{[p]} \otimes B_{[q]}$. The result of such tensor product, although it is a covariant tensor of order $p+q$, *is not* alternating. Meanwhile, the object $\mathrm{Alt}(A_{[p]} \otimes B_{[q]})$ is an alternating covariant tensor of order $p+q$, namely it is a $(p+q)$-vector.

Definition 2.1 ▶ *Let $A_{[p]} \in \bigwedge_p(V)$ be a p-vector and let $B_{[q]} \in \bigwedge_q(V)$ be a q-vector. The exterior product $\wedge : \bigwedge_p(V) \times \bigwedge_q(V) \to \bigwedge_{p+q}(V)$ is defined as*

$$\boxed{A_{[p]} \wedge B_{[q]} = \mathrm{Alt}(A_{[p]} \otimes B_{[q]})}. \tag{2.8}$$

Some consequences of this definition are now explored. First, the exterior product is *associative*, that is,

$$(A_{[p]} \wedge B_{[q]}) \wedge C_{[r]} = A_{[p]} \wedge (B_{[q]} \wedge C_{[r]}); \tag{2.9}$$

this associativity is an inherited property, elicited from the associativity of the tensor product. Obviously, the exterior product is also *bilinear*. If $a \in \bigwedge_0(V) = \mathbb{R}$ is a scalar, it follows that $a \wedge A_{[p]} = aA_{[p]}$. Equation (2.9) can be shown by starting with the action of the group of permutations S_p on the tensor space $\mathsf{T}_p(V)$. It may be either an action on the left side (such that $\sigma(\tau(A_{[p]})) = (\sigma\tau)(A_{[p]})$) or an action on the right side (such that $(A_{[p]}^{\sigma})^{\tau} = A_{[p]}^{\sigma\tau}$):

$$\begin{aligned} \text{either} \quad & \sigma(\mathbf{v}_1 \otimes \cdots \otimes \mathbf{v}_p) = \epsilon(\sigma)\mathbf{v}_{\sigma^{-1}(1)} \otimes \cdots \otimes \mathbf{v}_{\sigma^{-1}(p)}, \\ \text{or} \quad & (\mathbf{v}_1 \otimes \cdots \otimes \mathbf{v}_p)^{\sigma} = \epsilon(\sigma)\mathbf{v}_{\sigma(1)} \otimes \cdots \otimes \mathbf{v}_{\sigma(p)}\,. \end{aligned}$$

Thus, $\bigwedge^p(V)$ is the subspace of all $A_{[p]} \in \mathsf{T}_p(V)$ such that either $\sigma(A_{[p]}) = A_{[p]}$ or $A_{[p]}^{\sigma} = A_{[p]}$ for all $\sigma \in S_p$.

Moreover, the projector Alt is the classical projector from the space $\mathsf{T}_p(V)$ onto the subspace of invariant elements $\bigwedge^p(V)$. The definition of the exterior product is equivalent to

$$A_{[p]} \wedge B_{[q]} = \frac{p!q!}{(p+q)!} \sum_{\sigma \in S_{p,q}} \left(A_{[p]} \otimes B_{[q]}\right)^{\sigma},$$

where $S_{p,q}$ denotes the subset of S_{p+q} containing all σ such that $\sigma(i) < \sigma(i+1)$ if $0 < i < p$, or $p < i < p+q$. Subsequently, both sides of equality (2.9) prove to be equal to

$$\frac{p!q!r!}{(p+q+r)!} \sum_{\sigma \in S_{p,q,r}} \left(A_{[p]} \otimes B_{[q]} \otimes C_{[r]}\right),$$

where $S_{p,q,r}$ is a subset of S_{p+q+r} and contains all σ such that $\sigma(i) < \sigma(i+1)$ if (a) $0 < i < p$ or $p < i < p+q$; and (b) $0 < i < p+q$ or $p+q < i < p+q+r$.

The case involving the exterior product between two vectors is fundamental. According to the definition of an exterior product, the exterior product between two vectors reads

$$\mathbf{v} \wedge \mathbf{u} = \frac{1}{2}(\mathbf{v} \otimes \mathbf{u} - \mathbf{u} \otimes \mathbf{v}). \tag{2.10}$$

From this equation, it follows that

$$\boxed{\mathbf{v} \wedge \mathbf{u} = -\mathbf{u} \wedge \mathbf{v}}, \tag{2.11}$$

namely, that the exterior product involving two vectors is anti-commutative. In particular,

$$\mathbf{v} \wedge \mathbf{v} = 0. \tag{2.12}$$

It is straightforward to obtain the generalisation regarding eqn (2.11). Because of the bilinearity and associativity of the exterior product, it is enough to consider a p-vector $A_{[p]}$ and a q-vector $B_{[q]}$ in the form

$$A_{[p]} = \mathbf{v}_1 \wedge \cdots \wedge \mathbf{v}_p, \quad B_{[q]} = \mathbf{u}_1 \wedge \cdots \wedge \mathbf{u}_q. \tag{2.13}$$

The exterior product $A_{[p]} \wedge B_{[q]}$ can be thus written as

$$A_{[p]} \wedge B_{[q]} = \mathbf{v}_1 \wedge \cdots \wedge \mathbf{v}_p \wedge \mathbf{u}_1 \wedge \cdots \wedge \mathbf{u}_q. \tag{2.14}$$

Now using eqn (2.11), in order to reorder the vectors involved in the exterior products on the right-hand side, we find that

$$\mathbf{v}_1 \wedge \cdots \wedge \mathbf{v}_p \wedge \mathbf{u}_1 \wedge \cdots \wedge \mathbf{u}_q = (-1)^{pq} \mathbf{u}_1 \wedge \cdots \wedge \mathbf{u}_q \wedge \mathbf{v}_1 \wedge \cdots \wedge \mathbf{v}_p,$$

namely,

$$\boxed{A_{[p]} \wedge B_{[q]} = (-1)^{pq} B_{[q]} \wedge A_{[p]}}. \tag{2.15}$$

A p-vector that can be written as the exterior product of a p number of 1-vectors – as in eqn (2.13) – is called a *simple* p-vector. In vector spaces V such that $n = \dim V \leq 3$, every p-vector is simple. For $\dim V \geq 4$, not all p-vectors are simple. For instance, let V be a vector space of dimension 4, and $\mathfrak{B} = \{\mathbf{e}_1, \mathbf{e}_2, \mathbf{e}_3, \mathbf{e}_4\}$ a basis for V. Let $A_{[2]}$ be the 2-vector given by $A_{[2]} = \mathbf{e}_1 \wedge \mathbf{e}_2 + \mathbf{e}_3 \wedge \mathbf{e}_4$. There is no linear combination of the vectors $\{\mathbf{e}_i\}$ $(i = 1, 2, 3, 4)$ that allows us to write $A_{[2]} = \mathbf{v}_1 \wedge \mathbf{v}_2$. This prominent and useful result deserves a detailed exposition.

Let $0 \neq \psi \in \bigwedge_2(V)$. Then, ψ is simple if and only if $\psi \wedge \psi = 0 \in \bigwedge_4(V)$. Indeed, if $\psi = \mathbf{u} \wedge \mathbf{v}$, for $\mathbf{u}, \mathbf{v} \in V$, then $\psi \wedge \psi = \mathbf{u} \wedge \mathbf{v} \wedge \mathbf{u} \wedge \mathbf{v} = 0$. The reciprocal assertion can be shown using induction in the dimension of V. If dim $V = 0$ or 1, then $\bigwedge_2(V) = \{0\}$, and therefore the first case to be considered is when dim $V = 2$. In this case, dim $\bigwedge_2(V) = 1$, and $\mathbf{v}_1 \wedge \mathbf{v}_2$ is a non-trivial element if $\{\mathbf{v}_1, \mathbf{v}_2\}$ is a basis of V, and ψ is simple. Now let us consider the case where dim $V = 3$ is considered now. Given $0 \neq \psi \in \bigwedge_2(V)$, let us define a mapping $A : V \to \bigwedge_3(V)$ by $A(\mathbf{v}) = \psi \wedge \mathbf{v}$. Since $\dim \bigwedge_3(V) = 1$, therefore $\dim \ker A \geq 2$. Now let $\mathbf{u}_1$ and $\mathbf{u}_2$ be linearly independent vectors which are in ker A and can be extended to a basis $\{\mathbf{u}_1, \mathbf{u}_2, \mathbf{u}_3\}$ of V, so we can write $\psi = a\mathbf{u}_1 \wedge \mathbf{u}_2 + b\mathbf{u}_1 \wedge \mathbf{u}_3 + c\mathbf{u}_2 \wedge \mathbf{u}_3$. By definition, $A(\mathbf{u}_1) = 0$, and therefore $0 = \psi \wedge \mathbf{u}_1 = c\mathbf{u}_1 \wedge \mathbf{u}_2 \wedge \mathbf{u}_3$, which implies that $c = 0$. In the same way, $A(\mathbf{u}_2) = 0$, and so $0 = \psi \wedge \mathbf{u}_2 = -b\mathbf{u}_1 \wedge \mathbf{u}_2 \wedge \mathbf{u}_3$, which means that $b = 0$. It follows that $\psi = a\mathbf{u}_1 \wedge \mathbf{u}_2$, which is simple. Suppose now by induction that the assumption holds for dim $V \leq n - 1$, and consider the case where dim $V = n$. Using the basis $\{\mathbf{v}_1, \ldots, \mathbf{v}_n\}$, it follows that

$$\begin{aligned} \psi = \sum_{1 \leq i < j}^{n} a_{ij} \mathbf{v}_i \wedge \mathbf{v}_j &= \left(\sum_{i=1}^{n-1} a_{in} \mathbf{v}_i \right) \wedge \mathbf{v}_n + \sum_{1 \leq i < j}^{n-1} a_{ij} \mathbf{v}_i \wedge \mathbf{v}_j \\ &= \mathbf{u} \wedge \mathbf{v}_n + \psi', \end{aligned}$$

where U is the subspace generated by $\{\mathbf{v}_1, \ldots, \mathbf{v}_{n-1}\}$, $\mathbf{u} \in U$, and $\psi' \in \bigwedge_2(U)$. Now,

$$0 = \psi \wedge \psi = (\mathbf{u} \wedge \mathbf{v}_n + \psi') \wedge (\mathbf{u} \wedge \mathbf{v}_n + \psi') = 2\mathbf{u} \wedge \psi' \wedge \mathbf{v}_n + \psi' \wedge \psi',$$

but $\mathbf{v}_n$ appears neither in the expansion of $\mathbf{u} \wedge \psi'$ nor in the expansion of $\psi' \wedge \psi'$, and separately one obtains $\mathbf{u} \wedge \psi' = 0 = \psi' \wedge \psi'$. By induction, $0 = \psi' \wedge \psi'$ implies that

$\psi' = \mathbf{u}_1 \wedge \mathbf{u}_2$, and so $\mathbf{u} \wedge \mathbf{u}_1 \wedge \mathbf{u}_2 = 0$. Hence, there exists $\mu, \lambda_1, \lambda_2 \in \mathbb{K}$ such that $\mu\mathbf{u} + \lambda_2\mathbf{u}_2 + \lambda_1\mathbf{u}_1 = 0$. If $\mu = 0$, then $\mathbf{u}_1$ and $\mathbf{u}_2$ are linearly dependent, and therefore $\psi' = \mathbf{u}_1 \wedge \mathbf{u}_2 = 0$, which means that $\psi = \mathbf{u} \wedge \mathbf{v}_n$, and therefore ψ is simple. If $\mu \neq 0$, then $\mathbf{u} = -\frac{\lambda_2}{\mu}\mathbf{u}_2 - \frac{\lambda_1}{\mu}\mathbf{u}_1$, and

$$\psi = -\frac{\lambda_1}{\mu}\mathbf{u}_1 \wedge \mathbf{v}_n - \frac{\lambda_2}{\mu}\mathbf{u}_2 \wedge \mathbf{v}_n + \mathbf{u}_1 \wedge \mathbf{u}_2,$$

corresponding to the three-dimensional case, which was shown to be always simple.

Observation ☞ There is no consensus with respect to the definition of the exterior product. Some authors define the exterior product of vectors as in eqn (2.10), whereas other authors define it without the factor 1/2, namely, $\mathbf{v} \wedge \mathbf{u} = \mathbf{v} \otimes \mathbf{u} - \mathbf{u} \otimes \mathbf{v}$. The source of this difference lies in the definition of the alternating operator: some authors use the factor $1/p!$ as we use it in eqn (2.1), whereas other authors prefer not to use it. However, this difference is completely irrelevant when the necessary attention is taken and the corresponding adaptations made in the definitions. Without the factor $1/p!$, the operator Alt is not a projection operator. Therefore, in order for us to use it as a projection operators, we must employ a factor p in the definition of the contraction (or interior product; see eqn (2.34)). Authors that do not use the factor $1/p!$ in the definition of a p-vector do not need to use the factor p in eqn (2.34) but must use the factor $1/p!q!$ in eqn (2.8). For those mathematicians who need exterior algebras over arbitrary fields see, for instance, the work by Lam (1980), Lounesto (2001*a*), and Helmstetter and Micali (2008).

Bases

Suppose that $\dim V = n$ and let $\mathfrak{B} = \{\mathbf{e}_1, \ldots, \mathbf{e}_n\}$ be a basis for V. From this basis, we can then construct a basis for each one of the spaces $\bigwedge_p(V)$.

Consider at first the space $\bigwedge_2(V)$ and the exterior products of the form $\mathbf{e}_i \wedge \mathbf{e}_j$. Because of the anti-commutativity of the exterior product between basis vectors, the linearly independent set of 2-vectors is provided by

$$\begin{aligned}
&\mathbf{e}_1 \wedge \mathbf{e}_2, \mathbf{e}_1 \wedge \mathbf{e}_3, \mathbf{e}_1 \wedge \mathbf{e}_4, \ldots, \mathbf{e}_1 \wedge \mathbf{e}_n,\\
&\mathbf{e}_2 \wedge \mathbf{e}_3, \mathbf{e}_2 \wedge \mathbf{e}_4, \ldots, \mathbf{e}_2 \wedge \mathbf{e}_n,\\
&\cdots \quad \cdots \quad \cdots\\
&\mathbf{e}_{n-1} \wedge \mathbf{e}_n.
\end{aligned}$$

Therefore, there exist $(n-1)+(n-2)+\cdots+1 = n(n-1)/2$ elements, which is precisely $\dim \bigwedge_2(V)$. This set of 2-vectors forms a basis for $\bigwedge_2(V)$. An arbitrary 2-vector $A_{[2]}$ can thus be written as

$$A_{[2]} = \frac{1}{2}\sum_{i,j} A^{ij}\mathbf{e}_i \wedge \mathbf{e}_j = \sum_{i<j} A^{ij}\mathbf{e}_i \wedge \mathbf{e}_j. \tag{2.16}$$

Notice in the first expression the presence of the factor 1/2, which is absent in the second expression. Since, in the sum $\sum_{ij}$, we consider all the values for the indices

$i, j = 1, 2, \ldots, n$ and, as $A^{ij} = -A^{ji}$, and $\mathbf{e}_i \wedge \mathbf{e}_j = -\mathbf{e}_j \wedge \mathbf{e}_i$, we are indeed counting the same term twice. However, this factor is not necessary in the second expression, since we consider the sum $\sum_{i<j}$ restricted to all values of i and j such that $i < j$, so that no term is counted twice.

This result can be generalised for $\bigwedge_p(V)$. A basis for this space consists of elements of the form $\mathbf{e}_{\mu_1} \wedge \mathbf{e}_{\mu_2} \wedge \cdots \wedge \mathbf{e}_{\mu_p}$, and the number of distinct elements is the number of p-combinations of n elements, denoted by $\binom{n}{p}$. An arbitrary element $A_{[p]} \in \bigwedge_p(V)$ can be written as

$$\begin{aligned} A_{[p]} &= \frac{1}{p!} \sum_{\mu_1\mu_2\cdots\mu_p} A^{\mu_1\mu_2\cdots\mu_p} \mathbf{e}_{\mu_1} \wedge \mathbf{e}_{\mu_2} \wedge \cdots \wedge \mathbf{e}_{\mu_k} \\ &= \sum_{\mu_1<\mu_2<\cdots<\mu_p} A^{\mu_1\mu_2\cdots\mu_p} \mathbf{e}_{\mu_1} \wedge \mathbf{e}_{\mu_2} \wedge \cdots \wedge \mathbf{e}_{\mu_p}, \end{aligned} \tag{2.17}$$

where, in the first case, we consider the sum over all possible values for the indices $\mu_i = 1, 2, \ldots, n$ $(i = 1, 2, \ldots, p)$ and, in the second case, we consider the sum over the indices with the restriction that $\mu_1 < \mu_2 < \cdots < \mu_p$.

Observation ☞ Sum convention: In order to apply the sum convention to p-vectors as efficiently as possible, we impose the same restriction on the sum convention with respect to the indices, namely, that

$$A^{\mu_1\mu_2\cdots\mu_p} \mathbf{e}_{\mu_1} \wedge \mathbf{e}_{\mu_2} \wedge \cdots \wedge \mathbf{e}_{\mu_p} = \sum_{\mu_1<\mu_2<\cdots<\mu_p} A^{\mu_1\mu_2\cdots\mu_p} \mathbf{e}_{\mu_1} \wedge \mathbf{e}_{\mu_2} \wedge \cdots \wedge \mathbf{e}_{\mu_p}. \tag{2.18}$$

Thus, we avoid using expressions with the factor $p!$.

n-Vectors, or Pseudoscalars

The exterior product of m vectors is 0 whenever $m > n$, where $n = \dim V$. Indeed, let us consider the exterior product of $n+1$ vectors, $\mathbf{v}_1 \wedge \cdots \wedge \mathbf{v}_n \wedge \mathbf{v}_{n+1}$. If $\dim V = n$, there are at most n linearly independent vectors. Therefore, the $n+1$ given vectors are necessarily linearly dependent and we can write one of those vectors as a linear combination of the others. Without loss of generality, it is possible to choose such a vector as being $\mathbf{v}_{n+1}$. It follows that $\mathbf{v}_{n+1} = \sum_{i=1}^n a^i \mathbf{v}_i$. Using the anti-commutation relation (2.11), it reads

$$\begin{aligned} \mathbf{v}_1 \wedge \cdots \wedge \mathbf{v}_n \wedge \mathbf{v}_{n+1} =& \mathbf{v}_1 \wedge \mathbf{v}_2 \wedge \cdots \wedge \mathbf{v}_n \wedge (a^1\mathbf{v}_1 + a^2\mathbf{v}_2 + \cdots + a^n\mathbf{v}_n) \\ =& (-1)^{n-1} a^1 \mathbf{v}_1 \wedge \mathbf{v}_1 \wedge \mathbf{v}_2 \wedge \cdots \wedge \mathbf{v}_n \\ &+ (-1)^{n-2} a^2 \mathbf{v}_1 \wedge \mathbf{v}_2 \wedge \mathbf{v}_2 \wedge \mathbf{v}_3 \wedge \cdots \wedge \mathbf{v}_n \\ &+ \cdots + a^n \mathbf{v}_1 \wedge \cdots \wedge \mathbf{v}_{n-1} \wedge \mathbf{v}_n \wedge \mathbf{v}_n = 0, \end{aligned}$$

where $\mathbf{v}_i \wedge \mathbf{v}_i = 0$ is used. Consequently, we have

$$\mathbf{v}_1 \wedge \mathbf{v}_2 \wedge \cdots \wedge \mathbf{v}_m = 0 \quad \text{if} \quad m > n. \tag{2.19}$$

Indeed, a stronger result than that can be shown, namely, that

$$\mathbf{v}_1 \wedge \cdots \wedge \mathbf{v}_p = 0 \Longleftrightarrow \{\mathbf{v}_1, \ldots, \mathbf{v}_p\} \text{ is linearly dependent.}$$

This result shows that the vector space $\bigwedge_p(V)$ does not exist if $p > n$. The spaces that can be constructed are

$$\bigwedge{}_0(V), \bigwedge{}_1(V), \bigwedge{}_2(V), \ldots, \bigwedge{}_{n-1}(V), \bigwedge{}_n(V),$$

such that $\dim \bigwedge_p(V) = \dim \bigwedge_{n-p}(V)$. The space $\bigwedge_n(V)$ is particularly important.

The vector space $\bigwedge_n(V)$ has dimension $\binom{n}{n} = 1$. A basis for this space consists of the element $\mathbf{v}_1 \wedge \cdots \wedge \mathbf{v}_n$, where $\{\mathbf{v}_1, \ldots, \mathbf{v}_n\}$ is a set of linearly independent vectors. If $\mathfrak{B} = \{\mathbf{e}_1, \ldots, \mathbf{e}_n\}$ is a basis of V, it is natural to take as a basis for $\bigwedge_n(V)$ the exterior product of these n vectors of $\mathfrak{B}$. Whatever order these n vectors are taken, their product can be transformed into

$$\pm\mathbf{e}_1 \wedge \cdots \wedge \mathbf{e}_n,$$

because of the anti-commutativity associated with the exterior product. For instance, $\mathbf{e}_n \wedge \cdots \wedge \mathbf{e}_1 = (-1)^{n(n-1)/2}\mathbf{e}_1 \wedge \cdots \wedge \mathbf{e}_n$, where $(-1)^{n(n-1)/2} = \pm 1$, depending upon the value of n. For a basis for $\bigwedge_n(V)$, we shall consider the element

$$\mathbf{e}_1 \wedge \cdots \wedge \mathbf{e}_n,$$

and hence the exterior product of n linearly independent vectors can be always written as

$$\mathbf{v}_1 \wedge \cdots \wedge \mathbf{v}_n = p\mathbf{e}_1 \wedge \cdots \wedge \mathbf{e}_n, \tag{2.20}$$

where $p \in \mathbb{R}$. The elements of $\bigwedge_n(V)$ are called n-vectors, as before, and another usual denomination for the n-vectors is *pseudoscalars*.

2.4 The Exterior Algebra $\bigwedge(V)$

Let us consider the vector space $\bigwedge(V)$ defined by the direct sum of the vector spaces $\bigwedge_p(V)$, $(p = 0, 1, 2, \ldots, n)$:

$$\bigwedge(V) = \bigwedge{}_0(V) \oplus \bigwedge{}_1(V) \oplus \bigwedge{}_2(V) \oplus \cdots \oplus \bigwedge{}_n(V) = \bigoplus_{p=0}^{n} \bigwedge{}_p(V).$$

The exterior product of a p-vector and a q-vector is such that $\wedge : \bigwedge_p(V) \times \bigwedge_q(V) \to \bigwedge_{p+q}(V)$. The space $\bigwedge(V)$ is thereby closed with respect to the exterior product, when one extends by linearity what was discussed in section 2.3.

Definition 2.2 ▶ *The pair* $(\bigwedge(V), \wedge)$ *is called the exterior algebra associated with the vector space* V.

Arbitrary elements of $\bigwedge(V)$ will be called *multivectors*. An arbitrary multivector is therefore the sum of a scalar, a 1-vector, a 2-vector, ..., up to an n-vector or pseudoscalar. We can write in a general manner

$$\begin{aligned} A = \underbrace{a}_{\text{scalar}} + \underbrace{v^i \mathbf{e}_i}_{\text{vector}} + \underbrace{F^{ij}\mathbf{e}_i \wedge \mathbf{e}_j}_{\text{2-vector}} \\ + \underbrace{T^{ijk}\mathbf{e}_i \wedge \mathbf{e}_j \wedge \mathbf{e}_k}_{\text{3-vector}} + \cdots + \underbrace{p\mathbf{e}_1 \wedge \cdots \wedge \mathbf{e}_n}_{n\text{-vector}} \in \bigwedge(V). \end{aligned} \tag{2.21}$$

The dimension of $\bigwedge(V)$ can be straightforwardly calculated by using the well-known result

$$\binom{n}{0} + \binom{n}{1} + \binom{n}{2} + \cdots + \binom{n}{n-1} + \binom{n}{n} = \sum_{p=0}^{n}\binom{n}{p} = 2^n.$$

The dimension of $\bigwedge(V)$ is 2^n, since

$$\dim \bigwedge(V) = \sum_{p=0}^{n} \dim \bigwedge{}_p(V) = \sum_{p=0}^{n}\binom{n}{p} = 2^n. \tag{2.22}$$

We denote by $\langle\ \rangle_p$ the projector $\langle\ \rangle_p : \bigwedge(V) \to \bigwedge_p(V)$, in such a way that

$$\langle A\rangle_p = A_{[p]}, \tag{2.23}$$

where $A_{[p]} \in \bigwedge_p(V)$ is the p-vector component of the multivector A.

The operations grade involution, reversion, and conjugation, as defined in the tensor algebra, descend to the exterior algebra, since they leave the ideal $\mathcal{I}_E$ in eqn (2.29) invariant (see section 2.5). The grade involution is given by

$$\#(A_{[p]}) = \widehat{A}_{[p]} = (-1)^p A_{[p]}. \tag{2.24}$$

For the reversion, it follows that

$$(\widetilde{\mathbf{v}_1 \wedge \cdots \wedge \mathbf{v}_p}) = \mathbf{v}_p \wedge \cdots \wedge \mathbf{v}_1, \tag{2.25}$$

which implies that

$$\widetilde{A}_{[p]} = (-1)^{p(p-1)/2} A_{[p]}. \tag{2.26}$$

The conjugation is known to be the composition of the two operations

$$\bar{A}_{[p]} = \widetilde{\widehat{A}}_{[p]} = \widehat{\widetilde{A}}_{[p]}. \tag{2.27}$$

Now let us define the following notation:

$\bigwedge(V)$ = exterior algebra of V	$\bigwedge^*(V)$ = exterior algebra of V^*
$\mathbf{v}, \mathbf{u}, \ldots$ = vectors	$\alpha, \beta, \ldots$ = covectors
$A, B, \ldots$ = multivectors	$\Psi, \Phi, \ldots$ = multicovectors

One more word about this notation: whereas A_1 and A_2 denote two distinct multivectors, $A_{[1]}$ and $A_{[2]}$ respectively denote the 1-vector and 2-vector parts of the multivector A.

Example 2.3 Consider $V = \mathbb{R}^3$, and $A, B \in \bigwedge(\mathbb{R}^3)$ given by

$$A = 2 + 3\mathbf{e}_1 - \mathbf{e}_1 \wedge \mathbf{e}_2 \wedge \mathbf{e}_3,$$
$$B = -10 + \mathbf{e}_2 + 4\mathbf{e}_1 \wedge \mathbf{e}_2 + 3\mathbf{e}_2 \wedge \mathbf{e}_3.$$

The multivector A has scalar, 1-vector and 3-vector (or pseudoscalar) parts, and B has scalar, 1-vector and 2-vector components. For instance, the projector $\langle\ \rangle_2$ acting on those multivectors reads

$$\langle A\rangle_2 = 0, \qquad \langle B\rangle_2 = 4\mathbf{e}_1 \wedge \mathbf{e}_2 + 3\mathbf{e}_2 \wedge \mathbf{e}_3.$$

The exterior product $A \wedge B$ is then

$$-20 - 30\mathbf{e}_1 + 2\mathbf{e}_2 + 11\mathbf{e}_1 \wedge \mathbf{e}_2 + 6\mathbf{e}_2 \wedge \mathbf{e}_3 + 19\mathbf{e}_1 \wedge \mathbf{e}_2 \wedge \mathbf{e}_3,$$

whereas the exterior product $B \wedge A$ is

$$-20 - 30\mathbf{e}_1 + 2\mathbf{e}_2 + 5\mathbf{e}_1 \wedge \mathbf{e}_2 + 6\mathbf{e}_2 \wedge \mathbf{e}_3 + 19\mathbf{e}_1 \wedge \mathbf{e}_2 \wedge \mathbf{e}_3.$$

It is then clear that the exterior product between multivectors is neither commutative nor anti-commutative. Such properties generally only hold when the multivectors are simple, as shown in eqn (2.15). Furthermore,

$$A \wedge A = 4 + 12\mathbf{e}_1 - 4\mathbf{e}_1 \wedge \mathbf{e}_2 \wedge \mathbf{e}_3.$$

Indeed, $\mathbf{v} \wedge \mathbf{v} = 0$ always when $\mathbf{v} \in V \subset \bigwedge(V)$, since in this case the exterior product is anti-commutative and consequently the product of an element with itself must equal 0. If the exterior product of the multivectors is commutative or if it is neither commutative nor anti-commutative, in general the exterior product of an element with itself does not necessarily equal 0.

Example 2.4 Let W be a vector subspace of $V = \mathbb{R}^4$, generated by the vectors $\mathbf{v}_1 = (2,1,0,1)^\intercal$ and $\mathbf{v}_2 = (1,0,0,-1)^\intercal$. The 2-vector $I_W = \mathbf{v}_1 \wedge \mathbf{v}_2$ is the pseudoscalar in W. A vector $\mathbf{v}$ is in W if $\mathbf{v} \wedge I_W = 0$. Indeed, I_W is given by

$$\begin{aligned} I_W &= \mathbf{v}_1 \wedge \mathbf{v}_2 = (2\mathbf{e}_1 + \mathbf{e}_2 + \mathbf{e}_4) \wedge (\mathbf{e}_1 - \mathbf{e}_4) \\ &= -\mathbf{e}_1 \wedge \mathbf{e}_2 - 3\mathbf{e}_1 \wedge \mathbf{e}_4 - \mathbf{e}_2 \wedge \mathbf{e}_4. \end{aligned}$$

When $\mathbf{v} \wedge I_W$ is calculated, it follows that

$$\begin{aligned} \mathbf{v} \wedge I_W =& (-v^1 + 3v^2 - v^4)\mathbf{e}_1 \wedge \mathbf{e}_2 \wedge \mathbf{e}_4 + (-v^3)\mathbf{e}_1 \wedge \mathbf{e}_2 \wedge \mathbf{e}_3 \\ &+ (3v^3)\mathbf{e}_1 \wedge \mathbf{e}_3 \wedge \mathbf{e}_4 + (v^3)\mathbf{e}_2 \wedge \mathbf{e}_3 \wedge \mathbf{e}_4, \end{aligned}$$

where we write $\mathbf{v} = v^i\mathbf{e}_i$. The condition $\mathbf{v} \wedge I_W = 0$ then reveals that

$$v^1 - 3v^2 + v^4 = 0, \qquad v^3 = 0.$$

Taking $v^2 = a$, and $v^4 = a - b$, it follows that $v^1 = 2a + b$, $\mathbf{v}$ can be written as $\mathbf{v} = a\mathbf{v}_1 + b\mathbf{v}_2$, that is, $\mathbf{v} \in W$. The space W can be characterised by the 2-vector $\mathbf{v}_1 \wedge \mathbf{v}_2$. On the other hand, if $\mathbf{v} \wedge I_W \neq 0$, there exists another vector space W' characterised by its pseudoscalar $I_{W'} = \mathbf{v} \wedge I_W$. Let $\mathbf{v}_3 = (1,0,-1,1)^\intercal$ be such a vector. Indeed,

$$\begin{aligned} I_{W'} =& \mathbf{v}_3 \wedge I_W = \mathbf{v}_1 \wedge \mathbf{v}_2 \wedge \mathbf{v}_3 = -\mathbf{e}_1 \wedge \mathbf{e}_2 \wedge \mathbf{e}_3 \\ &- 2\mathbf{e}_1 \wedge \mathbf{e}_2 \wedge \mathbf{e}_4 + 3\mathbf{e}_1 \wedge \mathbf{e}_3 \wedge \mathbf{e}_4 + \mathbf{e}_2 \wedge \mathbf{e}_3 \wedge \mathbf{e}_4. \end{aligned}$$

The condition $\mathbf{v} \in W'$ is equivalent to $\mathbf{v} \wedge I_{W'}$ and implies that

$$\mathbf{v} \wedge I_{W'} = (v^1 + 3v^2 - 2v^3 - v^4)\mathbf{e}_1 \wedge \mathbf{e}_2 \wedge \mathbf{e}_3 \wedge \mathbf{e}_4.$$

Taking $v^2 = a$, together with $v^3 = -c$, and $v^4 = a - b + c$, it follows that $v^1 = 2a + b + c$, and $\mathbf{v} = a\mathbf{v}_1 + b\mathbf{v}_2 + c\mathbf{v}_3$.

Example 2.5 An interesting and prominent application of the exterior algebra is the solution of a linear system of equations. Consider a system of 3 equations and 3 variables (the generalisation for a system of n equations and n variables is trivial):

$$\begin{aligned} a_{11}x_1 + a_{12}x_2 + a_{13}x_3 &= y_1, \\ a_{21}x_1 + a_{22}x_2 + a_{23}x_3 &= y_2, \\ a_{31}x_1 + a_{32}x_2 + a_{33}x_3 &= y_3, \end{aligned}$$

that can be forthwith written as

$$\begin{pmatrix} a_{11} & a_{12} & a_{13} \\ a_{21} & a_{22} & a_{23} \\ a_{31} & a_{32} & a_{33} \end{pmatrix} \begin{pmatrix} x_1 \\ x_2 \\ x_3 \end{pmatrix} = \begin{pmatrix} y_1 \\ y_2 \\ y_3 \end{pmatrix}.$$

Now let us take a basis $\{\mathbf{e}_1, \mathbf{e}_2, \mathbf{e}_3\}$ of $\mathbb{R}^3$ and define the following vectors:

$$\begin{aligned} \mathbf{a}_1 &= a_{11}\mathbf{e}_1 + a_{21}\mathbf{e}_2 + a_{31}\mathbf{e}_3, \\ \mathbf{a}_2 &= a_{12}\mathbf{e}_1 + a_{22}\mathbf{e}_2 + a_{32}\mathbf{e}_3, \\ \mathbf{a}_3 &= a_{13}\mathbf{e}_1 + a_{23}\mathbf{e}_2 + a_{33}\mathbf{e}_3, \\ \mathbf{y} &= y_1\mathbf{e}_1 + y_2\mathbf{e}_2 + y_3\mathbf{e}_3. \end{aligned}$$

The linear systems of equations can thus be expressed as

$$x_1\mathbf{a}_1 + x_2\mathbf{a}_2 + x_3\mathbf{a}_3 = \mathbf{y}.$$

The exterior product between the vector equation and the 2-vector $\mathbf{a}_1 \wedge \mathbf{a}_2$ can be calculated. Since $\mathbf{a}_1 \wedge \mathbf{a}_2 \wedge \mathbf{a}_1 = \mathbf{a}_1 \wedge \mathbf{a}_2 \wedge \mathbf{a}_2 = 0$, it follows that

$$x_3\mathbf{a}_1 \wedge \mathbf{a}_2 \wedge \mathbf{a}_3 = \mathbf{a}_1 \wedge \mathbf{a}_2 \wedge \mathbf{y}.$$

Now let us calculate the exterior products. First,

$$\mathbf{a}_1 \wedge \mathbf{a}_2 = (a_{11}a_{22} - a_{21}a_{12})\mathbf{e}_1 \wedge \mathbf{e}_2 + (a_{11}a_{32} - a_{31}a_{12})\mathbf{e}_1 \wedge \mathbf{e}_3 + (a_{21}a_{32} - a_{31}a_{22})\mathbf{e}_2 \wedge \mathbf{e}_3.$$

The exterior product of this 2-vector and $\mathbf{a}_3$ is given by

$$\begin{aligned} \mathbf{a}_1 \wedge \mathbf{a}_2 \wedge \mathbf{a}_3 = &[a_{33}(a_{11}a_{22} - a_{21}a_{12}) - a_{23}(a_{11}a_{32} - a_{31}a_{12}) \\ &+ a_{13}(a_{21}a_{32} - a_{31}a_{22})]\mathbf{e}_1 \wedge \mathbf{e}_2 \wedge \mathbf{e}_3, \end{aligned}$$

which can be regarded, once the determinant function is introduced, as

$$\mathbf{a}_1 \wedge \mathbf{a}_2 \wedge \mathbf{a}_3 = \begin{vmatrix} a_{11} & a_{12} & a_{13} \\ a_{21} & a_{22} & a_{23} \\ a_{31} & a_{32} & a_{33} \end{vmatrix} \mathbf{e}_1 \wedge \mathbf{e}_2 \wedge \mathbf{e}_3.$$

On the other hand, the exterior product $\mathbf{a}_1 \wedge \mathbf{a}_2 \wedge \mathbf{y}$ implies that

$$\mathbf{a}_1 \wedge \mathbf{a}_2 \wedge \mathbf{y} = [y_3(a_{11}a_{22} - a_{21}a_{12}) - y_2(a_{11}a_{32} - a_{31}a_{12}) + y_1(a_{21}a_{32} - a_{31}a_{22})]\mathbf{e}_1 \wedge \mathbf{e}_2 \wedge \mathbf{e}_3,$$

which can be expressed as

$$\mathbf{a}_1 \wedge \mathbf{a}_2 \wedge \mathbf{y} = \begin{vmatrix} a_{11} & a_{12} & y_1 \\ a_{21} & a_{22} & y_2 \\ a_{31} & a_{32} & y_3 \end{vmatrix} \mathbf{e}_1 \wedge \mathbf{e}_2 \wedge \mathbf{e}_3.$$

Therefore, the vector equation $x_3\mathbf{a}_1 \wedge \mathbf{a}_2 \wedge \mathbf{a}_3 = \mathbf{a}_1 \wedge \mathbf{a}_2 \wedge \mathbf{y}$ has a solution if $\mathbf{a}_1 \wedge \mathbf{a}_2 \wedge \mathbf{a}_3 \neq 0$, that is, if

$$\begin{vmatrix} a_{11} & a_{12} & a_{13} \\ a_{21} & a_{22} & a_{23} \\ a_{31} & a_{32} & a_{33} \end{vmatrix} \neq 0,$$

where in this case x_3 is given by

$$x_3 = \frac{\begin{vmatrix} a_{11} & a_{12} & y_1 \\ a_{21} & a_{22} & y_2 \\ a_{31} & a_{32} & y_3 \end{vmatrix}}{\begin{vmatrix} a_{11} & a_{12} & a_{13} \\ a_{21} & a_{22} & a_{23} \\ a_{31} & a_{32} & a_{33} \end{vmatrix}}.$$

Repeating the same procedure but using instead the 2-vectors $\mathbf{a}_1 \wedge \mathbf{a}_3$ and $\mathbf{a}_2 \wedge \mathbf{a}_3$, we find that

$$x_2 = \frac{\begin{vmatrix} a_{11} & y_1 & a_{13} \\ a_{21} & y_2 & a_{23} \\ a_{31} & y_3 & a_{33} \end{vmatrix}}{\begin{vmatrix} a_{11} & a_{12} & a_{13} \\ a_{21} & a_{22} & a_{23} \\ a_{31} & a_{32} & a_{33} \end{vmatrix}}, \qquad x_1 = \frac{\begin{vmatrix} y_1 & a_{12} & a_{13} \\ y_2 & a_{22} & a_{23} \\ y_3 & a_{32} & a_{33} \end{vmatrix}}{\begin{vmatrix} a_{11} & a_{12} & a_{13} \\ a_{21} & a_{22} & a_{23} \\ a_{31} & a_{32} & a_{33} \end{vmatrix}}.$$

The final expressions for x_1, x_2, and x_3 comprise the well-known Cramer's rule. Clearly, this procedure can be immediately extended to high dimensions, corresponding to systems of linear equations of high order and can be straightforwardly computed.

2.5 The Exterior Algebra as the Quotient of the Tensor Algebra

In this section, we present another construction of the exterior algebra. Besides it being a prominent construction itself, a construction that is analogous to it can be used for Clifford algebras. Therefore, this approach is particularly interesting.

Equivalence Relations

An equivalence relation R in a set X is a relation that is (i) *reflexive*, (ii) *symmetric*, and (iii) *transitive*, namely, (i) xRx; (ii) if xRy then yRx; and (iii) if xRy and yRz, then xRz, $\forall x, y, z \in X$. The set of all elements that are equivalent to x constitutes an equivalence class of x, denoted by $[x] = \{y \in X \mid yRx\}$. The set of such (disjoint) equivalence classes is denoted by $\mathcal{X} = X/R = \{[x] \mid x \in X\}$. A notation that is frequently used for xRy is $x \sim y$.

Example 2.6 In the set of integers $\mathbb{Z}$, we can define the following equivalence relation:

$$m \sim n \Leftrightarrow m - n = kN, \quad k = 0, \pm 1, \pm 2, \ldots,$$

namely, that the integers m and n are considered equivalent if they differ by a multiple of a fixed natural number N. The equivalence class of m consists therefore of the set

$$[m] = \{n \in \mathbb{Z} | n = m + kN\} = \{m, m \pm N, m \pm 2N, m \pm 3N, \ldots\}.$$

The set of all equivalence classes is the set $\mathbb{Z}_N = \mathbb{Z}/\sim$, the set of integers modulo N.

Ideals

Let $\mathcal{A}$ denote an algebra. A set $I_L \subset \mathcal{A}$ is said to be a *left ideal* of $\mathcal{A}$ if, for all $a \in \mathcal{A}$ and for all $x \in I_L$, the relation $ax \in I_L$ holds. Analogously, $I_R \subset \mathcal{A}$ is said to be a *right ideal* of $\mathcal{A}$ if, for all $a \in \mathcal{A}$ and for all $x \in I_R$, it follows that $xa \in I_R$. The set $I \subset \mathcal{A}$ is said to be a *bilateral ideal* (or simply an *ideal*) of $\mathcal{A}$ if, $\forall a, b \in \mathcal{A}$ and $\forall x \in I$, it implies that $axb \in I$.

Given an algebra $\mathcal{A}$, let us write it as $\mathcal{A} = \mathcal{B} + \mathcal{C}$. Here, it is not supposed a priori that $\mathcal{B}$ and $\mathcal{C}$ are subalgebras. Notwithstanding, only the properties of the vector space which underlies the algebra $\mathcal{A}$ are emphasised. Moreover, this sum is not supposed to be a direct sum. Notice that only $\mathcal{A}$ and $\mathcal{C}$ are relevant, as it is always possible to set $\mathcal{B} = \mathcal{A}$. Of course, if $\mathcal{A} = \mathcal{B} \oplus \mathcal{C}$, then the quotient $\mathcal{A}/\mathcal{C}$ can be identified as $\mathcal{B}$. Given $a, b \in \mathcal{A}$, the following equivalence relation can be defined:

$$a \sim b \Longleftrightarrow a = b + x, \qquad x \in \mathcal{C}. \tag{2.28}$$

The set $\mathcal{A}/\sim$ of the equivalence classes has a natural vector space structure, with the sum of vectors defined by

$$[a] + [b] = [a + b]$$

and the multiplication by scalars $d \in \mathbb{K}$ defined by

$$d[a] = [da].$$

Now, the set $\mathcal{A}/\sim$ is also intended to present an algebra structure. A natural way to define the product between equivalence classes is

$$[a][b] = [ab].$$

According to the definition given in eqn (2.28), regarding the equivalence relation for $[a], [b] \in \mathcal{A}/\sim$, it follows that $[a] = [a+x]$, and $[b] = [b+y]$, where $x, y \in \mathcal{C}$; in addition,

$$[a][b] = [a+x][b+y] = [(a+x)(b+y)] = [ab + ay + xb + xy].$$

The last two equations result in

$$[ab] = [ab + ay + xb + xy],$$

namely

$$ay + xb + xy \in \mathcal{C}, \quad \forall a, b \in \mathcal{A}, \forall x, y \in \mathcal{C},$$

which only holds if $\mathcal{C}$ is an (bilateral) ideal. This algebra is called the quotient algebra of $\mathcal{A}$ by the ideal $\mathcal{C}$ and is denoted by $\mathcal{A}/\mathcal{C}$.

The Exterior Algebra as a Quotient of the Tensor Algebra

Let $\mathsf{T}(V)$ denote the algebra of contravariant tensors. Let $\mathcal{I}_E$ be the ideal of $\mathsf{T}(V)$, generated by the elements $\mathbf{v} \otimes \mathbf{v}$, $\mathbf{v} \in V$. In other words, the ideal $\mathcal{I}_E$ consists of the sums

$$\sum_i A_i \otimes \mathbf{v}_i \otimes \mathbf{v}_i \otimes B_i, \tag{2.29}$$

where $\mathbf{v}_i \in V$, and $A_i, B_i \in \mathsf{T}(V)$. Moreover, the ideal $\mathcal{I}_E$ is generated by the elements $\mathbf{v} \otimes \mathbf{u} + \mathbf{u} \otimes \mathbf{v}$, where $\mathbf{v}, \mathbf{u} \in V$. In addition,

$$\mathbf{v} \otimes \mathbf{u} + \mathbf{u} \otimes \mathbf{v} \in \ker \mathrm{Alt},$$

and $\ker(\mathrm{Alt}) = \mathcal{I}_E$. Let us now show that the *exterior algebra is isomorphic to the quotient algebra* $\mathsf{T}(V)/\mathcal{I}_E$. The respective equivalence class is given by eqn (2.28) with the identification $\mathcal{C} = \mathcal{I}_E$, namely

$$A \sim B \Longleftrightarrow A = B + x, \qquad x \in \mathcal{I}_E.$$

The equivalence class of A is denoted by $[A]$, and the product is denoted by

$$[A] \wedge [B] = [A \otimes B]. \tag{2.30}$$

For all $\mathbf{v}, \mathbf{u} \in V$, it is clear that $[\mathbf{v}] \sim \mathbf{v}$, and $[\mathbf{u}] \sim \mathbf{u}$. It is possible to calculate $\mathbf{v} \wedge \mathbf{u}$ according to the definition given in eqn (2.30), since

$$\mathbf{v} \otimes \mathbf{u} = \frac{1}{2}(\mathbf{v} \otimes \mathbf{u} - \mathbf{u} \otimes \mathbf{v}) + \frac{1}{2}(\mathbf{v} \otimes \mathbf{u} + \mathbf{u} \otimes \mathbf{v}),$$

where $\frac{1}{2}(\mathbf{v}\otimes\mathbf{u}+\mathbf{u}\otimes\mathbf{v})\in\mathcal{I}_E$. Indeed,

$$(\mathbf{v}\otimes\mathbf{u}+\mathbf{u}\otimes\mathbf{v})=(\mathbf{v}+\mathbf{u})\otimes(\mathbf{v}+\mathbf{u})-\mathbf{v}\otimes\mathbf{v}-\mathbf{u}\otimes\mathbf{u},$$

showing that this quantity is an element of $\mathcal{I}_E$. We conclude that

$$\mathbf{v}\otimes\mathbf{u}\sim\mathbf{v}\wedge\mathbf{u}=\frac{1}{2}(\mathbf{v}\otimes\mathbf{u}-\mathbf{u}\otimes\mathbf{v}), \tag{2.31}$$

or $[\mathbf{v}\otimes\mathbf{u}]=[\mathbf{v}\wedge\mathbf{u}]$, and $[\mathbf{v}]\wedge[\mathbf{u}]=[\mathbf{v}\wedge\mathbf{u}]$.

This result can be generalised as

$$\mathbf{v}_1\otimes\cdots\otimes\mathbf{v}_p\sim\mathrm{Alt}(\mathbf{v}_1\otimes\cdots\otimes\mathbf{v}_p)=\mathbf{v}_1\wedge\cdots\wedge\mathbf{v}_p\,. \tag{2.32}$$

In fact, $\mathbf{v}_1\otimes\cdots\otimes\mathbf{v}_p$ is congruent to every $(\mathbf{v}_1\otimes\cdots\otimes\mathbf{v}_p)^\sigma$, since this assertion holds when σ is the transposition of two consecutive elements of $\{1,2,\ldots,p\}$, and these transpositions generate the group S_p.

The congruence (2.32) has prominent consequences, summarised in the following theorem:

Theorem ▶ $\mathsf{T}(V)$ *is the direct sum of* $\bigwedge(V)$ *and* $\ker(\mathrm{Alt})$, *which is equal to the ideal* $\mathcal{I}_E$. *Moreover,* $A\wedge B\sim A\otimes B \mod \mathcal{I}_E$ *for all* $A,B\in\bigwedge(V)$.

Proof: It is clear that $\mathsf{T}(V)$ is the direct sum of $\bigwedge(V)$ (the image of Alt) and $\ker(\mathrm{Alt})$, and that $\mathcal{I}_E\subset\ker(\mathrm{Alt})$. Besides, eqn (2.32) proves that $\mathsf{T}(V)=\bigwedge(V)+\mathcal{I}_E$ because it shows that $A\sim\mathrm{Alt}(A)$ modulo $\mathcal{I}_E$ for all $A\in\mathsf{T}(V)$, proving the first assertion. The second assertion follows if we remember that $A_{[p]}\wedge B_{[q]}=\mathrm{Alt}(A_{[p]}\otimes B_{[q]})$. ✓

Theorem 2.1 allows to prove by induction on p that $\mathrm{Alt}(\mathbf{v}_1\otimes\cdots\otimes\mathbf{v}_p)=\mathbf{v}_1\wedge\cdots\wedge\mathbf{v}_p$. In addition, an immediate consequence of this theorem is that (up to an isomorphism) we have

$$\bigwedge(V)=\mathsf{T}(V)/\mathcal{I}_E. \tag{2.33}$$

The universal property of $\bigwedge(V)$ follows from eqn (2.33). In fact, since $\mathcal{I}_E$ is also the ideal of $\mathsf{T}(V)$, as generated by all $\mathbf{v}\otimes\mathbf{v}$, this property can be stated as follows: for every (associative unital) algebra $\mathcal{A}$, by mapping every algebra homomorphism $\bigwedge(V)\to\mathcal{A}$ to its restriction $V\to\mathcal{A}$, we obtain a bijection from the set of all such homomorphisms onto the set of all linear mappings $f:V\to\mathcal{A}$ such that $f(\mathbf{v})^2=0_{\mathcal{A}}$, for all $\mathbf{v}\in V$. Here, by definition, an algebra homomorphism must map the unit element to the unit element.

Besides the tensor algebra and, as we shall also see, the Clifford algebras and the exterior algebras, other objects have universal properties that are often useful in the study of Clifford algebras, such as, for example, the tensor product and the twisted tensor product of two (or several) algebras, the quotient of a vector space by a subspace, and the quotient of an algebra by an ideal. Moreover, the grade involution, the reversion, and consequently the conjugation preserve the ideal $\mathcal{I}_E$ and thus pass to the quotient, being also defined in $\bigwedge(V)$.

2.6 The Contraction, or Interior Product

It has already been seen that, given a vector $\mathbf{v} \in V$ and a covector $\alpha \in V^*$, there exists an associated scalar, $\alpha(\mathbf{v})$. Likewise, this property can be interpreted as representing the action of the linear functional α on $\mathbf{v}$ (in terms of the mapping $\alpha : V \to \mathbb{R}$), or the action of the linear functional $\xi_{\mathbf{v}}$ on α via the relation $\xi_{\mathbf{v}}(\alpha) = \alpha(\mathbf{v})$ (in this case, in terms of the mapping $\xi_{\mathbf{v}} : V^* \to \mathbb{R}$). The quantity $\alpha(\mathbf{v})$ can thus be interpreted as a kind of product between α and the vector $\mathbf{v}$, namely, the operation $V \times V^* \to \mathbb{R}$. Similarly, this operation can be defined in such a way that $V^* \times V \to \mathbb{R}$.

In the context of the exterior algebra, we can equivalently write $\alpha : \bigwedge_1(V) \to \bigwedge_0(V)$. The natural question arising is whether this viewpoint can be generalised, in order to define an operation such that $\bigwedge_p(V) \to \bigwedge_{p-1}(V)$, for all p such that $(0 \leq p \leq n)$. In what follows this generalisation is accomplished.

The Left Contraction

Let $A_{[p]}$ be a p-vector and let α be a covector. We aim to construct an operation that when acting on $A_{[p]} \in \bigwedge_p(V)$ yields an element of $\bigwedge_{p-1}(V)$, namely, a $(p-1)$-vector. The *left contraction* of a p-vector $A_{[p]}$ by a covector α, denoted from this point on by $\alpha\rfloor$, is defined as

$$\boxed{(\alpha\rfloor A_{[p]})(\alpha_1, \alpha_2, \ldots, \alpha_{p-1}) = pA_{[p]}(\alpha, \alpha_1, \alpha_2, \ldots, \alpha_{p-1})}, \tag{2.34}$$

where $\alpha_1, \ldots, \alpha_{p-1}$ are arbitrary covectors. Taking $A_{[p]} = \mathbf{v}_1 \wedge \cdots \wedge \mathbf{v}_p$ on the right-hand side of eqn (2.34) means that

$$\begin{aligned}&(\mathbf{v}_1 \wedge \cdots \wedge \mathbf{v}_p)(\alpha, \alpha_1, \ldots, \alpha_{p-1})\\ &\quad= \frac{1}{p!} \sum_{\sigma \in S_p} \varepsilon(\sigma)\alpha(\mathbf{v}_{\sigma(1)})\alpha_1(\mathbf{v}_{\sigma(2)}) \ldots \alpha_{p-1}(\mathbf{v}_{\sigma(p)}).\end{aligned}$$

This definition clearly shows that $\alpha\rfloor A_{[p]}$ is a $(p-1)$-vector. In the case where a covector $\mathbf{v}$ is considered, this definition obviously leads to

$$\alpha\rfloor\mathbf{v} = \alpha(\mathbf{v}). \tag{2.35}$$

For an element of $\bigwedge_0(V)$, it is assumed that

$$\alpha\rfloor 1 = 0. \tag{2.36}$$

Observation ☞ The left contraction is also called the *interior product* and can be denoted by $i(\alpha)$, for example, the interior product between $A_{[p]}$ and the covector α is denoted by $i(\alpha)(A_{[p]})$ (Frankel, 2012). However, this notation is not as suitable for our purposes as the other is. As will be shown, the right contraction emulates the left contraction. Using the notation $\alpha\rfloor$ to denote the left contraction makes this fact clear, so we will continue to use this notation.

Properties of the Left Contraction

Although the definition given in eqn (2.34) is appropriate for the formalism, it is not the best one to use for computation. However, an important and useful property can be introduced from this definition. First, let us consider the contraction of a 2-vector, which can be considered, without loss of generality, to be a simple 2-vector $\mathbf{v} \wedge \mathbf{u}$.[1] Using the definition given in eqn (2.34) as well as the definition of the exterior product, given $\alpha, \beta \in V^*$, it follows that

$$\begin{aligned}(\alpha \rfloor (\mathbf{v} \wedge \mathbf{u}))(\beta) &= 2(\mathbf{v} \wedge \mathbf{u})(\alpha, \beta) = \alpha(\mathbf{v})\beta(\mathbf{u}) - \alpha(\mathbf{u})\beta(\mathbf{v}) \\ &= \big(\alpha(\mathbf{v})\mathbf{u} - \alpha(\mathbf{u})\mathbf{v}\big)(\beta) = \big((\alpha \rfloor \mathbf{v})\mathbf{u} - (\alpha \rfloor \mathbf{u})\mathbf{v}\big)(\beta).\end{aligned}$$

Moreover, as this expression must hold for any covector β, it follows that

$$\boxed{\alpha \rfloor (\mathbf{v} \wedge \mathbf{u}) = (\alpha \rfloor \mathbf{v})\mathbf{u} - \mathbf{v}(\alpha \rfloor \mathbf{u}) = \alpha(\mathbf{v})\mathbf{u} - \mathbf{v}\alpha(\mathbf{u})}. \tag{2.37}$$

Now this result can be generalised for the 3-vector $\mathbf{v} \wedge \mathbf{u} \wedge \mathbf{w}$. Again, using eqn (2.34) together with the definition of the exterior product, we obtain

$$\begin{aligned}(\alpha \rfloor (\mathbf{v} \wedge \mathbf{u} \wedge \mathbf{w}))(\beta, \gamma) &= 3(\mathbf{v} \wedge \mathbf{u} \wedge \mathbf{w})(\alpha, \beta, \gamma) \\ &= (3/3!)[\alpha(\mathbf{v})\beta(\mathbf{u})\gamma(\mathbf{w}) + \beta(\mathbf{v})\gamma(\mathbf{u})\alpha(\mathbf{w}) + \gamma(\mathbf{v})\alpha(\mathbf{u})\beta(\mathbf{w}) \\ &\qquad - \gamma(\mathbf{v})\beta(\mathbf{u})\alpha(\mathbf{w}) - \beta(\mathbf{v})\alpha(\mathbf{u})\gamma(\mathbf{w}) - \alpha(\mathbf{v})\gamma(\mathbf{u})\beta(\mathbf{w})] \\ &= \alpha(\mathbf{v})(1/2!)[\beta(\mathbf{u})\gamma(\mathbf{w}) - \beta(\mathbf{w})\gamma(\mathbf{u}))] \\ &\qquad - \alpha(\mathbf{u})(1/2!)[(\beta(\mathbf{v})\gamma(\mathbf{w}) - \beta(\mathbf{w})\gamma(\mathbf{v})] \\ &\qquad + \alpha(\mathbf{w})(1/2!)[\beta(\mathbf{v})\gamma(\mathbf{u}) - \beta(\mathbf{u})\gamma(\mathbf{v})] \\ &= \alpha(\mathbf{v})(\mathbf{u} \wedge \mathbf{w})(\beta, \gamma) - \alpha(\mathbf{u})(\mathbf{v} \wedge \mathbf{w})(\beta, \gamma) + \alpha(\mathbf{w})(\mathbf{v} \wedge \mathbf{u})(\beta, \gamma).\end{aligned}$$

Since expression must hold for any covectors β and γ, it follows that

$$\begin{aligned}\alpha \rfloor (\mathbf{v} \wedge \mathbf{u} \wedge \mathbf{w}) &= (\alpha \rfloor \mathbf{v})\mathbf{u} \wedge \mathbf{w} - (\alpha \rfloor \mathbf{u})\mathbf{v} \wedge \mathbf{w} + (\alpha \rfloor \mathbf{w})\mathbf{v} \wedge \mathbf{u} \\ &= \alpha(\mathbf{v})\mathbf{u} \wedge \mathbf{w} - \alpha(\mathbf{u})\mathbf{v} \wedge \mathbf{w} + \alpha(\mathbf{w})\mathbf{v} \wedge \mathbf{u}.\end{aligned} \tag{2.38}$$

Using now eqns (2.37) and (2.38), together with the associativity of the exterior product, we obtain

$$\alpha \rfloor ((\mathbf{v} \wedge \mathbf{u}) \wedge \mathbf{w}) = (\alpha \rfloor (\mathbf{v} \wedge \mathbf{u})) \wedge \mathbf{w} + \mathbf{v} \wedge \mathbf{u}(\alpha \rfloor \mathbf{w}), \tag{2.39}$$

$$\alpha \rfloor (\mathbf{v} \wedge (\mathbf{u} \wedge \mathbf{w})) = (\alpha \rfloor \mathbf{v})\mathbf{u} \wedge \mathbf{w} - \mathbf{v} \wedge (\alpha \rfloor (\mathbf{u} \wedge \mathbf{w})). \tag{2.40}$$

where the contraction of the exterior product between a 1-vector and a 2-vector is considered. It is important to note on the right-hand side of these equations the presence of either a positive or negative sign, according to the presence of a 2-vector ($\mathbf{v} \wedge \mathbf{u}$ in eqn (2.39)) or a 1-vector ($\mathbf{v}$ in eqn (2.40)), respectively. The different signs can be taken into account by the use of the grade involution given by (2.24).

[1]It is enough to consider a simple 2-vector, because of the linearity of the contraction, which can be forthwith verified.

The generalisation of these equations for the exterior product of a p-vector and a q-vector is provided by

$$\alpha\rfloor(A_{[p]} \wedge B_{[q]}) = (\alpha\rfloor A_{[p]}) \wedge B_{[q]} + (-1)^p A_{[p]} \wedge (\alpha\rfloor B_{[q]}),$$

where the expression $\widehat{A}_{[p]} = \# A_{[p]} = (-1)^p A_{[p]}$ is used. Because of the linearity of the contraction, we can write in a general manner that

$$\boxed{\alpha\rfloor(A \wedge B) = (\alpha\rfloor A) \wedge B + \widehat{A} \wedge (\alpha\rfloor B)}, \tag{2.41}$$

where A and B are arbitrary multivectors. Notice that this expression resembles the Leibniz rule in differential calculus for the derivative of the product of functions. In fact, it is the *graded* Leibniz rule, so called because of the presence of the grade involution.

The Right Contraction

In eqn (2.34), which defines the left contraction, the covector α is the first argument of a p-vector. However, it would also be correct to locate the covector α in the last argument of that p-vector; there is no reason to prefer one or the other case. This observation gives rise to another definition: the *right contraction* of a p-vector $A_{[p]}$ by the covector α, denoted by $\lfloor\alpha$, is defined as

$$\boxed{(A_{[p]}\lfloor\alpha)(\alpha_1, \alpha_2, \ldots, \alpha_{p-1}) = pA_{[p]}(\alpha_1, \alpha_2, \ldots, \alpha_{p-1}, \alpha)}, \tag{2.42}$$

where $\alpha_1, \ldots, \alpha_{p-1}$ are arbitrary covectors.

Given the definition in eqn (2.42), in the case where $p = 1$, the contraction of a 1-vector $\mathbf{v}$ by the covector α is immediately reduced to

$$\mathbf{v}\lfloor\alpha = \alpha(\mathbf{v}) = \alpha\rfloor\mathbf{v}\,. \tag{2.43}$$

Analogously, for an element of $\bigwedge_0(V)$, the result equals 0:

$$1\lfloor\alpha = 0. \tag{2.44}$$

The differences between the left contraction and the right contraction can be seen when the right contraction of a 2-vector $\mathbf{v} \wedge \mathbf{u}$ is considered. Using the definition given in eqn (2.42), and a reasoning similar to that used to obtain eqn (2.37), we can conclude that

$$(\mathbf{v} \wedge \mathbf{u})\lfloor\alpha = \mathbf{v}(\mathbf{u}\lfloor\alpha) - \mathbf{u}(\mathbf{v}\lfloor\alpha) = \alpha(\mathbf{u})\mathbf{v} - \alpha(\mathbf{v})\mathbf{u}. \tag{2.45}$$

It is not difficult to demonstrate the analogue of eqn (2.41) for the right contraction. The result is

$$\boxed{(A \wedge B)\lfloor\alpha = A \wedge (B\lfloor\alpha) + (A\lfloor\alpha) \wedge \widehat{B}}, \tag{2.46}$$

where A and B are arbitrary multivectors.

The relationship between the left and right contractions can be straightforwardly obtained from their respective definitions. Indeed, it is enough to realise that

$$A_{[p]}(\alpha_1,\ldots,\alpha_{p-1},\alpha) = (-1)^{p-1}A_{[p]}(\alpha,\alpha_1,\ldots,\alpha_{p-1}),$$

which implies that

$$\alpha\rfloor A_{[p]} = (-1)^{p-1}A_{[p]}\lfloor\alpha. \tag{2.47}$$

In addition, the general expression

$$\boxed{\alpha\rfloor A = -\widehat{A}\lfloor\alpha} \tag{2.48}$$

can be obtained for an arbitrary multivector A.

Generalisation of the Left and the Right Contractions

It is well-known that the exterior product of covectors is a p-covector (or, in general terms, a multicovector). It is thus natural to generalise the definitions of the left and the right contractions for a p-covector.

First, we need to verify that $\alpha\rfloor\alpha\rfloor A_{[p]} = 0 = A_{[p]}\lfloor\alpha\lfloor\alpha$, for $\alpha \in V^*$ and for $A_{[p]} \in \bigwedge(V)$. As usual, let us consider covectors $\alpha_1,\ldots,\alpha_{p-2}$. Thus, eqn (2.34) implies that

$$\begin{aligned}(\alpha\rfloor(\alpha\rfloor A_{[p]}))(\alpha_1,\ldots,\alpha_{p-2}) &= (p-1)(\alpha\rfloor A_{[p]})(\alpha,\alpha_1,\alpha_2,\ldots,\alpha_{p-2})\\ &= p(p-1)A_{[p]}(\alpha,\alpha,\alpha_2,\ldots,\alpha_{p-2})\\ &= \frac{1}{(p-2)!}\sum_{\sigma\in S_p}\varepsilon(\sigma)\alpha(\mathbf{v}_{\sigma(1)})\ldots\alpha_{p-2}(\mathbf{v}_{\sigma(p)})\\ &= 0.\end{aligned} \tag{2.49}$$

The case $A_{[p]}\lfloor\alpha\lfloor\alpha$ can be proved analogously. Moreover,

$$1\lfloor A = A = A\rfloor 1\,,$$

where A denotes an arbitrary multivector, and we must be aware of the universal property of $\bigwedge(V^*)$. In fact, the mapping $V^* \to \mathrm{End}(\bigwedge(V))$, which maps every covector α to $A \mapsto \alpha\rfloor A$, extends to an algebra homomorphism $\bigwedge(V^*) \to \mathrm{End}(\bigwedge(V))$, where $\mathrm{End}(\bigwedge(V))$ stands for the space of endomorphisms of V. In addition, the unit element 1 of $\bigwedge(V^*)$ is mapped to the unit element of $\mathrm{End}(\bigwedge(V))$.

Hence, given a p-covector $\alpha_1\wedge\alpha_2\wedge\cdots\wedge\alpha_p$, we can define

$$\boxed{(\alpha_1\wedge\alpha_2\wedge\cdots\wedge\alpha_p)\rfloor = \alpha_1\rfloor\alpha_2\rfloor\cdots\alpha_p\rfloor}\,, \tag{2.50}$$

and

$$\boxed{\lfloor(\alpha_1\wedge\alpha_2\wedge\cdots\wedge\alpha_p) = \lfloor\alpha_1\lfloor\alpha_2\cdots\lfloor\alpha_p}\,. \tag{2.51}$$

From these definitions, it immediately follows that the (left or right) contraction of a p-vector by a q-covector equals 0 if $q > p$.

These definitions shall now be explored for the case of a 2-covector $\alpha \wedge \beta$. Let us consider a 2-vector $\mathbf{v} \wedge \mathbf{u}$. Equation (2.50) means that

$$(\alpha \wedge \beta) \rfloor (\mathbf{v} \wedge \mathbf{u}) = \alpha \rfloor (\beta \rfloor (\mathbf{v} \wedge \mathbf{u})), \tag{2.52}$$

and the result following from the use of eqn (2.37) is

$$(\alpha \wedge \beta) \rfloor (\mathbf{v} \wedge \mathbf{u}) = \alpha(\mathbf{u})\beta(\mathbf{v}) - \alpha(\mathbf{v})\beta(\mathbf{u})\,. \tag{2.53}$$

On the other hand, the action of the 2-covector $\alpha \wedge \beta$ on the 2-vector $\mathbf{v} \wedge \mathbf{u}$ can be defined. For the case involving covectors, the definition of tensor products of linear mappings, given at the end of chapter 1, reads

$$(\alpha \otimes \beta)(\mathbf{v} \otimes \mathbf{u}) = \alpha(\mathbf{v})\beta(\mathbf{u}).$$

Thus, using the definition of the exterior product, we can define $(\alpha \wedge \beta)(\mathbf{v} \wedge \mathbf{u})$ as follows:

$$(\alpha \wedge \beta)(\mathbf{v} \wedge \mathbf{u}) = \frac{1}{2}\big(\alpha(\mathbf{v})\beta(\mathbf{u}) - \alpha(\mathbf{u})\beta(\mathbf{v})\big). \tag{2.54}$$

Comparing eqns (2.53) and (2.54) yields

$$(\alpha \wedge \beta)(\mathbf{v} \wedge \mathbf{u}) = \frac{1}{2}(\beta \wedge \alpha) \rfloor (\mathbf{v} \wedge \mathbf{u}), \tag{2.55}$$

where the order of the exterior product of the covectors on the two sides of the equation must be taken into account. Moreover, the appropriate form for this expression is

$$(\alpha \wedge \beta)(\mathbf{v} \wedge \mathbf{u}) = \frac{1}{2}\widetilde{(\alpha \wedge \beta)} \rfloor (\mathbf{v} \wedge \mathbf{u}). \tag{2.56}$$

These results can be immediately generalised. If $A_{[p]}$ is a p-vector, and $\Psi^{[p]}$ a p-covector, this generalisation is provided by

$$\Psi^{[p]}(A_{[p]}) = \frac{1}{p!}\widetilde{\Psi}^{[p]} \rfloor A_{[p]}. \tag{2.57}$$

In the case involving a simple p-vector and a simple p-covector, this equation can be written as

$$\begin{aligned}(\alpha_1 \wedge \cdots \wedge \alpha_p)(\mathbf{v}_1 \wedge \cdots \wedge \mathbf{v}_p) &= \frac{1}{p!}(\alpha_p \wedge \cdots \wedge \alpha_1) \rfloor (\mathbf{v}_1 \wedge \cdots \wedge \mathbf{v}_p) \\ &= \frac{1}{p!}\begin{vmatrix} \alpha_1(\mathbf{v}_1) & \alpha_1(\mathbf{v}_2) & \ldots & \alpha_1(\mathbf{v}_p) \\ \alpha_2(\mathbf{v}_1) & \alpha_2(\mathbf{v}_2) & \ldots & \alpha_2(\mathbf{v}_p) \\ \vdots & \vdots & \ddots & \vdots \\ \alpha_p(\mathbf{v}_1) & \alpha_p(\mathbf{v}_2) & \ldots & \alpha_p(\mathbf{v}_p) \end{vmatrix}.\end{aligned} \tag{2.58}$$

The (left or right) contraction involving a p-covector and a q-vector can be similarly defined. This construction shall not be shown here, since it presents neither difficulties nor novelties. However, two related equations that are interesting are

$$\boxed{\Psi^{[p]} \rfloor A_{[q]} = (-1)^{p(q-1)} A_{[q]} \lfloor \Psi^{[p]}}, \tag{2.59}$$

where $p \leq q$, and

$$\boxed{A_{[q]} \rfloor \Psi^{[p]} = (-1)^{q(p-1)} \Psi^{[p]} \lfloor A_{[q]}}, \tag{2.60}$$

where $q \leq p$. Another result is given by

$$\widetilde{\Psi^{[p]} \rfloor A_{[q]}} = \tilde{A}_{[q]} \lfloor \tilde{\Psi}^{[p]}. \tag{2.61}$$

In addition, when $p = q$, it follows that

$$\Psi^{[p]} \rfloor A_{[p]} = A_{[p]} \lfloor \Psi^{[p]} = A_{[p]} \rfloor \Psi^{[p]} = \Psi^{[p]} \lfloor A_{[p]}, \tag{2.62}$$

which can be used to write eqn (2.57) as other, equivalent expressions.

Example 2.7 Let α and β be the covectors

$$\alpha = 5\mathbf{e}^1 - 2\mathbf{e}^2, \qquad \beta = \mathbf{e}^2 + 3\mathbf{e}^3 - \mathbf{e}^4,$$

and let A be the multivector

$$A = \mathbf{e}_1 \wedge \mathbf{e}_2 \wedge \mathbf{e}_3 + 2\mathbf{e}_1 \wedge \mathbf{e}_4.$$

First, $\alpha \rfloor A$ is calculated:

$$\begin{aligned}
\alpha \rfloor A &= (5\mathbf{e}^1 - 2\mathbf{e}^2) \rfloor (\mathbf{e}_1 \wedge \mathbf{e}_2 \wedge \mathbf{e}_3) + (5\mathbf{e}^1 - 2\mathbf{e}^2) \rfloor (2\mathbf{e}_1 \wedge \mathbf{e}_4) \\
&= 5\mathbf{e}^1 \rfloor (\mathbf{e}_1 \wedge \mathbf{e}_2 \wedge \mathbf{e}_3) - 2\mathbf{e}^2 \rfloor (\mathbf{e}_1 \wedge \mathbf{e}_2 \wedge \mathbf{e}_3) + 10\mathbf{e}^1 \rfloor (\mathbf{e}_1 \wedge \mathbf{e}_4) - 4\mathbf{e}^2 \rfloor (\mathbf{e}_1 \wedge \mathbf{e}_4) \\
&= 5\mathbf{e}_2 \wedge \mathbf{e}_3 + 2\mathbf{e}_1 \wedge \mathbf{e}_3 + 10\mathbf{e}_4,
\end{aligned}$$

since $\mathbf{e}^2 \rfloor (\mathbf{e}_1 \wedge \mathbf{e}_4) = 0$. On the other hand, $\beta \rfloor A$ is given by

$$\begin{aligned}
\beta \rfloor A &= (\mathbf{e}^2 + 3\mathbf{e}^3 - \mathbf{e}^4) \rfloor (\mathbf{e}_1 \wedge \mathbf{e}_2 \wedge \mathbf{e}_3) + (\mathbf{e}^2 + 3\mathbf{e}^3 - \mathbf{e}^4) \rfloor (2\mathbf{e}_1 \wedge \mathbf{e}_4) \\
&= \mathbf{e}^2 \rfloor (\mathbf{e}_1 \wedge \mathbf{e}_2 \wedge \mathbf{e}_3) + 3\mathbf{e}^3 \rfloor (\mathbf{e}_1 \wedge \mathbf{e}_2 \wedge \mathbf{e}_3) - \mathbf{e}^4 \rfloor (\mathbf{e}_1 \wedge \mathbf{e}_2 \wedge \mathbf{e}_3) \\
&\quad + \mathbf{e}^2 \rfloor (2\mathbf{e}_1 \wedge \mathbf{e}_4) + 3\mathbf{e}^3 \rfloor (2\mathbf{e}_1 \wedge \mathbf{e}_4) - \mathbf{e}^4 \rfloor (2\mathbf{e}_1 \wedge \mathbf{e}_4) \\
&= -\mathbf{e}_1 \wedge \mathbf{e}_3 + 3\mathbf{e}_1 \wedge \mathbf{e}_2 + 2\mathbf{e}_1,
\end{aligned}$$

where we use $\mathbf{e}^4 \rfloor (\mathbf{e}_1 \wedge \mathbf{e}_2 \wedge \mathbf{e}_3) = \mathbf{e}^2 \rfloor (2\mathbf{e}_1 \wedge \mathbf{e}_4) = \mathbf{e}^3 \rfloor (2\mathbf{e}_1 \wedge \mathbf{e}_4) = 0$. In addition,

$$\beta \rfloor (\alpha \rfloor A) = 5\mathbf{e}_3 - 15\mathbf{e}_2 - 6\mathbf{e}_1 - 10,$$

and

$$\alpha \rfloor (\beta \rfloor A) = -5\mathbf{e}_3 + 15\mathbf{e}_2 + 10 + 6\mathbf{e}_1.$$

These two last results are expected, since

$$\beta \rfloor (\alpha \rfloor A) = (\beta \wedge \alpha) \rfloor A = -(\alpha \wedge \beta) \rfloor A = -\alpha \rfloor (\beta \rfloor A).$$

Finally, with respect to the right contraction, it follows that

$$A \lfloor \alpha = 5\mathbf{e}_2 \wedge \mathbf{e}_3 - 10\mathbf{e}_4 + 2\mathbf{e}_1 \wedge \mathbf{e}_3.$$

The relationship between $\alpha \rfloor A$ and $A \lfloor \alpha$ is obvious, and, in this case, given by eqn (2.48).

2.7 Orientation, and Quasi-Hodge Isomorphisms

Before discussing the Hodge isomorphism, let us define an isomorphism which is called the 'quasi-Hodge' isomorphism. In order to introduce it, it is necessary to *choose an n-vector (pseudoscalar)*.

Orientation

There is no criteria that would induce us to prefer the choice of some specific n-vector. If a basis $\mathfrak{B} = \{\mathbf{e}_1, \ldots, \mathbf{e}_n\}$ of V is taken into account, it is natural to choose the n-vector given by the exterior product of the basis vectors $\mathfrak{B}$. No matter the order that such exterior products are taken – for instance, $\mathbf{e}_1 \wedge \cdots \wedge \mathbf{e}_n$, or $\mathbf{e}_n \wedge \cdots \wedge \mathbf{e}_1$, or even $\mathbf{e}_{\sigma(1)} \wedge \cdots \wedge \mathbf{e}_{\sigma(n)}, \forall \sigma \in S_n$ – the result can be always written as

$$\pm \mathbf{e}_1 \wedge \mathbf{e}_2 \wedge \cdots \wedge \mathbf{e}_n.$$

Indeed, much more information can be extracted from it. Given any set of n linearly independent vectors $\{\mathbf{v}_1, \ldots, \mathbf{v}_n\}$, it follows that

$$\mathbf{v}_1 \wedge \cdots \wedge \mathbf{v}_n = J \mathbf{e}_1 \wedge \cdots \wedge \mathbf{e}_n,$$

where $J > 0$, or $J < 0$.[2] This result implies that, once an n-vector has been chosen, any other set of linearly independent n vectors can be represented by one of the two following classes: (i) that one for which $J > 0$, and (ii) that one such that $J < 0$.

In other words, when an n-vector Ω has been chosen, we can define two equivalence classes, $[\Omega]$ and $[-\Omega]$, as

$$[\Omega] = \{A_{[n]} \in \textstyle\bigwedge_n(V) \mid A_{[n]} = J\Omega, \quad J > 0\},$$
$$[-\Omega] = \{A_{[n]} \in \textstyle\bigwedge_n(V) \mid A_{[n]} = J\Omega, \quad J < 0\}.$$

These two equivalence classes comprise the *vector space orientation.* There exists two possible orientations for a vector space, completely determined once an n-vector has been selected. The choice of an n-vector Ω_V in an equivalence class defines a pair constituted by a vector space V endowed with an n-vector Ω_V.

We can immediately define an orientation for the dual space V^* via the n-covector Ω_{V^*}. The question to be posed is whether the orientation of V^* can be related to the orientation of V. In order to answer this question, we write

$$\Omega_V = J \mathbf{e}_1 \wedge \cdots \wedge \mathbf{e}_n, \qquad \Omega_{V^*} = J' \mathbf{e}^1 \wedge \cdots \wedge \mathbf{e}^n.$$

Using eqn (2.57) and (2.58), we can see that

$$\Omega_{V^*}(\Omega_V) = \frac{1}{n!} \widetilde{\Omega}_{V^*} \rfloor \Omega_V = \frac{1}{n!} J J' (\mathbf{e}^n \wedge \cdots \wedge \mathbf{e}^1) \rfloor (\mathbf{e}_1 \wedge \cdots \wedge \mathbf{e}_n) = \frac{1}{n!} J J',$$

in such a way that

$$0 \neq \Omega_{V^*}(\Omega_V) = \begin{cases} > 0 & \text{if } J > 0 \text{ and } J' > 0; \text{ or } J < 0 \text{ and } J' < 0, \\ < 0 & \text{if } J > 0 \text{ and } J' < 0; \text{ or } J < 0 \text{ and } J' > 0. \end{cases}$$

As $\Omega_{V^*}(\Omega_V) > 0$, or $\Omega_{V^*}(\Omega_V) < 0$, we can use this fact to relate both the orientations of V and V^*.

[2] Let us remember that, if these n vectors are not linearly independent, the exterior product among them equals 0.

The orientations of V and V^* determined by the n-vector Ω_V and the n-covector Ω_{V^*}, respectively, are called *compatible* if $\Omega_{V^*}(\Omega_V) > 0$. Hence, if the orientations of V and V^* are compatible, once the orientation for one of those spaces are chosen, the orientation for the other space is forthwith defined.

In order to define Ω_{V^*} given Ω_V, in such a way that both orientations are compatible, let us choose the most natural way. If with respect to a basis $\mathfrak{B} = \{\mathbf{e}_i\}$ we have

$$\Omega_V = \mathbf{e}_1 \wedge \cdots \wedge \mathbf{e}_n, \tag{2.63}$$

then, in terms of a dual basis $\mathfrak{B} = \{\mathbf{e}^i\}$,

$$\Omega_{V^*} = \mathbf{e}^1 \wedge \cdots \wedge \mathbf{e}^n. \tag{2.64}$$

It follows that

$$\widetilde{\Omega}_{V^*} \rfloor \Omega_V = 1. \tag{2.65}$$

Quasi-Hodge Isomorphisms

Given the pair (V, Ω_V), it is possible to define the two operators $\bar{\star}$ and $\underline{\star}$, which will be called quasi-Hodge isomorphisms:

$$\bar{\star} : \bigwedge_p(V) \to \bigwedge^{n-p}(V), \qquad \underline{\star} : \bigwedge^p(V) \to \bigwedge_{n-p}(V).$$

Given $A_{[p]} \in \bigwedge_p(V)$, we define $\bar{\star}A_{[p]}$ as

$$\Psi^{[p]} \wedge \bar{\star}A_{[p]} = p!\Psi^{[p]}(A_{[p]})\Omega_{V^*}, \qquad \forall \Psi^{[p]} \in \bigwedge^p(V). \tag{2.66}$$

Similarly, given $\Psi^{[p]} \in \bigwedge^p(V)$, we define $\underline{\star}\Psi^{[p]}$ as

$$A_{[p]} \wedge \underline{\star}\Psi^{[p]} = p!\Psi^{[p]}(A_{[p]})\Omega_V, \qquad \forall A_{[p]} \in \bigwedge_p(V). \tag{2.67}$$

Now let us focus on some of the properties and applications of these definitions, considering only the case for $\bar{\star}$, since both cases can be similarly analysed. In eqns (2.66) and (2.67), when $p = 0$, it is implicit that $a(b) = ab$ for $a \in \bigwedge^0(V) = \mathbb{R}$, and $b \in \bigwedge_0(V) = \mathbb{R}$. Taking this fact into account, for $1 \in \bigwedge_0(V)$, eqn (2.66) reads

$$a \wedge \bar{\star}1 = a(1)\Omega_{V^*} = a\Omega_{V^*},$$

so

$$\bar{\star}1 = \Omega_{V^*}. \tag{2.68}$$

For $\mathbf{v} \in V$, let us calculate $\bar{\star}\mathbf{v}$. According to eqn (2.66), it follows that

$$\alpha \wedge \bar{\star}\mathbf{v} = \alpha(\mathbf{v})\Omega_{V^*} = (\alpha \rfloor \mathbf{v})\Omega_{V^*} = (\mathbf{v} \rfloor \alpha)\Omega_{V^*}, \tag{2.69}$$

where eqn (2.62) was used in order to express the right-hand side in terms of the contraction of a covector by a vector; the reason for this approach will be made clear in what follows. Since $\Omega_{V^*} \in \bigwedge^n(V)$, thus $\alpha \wedge \Omega_{V^*} = 0$, $\forall \alpha \in \bigwedge^1(V)$. The contraction $\mathbf{v}\rfloor$ satisfies the graded Leibniz rule

$$\mathbf{v}\rfloor(\Psi \wedge \Phi) = (\mathbf{v}\rfloor\Psi) \wedge \Phi + \widehat{\Psi} \wedge (\mathbf{v}\rfloor\Phi),$$

where Ψ and Φ are multicovectors, and it implies that

$$0 = \mathbf{v}\rfloor(\alpha \wedge \Omega_{V^*}) = (\mathbf{v}\rfloor\alpha)\Omega_{V^*} - \alpha \wedge (\mathbf{v}\rfloor\Omega_{V^*}).$$

Hence, eqn (2.69) can be written as

$$\alpha \wedge \bar{\star}\mathbf{v} = \alpha \wedge (\mathbf{v}\rfloor\Omega_{V^*}). \tag{2.70}$$

Since this expression holds for all $\alpha \in \bigwedge^1(V)$, it reads

$$\bar{\star}\mathbf{v} = \mathbf{v}\rfloor\Omega_{V^*}. \tag{2.71}$$

Now let us consider the p-vector $A_{[p-1]} \wedge \mathbf{v}$. According to eqn (2.66),

$$\Psi^{[p]} \wedge \bar{\star}(A_{[p-1]} \wedge \mathbf{v}) = p!\Psi^{[p]}(A_{[p-1]} \wedge \mathbf{v})\Omega_{V^*} = ((\widetilde{A_{[p-1]} \wedge \mathbf{v}})\rfloor\Psi^{[p]})\Omega_{V^*}, \tag{2.72}$$

where eqns (2.57) and (2.62) were employed; consequently, we have

$$\begin{aligned}\Psi^{[p]} \wedge \bar{\star}(A_{[p-1]} \wedge \mathbf{v}) &= ((\mathbf{v} \wedge \widetilde{A}_{[p-1]}))\rfloor\Psi^{[p]})\Omega_{V^*} \\ &= (-1)^{p-1}((\widetilde{A}_{[p-1]} \wedge \mathbf{v})\rfloor\Psi^{[p]})\Omega_{V^*} \\ &= (-1)^{p-1}(\widetilde{A}_{[p-1]}\rfloor(\mathbf{v}\rfloor\Psi^{[p]}))\Omega_{V^*},\end{aligned} \tag{2.73}$$

where eqn (2.15) and the analogue for eqn (2.50) for the case involving vectors were used. Again by eqn (2.66), the last term of eqn (2.73) reads

$$(-1)^{p-1}(\widetilde{A}_{[p-1]}\rfloor(\mathbf{v}\rfloor\Psi^{[p]}))\Omega_{V^*} = (-1)^{p-1}(\mathbf{v}\rfloor\Psi^{[p]}) \wedge \bar{\star}A_{[p-1]}. \tag{2.74}$$

On the other hand, according to the graded Leibniz rule,

$$\mathbf{v}\rfloor(\Psi^{[p]} \wedge \bar{\star}A_{[p-1]}) = (\mathbf{v}\rfloor\Psi^{[p]}) \wedge \bar{\star}A_{[p-1]} + (-1)^p\Psi^{[p]} \wedge (\mathbf{v}\rfloor\bar{\star}A_{[p-1]}). \tag{2.75}$$

But $\bar{\star}A_{[p-1]} \in \bigwedge^{n-(p-1)}(V) = \bigwedge^{n-p+1}(V)$; consequently, it follows that

$$\Psi^{[p]} \wedge \bar{\star}A_{[p-1]} = 0.$$

Finally, the expression

$$(\mathbf{v}\rfloor\Psi^{[p]}) \wedge \bar{\star}A_{[p-1]} = -(-1)^p\Psi^{[p]} \wedge (\mathbf{v}\rfloor\bar{\star}A_{[p-1]}) \tag{2.76}$$

is obtained, and we can substitute it into eqn (2.74) and subsequently eqn (2.73), obtaining

$$\Psi^{[p]} \wedge \bar{\star}(A_{[p-1]} \wedge \mathbf{v}) = \Psi^{[p]} \wedge (\mathbf{v}\rfloor\bar{\star}A_{[p-1]}). \tag{2.77}$$

As it must hold $\forall \Psi^{[p]} \in \bigwedge^p(V)$, it then reads

$$\bar{\star}(A_{[p-1]} \wedge \mathbf{v}) = (\mathbf{v} \rfloor \bar{\star} A_{[p-1]}). \tag{2.78}$$

Now eqn (2.71) can be used recursively, together with eqn (2.78), in order to obtain

$$\bar{\star}(\mathbf{v}_1 \wedge \cdots \wedge \mathbf{v}_p) = (\mathbf{v}_p \wedge \cdots \wedge \mathbf{v}_1) \rfloor \Omega_{V^*}, \tag{2.79}$$

and, when extended by linearity, it yields

$$\boxed{\bar{\star} A_{[p]} = \widetilde{A}_{[p]} \rfloor \Omega_{V^*}}. \tag{2.80}$$

This equation and eqn (2.68) completely determine the quasi-Hodge isomorphism $\bar{\star}$.

The calculations for $\underline{\star}$ are similar and they imply that

$$\underline{\star} 1 = \Omega_V, \tag{2.81}$$

and

$$\boxed{\underline{\star} \Psi^{[p]} = \widetilde{\Psi}^{[p]} \rfloor \Omega_V}. \tag{2.82}$$

It is possible to show that the isomorphisms $\bar{\star}$ and $\underline{\star}$ are the inverse of each other, up to a sign. Specifically:

$$\underline{\star}\bar{\star} = \bar{\star}\underline{\star} = (-1)^{p(n-p)} 1, \tag{2.83}$$

where 1 denotes the identity operator, and it is natural to define

$$\bar{\star}^{-1} = (-1)^{p(n-p)} \underline{\star}, \qquad \underline{\star}^{-1} = (-1)^{p(n-p)} \bar{\star}. \tag{2.84}$$

Remembering the assumption about the compatibility between the orientations of V and V^*, and therefore the validity of the relation $\tilde{\Omega}_{V^*} \rfloor \Omega_V = \tilde{\Omega}_V \rfloor \Omega_{V^*} = 1$, by using eqn (2.66) together with eqn (2.57 and eqn (2.62), we obtain

$$(\widetilde{\Psi^{[p]} \wedge \bar{\star} A_{[p]}}) \rfloor \Omega_V = \tilde{A}_{[p]} \rfloor \Psi^{[p]} (\tilde{\Omega}_{V^*} \rfloor \Omega_V) = \tilde{A}_{[p]} \rfloor \Psi^{[p]}. \tag{2.85}$$

Alternatively, from eqn (2.82), we obtain

$$\underline{\star}(\Psi^{[p]} \wedge \bar{\star} A_{[p]}) = \tilde{A}_{[p]} \rfloor \Psi^{[p]}. \tag{2.86}$$

On the other hand, the definition given in eqn (2.67) implies that

$$(\widetilde{A_{[p]} \wedge \underline{\star} \Psi^{[p]}}) \rfloor \Omega_{V^*} = \tilde{A}_{[p]} \rfloor \Psi^{[p]} (\tilde{\Omega}_{V^*} \rfloor \Omega_{V^*}) = \tilde{A}_{[p]} \rfloor \Psi^{[p]}, \tag{2.87}$$

and, by using eqn (2.80), we obtain

$$\bar{\star}(A_{[p]} \wedge \underline{\star} \Psi^{[p]}) = \tilde{A}_{[p]} \rfloor \Psi^{[p]}. \tag{2.88}$$

When we compare eqns (2.86) and (2.88), we can see that

$$\underline{\star}(\Psi^{[p]} \wedge \bar{\star} A_{[p]}) = \bar{\star}(A_{[p]} \wedge \underline{\star} \Psi^{[p]}). \tag{2.89}$$

Example 2.8 Let us prove eqn (2.83). Because of linearity, it is enough to prove this equation for $A_{[p]}$ being a simple p-vector. Given $\mathfrak{B} = \{\mathbf{e}_i\}$ $(i = 1, \ldots, n)$, let us write an arbitrary simple p-vector as

$A_{[p]} = \mathbf{e}_{\mu_1} \wedge \cdots \wedge \mathbf{e}_{\mu_p}$, where $\mu_i = 1, \ldots, n$; $\mu_i \neq \mu_j$ $(i \neq j)$; $i, j = 1, \ldots, p$. Let σ be a permutation of n elements $\sigma : \{1, \ldots, n\} \to \{\sigma(1), \ldots, \sigma(n)\}$ such that $\sigma(i) = \mu_i$ $(i = 1, \ldots, p)$. In this way, we can write this arbitrary simple p-vector as

$$A_{[p]} = \mathbf{e}_{\sigma(1)} \wedge \cdots \wedge \mathbf{e}_{\sigma(p)}.$$

The n-vectors Ω_V and Ω_{V^*} can be written as

$$\begin{aligned}\Omega_V &= \mathbf{e}_1 \wedge \cdots \wedge \mathbf{e}_p \wedge \mathbf{e}_{p+1} \wedge \cdots \wedge \mathbf{e}_n \\ &= \epsilon(\sigma)\, \mathbf{e}_{\sigma(1)} \wedge \cdots \wedge \mathbf{e}_{\sigma(p)} \wedge \mathbf{e}_{\sigma(p+1)} \wedge \cdots \wedge \mathbf{e}_{\sigma(p)}, \\ \Omega_{V^*} &= \mathbf{e}^1 \wedge \cdots \wedge \mathbf{e}^p \wedge \mathbf{e}^{p+1} \wedge \cdots \wedge \mathbf{e}^n \\ &= \epsilon(\sigma)\, \mathbf{e}^{\sigma(1)} \wedge \cdots \wedge \mathbf{e}^{\sigma(p)} \wedge \mathbf{e}^{\sigma(p+1)} \wedge \cdots \wedge \mathbf{e}^{\sigma(p)},\end{aligned}$$

where $\epsilon(\sigma)$ is the sign of the permutation σ. Now for $\overline{\star} A_{[p]}$ we have

$$\begin{aligned}\overline{\star} A_{[p]} &= (\mathbf{e}_{\sigma(p)} \wedge \cdots \wedge \mathbf{e}_{\sigma(1)}) \rfloor \left(\epsilon(\sigma)\, \mathbf{e}^{\sigma(1)} \wedge \cdots \wedge \mathbf{e}^{\sigma(p)} \wedge \mathbf{e}^{\sigma(p+1)} \wedge \cdots \wedge \mathbf{e}^{\sigma(n)}\right) \\ &= \epsilon(\sigma)\, \mathbf{e}^{\sigma(p+1)} \wedge \cdots \wedge \mathbf{e}^{\sigma(n)}.\end{aligned}$$

In a similar way,

$$\begin{aligned}\underline{\star}(\overline{\star} A_{[p]}) &= \left(\epsilon(\sigma)\, \mathbf{e}^{\sigma(n)} \wedge \cdots \wedge \mathbf{e}^{\sigma(p+1)}\right) \rfloor \left(\epsilon(\sigma)\, \mathbf{e}_{\sigma(1)} \wedge \cdots \wedge \mathbf{e}_{\sigma(p)} \wedge \mathbf{e}_{\sigma(p+1)} \wedge \cdots \wedge \mathbf{e}_{\sigma(n)}\right) \\ &= \left(\mathbf{e}^{\sigma(n)} \wedge \cdots \wedge \mathbf{e}^{\sigma(p+1)}\right) \rfloor \left((-1)^{n(p-n)} (\mathbf{e}_{\sigma(p+1)} \wedge \cdots \wedge \mathbf{e}_{\sigma(n)}) \wedge (\mathbf{e}_{\sigma(1)} \wedge \cdots \mathbf{e}_{\sigma(p)})\right) \\ &= (-1)^{p(n-p)} \mathbf{e}_{\sigma(1)} \wedge \cdots \wedge \mathbf{e}_{\sigma(p)} = (-1)^{p(n-p)} A_{[p]}.\end{aligned}$$

The proof that $\overline{\star}\underline{\star} = (-1)^{p(n-p)}$ is analogous.

Example 2.9 In what follows, quasi-Hodge isomorphisms are used to find the inverse of a matrix. Specifically, consider the matrix A given by

$$A = \begin{pmatrix} 1 & 2 & 0 \\ 0 & 4 & 1 \\ 1 & -1 & 1 \end{pmatrix}.$$

Take the entries a_i^j of this matrix as the vector components $\mathbf{a}_i = a_i^j \mathbf{e}_j$, namely

$$\begin{aligned}\mathbf{a}_1 &= \mathbf{e}_1 + \mathbf{e}_3, \\ \mathbf{a}_2 &= 2\mathbf{e}_1 + 4\mathbf{e}_2 - \mathbf{e}_3, \\ \mathbf{a}_3 &= \mathbf{e}_2 + \mathbf{e}_3.\end{aligned}$$

In this way, the matrix A is defined as

$$A = \begin{pmatrix} \mathbf{a}_1 & \mathbf{a}_2 & \mathbf{a}_3 \end{pmatrix}.$$

By using quasi-Hodge isomorphisms, we can define a new set of covectors $\{\overline{\star}(\mathbf{a}_i \wedge \mathbf{a}_j)\}$ $(i \neq j)$:

$$\overline{\star}(\mathbf{a}_i \wedge \mathbf{a}_j) = (\mathbf{a}_j \wedge \mathbf{a}_i) \rfloor \epsilon,$$

where the n-covector ϵ is given by

$$\epsilon = \mathbf{e}^1 \wedge \mathbf{e}^2 \wedge \mathbf{e}^3.$$

Now the vectors $\bar{\mathbf{a}}^1$, $\bar{\mathbf{a}}^2$, and $\bar{\mathbf{a}}^3$ are defined as

$$\bar{\mathbf{a}}^1 = \overline{\star}(\mathbf{a}_2 \wedge \mathbf{a}_3), \quad \bar{\mathbf{a}}^2 = \overline{\star}(\mathbf{a}_3 \wedge \mathbf{a}_1), \quad \bar{\mathbf{a}}^3 = \overline{\star}(\mathbf{a}_1 \wedge \mathbf{a}_2).$$

Computing these quantities, we obtain

$$\begin{aligned}\bar{\mathbf{a}}^1 &= \mathbf{a}_3 \rfloor (\mathbf{a}_2 \rfloor (\mathbf{e}^1 \wedge \mathbf{e}^2 \wedge \mathbf{e}^3)) = 5\mathbf{e}^1 - 2\mathbf{e}^2 + 2\mathbf{e}^3, \\ \bar{\mathbf{a}}^2 &= \mathbf{a}_1 \rfloor (\mathbf{a}_3 \rfloor (\mathbf{e}^1 \wedge \mathbf{e}^2 \wedge \mathbf{e}^3)) = \mathbf{e}^1 + \mathbf{e}^2 - \mathbf{e}^3, \\ \bar{\mathbf{a}}^3 &= \mathbf{a}_2 \rfloor (\mathbf{a}_1 \rfloor (\mathbf{e}^1 \wedge \mathbf{e}^2 \wedge \mathbf{e}^3)) = -4\mathbf{e}^1 + 3\mathbf{e}^2 + 4\mathbf{e}^3.\end{aligned}$$

It straightforwardly follows that

$$\bar{\mathbf{a}}^i(\mathbf{a}_j) = 7\delta^i_j, \quad (i, j = 1, 2, 3),$$

and that

$$\det A = 7.$$

These results can be used to obtain the inverse matrix of A, as follows. First, place the vector components (according to our convention) into a matrix column; then, the covector components must be placed in the corresponding rows of the matrix. Thus, using the covectors $\bar{\mathbf{a}}^1$, $\bar{\mathbf{a}}^2$, and $\bar{\mathbf{a}}^3$, the matrix $\bar{A}$ can be defined as

$$\bar{A} = \begin{pmatrix} \bar{\mathbf{a}}^1 \\ \bar{\mathbf{a}}^2 \\ \bar{\mathbf{a}}^3 \end{pmatrix} = \begin{pmatrix} 5 & -2 & 2 \\ 1 & 1 & -1 \\ -4 & 3 & 4 \end{pmatrix}.$$

As can be immediately verified,

$$\bar{A}A = A\bar{A} = 7\mathbf{1} = (\det A)\mathbf{1}.$$

Hence, the inverse matrix of A is given by

$$A^{-1} = \bar{A}/(\det A),$$

and the determinant can be implicitly written as

$$\mathbf{a}_1 \wedge \mathbf{a}_2 \wedge \mathbf{a}_3 = (\det A)\mathbf{e}_1 \wedge \mathbf{e}_2 \wedge \mathbf{e}_3.$$

2.8 The Regressive Product

A close look at the *Ausdehnungslehre* of 1844 (Grassmann, 1844) shows that Grassmann actually used the Hodge star operator (which he called the *Ergänzung*) to define the regressive product. He also used it in the revised *Ausdehnungslehre*, which was published in 1862 (Grassmann, 1862), and indeed his *Ergänzung* depended on the determinant of an otherwise arbitrary metric (Grassmann, 1894; Fearnley-Sandre and Stokes, 1997). Here, by using $\overline{\star}$ and $\underline{\star}$, we can define a new product among elements of $\bigwedge(V)$ from the product $\wedge$. Let us define the so-called *regressive product*, which we shall denote by $\vee$, as

$$\boxed{A_{[p]} \vee B_{[q]} = \overline{\star}^{-1}(\overline{\star}A_{[p]} \wedge \overline{\star}B_{[q]})}. \tag{2.90}$$

The combination of these operations shows that $\overline{\star}A_{[p]} \in \bigwedge^{n-p}(V)$ and $\overline{\star}B_{[q]} \in \bigwedge^{n-q}(V)$. Hence, the exterior product in eqn (2.90) belongs to $\bigwedge^{2n-p-q}(V)$. Since $\overline{\star}^{-1}$ differs from $\underline{\star}$ only by a sign (see eqn (2.84)) it follows that $A_{[p]} \vee B_{[q]} \in \bigwedge_{p+q-n}(V)$. This result leads us to define

$$A_{\{p\}} = A_{[n-p]}, \tag{2.91}$$

and

$$\bigvee{}_p(V) = \bigwedge{}_{n-p}(V). \tag{2.92}$$

Thus, the regressive product between elements of $\bigvee_p(V)$ and $\bigvee_q(V)$ is an element of $\bigwedge_{(n-p)+(n-q)-n}(V) = \bigwedge_{n-p-q}(V) = \bigvee_{p+q}(V)$, that is,

$$\vee : \bigvee_p(V) \times \bigvee_q(V) \to \bigvee_{p+q}(V).$$

The expression involving p-covectors is similar:

$$\boxed{\Psi^{[p]} \vee \Phi^{[q]} = \underline{\star}^{-1}(\underline{\star}\Psi^{[p]} \wedge \underline{\star}\Phi^{[q]})}. \tag{2.93}$$

The regressive product is associative. Indeed,

$$
\begin{aligned}
(A_{[p]} \vee B_{[q]}) \vee C_{[r]} &= \bar{\star}^{-1}(\bar{\star}(A_{[p]} \vee B_{[q]}) \wedge \bar{\star}C_{[r]}) \\
&= \bar{\star}^{-1}(\bar{\star}\bar{\star}^{-1}(\bar{\star}A_{[p]} \wedge \bar{\star}B_{[q]}) \wedge \bar{\star}C_{[r]}) = \bar{\star}^{-1}((\bar{\star}A_{[p]} \wedge \bar{\star}B_{[q]}) \wedge \bar{\star}C_{[r]}) \\
&= \bar{\star}^{-1}(\bar{\star}A_{[p]} \wedge (\bar{\star}B_{[q]} \wedge \bar{\star}C_{[r]})) = \bar{\star}^{-1}(\bar{\star}A_{[p]} \wedge \bar{\star}\bar{\star}^{-1}(\bar{\star}B_{[q]} \wedge \bar{\star}C_{[r]})) \\
&= \bar{\star}^{-1}(\bar{\star}A_{[p]} \wedge \bar{\star}(B_{[q]} \vee C_{[r]})) \\
&= A_{[p]} \vee (B_{[q]} \vee C_{[r]}).
\end{aligned}
$$

In addition, the regressive product can be related to the contraction as follows:

$$
\begin{aligned}
A_{[p]} \vee \underline{\star}\Psi^{[q]} &= \bar{\star}^{-1}(\bar{\star}A_{[p]} \wedge \bar{\star}\underline{\star}\Psi^{[q]}) = (-1)^{q(n-q)}\bar{\star}^{-1}(\bar{\star}A_{[p]} \wedge \Psi^{[q]}) \\
&= (-1)^{q(n-q)}(-1)^{(n+q-p)(p-q)}\underline{\star}(\bar{\star}A_{[p]} \wedge \Psi^{[q]}) \\
&= (-1)^{p(n-p)}(\bar{\star}\widetilde{A_{[p]} \wedge \Psi^{[q]}})\rfloor\Omega_V = (-1)^{p(n-p)}\tilde{\Psi}^{[q]}\rfloor(\widetilde{\bar{\star}A_{[p]}}\rfloor\Omega_V) \\
&= (-1)^{p(n-p)}\tilde{\Psi}^{[q]}\rfloor(\underline{\star}\bar{\star}A_{[p]}) \\
&= \tilde{\Psi}^{[q]}\rfloor A_{[p]},
\end{aligned}
$$

namely,

$$
\tilde{\Psi}^{[q]}\rfloor A_{[p]} = A_{[p]} \vee \underline{\star}\Psi^{[q]}. \tag{2.94}
$$

Analogously,

$$
\tilde{A}_{[p]}\rfloor\Psi^{[q]} = \Psi^{[q]} \vee \bar{\star}A_{[p]}. \tag{2.95}
$$

2.9 The Grassmann Algebra

The exterior algebra is defined as being the pair $(\bigwedge(V), \wedge)$, consisting of the vector space $\bigwedge(V)$ endowed with the exterior product $\wedge$. In addition, we can define a symmetric bilinear functional $g : V \times V \to \mathbb{R}$ – or, equivalently, a symmetric correlation $\flat : V \to V^*$. Once the space $\bigwedge(V)$ is constructed from V, if V is endowed with a bilinear functional g, it is natural to ask whether it would be possible to generalise the definition of g for the space $\bigwedge(V)$. The Grassmann algebra, in this context, is the *exterior algebra endowed with the extension* of g, which we intend to define in what follows. The bilinear functional g is often called a *metric*, so the pair (V, g) is then a metric space.

Instead of taking the bilinear functional g, we prefer to take into account the correlation, since the expressions seem straightforward in this case. Let us therefore consider the correlation $\flat : V \to V^*$ and define its extension, which is denoted by the same symbol,[3] as the mapping $\flat : \bigwedge_p(V) \to \bigwedge^p(V)$ given by

$$
(\mathbf{v}_1 \wedge \mathbf{v}_2 \wedge \cdots \wedge \mathbf{v}_p)_\flat = \mathbf{v}_{1\flat} \wedge \mathbf{v}_{2\flat} \wedge \cdots \wedge \mathbf{v}_{p\flat}. \tag{2.96}
$$

By the extension of $\flat$, we can immediately define the extension of the bilinear functional $g : V \times V \to \mathbb{R}$, namely $g(\mathbf{v}, \mathbf{u}) = \mathbf{v}_\flat(\mathbf{u})$. Let us denote this extension by G. We define $G : \bigwedge_p(V) \times \bigwedge_p(V) \to \mathbb{R}$ for the case of simple p-vectors as

[3]Evidently, it is a usual abuse of notation. Instead of writing this extension as $\flat$, we could write $\flat_p$; however, writing it in this way seems a preciousness that simply overloads the notation.

$$G(\mathbf{v}_1 \wedge \cdots \wedge \mathbf{v}_p, \mathbf{u}_1 \wedge \cdots \wedge \mathbf{u}_p) = p!(\mathbf{v}_1 \wedge \cdots \wedge \mathbf{v}_p)_\flat(\mathbf{u}_1 \wedge \cdots \wedge \mathbf{u}_p), \tag{2.97}$$

and generalise this expression for all $\bigwedge_p(V)$ by bilinearity. The presence of the factor $p!$ is a convention. Previously, the duality mapping $\bigwedge(V^*) \times \bigwedge(V) \to \mathbb{K}$ was defined by means of the duality mapping $\mathsf{T}(V^*) \times \mathsf{T}(V) \to \mathbb{K}$, as well as by the inclusions $\bigwedge(V) \subset \mathsf{T}(V)$ and $\bigwedge(V^*) \subset \mathsf{T}(V^*)$. However, this method generates the inopportune factor $p!$ when elements of degree p are taken into account. The bilinear functional G just defined is symmetric:

$$G(\mathbf{v}_1 \wedge \cdots \wedge \mathbf{v}_p, \mathbf{u}_1 \wedge \cdots \wedge \mathbf{u}_p) = G(\mathbf{u}_1 \wedge \cdots \wedge \mathbf{u}_p, \mathbf{v}_1 \wedge \cdots \wedge \mathbf{v}_p). \tag{2.98}$$

Equivalently, we can write

$$G(\mathbf{v}_1 \wedge \cdots \wedge \mathbf{v}_p, \mathbf{u}_1 \wedge \cdots \wedge \mathbf{u}_p) = (\mathbf{v}_p \wedge \cdots \wedge \mathbf{v}_1)_\flat \rfloor (\mathbf{u}_1 \wedge \cdots \wedge \mathbf{u}_p), \tag{2.99}$$

or even

$$G(\mathbf{v}_1 \wedge \cdots \wedge \mathbf{v}_p, \mathbf{u}_1 \wedge \cdots \wedge \mathbf{u}_p) = \begin{vmatrix} g(\mathbf{v}_1, \mathbf{u}_1) & g(\mathbf{v}_1, \mathbf{u}_2) & \dots & g(\mathbf{v}_1, \mathbf{u}_p) \\ g(\mathbf{v}_2, \mathbf{u}_1) & g(\mathbf{v}_2, \mathbf{u}_2) & \dots & g(\mathbf{v}_2, \mathbf{u}_p) \\ \vdots & \vdots & \ddots & \vdots \\ g(\mathbf{v}_p, \mathbf{u}_1) & g(\mathbf{v}_p, \mathbf{u}_2) & \dots & g(\mathbf{v}_p, \mathbf{u}_p) \end{vmatrix}. \tag{2.100}$$

Equation (2.99) is still valid over fields of non-zero characteristic.

Finally, these definitions can be generalised for the vector space $\bigwedge(V)$. In order to do so, we just need to consider the case of $\bigwedge_p(V)$ and $\bigwedge_q(V)$ with $p \neq q$, as

$$G(A_{[p]}, B_{[q]}) = 0, \quad \text{when } p \neq q, \tag{2.101}$$

where $A_{[p]} \in \bigwedge_p(V)$, and $B_{[q]} \in \bigwedge_q(V)$.

Definition 2.3 ▶ *The exterior algebra $(\bigwedge(V), \wedge)$ endowed with the extension G of g for all $\bigwedge(V)$ is the Grassmann algebra of the vector space V and is denoted by $\mathcal{G}(V)$.*

Example 2.10 Let $\flat$ be the correlation defined in chapter 1, example 1.3, namely, $\mathbf{e}_{1\flat} = 3\mathbf{e}^1 + \mathbf{e}^2$; $\mathbf{e}_{2\flat} = \mathbf{e}^1 + 3\mathbf{e}^2$; and $\mathbf{e}_{3\flat} = 2\mathbf{e}^3$. Equation (2.96) provides the extension of $\flat$ to $\bigwedge_p(V)$ $(p = 2, 3)$. For $\bigwedge_2(V)$, it yields

$$(\mathbf{e}_1 \wedge \mathbf{e}_2)_\flat = 8\mathbf{e}^1 \wedge \mathbf{e}^2, \quad (\mathbf{e}_1 \wedge \mathbf{e}_3)_\flat = 6\mathbf{e}^1 \wedge \mathbf{e}^3 + 2\mathbf{e}^2 \wedge \mathbf{e}^3,$$
$$(\mathbf{e}_2 \wedge \mathbf{e}_3)_\flat = 2\mathbf{e}^1 \wedge \mathbf{e}^3 + 6\mathbf{e}^2 \wedge \mathbf{e}^3,$$

and for $\bigwedge^3(V)$ it reads

$$(\mathbf{e}_1 \wedge \mathbf{e}_2 \wedge \mathbf{e}_3)_\flat = 16\mathbf{e}^1 \wedge \mathbf{e}^2 \wedge \mathbf{e}^3.$$

For the symmetric bilinear functional G, one finds

$$\begin{aligned} &G(\mathbf{e}_1 \wedge \mathbf{e}_2, \mathbf{e}_1 \wedge \mathbf{e}_2) = 8, \quad G(\mathbf{e}_1 \wedge \mathbf{e}_2, \mathbf{e}_1 \wedge \mathbf{e}_3) = 0, \quad G(\mathbf{e}_1 \wedge \mathbf{e}_2, \mathbf{e}_2 \wedge \mathbf{e}_3) = 0, \\ &G(\mathbf{e}_1 \wedge \mathbf{e}_3, \mathbf{e}_1 \wedge \mathbf{e}_2) = 0, \quad G(\mathbf{e}_1 \wedge \mathbf{e}_3, \mathbf{e}_1 \wedge \mathbf{e}_3) = 6, \quad G(\mathbf{e}_1 \wedge \mathbf{e}_3, \mathbf{e}_2 \wedge \mathbf{e}_3) = 2, \\ &G(\mathbf{e}_2 \wedge \mathbf{e}_3, \mathbf{e}_1 \wedge \mathbf{e}_2) = 0, \quad G(\mathbf{e}_2 \wedge \mathbf{e}_3, \mathbf{e}_1 \wedge \mathbf{e}_3) = 2, \quad G(\mathbf{e}_2 \wedge \mathbf{e}_3, \mathbf{e}_2 \wedge \mathbf{e}_3) = 6, \end{aligned}$$

and

$$G(\mathbf{e}_1 \wedge \mathbf{e}_2 \wedge \mathbf{e}_3, \mathbf{e}_1 \wedge \mathbf{e}_2 \wedge \mathbf{e}_3) = 16.$$

2.10 The Hodge Isomorphism

Previously, the vector spaces $\bigwedge_p(V)$ and $\bigwedge_{n-p}(V)$ were shown to have the same dimension and therefore to be isomorphic, although this isomorphism is not canonical. We must therefore define this isomorphism and, among numerous possible choices, the *Hodge isomorphism* has prominent importance. The Hodge isomorphism is obviously closely related to the quasi-Hodge isomorphisms. In order to define the quasi-Hodge isomorphisms, an orientation must be chosen for the vector space V. In order to define the Hodge isomorphism, one additional structure is demanded: a *symmetric bilinear functional* on V. Consequently, the Hodge isomorphism is defined *only in the context of the Grassmann algebra* and not in the exterior algebra, as only quasi-Hodge isomorphisms exist in the latter.

The presence of a symmetric bilinear functional g makes it possible to relate the spaces V and V^* via the correlation $\flat : V \to V^*$. This correlation can be generalised via $\flat : \bigwedge_p(V) \to \bigwedge^p(V)$ from eqn (2.96). The same reasoning applies to the correlation $\sharp = \flat^{-1}$, which can be similarly generalised (see eqn (2.96)), so that $\sharp : \bigwedge^p(V) \to \bigwedge_p(V)$.

Now let us consider the quasi-Hodge isomorphism $\overline{\star} : \bigwedge_p(V) \to \bigwedge^{n-p}(V)$. Since $\sharp : \bigwedge^{n-p}(V) \to \bigwedge_{n-p}(V)$, the composition $\sharp \circ \overline{\star}$ satisfies

$$\sharp \circ \overline{\star} : \bigwedge{}_p(V) \to \bigwedge{}_{n-p}(V). \tag{2.102}$$

Furthermore, let us consider the composition $\overline{\star} \circ \sharp$ such that

$$\overline{\star} \circ \sharp : \bigwedge{}^p(V) \to \bigwedge{}^{n-p}(V). \tag{2.103}$$

With respect to the quasi-Hodge isomorphism $\underline{\star} : \bigwedge^p(V) \to \bigwedge_{n-p}(V)$, we can consider two compositions: (i) $\flat \circ \underline{\star}$, and (ii) $\underline{\star} \circ \flat$. The first composition is provided by

$$\flat \circ \underline{\star} : \bigwedge{}^p(V) \to \bigwedge{}^{n-p}(V), \tag{2.104}$$

whereas the second one reads

$$\underline{\star} \circ \flat : \bigwedge{}_p(V) \to \bigwedge{}_{n-p}(V). \tag{2.105}$$

Let us now define the isomorphism $\bigwedge_p(V) \to \bigwedge_{n-p}(V)$. From what we have previously discussed, there exist two isomorphisms which can be used to define it: (i) $\sharp \circ \overline{\star}$, and (ii) $\underline{\star} \circ \flat$. Which one to choose? In order to settle this question, we demanded that

$$\sharp \circ \overline{\star} = \underline{\star} \circ \flat. \tag{2.106}$$

However, which conditions must hold in order for this expression to be valid?

Let us first consider the isomorphism $\overline{\star}$ defined by eqn (2.66). By applying the correlation $\sharp$ at both sides of this equation and using $\sharp(\Psi \wedge \Phi) = \Psi^\sharp \wedge \Phi^\sharp$, we can see that

$$\Psi^{[p]\sharp} \wedge (\sharp \circ \overline{\star}) A_{[p]} = p! \Psi^{[p]}(A_{[p]}) \Omega^\sharp_{V^*}, \tag{2.107}$$

which can be written as

$$B_{[p]} \wedge (\sharp \circ \overline{\star}) A_{[p]} = p! A_{[p]\flat}(B_{[p]}) \Omega^{\sharp}_{V^*}. \tag{2.108}$$

On the other hand, using the definition of $\underline{\star}$ given in eqn (2.67), we find that

$$A_{[p]} \wedge (\underline{\star} \circ \flat) B_{[p]} = p! B_{[p]\flat}(A_{[p]}) \Omega_V. \tag{2.109}$$

Comparing those two last equations, if the equality $\sharp \circ \overline{\star} = \underline{\star} \circ \flat$ holds, then $\Omega^{\sharp}_{V^*} = \Omega_V$ or equivalently

$$\Omega_{V^*} = \Omega^{\flat}_V. \tag{2.110}$$

The relationship between Ω_{V^*} and Ω_V is given by eqn (2.65). With this last condition, it therefore follows that

$$\widetilde{\Omega^{\flat}_V}(\Omega_V) = 1. \tag{2.111}$$

In order to understand the meaning of this equation, the definition of G is used together with eqn (2.99), implying that

$$G(\Omega_V, \Omega_V) = 1. \tag{2.112}$$

Then $\sharp \circ \overline{\star} = \underline{\star} \circ \flat$ holds if the n-vector Ω_V is *unitary*. The *Hodge isomorphism*, denoted by $\star$, is defined by

$$\boxed{\star = \sharp \circ \overline{\star} = \underline{\star} \circ \flat}. \tag{2.113}$$

Using eqn (2.97), we can thus define $\star$ as

$$A_{[p]} \wedge \star B_{[p]} = G(A_{[p]}, B_{[p]}) \Omega_V, \tag{2.114}$$

where $G(\Omega_V, \Omega_V) = 1$; or, even better than that, as

$$\boxed{A \wedge \star B = G(A, B) \Omega_V}, \tag{2.115}$$

where $A, B \in \bigwedge(V)$. Here we used the fact that $G(A_{[p]}, B_{[q]}) = 0$ if $p \neq q$. The Hodge isomorphism between the spaces $\bigwedge^p(V)$ and $\bigwedge^{n-p}(V)$ is similarly defined. In fact, the mapping $\overset{*}{\star} : \bigwedge^p(V) \to \bigwedge^{n-p}(V)$ can be written as

$$\overset{*}{\star} = \flat \circ \star \circ \sharp. \tag{2.116}$$

Evidently, the calculations for $\overline{\star}$ and $\underline{\star}$ can be applied to $\star$ by careful considering the correlations $\flat$ and $\sharp$, in the context of either eqns (2.68) and (2.80), or eqns (2.81) and (2.82). For the case of the Hodge isomorphism $\star$, it is immediately clear that

$$\boxed{\begin{aligned} \star 1 &= \Omega_V, \\ \star A &= \widetilde{A}_{\flat} \rfloor \Omega_V \end{aligned}}, \tag{2.117}$$

where $A \in \bigwedge(V)$.

The *regressive product* can be written using the Hodge isomorphism. From eqn (2.113), it reads

$$\underline{\star} = \star \circ \sharp, \qquad \overline{\star} = \flat \circ \star. \tag{2.118}$$

Taking it and $\overline{\star}^{-1} = \star \circ \flat^{-1} = \star \circ \sharp$ into account, from eqn (2.90) we find that

$$\boxed{A_{[p]} \vee B_{[q]} = \star^{-1}(\star A_{[p]} \wedge \star B_{[q]})}. \tag{2.119}$$

To conclude this discussion, let us explicitly write the unit pseudoscalar Ω_V. Let $\mathfrak{B} = \{\mathbf{e}_1, \dots, \mathbf{e}_n\}$ be a basis of V and let g be the symmetric bilinear functional $g(\mathbf{v}, \mathbf{u}) = \mathbf{v}_\flat(\mathbf{u}) = g_{ij} v^i u^j$, where $g_{ij} = g(\mathbf{e}_i, \mathbf{e}_j) = g_{ji}$. Let $\mathfrak{B}' = \{\mathbf{e}'_1, \dots, \mathbf{e}'_n\}$ be an orthonormal basis,[4]

$$g(\mathbf{e}'_i, \mathbf{e}'_j) = \lambda_i \delta_{ij}, \tag{2.120}$$

where

$$\lambda_i = \begin{cases} 1, & (i = 1, \dots, p), \\ -1, & (i = p+1, \dots, n), \end{cases} \tag{2.121}$$

defining the quadratic space $\mathbb{R}^{p,q}$ $(p + q = n)$.

Since the vectors $\{\mathbf{e}'_i\}$ are unitary, from the definition of G, it immediately follows that the n-vector

$$\Omega_V = \mathbf{e}'_1 \wedge \cdots \wedge \mathbf{e}'_n \tag{2.122}$$

is unitary. We can then use this expression for Ω_V with respect to the basis $\mathfrak{B}'$ in order to explicitly calculate the Hodge isomorphism. Sometimes it is appropriate not to take into account an orthonormal basis $\mathfrak{B}'$ but rather an arbitrary basis $\mathfrak{B}$. Let us then express Ω_V with respect to the basis $\mathfrak{B}$, writing the relation between $\mathfrak{B}$ and $\mathfrak{B}'$ as

$$\mathbf{e}'_i = h(\mathbf{e}_i) = h^j_i \mathbf{e}_j, \tag{2.123}$$

where the demonstration of

$$\mathbf{e}'_1 \wedge \cdots \wedge \mathbf{e}'_n = (\det h)\mathbf{e}_1 \wedge \cdots \wedge \mathbf{e}_n \tag{2.124}$$

is left as an exercise, and h denotes the matrix

$$h = \begin{pmatrix} h^1_1 & h^1_2 & \cdots & h^1_n \\ h^2_1 & h^2_2 & \cdots & h^2_n \\ \vdots & \vdots & \ddots & \vdots \\ h^n_1 & h^n_2 & \cdots & h^n_n \end{pmatrix}. \tag{2.125}$$

On the other hand, we can write from eqn (2.120) that

$$\lambda_i \delta_{ij} = h^k_i g_{kl} h^l_i. \tag{2.126}$$

This equation can be written in matrix form as

$$\lambda = h^\intercal g h, \tag{2.127}$$

[4]Equation (2.120) *does not* denote the sum convention, since the index i, although repeated, is not placed as a raised or lowered index, which would be the case if it did.

where g denotes the matrix

$$g = \begin{pmatrix} g_{11} & g_{12} & \cdots & g_{1n} \\ g_{21} & g_{22} & \cdots & g_{2n} \\ \vdots & \vdots & \ddots & \vdots \\ g_{n1} & g_{n2} & \cdots & g_{nn} \end{pmatrix}, \tag{2.128}$$

$h^\intercal$ denotes the transposed matrix associated with h, and λ is the diagonal matrix with entries $\lambda_{ii} = \lambda_i$ given by eqn (2.121). Using the well-known properties $\det(AB) = \det A \det B$, and $\det A^\intercal = \det A$, as well as the fact that $\det \lambda = (-1)^q$, where $q = n-p$, it follows that $(-1)^q = (\det h)^2 \det g$ or

$$1 = (\det h)^2 |\det g|. \tag{2.129}$$

There are now two possibilities: (i) $\det h = +|\det g|^{-1/2}$, or (ii) $\det h = -|\det g|^{-1/2}$. Since the set of matrices h must include the identity matrix 1, the first possibility must be chosen:

$$\det h = |\det g|^{-1/2}. \tag{2.130}$$

Thus, the unit n-vector Ω_V can be written in terms of the basis $\mathfrak{B}$ as

$$\Omega_V = \frac{1}{\sqrt{|\det g|}} \mathbf{e}_1 \wedge \cdots \wedge \mathbf{e}_n. \tag{2.131}$$

Another useful expression is

$$\Omega_V = \sqrt{|\det g^{-1}|}\mathbf{e}_1 \wedge \cdots \wedge \mathbf{e}_n, \tag{2.132}$$

where g^{-1} is the inverse matrix related to g,

$$g^{-1} = \begin{pmatrix} g^{11} & g^{12} & \cdots & g^{1n} \\ g^{21} & g^{22} & \cdots & g^{2n} \\ \vdots & \vdots & \ddots & \vdots \\ g^{n1} & g^{n2} & \cdots & g^{nn} \end{pmatrix}, \tag{2.133}$$

where $g^{ij} = g(\mathbf{e}^i, \mathbf{e}^j)$, $g^{ij} g_{jk} = \delta^i_k$.

Example 2.11 Let us establish the Hodge isomorphism for the case where the space $\mathbb{R}^3$ is endowed with the usual Euclidean scalar product. The standard basis $\{\mathbf{e}_i\}$ satisfies $g(\mathbf{e}_i, \mathbf{e}_j) = \delta_{ij}$. The unit 3-vector is given by $\Omega_{\mathbb{R}^3}$, corresponding to the usual orientation of $\mathbb{R}^3$ given by $\Omega_{\mathbb{R}^3} = \mathbf{e}^1 \wedge \mathbf{e}^2 \wedge \mathbf{e}^3$. Using eqn (2.117), we find that

$$\begin{aligned} \star 1 &= \mathbf{e}^1 \wedge \mathbf{e}^2 \wedge \mathbf{e}^3, & \star \mathbf{e}^1 \wedge \mathbf{e}^2 \wedge \mathbf{e}^3 &= 1, \\ \star \mathbf{e}^1 &= \mathbf{e}^2 \wedge \mathbf{e}^3, & \star \mathbf{e}^2 \wedge \mathbf{e}^3 &= \mathbf{e}^1, \\ \star \mathbf{e}^2 &= \mathbf{e}^3 \wedge \mathbf{e}^1, & \star \mathbf{e}^3 \wedge \mathbf{e}^1 &= \mathbf{e}^2, \\ \star \mathbf{e}^3 &= \mathbf{e}^1 \wedge \mathbf{e}^2, & \star \mathbf{e}^1 \wedge \mathbf{e}^2 &= \mathbf{e}^3. \end{aligned}$$

Example 2.12 This example aims to establish the Hodge isomorphism for the space $\mathbb{R}^3$ endowed with the correlation investigated in example 1.3. The results in example 2.9 show that the unit 3-vector $\Omega_{\mathbb{R}^3}$ with the usual orientation of $\mathbb{R}^3$ reads

$$\Omega_{\mathbb{R}^3} = \frac{1}{4}\mathbf{e}_1 \wedge \mathbf{e}_2 \wedge \mathbf{e}_3.$$

Using the definition of $\flat$ for this case and its extension for $\bigwedge_p(V)$ calculated in the example 2.8 in eqn (2.117) , we find that

$$\star 1 = \frac{1}{4}\mathbf{e}_1 \wedge \mathbf{e}_2 \wedge \mathbf{e}_3, \qquad \star\mathbf{e}_1 = \frac{3}{4}\mathbf{e}_2 \wedge \mathbf{e}_3 + \frac{1}{4}\mathbf{e}_3 \wedge \mathbf{e}_1,$$

$$\star\,\mathbf{e}_2 = \frac{1}{4}\mathbf{e}_2 \wedge \mathbf{e}_3 + \frac{3}{4}\mathbf{e}_3 \wedge \mathbf{e}_1, \qquad \star\mathbf{e}_3 = \frac{1}{2}\mathbf{e}_1 \wedge \mathbf{e}_2,$$

$$\star\,\mathbf{e}_1 \wedge \mathbf{e}_2 = 2\mathbf{e}_3, \qquad \star\mathbf{e}_3 \wedge \mathbf{e}_1 = \frac{3}{2}\mathbf{e}_2 - \frac{1}{2}\mathbf{e}_1,$$

$$\star\,\mathbf{e}_2 \wedge \mathbf{e}_3 = \frac{3}{2}\mathbf{e}_1 - \frac{1}{2}\mathbf{e}_2, \qquad \star\mathbf{e}_1 \wedge \mathbf{e}_2 \wedge \mathbf{e}_3 = 4.$$

2.11 Additional Readings

A thorough discussion of the exterior algebra can be found in Greub's (1978) book, whereas some results obtained by using the exterior product in linear algebra are revisited in Winitzki's (2010) book. The exterior algebra is the cornerstone of the calculus of differential forms. A good introduction to the formalism of differential forms is provided by do Carmo (1994). For the use of such forms in differential geometry via Cartan's moving frame method, a great introductory text has been provided by O'Neill (2006) whereas, for a modern treatment, including vector bundles, a nice introduction has been given by Darling (1994). For applications of differential forms in physics, the classic work is Flanders (1963), whereas Frankel (2012) is modern and complete. Another text with an emphasis on applications in thermodynamics, electromagnetism, and gauge theories is the one by Edelen (1985). The exterior calculus without a metric g is fundamental for a metric-free formulation of classical electrodynamics (Hehl and Obukhov, 2003), with subsequent applications (da Rocha and Rodrigues, 2010). The Grassmann algebra also underlies supermanifolds and supersymmetry; for an introduction to this subject, from the geometric point of view, we suggest the work by Rodrigues Jr, da Rocha, Bernardini and Vaz Jr (2005) as well as that by Rogers (2007).

2.12 Exercises

(1) Let $\mathfrak{B} = \{\mathbf{e}^1, \mathbf{e}^2, \mathbf{e}^3, \mathbf{e}^4, \mathbf{e}^5\}$ be a basis of V^* and let α, β, and γ the following multicovectors:

$$\alpha = \mathbf{e}^1 + 2\mathbf{e}^1 \wedge \mathbf{e}^2 + 5\mathbf{e}^3 \wedge \mathbf{e}^4,$$

$$\beta = \mathbf{e}^3 - \mathbf{e}^1 \wedge \mathbf{e}^2 + 3\mathbf{e}^1 \wedge \mathbf{e}^3 \wedge \mathbf{e}^4 - \mathbf{e}^3 \wedge \mathbf{e}^4 \wedge \mathbf{e}^5,$$

$$\gamma = \mathbf{e}^5 - 2\mathbf{e}^4 \wedge \mathbf{e}^5.$$

Calculate the following: (a) $\alpha \wedge \alpha$; (b) $\beta \wedge \beta$; (c) $\gamma \wedge \gamma$; (d) $\alpha \wedge \beta$; (e) $\beta \wedge \alpha$; (f) $\alpha \wedge \gamma$; (g) $\gamma \wedge \alpha$; (h) $\beta \wedge \gamma$; (i) $\gamma \wedge \beta$; (j) $\alpha \wedge \alpha \wedge \alpha$; (k)$\alpha \wedge \alpha \wedge \beta$; (l) $\alpha \wedge \alpha \wedge \gamma$; (m) $\beta \wedge \beta \wedge \alpha$; (n) $\beta \wedge \beta \wedge \beta$; and (o) $\beta \wedge \beta \wedge \gamma$.

(2) Show that an arbitrary 2-vector $F \in \bigwedge^2(V)$ can be written as the exterior product of two 1-vectors $\mathbf{v}, \mathbf{u} \in V = \bigwedge^1(V)$, that is, $F = \mathbf{v} \wedge \mathbf{u}$, only if $\dim V \leq 3$.

(3) Use the method described in example 2.8 to find the inverse of the matrix

$$\begin{pmatrix} 1 & 2 & 1 & -1 \\ 0 & 3 & -1 & 1 \\ 2 & 1 & 2 & 0 \\ 1 & 1 & 0 & -1 \end{pmatrix}.$$

(4) Let $\mathfrak{B} = \{\mathbf{e}_i\}$, and $\mathfrak{B}' = \{\mathbf{e}'_i\}$, be two bases of V related by $\mathbf{e}_i = B_i^j \mathbf{e}_j$ and let $\mathbf{B} = \{B_j^i\}$ be the matrix that changes the two bases, where B_j^i corresponds to the element of the jth row and the ith column $(i, j = 1, \ldots, n)$. Such change of basis obviously induces a change of basis for the space of the k-vectors. (a) Show that

$$\mathbf{e}'_i \wedge \mathbf{e}'_j = \sum_{k<l} (\det \Delta_{ij}^{kl}) \mathbf{e}_k \wedge \mathbf{e}_l,$$

where Δ_{ij}^{kl} denotes the matrix of order 2 obtained from the matrix $\mathbf{B}$ in the following way: the first row of Δ_{ij}^{kl} is given by the ith row of $\mathbf{B}$, and the second row of Δ_{ij}^{kl} by the jth row of $\mathbf{B}$; and the first column of Δ_{ij}^{kl} is given by the kth column of $\mathbf{B}$, and the second column of Δ_{ij}^{kl} by the lth column of $\mathbf{B}$. In this expression, $\sum_{j<l}$ denotes the sum over all j and l such that $k < l$. (b) Generalise the previous result, showing that, in the space of k-vectors $(k < n)$, the property

$$\mathbf{e}'_{\mu_1} \wedge \cdots \wedge \mathbf{e}'_{\mu_k} = \sum_{\nu_1 < \cdots < \nu_k} (\det \Delta_{\mu_1 \cdots \mu_k}^{\nu_1 \cdots \nu_k}) \mathbf{e}_{\nu_1} \wedge \cdots \wedge \mathbf{e}_{\nu_k}$$

holds, where $\Delta_{\mu_1 \cdots \mu_k}^{\nu_1 \cdots \nu_k}$ denotes the matrix of order k obtained from the matrix $\mathbf{B}$ when we take the ith row of $\Delta_{\mu_1 \cdots \mu_k}^{\nu_1 \cdots \nu_k}$ as the μ_i^{th} row of $\mathbf{B}$ and the jth column of $\Delta_{\mu_1 \cdots \mu_k}^{\nu_1 \cdots \nu_k}$ as being the ν_j^{th} column of $\mathbf{B}$. (c) Show that, as a consequence of these results, for pseudoscalars, the expression

$$\mathbf{e}'_1 \wedge \cdots \wedge \mathbf{e}'_n = (\det \mathbf{B}) \mathbf{e}_1 \wedge \cdots \wedge \mathbf{e}_n$$

holds.

(5) From this last expression (which can be taken as the definition of the determinant of a linear transformation), deduce the Laplace rule for the calculation of determinants.

(6) Let V be a vector space $(\dim V = n)$ and let W be the k-dimensional subspace generated by $\{\mathbf{v}_1, \ldots, \mathbf{v}_k\}$. The k-vector $I_W = \mathbf{v}_1 \wedge \cdots \wedge \mathbf{v}_k$ completely defines this subspace, and a vector $\mathbf{v}$ is in W if and only if $\mathbf{v} \wedge I_W = 0$ (see example 2.4). If $\mathfrak{B} = \{\mathbf{e}_i\}$ $(i = 1, \ldots, n)$ is a basis of V, the components of the k-vector I_W with respect to the basis $\{\mathbf{e}_{\mu_1} \wedge \cdots \wedge \mathbf{e}_{\mu_k} \mid \mu_1 < \cdots < \mu_k\}$ of $\bigwedge_k(V)$ are called the Plücker coordinates of the subspace W in the basis $\mathfrak{B}$, that is, the Plücker coordinates $V^{\mu_1 \cdots \mu_k}$ are given by

$$I_W = \mathbf{v}_1 \wedge \cdots \wedge \mathbf{v}_k = V^{\mu_1 \cdots \mu_k} \mathbf{e}_{\mu_1} \wedge \cdots \wedge \mathbf{e}_{\mu_k}.$$

(a) Show that, in general, the Plücker coordinates are not all independent but satisfy a set of identities called the Plücker correlations and given by

$$V^{[\mu_1\cdots\mu_k}V^{\nu_1]\nu_2\cdots\nu_k} = 0,$$

where the brackets denote antisymmetrisation of the indices, that is,

$$\begin{aligned}
&V^{[\mu_1\cdots\mu_k}V^{\nu_1]\nu_2\cdots\nu_k}\\
&\quad = V^{\mu_1\cdots\mu_k}V^{\nu_1\nu_2\cdots\nu_k} - V^{\mu_1\cdots\mu_{k-1}\nu_1}V^{\mu_k\nu_2\cdots\nu_k}\\
&\qquad + V^{\mu_1\cdots\mu_{k-2}\nu_1}V^{\mu_{k-1}\nu_2\cdots\nu_k} - V^{\mu_1\cdots\mu_{k-3}\mu_{k-1}\nu_1}V^{\mu_{k-2}\nu_2\cdots\nu_k} + \cdots\\
&\qquad + (-1)^{k-1}V^{\mu_1\mu_3\cdots\mu_{k-1}\nu_1}V^{\mu_2\nu_2\cdots\nu_k} + (-1)^k V^{\mu_2\mu_3\cdots\mu_{k-1}\nu_1}V^{\mu_1\nu_2\cdots\nu_k}.
\end{aligned}$$

(b) Not all Plücker correlations are trivial. Some of them, for instance, follow from the anti-commutativity of the components $V^{\mu_1\cdots\mu_k} = V^{[\mu_1\cdots\mu_k]}$ of the k-vector I_W. Show that, for $k = n$, and $k = n-1$, not all Plücker correlations are trivial. (c) Consider the case where $n = 4$, and $k = 2$. Show that, for this case, there exists only one non-trivial Plücker correlation.

(7) Show that $\widetilde{(\alpha\rfloor A)} = \widetilde{A}\lfloor\alpha$, for all $\alpha \in V^*$, and $A \in \bigwedge(V)$. This result implies that $\widetilde{(\Psi\rfloor A)} = \widetilde{A}\lfloor\tilde{\Psi}$, for all $\Psi \in \bigwedge(V^*)$, and $A \in \bigwedge(V)$.

(8) Consider $T_{[p]} \in \bigwedge_p(V)$, and $S^{[q]} \in \bigwedge^q(V)$. Show that

$$T_{[p]}\lfloor S^{[q]} = (-1)^{q(p+1)}\left(S^{[q]}\rfloor T_{[p]}\right).$$

(9) Let $\mathbb{R}^3$ be endowed with the usual Euclidean scalar product. Let $\dot{\times}$ denote the usual vector product between vectors (which defines the Gibbs–Heaviside vector algebra) and the vector product $\times$ defined as

$$\mathbf{v} \times \mathbf{u} = \star(\mathbf{v} \wedge \mathbf{u}),$$

where $\mathbf{v}, \mathbf{u} \in \mathbb{R}^3$. (a) Show that the objects defined by $\mathbf{v} \times \mathbf{u}$ and $\mathbf{v} \dot{\times} \mathbf{u}$ present the same components. (b) Show that the object defined by $\mathbf{v} \dot{\times} \mathbf{u}$ corresponds to what sometimes is called the axial vector or pseudovector (i. e. it is an object that does not change sign under the coordinate system inversion), and that, in contrast, the object defined by $\mathbf{v} \times \mathbf{u}$ is indeed a vector (sometimes called polar vector, in order to be distinguished from the axial vector, since in this case there is a change sign under the coordinate system inversion).

(10) Consider $\{\mathbf{u}_1, \mathbf{u}_2, \ldots, \mathbf{u}_{2r-1}, \mathbf{u}_{2r}\}$ vectors in V. Given $\mathbf{v} = \mathbf{u}_1 \wedge \mathbf{u}_2 + \mathbf{u}_3 \wedge \mathbf{u}_4 + \mathbf{u}_5 \wedge \mathbf{u}_6 + \cdots + \mathbf{u}_{2r-1} \wedge \mathbf{u}_{2r}$, show that

$$\underbrace{\mathbf{v} \wedge \mathbf{v} \wedge \cdots \wedge \mathbf{v}}_{r\ \text{times}} = r!\,(\mathbf{u}_1 \wedge \mathbf{u}_2 \wedge \cdots \wedge \mathbf{u}_{2r-1} \wedge \mathbf{u}_{2r}).$$

3
Clifford, or Geometric, Algebra

In this chapter, we introduce the so-called Clifford, or geometric, algebras. We are here interested in the universal Clifford algebras and, in this text, except for the first section in this chapter, 'Clifford algebra' will be a synonym for 'universal Clifford algebra'. Besides the standard definition of a Clifford algebra via the so-called Clifford mapping, we present an explicit construction of the (universal) Clifford algebra associated with a quadratic space as a quotient of the tensor algebra. Moreover, Clifford algebras are also presented in the context of Grassmann algebras (Clifford, 1878). The prominent features of Clifford algebras are presented and creation operators and annihilation operators are introduced. For a discussion regarding Clifford algebras over infinite-dimensional spaces, see the book by Plymen and Robinson (1990).

3.1 Definition of a Clifford Algebra

Let V be a vector space over $\mathbb{R}$, endowed with a symmetric bilinear form g. Let $\mathcal{A}$ be an associative algebra with unity $1_{\mathcal{A}}$ and let γ be the linear mapping $\gamma : V \to \mathcal{A}$.

Definition 3.1 ▶ *The pair $(\mathcal{A}, \gamma)$ is a Clifford algebra for the quadratic space (V, g) when $\mathcal{A}$ is generated as an algebra by $\{\gamma(\mathbf{v}) \mid \mathbf{v} \in V\}$ and $\{a1_{\mathcal{A}} \mid a \in \mathbb{R}\}$, and γ satisfies*

$$\boxed{\gamma(\mathbf{v})\gamma(\mathbf{u}) + \gamma(\mathbf{u})\gamma(\mathbf{v}) = 2g(\mathbf{v}, \mathbf{u})1_{\mathcal{A}}} \tag{3.1}$$

for all $\mathbf{v}, \mathbf{u} \in V$.

Equation (3.1) holds for all $\mathbf{u}, \mathbf{v} \in V$ if and only if $\gamma(\mathbf{v})^2 = Q(\mathbf{v})$ holds for all $\mathbf{v} \in V$. In many cases, it is easier to verify the equality $\gamma(\mathbf{v})^2 = Q(\mathbf{v})$ than the equality in eqn (3.1).

When g is not degenerated, the Clifford algebra for the quadratic space (V^*, g^{-1}) is defined in a completely analogous way, where now the linear mapping $\gamma : V^* \to \mathcal{A}$ is defined, satisfying

$$\gamma(\alpha)\gamma(\beta) + \gamma(\beta)\gamma(\alpha) = 2g^{-1}(\alpha, \beta)1_{\mathcal{A}}, \tag{3.2}$$

for all $\alpha, \beta \in V^*$, and $g^{-1} : V^* \times V^* \to \mathbb{R}$.

The mapping γ is a kind of 'square root' of the quadratic form $Q(\mathbf{v}) = g(\mathbf{v}, \mathbf{v})$, since

$$\boxed{(\gamma(\mathbf{v}))^2 = Q(\mathbf{v}) = g(\mathbf{v}, \mathbf{v})}. \tag{3.3}$$

Such a mapping γ is said to be a Clifford mapping.

An Introduction to Clifford Algebras and Spinors. First Edition. Jayme Vaz, Jr. and Roldão da Rocha, Jr.

Now let us consider an orthonormal basis $\mathfrak{B} = \{\mathbf{e}_1, \ldots, \mathbf{e}_n\}$ of V. In the Clifford algebra $(\mathcal{A}, \gamma)$ for (V, g), we have

$$\gamma(\mathbf{e}_i)\gamma(\mathbf{e}_j) + \gamma(\mathbf{e}_j)\gamma(\mathbf{e}_i) = 0_{\mathcal{A}}, \quad i \neq j, \tag{3.4}$$

and

$$\gamma(\mathbf{e}_i)^2 = Q(\mathbf{e}_i)1_{\mathcal{A}}, \tag{3.5}$$

where $Q(\mathbf{e}_i) = g(\mathbf{e}_i, \mathbf{e}_i)$. By using such relations, any product involving $\gamma(\mathbf{e}_i)$ ($i = 1, \ldots, n$), and their powers as well, can be reordered to yield

$$\gamma(\mathbf{e}_1)^{\mu_1}\gamma(\mathbf{e}_2)^{\mu_2}\cdots\gamma(\mathbf{e}_n)^{\mu_n}, \qquad \mu_i = 0, 1, \ (i = 1, \ldots, n),$$

where $\gamma(\mathbf{e}_1)^0\gamma(\mathbf{e}_2)^0\cdots\gamma(\mathbf{e}_n)^0$ is the identity $1_{\mathcal{A}}$ of $\mathcal{A}$. As $\mathcal{A}$ is generated as an algebra by $\{\gamma(\mathbf{v}) \mid \mathbf{v} \in V\}$ and $\{a1_{\mathcal{A}} \mid a \in \mathbb{R}\}$, it is generated by these products:

$$\mathcal{A} = \operatorname{span}\{\gamma(\mathbf{e}_1)^{\mu_1}\gamma(\mathbf{e}_2)^{\mu_2}\cdots\gamma(\mathbf{e}_n)^{\mu_n} \mid \mu_i = 0, 1\}. \tag{3.6}$$

However, since the number of elements of type $\gamma(\mathbf{e}_1)^{\mu_1}\gamma(\mathbf{e}_2)^{\mu_2}\cdots\gamma(\mathbf{e}_n)^{\mu_n}$, where $\mu_i = 0, 1$ is 2^n, therefore $\dim \mathcal{A} \leq 2^n$ and so the maximal dimension of a Clifford algebra is 2^n. Although there exist examples of Clifford algebras with fewer than 2^n-dimensions(Porteous, 1995), Clifford algebras of maximal dimension will be the focus here. Indeed, such algebras have a prominent property that distinguishes them from algebras with fewer than 2^n dimensions. This property is called *universality*, and the Clifford algebras that have such property have dimension 2^n.

Definition 3.2 ▶ *A Clifford algebra $(\mathcal{A}, \gamma)$ for the quadratic space (V, g) is said to be a universal Clifford algebra if, for each Clifford algebra $(\mathcal{B}, \rho)$ for (V, g), there exists an isomorphism $\phi : \mathcal{A} \to \mathcal{B}$ such that $\rho = \phi \circ \gamma$, and $\phi(1_{\mathcal{A}}) = 1_{\mathcal{B}}$. A universal Clifford algebra for the quadratic space (V, g) is denoted by $C\ell(V, g)$.*

The universal Clifford algebra $C\ell(V, g)$, if it exists, is unique up to a unique isomorphism. In fact, since $C\ell(V, g)$ is a Clifford algebra, there is a unique isomorphism such that $\phi : \mathcal{A} \to \mathcal{B}$ such that $\rho = \phi \circ \gamma$, and $\phi(1_{\mathcal{A}}) = 1_{\mathcal{B}}$ whereas, since $(\mathcal{B}, \rho)$ is also a Clifford algebra, there is a unique isomorphism such that $\phi' : \mathcal{B} \to \mathcal{A}$ such that $\gamma = \phi' \circ \rho$, and $\phi'(1_{\mathcal{B}}) = 1_{\mathcal{A}}$. Now, the composition $\phi' \circ \phi : C\ell(V, g) \to C\ell(V, g)$ is such that $\gamma = (\phi' \circ \phi) \circ \gamma$ and the identity $1_{\mathcal{A}} : C\ell(V, g) \to C\ell(V, g)$, whence $\phi' \circ \phi = 1_{\mathcal{A}}$. A similar argument shows that $\phi \circ \phi' = 1_{\mathcal{B}}$, whence $\phi : \mathcal{A} \to \mathcal{B}$ is an isomorphism.

Theorem 3.1 ▶ *The Clifford algebra $(\mathcal{A}, \gamma)$ for the quadratic space (V, g) is universal when* $\dim \mathcal{A} = 2^n$, *where* $n = \dim V$.

Proof: Consider an orthonormal basis $\mathfrak{B} = \{\mathbf{e}_1, \ldots, \mathbf{e}_n\}$ of V. For the Clifford algebra $(\mathcal{A}, \gamma)$, it follows that $\gamma(\mathbf{e}_i)\gamma(\mathbf{e}_j) + \gamma(\mathbf{e}_j)\gamma(\mathbf{e}_i) = 0_{\mathcal{A}}$, $(i \neq j)$ and $\gamma(\mathbf{e}_i)^2 = Q(\mathbf{e}_i)1_{\mathcal{A}}$. It is supposed that $\dim \mathcal{A} = 2^n$. In this case, the set $\{\gamma(\mathbf{e}_1)^{\mu_1}\gamma(\mathbf{e}_2)^{\mu_2}\cdots\gamma(\mathbf{e}_n)^{\mu_n}\}$ with $\mu_i = 0, 1$ for $i = 1, \ldots, n$ does not only generate $\mathcal{A}$ but is in addition a basis for $\mathcal{A}$. Let now $(\mathcal{B}, \rho)$ be an arbitrary Clifford algebra. Then $\rho(\mathbf{e}_i)\rho(\mathbf{e}_j) + \rho(\mathbf{e}_j)\rho(\mathbf{e}_i) = 0_{\mathcal{B}}$ (where

$i \neq j$), and $\rho(\mathbf{e}_i)^2 = Q(\mathbf{e}_i)1_{\mathcal{B}}$, and the set $\{\rho(\mathbf{e}_1)^{\mu_1}\rho(\mathbf{e}_2)^{\mu_2}\cdots\rho(\mathbf{e}_n)^{\mu_n} \mid \mu_i = 0,1\}$ generates $\mathcal{B}$. Define the mapping ϕ as a linear mapping $\phi : \mathcal{A} \to \mathcal{B}$ such that

$$\phi(\gamma(\mathbf{e}_1)^{\mu_1}\gamma(\mathbf{e}_2)^{\mu_2}\cdots\gamma(\mathbf{e}_n)^{\mu_n}) = \rho(\mathbf{e}_1)^{\mu_1}\rho(\mathbf{e}_2)^{\mu_2}\cdots\rho(\mathbf{e}_n)^{\mu_n}.$$

It is straightforward to see that ϕ as defined is an algebra isomorphism, that is, $\phi(aa') = \phi(a)\phi(a')$, $\forall\, a, a' \in \mathcal{A}$, satisfying $\phi(\mathbf{e}_i)\phi(\mathbf{e}_j) + \phi(\mathbf{e}_j)\phi(\mathbf{e}_i) = 2g(\mathbf{e}_i, \mathbf{e}_j)1_{\mathcal{B}}$. From the definition of a Clifford algebra, $(\mathcal{A}, \gamma)$ is therefore a universal Clifford algebra $\mathcal{C}\ell(V, g)$. ✓

Theorem 3.2 ► *For all quadratic space* (V, g), *there exist a universal Clifford algebra, and every universal Clifford algebra has dimension* 2^n.

This theorem is demonstrated from the universal Clifford algebra associated with a quadratic space. Its explicit construction is accomplished by a quotient of the tensor algebra by a specific ideal. However, this discussion is reserved for section 3.2, as a point that deserves special attention. Since the universal Clifford algebra has been shown to be unique up to a uniquely given isomorphism, it is legitimate to say *the* universal Clifford algebra associated with a quadratic space.

Example 3.1 Let V be a 1-dimensional vector space. In this case, any vector $\mathbf{v} \in V$ can be written as $\mathbf{v} = y\mathbf{e}$, where $\{\mathbf{e}\}$ is a basis of V. Let now g be the symmetric bilinear functional such that $g(\mathbf{e}, \mathbf{e}) = -1$. Then, $g(\mathbf{v}, \mathbf{v}) = -y^2$. Consider now the subalgebra of the matrix algebra defined by

$$\mathcal{A} = \left\{ \begin{pmatrix} x & y \\ -y & x \end{pmatrix} \mid x, y \in \mathbb{R} \right\}.$$

Obviously, this algebra is generated by

$$\left\{ y \begin{pmatrix} 0 & 1 \\ -1 & 0 \end{pmatrix} \mid y \in \mathbb{R} \right\} \quad \text{and} \quad \left\{ x \begin{pmatrix} 1 & 0 \\ 0 & 1 \end{pmatrix} = x1_{\mathcal{A}} \mid x \in \mathbb{R} \right\}.$$

Now let us define the mapping $\gamma : V \to \mathcal{A}$ as

$$\gamma(\mathbf{e}) = \begin{pmatrix} 0 & 1 \\ -1 & 0 \end{pmatrix}.$$

Consequently,

$$\gamma(\mathbf{v}) = \begin{pmatrix} 0 & y \\ -y & 0 \end{pmatrix}.$$

The mapping γ is a Clifford mapping, since

$$[\gamma(\mathbf{v})]^2 = \begin{pmatrix} 0 & y \\ -y & 0 \end{pmatrix} \begin{pmatrix} 0 & y \\ -y & 0 \end{pmatrix} = -y^2 \begin{pmatrix} 1 & 0 \\ 0 & 1 \end{pmatrix} = g(\mathbf{v}, \mathbf{v})1_{\mathcal{A}}.$$

The algebra $\mathcal{A}$ is isomorphic to the algebra of complex numbers algebra $\mathbb{C}$, and $\mathbb{C}$ is an example of a Clifford algebra.

Example 3.2 Let $V = \mathbb{R}^3$ be endowed with the usual scalar product $g(\mathbf{v}, \mathbf{u}) = v_1u_1 + v_2u_2 + v_3u_3$, where $\mathbf{v} = (v_1, v_2, v_3)^\intercal$, and $\mathbf{u} = (u_1, u_2, u_3)^\intercal$. Consider now the algebra of the 2×2 complex matrices:

$$\mathcal{A} = \left\{ \begin{pmatrix} z_1 & z_2 \\ z_3 & z_4 \end{pmatrix} \mid z_i \in \mathbb{C} (i = 1, 2, 3, 4) \right\},$$

with the matrix product. This algebra is generated by $\{x_1\sigma_1, x_2\sigma_2, x_3\sigma_3 \mid x_i \in \mathbb{R}\}$ and $\{x_0 1 \mid x_0 \in \mathbb{R}\}$, where 1 denotes the identity matrix, and σ_i denotes the matrices

$$\sigma_1 = \begin{pmatrix} 0 & 1 \\ 1 & 0 \end{pmatrix}, \qquad \sigma_2 = \begin{pmatrix} 0 & -i \\ i & 0 \end{pmatrix}, \qquad \sigma_3 = \begin{pmatrix} 1 & 0 \\ 0 & -1 \end{pmatrix}.$$

These matrices are called the Pauli matrices. In order to realise that this algebra is indeed generated by these matrices, it is enough to use the fact that

$$\sigma_1\sigma_2 = i\sigma_3, \qquad \sigma_2\sigma_3 = i\sigma_1, \qquad \sigma_3\sigma_1 = i\sigma_2,$$

which can be straightforwardly verified. Let us now define the mapping $\gamma : \mathbb{R}^3 \to \mathcal{A}$:

$$\gamma(1,0,0) = \sigma_1, \qquad \gamma(0,1,0) = \sigma_2, \qquad \gamma(0,0,1) = \sigma_3.$$

The mapping γ is a Clifford mapping, and the algebra $\mathcal{A}$ is a Clifford algebra. Indeed,

$$\gamma(\mathbf{v}) = \begin{pmatrix} v_3 & v_1 - iv_2 \\ v_1 + iv_2 & -v_3 \end{pmatrix},$$

which implies that

$$\gamma(\mathbf{v})\gamma(\mathbf{u}) + \gamma(\mathbf{u})\gamma(\mathbf{v}) = 2(v_1u_1 + v_2u_2 + v_3u_3)\begin{pmatrix} 1 & 0 \\ 0 & 1 \end{pmatrix} = 2g(\mathbf{v},\mathbf{u})1_{\mathcal{A}}.$$

3.2 Universal Clifford Algebra as a Quotient of the Tensor Algebra

This construction was proposed by Chevalley in 1954 (Chevalley, 1954) as well as by Bourbaki in 1959 (Bourbaki, 1989); In it, the Clifford algebra is presented as a quotient of the tensor algebra by a two-sided ideal. This approach provides a proof of existence by construction, and it is very suitable to a fast access to the main properties of Clifford algebras over commutative rings (Helmstetter and Micali, 2008).

Let (V, g) be a quadratic space and let $\mathsf{T}(V)$ be the algebra of the contravariant tensors. Let us consider the ideal $\mathcal{I}_C$ of $\mathsf{T}(V)$ generated by elements of type

$$\mathbf{v} \otimes \mathbf{v} - Q(\mathbf{v})1,$$

where $Q(\mathbf{v}) = g(\mathbf{v},\mathbf{v})$, and 1 is the identity of the tensor algebra. The ideal $\mathcal{I}_C$ consists therefore of all sums

$$\sum_i A_i \otimes (\mathbf{v} \otimes \mathbf{v} - Q(\mathbf{v})1) \otimes B_i,$$

where $A_i, B_i \in \mathsf{T}(V)$. We can also realise that the ideal $\mathcal{I}_C$ is generated by the elements

$$\mathbf{v} \otimes \mathbf{u} + \mathbf{u} \otimes \mathbf{v} - 2g(\mathbf{v},\mathbf{u})1.$$

In order to construct the quotient algebra $\mathsf{T}(V)/\mathcal{I}_C$, we must consider the equivalence relation

$$A \sim B \Leftrightarrow A = B + x, \qquad x \in \mathcal{I}_C.$$

Let us denote the product of equivalence classes by $\diamond$:

$$[A] \diamond [B] = [A \otimes B].$$

Let $\mathbf{v}, \mathbf{u} \in V$ and consider the tensor product $\mathbf{v} \otimes \mathbf{u}$ suitably expressed by

$$\begin{aligned} \mathbf{v} \otimes \mathbf{u} =& \frac{1}{2}(\mathbf{v} \otimes \mathbf{u} - \mathbf{u} \otimes \mathbf{v}) + g(\mathbf{v},\mathbf{u}) + \frac{1}{2}\big[(\mathbf{v}+\mathbf{u}) \otimes (\mathbf{v}+\mathbf{u}) \\ &- g(\mathbf{v}+\mathbf{u},\mathbf{v}+\mathbf{u}) - \mathbf{v} \otimes \mathbf{v} + g(\mathbf{v},\mathbf{v}) - \mathbf{u} \otimes \mathbf{u} + g(\mathbf{u},\mathbf{u})\big]. \end{aligned}$$

The term in the square brackets is an element of the ideal $\mathcal{I}_C$, so

$$\mathbf{v} \otimes \mathbf{u} \sim \frac{1}{2}(\mathbf{v} \otimes \mathbf{u} - \mathbf{u} \otimes \mathbf{v}) + g(\mathbf{v}, \mathbf{u}), \tag{3.7}$$

or, equivalently,

$$\mathbf{v} \otimes \mathbf{u} \sim \mathbf{v} \wedge \mathbf{u} + g(\mathbf{v}, \mathbf{u}). \tag{3.8}$$

Hence, by denoting the quotient mapping $\pi : \mathsf{T} \to \mathsf{T}(V)/\mathcal{I}_C \simeq C\ell(V, g)$, the product of vectors in the quotient algebra $\mathsf{T}(V)/\mathcal{I}_C$ can be written as

$$\boxed{\mathbf{v} \diamond \mathbf{u} = \pi(\mathbf{v} \wedge \mathbf{u}) + g(\mathbf{v}, \mathbf{u})}. \tag{3.9}$$

In eqn (3.8), $\mathbf{v}\wedge\mathbf{u}$ is the element mentioned in eqn (3.7), since we have already identified $\bigwedge(V)$ with a subspace of $\mathsf{T}(V)$. Although in eqn (3.9) we have come down into the quotient $\mathsf{T}(V)/\mathcal{I}_C$, the element $\pi(\mathbf{v}\wedge\mathbf{u})$ can be interpreted as $\mathbf{v}\wedge\mathbf{u}$, when every element of $\bigwedge(V)$ is identified with its image by π. Nevertheless, this interpretation is allowed only if the mapping π is injective on $\bigwedge(V)$, and we must prove that $\mathcal{I}_C \cap \bigwedge(V) = \{0\}$. In other words, if A is an element of $\mathcal{I}_C$ and is other than the zero multivector, then $\mathrm{Alt}(A) \neq A$. In fact, $\mathrm{Alt}[\sum_{i=1}^{n} A_i \otimes \mathbf{v} \otimes \mathbf{v} \otimes B_i] = 0$ since, for each permutation σ, there exists a permutation σ' with opposite sign (that exchanges $\mathbf{v}$ in $\sigma(j)$ and $\mathbf{v}$ in $\sigma(j+1)$), so the two cancel each other out. Thus,

$$\begin{aligned}
&\mathrm{Alt}\left[\sum_{i=1}^{n} A_i \otimes (\mathbf{v} \otimes \mathbf{v} - Q(\mathbf{v})1) \otimes B_i\right] \\
&= \mathrm{Alt}\left[\sum_{i=1}^{n} A_i \otimes \mathbf{v} \otimes \mathbf{v} \otimes B_i\right] - \mathrm{Alt}\left[\sum_{i=1}^{n} A_i \otimes Q(\mathbf{v})1 \otimes B_i\right] \\
&= -Q(\mathbf{v})\mathrm{Alt}\left[\sum_{i=1}^{n} A_i \otimes 1 \otimes B_i\right] = 0 \quad \text{if and only if} \quad Q(\mathbf{v}) = 0\,.
\end{aligned}$$

Therefore, when $Q(\mathbf{v}) \neq 0$, then $\mathrm{Alt}(A) \neq A$.

Keeping in mind that every element of $\bigwedge(V)$ is identified with its image by π, we can thus simplify the notation in what follows, writing in particular eqn (3.9) as

$$\mathbf{v} \diamond \mathbf{u} = \mathbf{v} \wedge \mathbf{u} + g(\mathbf{v}, \mathbf{u}). \tag{3.10}$$

Moreover, $\bigwedge(V)$ is identified with $C\ell(V, g)$. In fact, the quotient mapping $\pi : \mathsf{T}(V) \to C\ell(V, g)$ induces a mapping $\bigwedge(V) \to C\ell(V, g)$ which, by a reasoning similar to that presented after eqn (3.9), can be shown to be a bijection.

The next step is to generalise these expressions, considering $\mathbf{v}, \mathbf{u}, \mathbf{w} \in V$. The exterior product $\mathbf{v} \wedge \mathbf{u} \wedge \mathbf{w}$ can be written as (see chapter 2, examples 2.1 and 2.2)

$$\mathbf{v} \wedge \mathbf{u} \wedge \mathbf{w} = \frac{1}{3}(\mathbf{v} \otimes (\mathbf{u} \wedge \mathbf{w}) - \mathbf{u} \otimes (\mathbf{v} \wedge \mathbf{w}) + \mathbf{w} \otimes (\mathbf{v} \wedge \mathbf{u})). \tag{3.11}$$

Now eqn (3.8) implies that the two last terms in eqn (3.11) respectively read

$$\begin{aligned} \mathbf{u} \otimes (\mathbf{v} \wedge \mathbf{w}) &\sim \mathbf{u} \otimes \mathbf{v} \otimes \mathbf{w} - g(\mathbf{v}, \mathbf{w})\mathbf{u}, \\ \mathbf{w} \otimes (\mathbf{v} \wedge \mathbf{u}) &\sim \mathbf{w} \otimes \mathbf{v} \otimes \mathbf{u} - g(\mathbf{v}, \mathbf{u})\mathbf{w}. \end{aligned} \tag{3.12}$$

By using the equivalence classes

$$\begin{aligned} \mathbf{u} \otimes \mathbf{v} + \mathbf{v} \otimes \mathbf{u} &\sim 2g(\mathbf{u}, \mathbf{v}), \\ \mathbf{w} \otimes \mathbf{v} + \mathbf{w} \otimes \mathbf{u} &\sim 2g(\mathbf{w}, \mathbf{v}), \end{aligned}$$

eqn (3.12) yields

$$\begin{aligned} \mathbf{u} \otimes (\mathbf{v} \wedge \mathbf{w}) &\sim -\mathbf{v} \otimes \mathbf{u} \otimes \mathbf{w} + 2g(\mathbf{u}, \mathbf{v})\mathbf{w} - g(\mathbf{v}, \mathbf{w})\mathbf{u}, \\ \mathbf{w} \otimes (\mathbf{v} \wedge \mathbf{u}) &\sim -\mathbf{v} \otimes \mathbf{w} \otimes \mathbf{u} + 2g(\mathbf{w}, \mathbf{v})\mathbf{u} - g(\mathbf{v}, \mathbf{u})\mathbf{w}. \end{aligned} \tag{3.13}$$

Now, substituting eqn (3.13) into eqn (3.11), we obtain

$$\mathbf{v} \wedge \mathbf{u} \wedge \mathbf{w} = \mathbf{v} \otimes (\mathbf{u} \wedge \mathbf{w}) + g(\mathbf{v}, \mathbf{w})\mathbf{u} - g(\mathbf{v}, \mathbf{u})\mathbf{w}. \tag{3.14}$$

It is worth emphasising that, by using the expression for $\mathbf{v}_\flat \rfloor (\mathbf{u} \wedge \mathbf{w})$ (see e.g. eqn (2.37)) we can finally express

$$\mathbf{v} \otimes (\mathbf{u} \wedge \mathbf{w}) \sim \mathbf{v} \wedge \mathbf{u} \wedge \mathbf{w} + \mathbf{v}_\flat \rfloor (\mathbf{u} \wedge \mathbf{w}), \tag{3.15}$$

namely, the product between a 1-vector and a 2-vector in the quotient algebra $\mathsf{T}(V)/\mathcal{I}_C$ can be written as

$$\mathbf{v} \diamond (\mathbf{u} \wedge \mathbf{w}) = \mathbf{v} \wedge \mathbf{u} \wedge \mathbf{w} + \mathbf{v}_\flat \rfloor (\mathbf{u} \wedge \mathbf{w}), \tag{3.16}$$

which is a clear generalisation of eqn (3.9).

Equation (3.16) can be generalised as

$$\mathbf{v} \diamond A_{[p]} = \mathbf{v} \wedge A_{[p]} + \mathbf{v}_\flat \rfloor A_{[p]}. \tag{3.17}$$

This formula can be demonstrated by induction, and the cases for $p = 1$ and for $p = 2$ have already been shown.

The quotient algebra $\mathsf{T}(V)/\mathcal{I}_C$ is a *Clifford algebra*. Indeed, from eqn (3.9), it follows that

$$\mathbf{v} \diamond \mathbf{u} + \mathbf{u} \diamond \mathbf{v} = 2g(\mathbf{v}, \mathbf{u}). \tag{3.18}$$

Moreover, Clifford algebra is *universal*, as will be shown in 'Universality'. Hence,

$$\boxed{\mathcal{C}\ell(V, g) = \mathsf{T}(V)/\mathcal{I}_C}. \tag{3.19}$$

Example 3.3 Let us prove eqn (3.17) by induction. Because of linearity, it is sufficient to prove this equation for the case where $A_{[p]}$ is a simple p-vector. First, we slightly change the notation, so that it shall be suitable for this purpose. Let us denote a simple p-vector by $A_{[1\cdots p]}$, that is,

$$A_{[1\cdots p]} = \mathbf{v}_1 \wedge \cdots \wedge \mathbf{v}_p.$$

In addition, we introduce the notation

$$A_{[1\cdots\hat{\imath}\cdots p]} = \mathbf{v}_1 \wedge \cdots \wedge \mathbf{v}_{i-1} \wedge \mathbf{v}_{i+1} \wedge \cdots \wedge \mathbf{v}_p$$

to denote the $(p-1)$-vector obtained when the vector $\mathbf{v}_i$ is taken out of the product. Thus, the 'hat' here *does not* denote the grade involution but rather the omission of the corresponding vector in the wedge product.

Using this notation and the definition of the wedge product, we can write

$$\mathbf{u}\wedge A_{[1\cdots p]}=\frac{1}{p+1}\left[\mathbf{u}\otimes A_{[1\cdots p]}+\sum_{i=1}^{p}(-1)^{i}\mathbf{v}_i\otimes\left(\mathbf{u}\wedge A_{[1\cdots\hat{\imath}\cdots p]}\right)\right].$$

Assuming that the inductive hypothesis holds for $(p-1)$-vector, this equation implies that

$$\mathbf{u}\wedge A_{[1\cdots\hat{\imath}\cdots p]}\sim\mathbf{u}\otimes A_{[1\cdots\hat{\imath}\cdots p]}-\mathbf{u}_\flat\rfloor A_{[1\cdots\hat{\imath}\cdots p]}.$$

Hence, we have

$$\begin{aligned}\mathbf{v}_i\otimes\left(\mathbf{u}\wedge A_{[1\cdots\hat{\imath}\cdots p]}\right)&\sim\mathbf{v}_i\otimes\mathbf{u}\otimes A_{[1\cdots\hat{\imath}\cdots p]}-\mathbf{v}_i\otimes\left(\mathbf{u}_\flat\rfloor A_{[1\cdots\hat{\imath}\cdots p]}\right)\\&\sim 2g(\mathbf{v}_i,\mathbf{u})A_{[1\cdots\hat{\imath}\cdots p]}-\mathbf{u}\otimes\mathbf{v}_i\otimes A_{[1\cdots\hat{\imath}\cdots p]}-\mathbf{v}_i\otimes\left(\mathbf{u}_\flat\rfloor A_{[1\cdots\hat{\imath}\cdots p]}\right).\end{aligned}$$

Now, keeping in mind the definition of the wedge product, we obtain

$$\sum_{i=1}^{p}(-1)^{i+1}\mathbf{v}_i\otimes A_{[1\cdots\hat{\imath}\cdots p]}=pA_{[1\cdots p]},$$

which yields

$$\begin{aligned}\mathbf{u}\wedge A_{[1\cdots p]}\sim\frac{1}{p+1}\Bigg[&u\otimes A_{[1\cdots p]}+\mathbf{u}\otimes\left(pA_{[1\cdots p]}\right)\\&-\sum_{i=1}^{p}(-1)^{i+1}2g(\mathbf{v}_i,\mathbf{u})A_{[1\cdots\hat{\imath}\cdots p]}-\sum_{i=1}^{p}(-1)^{i}\mathbf{v}_i\otimes\left(\mathbf{u}_\flat\rfloor A_{[1\cdots\hat{\imath}\cdots p]}\right)\Bigg].\end{aligned}$$

However, we can also write

$$\begin{aligned}&\sum_{i=1}^{p}(-1)^{i}\mathbf{v}_i\otimes\left(\mathbf{u}_\flat\rfloor A_{[1\cdots\hat{\imath}\cdots p]}\right)\\&=\sum_{i=1}^{p}(-1)^{i}\mathbf{v}_i\otimes\left[\sum_{j=1}^{i-1}(-1)^{j+1}g(\mathbf{u},\mathbf{v}_j)A_{[1\cdots\hat{\jmath}\cdots\hat{\imath}\cdots p]}+\sum_{j=i+1}^{p}(-1)^{j}g(\mathbf{u},\mathbf{v}_j)A_{[1\cdots\hat{\imath}\cdots\hat{\jmath}\cdots p]}\right]\\&=\sum_{j=1}^{p}\sum_{i=j+1}^{p}(-1)^{i}(-1)^{j+1}g(\mathbf{u},\mathbf{v}_j)\mathbf{v}_i\otimes A_{[1\cdots\hat{\jmath}\cdots\hat{\imath}\cdots p]}\\&\quad+\sum_{j=1}^{p}\sum_{i=1}^{j-1}(-1)^{i}(-1)^{j}g(\mathbf{u},\mathbf{v}_j)\mathbf{v}_i\otimes A_{[1\cdots\hat{\imath}\cdots\hat{\jmath}\cdots p]}\\&=\sum_{j=1}^{p}(-1)^{j+1}g(\mathbf{u},\mathbf{v}_j)\left[\sum_{i=1}^{j-1}(-1)^{i+1}\mathbf{v}_i\otimes A_{[1\cdots\hat{\imath}\cdots\hat{\jmath}\cdots p]}+\sum_{i=j+1}^{p}(-1)^{i}\mathbf{v}_i\otimes A_{[1\cdots\hat{\jmath}\cdots\hat{\imath}\cdots p]}\right]\\&=\sum_{j=1}^{p}(-1)^{j+1}g(\mathbf{u},\mathbf{v}_j)(p-1)A_{[1\cdots\hat{\jmath}\cdots p]},\end{aligned}$$

where in the last equality we have used again the definition of the exterior product. Finally, we obtain

$$\begin{aligned}\mathbf{u}\wedge A_{[1\cdots p]}&\sim\mathbf{u}\otimes A_{[1\cdots p]}-\frac{1}{p+1}\Bigg[2\sum_{i=1}^{p}(-1)^{i+1}g(\mathbf{v}_i,\mathbf{u})A_{[1\cdots\hat{\imath}\cdots p]}\\&\qquad+(p-1)\sum_{j=1}^{p}(-1)^{j+1}g(\mathbf{v}_j,\mathbf{u})A_{[1\cdots\hat{\jmath}\cdots p]}\Bigg]\\&\sim\mathbf{u}\otimes A_{[1\cdots p]}-\mathbf{u}_\flat\rfloor A_{[1\cdots p]},\end{aligned}$$

which proves eqn (3.17).

Observation ☞ The Clifford mapping γ is obviously $\gamma = \pi \circ \imath$, where $\imath$ denotes the inclusion $\imath : V \hookrightarrow \mathsf{T}(V)$, and $\pi : \mathsf{T}(V) \to \mathcal{C}\ell(V,g) = \mathsf{T}(V)/\mathcal{I}_C$, in such a way that $\gamma(\mathbf{v}) = [\mathbf{v}]$. In these expressions, we did not explicitly write $[\mathbf{v}]$ or $\gamma(\mathbf{v})$, since doing so would have overloaded the notation. Although there are situations where the notation $\gamma(\mathbf{v})$ is useful and where using it can help detailed understanding, in this text, except when it is convenient, we avoid explicitly writing the Clifford mapping γ.

Observation ☞ **Notation:** At first glance, using the notation $\diamond$ for the product in the Clifford algebra $\mathcal{C}\ell(V,g)$ seems convenient. However, the side effect of using $\diamond$ is that it would have to be written on every single page in this text, to denote the Clifford product. Hence, in order to employ a straightforward notation, we shall abandon the use of $\diamond$ and simply denote the product in $\mathcal{C}\ell(V,g)$ by *juxtaposition*. Equations (3.17) and (3.18) then read

$$\boxed{\mathbf{vu} + \mathbf{uv} = 2g(\mathbf{v},\mathbf{u})}, \tag{3.20}$$

and

$$\boxed{\mathbf{v}A_{[p]} = \mathbf{v} \wedge A_{[p]} + \mathbf{v}_\flat \rfloor A_{[p]}}, \tag{3.21}$$

where $A_{[p]}$ is a p-vector.

Similarly, we can write

$$\boxed{A_{[p]}\mathbf{v} = A_{[p]} \wedge \mathbf{v} + A_{[p]} \lfloor \mathbf{v}_\flat}. \tag{3.22}$$

Using eqn (2.15) for the exterior product, namely $A_{[p]} \wedge \mathbf{v} = (-1)^p \mathbf{v} \wedge A_{[p]}$, as well as eqn (2.47) which relates the left and the right contractions, that is $A_{[p]} \lfloor \mathbf{v}_\flat = -(-1)^p \mathbf{v}_\flat \rfloor A_{[p]}$, we obtain

$$A_{[p]}\mathbf{v} = (-1)^p \mathbf{v} \wedge A_{[p]} - (-1)^p \mathbf{v}_\flat \rfloor A_{[p]}. \tag{3.23}$$

When we now compare eqns (3.21) and (3.23), we find that

$$\boxed{\mathbf{v} \wedge A_{[p]} = \frac{1}{2}(\mathbf{v}A_{[p]} + (-1)^p A_{[p]}\mathbf{v})}, \tag{3.24}$$

and

$$\boxed{\mathbf{v}_\flat \rfloor A_{[p]} = \frac{1}{2}(\mathbf{v}A_{[p]} - (-1)^p A_{[p]}\mathbf{v})}. \tag{3.25}$$

These equations can be further generalised into expressions involving an arbitrary multivector A, as, respectively,

$$\boxed{\mathbf{v} \wedge A = \frac{1}{2}(\mathbf{v}A + \widehat{A}\mathbf{v})}, \tag{3.26}$$

and

$$\boxed{\mathbf{v}_\flat \rfloor A = \frac{1}{2}(\mathbf{v}A - \widehat{A}\mathbf{v})}. \tag{3.27}$$

Observation ☞ The deep meaning of these last two equations is much more subtle that what it might seem at first sight: they illustrate the relationship between the

Grassmann and the Clifford algebras. This fact is a prominent point that requires a special discussion. Nevertheless, the relationship between these two algebras can be interpreted incorrectly when the notation is used carelessly. We will see how this can happen in what follows!

Universality

Let us now introduce the result that ensures that the Clifford algebra constructed as the quotient of the tensor algebra is indeed a universal Clifford algebra.

Theorem 3.3 ▶ *Let (V, g) be a quadratic space, let $C\ell(V, g)$ be the Clifford algebra $C\ell(V, g) = \mathsf{T}(V)/\mathcal{I}_C$, and let $(\mathcal{B}, \rho)$ a Clifford algebra for (V, g). Then there exists a homomorphism $\phi : C\ell(V, g) \to \mathcal{B}$ such that $\rho : \phi \circ \gamma$, where γ is a Clifford mapping $\gamma : V \to C\ell(V, g)$.*

Proof: For the Clifford algebra $(\mathcal{B}, \rho)$, let us consider the function $\rho : V \to \mathcal{B}$, where $\rho(\mathbf{v})^2 = Q(\mathbf{v})$. This mapping ρ can be extended to $\mathsf{T}(V)$ as the linear mapping $\rho' : \mathsf{T}(V) \to \mathcal{B}$ given by

$$\rho'(\mathbf{v}_1 \otimes \cdots \otimes \mathbf{v}_k) = \rho'(\mathbf{v}_1) \cdots \rho'(\mathbf{v}_k) = \rho(\mathbf{v}_1) \cdots \rho(\mathbf{v}_k).$$

There is a linear bijection from $\mathrm{Lin}(\mathsf{T}_k(V), \mathcal{B})$ onto $\mathrm{Lin}_{(k)}(V, V, \ldots, V; \mathcal{B})$; the restriction $\rho'_k : \mathsf{T}_k(V) \to A$ is an element of the first space and corresponds to the element defined by $(\mathbf{v}_1, \ldots, \mathbf{v}_k) \mapsto \rho(\mathbf{v}_1) \cdots \rho(\mathbf{v}_k)$ in the second space.

Let us consider now the quotient space $\mathsf{T}(V)/\ker\rho'$, where the elements of $\mathsf{T}(V)/\ker\rho'$ consist of the equivalence classes $[x]$ constructed from $x \sim y \Leftrightarrow x - y \in \ker\rho'$. We can express $[x] = \pi(x)$, where $x \in \mathsf{T}(V)$, and $\pi : \mathsf{T}(V) \to \mathsf{T}/\ker\rho'$. Hence, in this case, there exists a mapping $\phi : \mathsf{T}/\ker\rho' \to \mathcal{B}$ given by

$$\phi([x]) = \rho'(x), \quad \forall x \in \mathsf{T}(V),$$

which is the homomorphism

$$\phi([x][y]) = \phi([x \otimes y]) = \rho'(x \otimes y) = \rho'(x)\rho'(y) = \phi([x])\phi([y]).$$

On the other hand, looking at ρ', we can immediately see that

$$\rho'(\mathbf{v} \otimes \mathbf{v} - Q(\mathbf{v})) = 0,$$

namely, $\mathcal{I}_C \subseteq \ker\rho'$. This result implies that there is a surjective homomorphism $\mathsf{T}(V)/\mathcal{I}_C \to \mathsf{T}(V)/\ker\rho'$. Indeed, if U_i is a subspace of a vector space W_i (for $i = 1, 2$) and if $f : W_1 \to W_2$ is a linear mapping such that $f(U_1) \subset U_2$, then f induces a linear mapping $f' : W_1/U_1 \to W_2/U_2$. Moreover, when f is surjective, then f' is surjective too. The surjective homomorphism $\mathsf{T}(V)/\mathcal{I}_C \to \mathsf{T}(V)/\ker\rho'$ shows that $\dim \mathsf{T}(V)/\mathcal{I}_C \geq \dim \mathsf{T}(V)/\ker\rho'$.

The homomorphism $\mathcal{C}\ell(V,g) \to \mathcal{B}$ follows from

$$\mathcal{C}\ell(V,g) = \mathsf{T}(V)/\mathcal{I}_C \to \mathsf{T}(V)/\ker\rho' \xrightarrow{\phi} \mathcal{B}\,.$$

The homomorphism ϕ is injective and even bijective if $\mathcal{B}$ is generated by $\rho(V)$. For $\mathbf{v} \in V$, it follows that $\phi([\mathbf{v}]) = \rho'(\mathbf{v}) = \rho(\mathbf{v})$. Since $[\mathbf{v}] = \gamma(\mathbf{v})$, we obtain

$$\phi \circ \gamma = \rho,$$

which demonstrates the universality of $\mathcal{C}\ell(V,g)$. ✓

Observation ☞ Although we refer to $\mathcal{C}\ell(V,g)$ simply as the Clifford algebra, we mean that it is the universal Clifford algebra.

3.3 Some General Considerations

Let $\mathfrak{B} = \{\mathbf{e}_1, \ldots, \mathbf{e}_n\}$ be an orthogonal basis of V. Equation (3.21) yields

$$\mathbf{e}_i\mathbf{e}_j = \mathbf{e}_i \wedge \mathbf{e}_j, \qquad (i \neq j), \tag{3.28}$$

which can be used recursively to show that

$$\mathbf{e}_{\mu_1}\mathbf{e}_{\mu_2}\cdots\mathbf{e}_{\mu_p} = \mathbf{e}_{\mu_1} \wedge \mathbf{e}_{\mu_2} \wedge \cdots \wedge \mathbf{e}_{\mu_p} \quad (\mu_1 \neq \mu_2 \neq \cdots \neq \mu_p). \tag{3.29}$$

The dimension of a universal $\mathcal{C}\ell(V,g)$ is $\sum_{p=0}^{n}\binom{n}{p}$, namely,

$$\boxed{\dim \mathcal{C}\ell(V,g) = 2^{\dim V}}. \tag{3.30}$$

There exists an isomorphism between the Clifford algebra $\mathcal{C}\ell(V,g)$ and the exterior algebra $\bigwedge(V)$ or the Grassmann algebra $\mathcal{G}(V)$. Clearly, this isomorphism is not an algebra isomorphism (we will discuss this point in more details later on), but a *vector space isomorphism*,

$$\mathcal{C}\ell(V,g) \underset{V}{\simeq} \bigwedge(V). \tag{3.31}$$

Hence, for the operations of the vector space structure, it is natural to use for the Clifford algebra $\mathcal{C}\ell(V,g)$ the same notation used for the exterior algebra, although the subspace of the p-vectors will still be denoted by $\bigwedge_p(V)$. Then the following equality must be realised as an equality of vector spaces:

$$\mathcal{C}\ell(V,g) = \bigoplus_{p=0}^{n} \bigwedge{}_p(V). \tag{3.32}$$

The projection operators are equivalently denoted by

$$\langle\ \rangle_p : \mathcal{C}\ell(V,g) \to \bigwedge{}_p(V). \tag{3.33}$$

The projection operator into the scalar part $\langle\ \rangle_0$ has prominent importance in the formalism of Clifford algebras. It can be shown that, for $A, B \in \mathcal{C}\ell(V,g)$, the scalar part is such that

$$\langle AB\rangle_0 = \langle BA\rangle_0. \tag{3.34}$$

Observation ☞ The underlying multivector structure of the Clifford algebra $\mathcal{C}\ell(V,g)$ is immediately seen in eqn (3.29), where an orthogonal basis is taken into account.

When the basis is not orthogonal, the multivector structure is not as straightforward to realise: a Clifford algebra is a $\mathbb{Z}_2$-graded algebra, whereas the exterior algebra and Grassmann algebra are $\mathbb{Z}_n$-graded algebras. However, as the multivector structure is basis independent, there is no reason to encounter this problem. Nonetheless, we want to emphasise that calculations with Clifford algebras are more straightforward when we work with an orthogonal basis than when we work with a non-orthogonal basis. Hence, in most mappings, it is preferred to take an orthogonal basis into account – in some cases, indeed, it is indispensable.

Let g be a symmetric bilinear form in $\mathbb{R}^n$ of signature (p,q), where $p+q=n$, such that $\mathfrak{B}=\{\mathbf{e}_1,\dots,\mathbf{e}_n\}$ is an orthonormal basis. On any vector $\mathbf{v}=v^i\mathbf{e}_i\in V$, the symmetric bilinear form can be evaluated as

$$g(\mathbf{v},\mathbf{v})=(v^1)^2+\cdots+(v^p)^2-(v^{p+1})^2-\cdots-(v^n)^2. \tag{3.35}$$

We denote this quadratic space by $\mathbb{R}^{p,q}$, and the corresponding Clifford algebra by $\mathcal{C}\ell_{p,q}$:

$$\mathcal{C}\ell_{p,q}=\mathcal{C}\ell(\mathbb{R}^{p,q}). \tag{3.36}$$

The *centre* $\mathrm{Cen}(\mathcal{C}\ell_{p,q})$ of the Clifford algebra $\mathcal{C}\ell_{p,q}$ is defined as being the set of elements in $\mathcal{C}\ell_{p,q}$ that commute with all elements of $\mathcal{C}\ell_{p,q}$:

$$\mathrm{Cen}(\mathcal{C}\ell_{p,q})=\{a\in\mathcal{C}\ell_{p,q}\mid ax=xa,\forall x\in\mathcal{C}\ell_{p,q}\}.$$

When a basis $\{\mathbf{e}_1,\dots,\mathbf{e}_n\}$ is taken into account, every $\mathbf{e}_i$ either commutes or anti-commutes with $\mathbf{e}_{\mu_1}\mathbf{e}_{\mu_2}\cdots\mathbf{e}_{\mu_p}$, according to whether p is even or odd, respectively, and whether i belongs or does not belong, respectively, to the set $\{\mu_1,\dots,\mu_p\}$. This property provides the elements of the centre in $\mathcal{C}\ell(V,g)$. It is left as an exercise to show that

$$\mathrm{Cen}(\mathcal{C}\ell_{p,q})=\begin{cases}\bigwedge_0(\mathbb{R}^{p,q}), & \text{if}\dim\mathbb{R}^{p,q}\ \text{is even},\\ \bigwedge_0(\mathbb{R}^{p,q})\oplus\bigwedge_n(\mathbb{R}^{p,q}), & \text{if}\dim\mathbb{R}^{p,q}\ \text{is odd}.\end{cases} \tag{3.37}$$

The grade involution, the reversion, and the conjugation preserve the ideal $\mathcal{I}_C$. Then, as in the exterior algebras, such operations pass to the quotient. Hence, for the grade involution, it follows that

$$\#A_{[p]}=\widehat{A}_{[p]}=(-1)^pA_{[p]}\,. \tag{3.38}$$

For the reversion, we have the property

$$(\widetilde{A_{[p]}B_{[q]}})=\widetilde{B}_{[q]}\widetilde{A}_{[p]}, \tag{3.39}$$

with $\widetilde{A}_{[0]}=A_{[0]}$, $\widetilde{A}_{[1]}=A_{[1]}$. This result implies that

$$\widetilde{A}_{[p]}=(-1)^{p(p-1)/2}A_{[p]}, \tag{3.40}$$

and, for the conjugation,

$$\bar{A}_{[p]}=\widetilde{\widehat{A}}_{[p]}=\widehat{\widetilde{A}}_{[p]}. \tag{3.41}$$

Example 3.4 Equation (3.39) is straightforward to prove when $A_{[p]}$ and $B_{[q]}$ are products of vectors of the orthogonal basis $\{\mathbf{e}_1,\mathbf{e}_2,\dots,\mathbf{e}_n\}$. When some vectors $\mathbf{e}_i$ are factors both in $A_{[p]}$ and in $B_{[q]}$, we can, without loss of generality, choose $\mathbf{e}_p$ as the common vector that composes $A_{[p]}$ and $B_{[q]}$. Up to a sign, we can write $A_{[p]}$ and $B_{[q]}$ as

$$A_{[p]} = \mathbf{e}_1 \wedge \cdots \wedge \mathbf{e}_p, \qquad B_{[q]} = \mathbf{e}_p \wedge \cdots \wedge \mathbf{e}_{p+q}\,.$$

Then we obtain

$$\begin{aligned}\widetilde{A_{[p]}B_{[q]}} &= [(\mathbf{e}_1 \wedge \cdots \wedge \mathbf{e}_p)(\mathbf{e}_p \wedge \mathbf{e}_{p+q})]^\sim \\ &= [[(\mathbf{e}_1 \wedge \cdots \wedge \mathbf{e}_{p-1})\mathbf{e}_p - (\mathbf{e}_1 \wedge \cdots \wedge \mathbf{e}_{p-1})\lfloor \mathbf{e}_p](\mathbf{e}_p \wedge \cdots \wedge \mathbf{e}_{p+q})]^\sim\,.\end{aligned}$$

Nevertheless, $(\mathbf{e}_1 \wedge \cdots \wedge \mathbf{e}_{p-1})\lfloor \mathbf{e}_p = 0$, and hence

$$\begin{aligned}\widetilde{A_{[p]}B_{[q]}} &= [(\mathbf{e}_1 \wedge \cdots \wedge \mathbf{e}_{p-1})\mathbf{e}_p(\mathbf{e}_p \wedge \cdots \wedge \mathbf{e}_{p+q})]^\sim \\ &= [(\mathbf{e}_1 \wedge \cdots \wedge \mathbf{e}_{p-1})[\mathbf{e}_p \rfloor (\mathbf{e}_p \wedge \cdots \wedge \mathbf{e}_{p+q}) + \mathbf{e}_p \wedge (\mathbf{e}_p \wedge \cdots \wedge \mathbf{e}_{p+q})]]^\sim \\ &= [(\mathbf{e}_1 \wedge \cdots \wedge \mathbf{e}_{p-1})[\mathbf{e}_p \rfloor (\mathbf{e}_p \wedge \cdots \wedge \mathbf{e}_{p+q})]]^\sim \\ &= [(\mathbf{e}_1 \wedge \cdots \wedge \mathbf{e}_{p-1})(\mathbf{e}_{p+1} \wedge \cdots \wedge \mathbf{e}_{p+q})]^\sim \\ &= [(\mathbf{e}_1 \wedge \cdots \wedge \mathbf{e}_{p-1}) \wedge (\mathbf{e}_{p+1} \wedge \cdots \wedge \mathbf{e}_{p+q})]^\sim \\ &= (\mathbf{e}_{p+q} \wedge \cdots \wedge \mathbf{e}_{p+1}) \wedge (\mathbf{e}_{p-1} \wedge \cdots \wedge \mathbf{e}_1) \\ &= [(\mathbf{e}_{p+q} \wedge \cdots \wedge \mathbf{e}_{p+1})(\mathbf{e}_{p-1} \wedge \cdots \wedge \mathbf{e}_1)] \\ &= [(\mathbf{e}_{p+q} \wedge \cdots \wedge \mathbf{e}_{p+1})(\mathbf{e}_p)^2(\mathbf{e}_{p-1} \wedge \cdots \wedge \mathbf{e}_1)] \\ &= [(\mathbf{e}_{p+q} \wedge \cdots \wedge \mathbf{e}_{p+1})\mathbf{e}_p][\mathbf{e}_p(\mathbf{e}_{p-1} \wedge \cdots \wedge \mathbf{e}_1)] \\ &= [(\mathbf{e}_{p+q} \wedge \cdots \wedge \mathbf{e}_{p+1}) \wedge \mathbf{e}_p][\mathbf{e}_p \wedge (\mathbf{e}_{p-1} \wedge \cdots \wedge \mathbf{e}_1)] \\ &= \widetilde{B}_{[q]}\widetilde{A}_{[p]}\,.\end{aligned} \tag{3.42}$$

The Clifford algebra $\mathcal{C}\ell(V,g)$ is clearly a $\mathbb{Z}_2$-graded algebra. We can write

$$\mathcal{C}\ell(V,g) = \mathcal{C}\ell^+(V,g) \oplus \mathcal{C}\ell^-(V,g), \tag{3.43}$$

where

$$\mathcal{C}\ell^\pm(V,g) = \Pi_\pm(\mathcal{C}\ell(V,g)) = \frac{1}{2}(1 \pm \#)(\mathcal{C}\ell(V,g)). \tag{3.44}$$

Consequently, it reads

$$\mathcal{C}\ell^\pm(V,g)\,\mathcal{C}\ell^\pm(V,g) \subset \mathcal{C}\ell^+(V,g), \quad \mathcal{C}\ell^\pm(V,g)\,\mathcal{C}\ell^\mp(V,g) \subset \mathcal{C}\ell^-(V,g), \tag{3.45}$$

Since $\mathcal{C}\ell^\pm(V,g)\,\mathcal{C}\ell^\pm(V,g) \subset \mathcal{C}\ell^+(V,g)$, the set $\mathcal{C}\ell^+(V,g)$ is a *subalgebra* of the Clifford algebra $\mathcal{C}\ell(V,g)$. This subalgebra is prominent in the Clifford algebra formalism and applications. It is called the *even subalgebra*:

$$\boxed{\mathcal{C}\ell^+(V,g) = \{A \in \mathcal{C}\ell(V,g) \mid A = \#A = \widehat{A}\}}\,. \tag{3.46}$$

The assertion that $\mathcal{C}\ell^+(V,g)$ is a subalgebra can also be proved in a different way. Let $A_{[p]}$ and $B_{[q]}$ be a p-vector and a q-vector, respectively. The repeated use of eqn (3.21) permits us to conclude that

$$A_{[p]}B_{[q]} = \langle A_{[p]}B_{[q]}\rangle_{|p-q|} + \langle A_{[p]}B_{[q]}\rangle_{|p-q|+2} + \cdots + \langle A_{[p]}B_{[q]}\rangle_{p+q}. \tag{3.47}$$

In general, the product of a p-vector and a q-vector in $\mathcal{C}\ell^+(V,g)$ is not a graded-defined multivector but a sum involving a $|p-q|$-vector, a $(|p-q|+2)$-vector, and so on up to a $(p+q)$-vector. From this equation, we obtain the result explicitly described by eqn (3.45).

It is left as an exercise to show that, for $A, B \in \mathcal{C}\ell(V, g)$, the following expression holds:

$$\boxed{\langle \widetilde{A}B \rangle_0 = G(A, B)}, \tag{3.48}$$

where G is the extension of g for the Grassmann algebra as given by eqns (2.99–2.101). Using eqn (3.34) and the definition of the reversion, we can realise that $\langle \widetilde{A}B \rangle_0 = \langle \widetilde{B}A \rangle_0$, according to the symmetry of G.

For elements of the Clifford algebra, if we try to define a norm $|A|^2$ by

$$|A|^2 = \langle \widetilde{A}A \rangle_0 \,, \tag{3.49}$$

it would not be the unique norm that we might define in $\mathcal{C}\ell(V, g)$. Besides, it can have negative values when g is not Euclidean. We can however define

$$|A|'^2 = \langle \bar{A}A \rangle_0. \tag{3.50}$$

Since $\bar{A}_{[p]} = (-1)^p \widetilde{A}_{[p]}$ for $A_{[p]} \in \bigwedge_p(V)$,

$$|A_{[p]}|'^2 = (-1)^p |A_{[p]}|^2. \tag{3.51}$$

Nevertheless, these definitions do not ensure that the functions are non-negative. Therefore, we opt to define, for simple multivectors, the norm as

$$|A|^2 = |\langle \bar{A}A \rangle_0| \,. \tag{3.52}$$

When $\mathbb{K}$ is either $\mathbb{R}$ or $\mathbb{C}$, the positive root is chosen to be the norm.

In the case where the multivectors are not simple, we have two possible definitions for the norm:

$$|A|^2 = |\langle \widetilde{A}A \rangle_0| \,, \tag{3.53}$$

$$|A|'^2 = |\langle \bar{A}A \rangle_0|. \tag{3.54}$$

We can show that, for multivectors that are not simple, the norms in eqn (3.53) and in eqn (3.54) can be considered, as they are in general, indeed, distinct. This fact is illustrated in the following example.

Example 3.5 Let us consider the Clifford algebra $\mathcal{C}\ell_{1,2}$ generated by $\{\mathrm{e}_2, \mathrm{e}_2, \mathrm{e}_3\}$ (here we adopt the convention that $\mathrm{e}_1^2 = -1$), and take $u = \mathrm{e}_1 - \mathrm{e}_2\mathrm{e}_3$. The expressions for $\tilde{u}u$ and $\bar{u}u$ then read

$$\tilde{u}u = (\mathrm{e}_1 + \mathrm{e}_2\mathrm{e}_3)(\mathrm{e}_1 - \mathrm{e}_2\mathrm{e}_3) = -1 + \mathrm{e}_2^2\mathrm{e}_3^2 = 0,$$
$$\bar{u}u = (\mathrm{e}_1 - \mathrm{e}_2\mathrm{e}_3)(-\mathrm{e}_1 + \mathrm{e}_2\mathrm{e}_3) = 2 + 2\mathrm{e}_1\mathrm{e}_2\mathrm{e}_3 \neq 0.$$

Hence, $|u|^2 = 0$, and $|u|'^2 = 2$.

The Hodge isomorphism $\star$ can be written in terms of $\mathcal{C}\ell(V,g)$ simply as

$$\boxed{\star A = \widetilde{A}\Omega_V}. \tag{3.55}$$

In order to show this result, we first observe from eqn (3.47) that it is possible to write

$$A_{[p]} \wedge B_{[q]} = \langle A_{[p]} B_{[q]} \rangle_{p+q}. \tag{3.56}$$

Equation (3.47) allows us to conclude that, as Ω_V is an n-vector, therefore

$$\langle A_{[p]} B_{[q]} \rangle_0 \Omega_V = \langle A_{[p]} B_{[q]} \Omega_V \rangle_n. \tag{3.57}$$

Using the definition of the Hodge isomorphism and eqns (3.48), (3.56), and (3.57) we find that

$$\begin{aligned} \langle B_{[p]} \star A_{[p]} \rangle_n &= B_{[p]} \wedge \star A_{[p]} = G(B_{[p]}, A_{[p]})\Omega_V \\ &= \langle B_{[p]} \widetilde{A}_{[p]} \rangle_0 \Omega_V = \langle B_{[p]} \widetilde{A}_{[p]} \Omega_V \rangle_n, \end{aligned} \tag{3.58}$$

from which eqn (3.55) ensues.

Example 3.6 Let $V = \mathbb{R}^3$ be endowed with the usual Euclidean scalar product. Consider the Clifford algebra $\mathcal{C}\ell_3$, which is generated by 1 and $\{\gamma_i = \gamma(\mathbf{e}_i)\}$, where we take $\{\mathbf{e}_i\}$ as the standard basis of $\mathbb{R}^3$. Take two vectors, for example, $\mathbf{v} = (3, 2, -1)^\intercal$, and $\mathbf{u} = (1, -1, 3)^\intercal$. In $\mathcal{C}\ell_3$, we write $\mathbf{v} = 3\gamma_1 + 2\gamma_2 - \gamma_3$, and $\mathbf{u} = \gamma_1 - \gamma_2 + 3\gamma_3$. The products $\mathbf{vu}$ and $\mathbf{uv}$ provide

$$\begin{aligned} \mathbf{vu} &= -2 - 5\gamma_1\gamma_2 + 10\gamma_1\gamma_3 + 5\gamma_2\gamma_3, \\ \mathbf{uv} &= -2 + 5\gamma_1\gamma_2 - 10\gamma_1\gamma_3 - 5\gamma_2\gamma_3. \end{aligned}$$

The symmetric part of the geometric product corresponds to the scalar product, and the alternating part corresponds to the exterior product. Indeed,

$$\frac{1}{2}(\mathbf{vu} + \mathbf{uv}) = -2 = g(\mathbf{v}, \mathbf{u}),$$

and

$$\frac{1}{2}(\mathbf{vu} - \mathbf{uv}) = -5\gamma_1\gamma_2 + 10\gamma_1\gamma_3 - 5\gamma_2\gamma_3.$$

In particular, since $g(\mathbf{e}_i, \mathbf{e}_j) = 0$ for $i \neq j$,

$$\gamma_i\gamma_j = -\gamma_j\gamma_i, \qquad (i \neq j), \qquad \gamma_i^2 = 1.$$

An arbitrary element of $\mathcal{C}\ell_3$ can be written as

$$A = a^0 + a^1\gamma_1 + a^2\gamma_2 + a^3\gamma_3 + a^{12}\gamma_1\gamma_2 + a^{13}\gamma_1\gamma_3 + a^{23}\gamma_2\gamma_3 + a^{123}\gamma_1\gamma_2\gamma_3.$$

Since $\gamma_i\gamma_j = -\gamma_j\gamma_i$, if $i \neq j$, then the element $\gamma_1\gamma_2\gamma_3$ commutes with all elements of $\mathcal{C}\ell_3$. In addition, $(\gamma_1\gamma_2\gamma_3)^2 = -1$, so $\gamma_1\gamma_2\gamma_3$ can be denoted by $\mathbf{i}$:

$$\mathbf{i} = \gamma_1\gamma_2\gamma_3\,.$$

Hence, $\mathbf{i}A = A\mathbf{i}$ ($\forall\, A \in \mathcal{C}\ell_3$), and $\mathbf{i}^2 = -1$. The Hodge isomorphism applied to the multivector A then reads

$$\star A = \widetilde{A}\mathbf{i},$$

and, consequently,

$$\begin{aligned} \gamma_1\gamma_2 &= \gamma_1\gamma_2\gamma_3\gamma_3 = \mathbf{i}\gamma_3 = \star\gamma_3, \\ \gamma_1\gamma_3 &= \gamma_1\gamma_2\gamma_2\gamma_3 = -\mathbf{i}\gamma_2 = -\star\gamma_2, \\ \gamma_2\gamma_3 &= \gamma_1\gamma_1\gamma_2\gamma_3 = \gamma_1\mathbf{i} = \star\gamma_1, \end{aligned}$$

Finally, an arbitrary element $\psi \in \mathcal{C}\ell_3$ can be written as

$$A = (a^0 + \mathbf{i}a^{123}) + (a^1 + \mathbf{i}a^{23})\gamma_1 + (a^2 + \mathbf{i}a^{31})\gamma_2 + (a^3 + \mathbf{i}a^{12})\gamma_3.$$

As we remember from example 3.3, the Clifford algebra $\mathcal{C}\ell_3$ is isomorphic to the algebra of complex matrices of order 2. The isomorphism ρ is explicitly given by $\rho(1) = 1$ (the matrix identity) and $\rho(\gamma_i) = \sigma_i$ (the Pauli matrices). An arbitrary element $A \in \mathcal{C}\ell_3$ is mapped in $\mathsf{A} = \rho(A)$ given by

$$\rho(A) = \mathsf{A} = \begin{pmatrix} (a^0 + a^3) + i(a^{12} + a^{123}) & (a^1 + a^{31}) + i(a^{23} - a^2) \\ (a^1 - a^{31}) + i(a^{23} + a^2) & (a^0 - a^3) + i(a^{123} - a^{12}) \end{pmatrix} = \begin{pmatrix} z_1 & z_2 \\ z_3 & z_4 \end{pmatrix}.$$

In terms of the matrix representation, the reversion, the grade involution, and the conjugation correspond to

$$\rho(\widetilde{A}) = \mathsf{A}^\dagger = \begin{pmatrix} z_1^* & z_3^* \\ z_2^* & z_4^* \end{pmatrix},$$

$$\rho(\widehat{A}) = \mathrm{adj}(\mathsf{A}^\dagger) = \begin{pmatrix} z_4^* & -z_3^* \\ -z_2^* & z_1^* \end{pmatrix},$$

$$\rho(\bar{A}) = \mathrm{adj}(\mathsf{A}) = \begin{pmatrix} z_4 & -z_2 \\ -z_3 & z_1 \end{pmatrix},$$

where the asterisk denotes complex conjugation.

Example 3.7 The Clifford algebra structure is very rich and exhibits a large number of particularities because of the dimension and the signature associated with the quadratic space. For instance, the equality $u\bar{u} = \bar{u}u$ does not always hold. Indeed, consider the Clifford algebra $\mathcal{C}\ell_{3,1} \simeq \mathcal{M}(4, \mathbb{R})$ of $\mathbb{R}^{3,1}$. An element

$$u = (1 + \mathbf{e}_1)(1 + \mathbf{e}_{234}) = 1 + \mathbf{e}_1 + \mathbf{e}_{234} + \mathbf{e}_{1234},$$

presents a Clifford conjugate

$$\bar{u} = (1 + \mathbf{e}_{234})(1 - \mathbf{e}_1) = 1 - \mathbf{e}_1 + \mathbf{e}_{234} + \mathbf{e}_{1234}.$$

By calculating the products between u and $\bar{u}$, we obtain

$$\begin{aligned} u\bar{u} &= 4(\mathbf{e}_{234} + \mathbf{e}_{1234}), \\ \bar{u}u &= 0. \end{aligned}$$

In the Clifford algebra associated with finite-dimensional spaces, the norm can be naturally defined as

$$\begin{aligned} |u| &= \sqrt{\langle u\tilde{u}\rangle_0}, && \text{for } u \in \mathcal{C}\ell_{n,0}, \\ |u| &= \sqrt{\langle u\bar{u}\rangle_0}, && \text{for } u \in \mathcal{C}\ell_{0,n}, \end{aligned} \tag{3.59}$$

where we used eqn (3.49) and eqn (3.50). Indeed, in $\mathcal{C}\ell_{0,2} \simeq \mathbb{H}$, the corresponding conjugate $\bar{q} = w - ix - jy - kz$ is associated with $q = w + ix + jy + kz \in \mathbb{H}$. The norm is given by $|q| = \sqrt{q\bar{q}}$, and the inverse is $q^{-1} = \bar{q}/|q|^2$. Nevertheless, it is not always possible to calculate the inverse of an arbitrary element u.

Example 3.8 Considering the Clifford algebra $\mathcal{C}\ell_{3,1}$, take $u = \mathbf{e}_1 - \mathbf{e}_2\mathbf{e}_3$. Computing $u\tilde{u}$ and $u\bar{u}$ we obtain

$$\begin{aligned} u\tilde{u} &= (\mathbf{e}_1 - \mathbf{e}_2\mathbf{e}_3)(\mathbf{e}_1 + \mathbf{e}_2\mathbf{e}_3) = -1 + \mathbf{e}_2^2\mathbf{e}_3^2 = 0, \\ u\bar{u} &= (\mathbf{e}_1 - \mathbf{e}_2\mathbf{e}_3)(-\mathbf{e}_1 + \mathbf{e}_2\mathbf{e}_3) = 2 + 2\mathbf{e}_1\mathbf{e}_2\mathbf{e}_3 \neq 0. \end{aligned}$$

Since $u\tilde{u} = 0$, and $u\bar{u} \in \bigwedge_0 \oplus \bigwedge_3$, no inverse element can be constructed for $u = \mathbf{e}_1 - \mathbf{e}_2\mathbf{e}_3$ by using an anti-automorphism of the Clifford algebra. Some particular and illustrative examples are investigated in the work by Lounesto (1996).

In order to finish those considerations, now two general results concerning Clifford algebras are presented.

Theorem 3.4 ▶ (M. Riesz) *An orthonormal basis* $\mathfrak{B} = \{\mathbf{e}_1, \mathbf{e}_2, \ldots, \mathbf{e}_n\}$ *generates a* 2^n*-dimensional algebra, unless the pseudoscalar* $\mathbf{e}_1\mathbf{e}_2 \ldots \mathbf{e}_n$ *is a scalar multiple of the identity.*

Proof: At first, with at most one exception, the product involving one or more elements of the basis $\mathfrak{B}$ anti-commutes with at least one element of $\mathfrak{B}$. This property holds only if the product of an element of the basis anti-commutes with any element appearing in the product. Moreover, the product of an odd number of such elements anti-commutes with any element that does not appear in the product. The unique product that commutes with any element of the set $\{\mathbf{e}_i\}$ is the pseudoscalar $\mathbf{e}_1\mathbf{e}_2 \ldots \mathbf{e}_n$ when n is odd, according to eqn (3.37). Now the proof is by *reductio ad absurdum*. Consider the products that are not linearly independent:

$$a^0 + a^i\mathbf{e}_i + a^{ij}\mathbf{e}_i\mathbf{e}_j + \cdots + p\mathbf{e}_1\mathbf{e}_2 \cdots \mathbf{e}_n = 0,$$

which can be also written in the notation of multiple indices. In this case, there exists a set of coefficients $a^{i_1 i_2 \ldots i_k}$, with some of them non-zero, such that

$$\sum_{k=0}^{n} a^{i_1 i_2 \ldots i_k} \mathbf{e}_{i_1}\mathbf{e}_{i_2} \cdots \mathbf{e}_{i_k} = 0\,. \tag{3.60}$$

If the coefficient of 1 does not equal zero in eqn (3.60), we can divide the equation by this coefficient, obtaining

$$1 + \sum_{k=1}^{n} b^{i_1 i_2 \ldots i_k} \mathbf{e}_{i_1}\mathbf{e}_{i_2} \cdots \mathbf{e}_{i_k} = 0, \tag{3.61}$$

where the sum does not include the identity. On the other hand, if the coefficient of 1 in eqn (3.61) equals zero, we can take a term with a non-zero coefficient, let us say, $\mathbf{e}_{m_1}\mathbf{e}_{m_2} \cdots \mathbf{e}_{m_k} \equiv \mathbf{e}_{m_1 m_2 \ldots m_k}$, out and then multiply the equation by $\mathbf{e}^{-1}_{m_1 m_2 \ldots m_k} = \pm\mathbf{e}_{m_1 m_2 \ldots m_k}$. In this way, it is always possible to obtain an equation of the same type as eqn (3.61) from eqn (3.60). If all the coefficients $b^{i_1 i_2 \ldots i_k}$ are equal to zero, the equality $1 = 0$ is already a contradiction. If the sum in eqn (3.61) contains a product, let us say, $\mathbf{e}_{m_1 m_2 \ldots m_k}$, which anti-commutes with some $\mathbf{e}_m$, then we can multiply eqn (3.61) from the left by $\mathbf{e}_m$ and from the right by $\mathbf{e}_m^{-1}$ and thus obtain

$$1 + \sum_{k=0}^{2^n-1} b^{i_1 i_2 \ldots i_k} \mathbf{e}_m \mathbf{e}_{i_1 i_2 \ldots i_k} \mathbf{e}_m^{-1} = 0. \tag{3.62}$$

Notice that

$$\mathbf{e}_m \mathbf{e}_{m_1 m_2 \ldots m_k} \mathbf{e}_m^{-1} = -\mathbf{e}_{m_1 m_2 \ldots m_k} \mathbf{e}_m \mathbf{e}_m^{-1} = -\mathbf{e}_{m_1 m_2 \ldots m_k}.$$

Therefore, eqn (3.62) can be added to eqn (3.61), yielding a new equation that up to a factor 2 is identical to eqn (3.60), except that now at least one term of type $(\mathbf{e}_{m_1 m_2 \ldots m_k})$ is not in the sum. If n is even, the process can be repeated until the sum is reduced to 0, and we obtain the contradiction $1 = 0$. If n is odd, then the procedure can be repeated until

$$1 + a\mathbf{e}_1\mathbf{e}_2 \cdots \mathbf{e}_n = 0,$$

for some scalar $a \neq 0$. Hence, we again obtain a contradiction, unless $\mathbf{e}_1\mathbf{e}_2 \cdots \mathbf{e}_n$ is a scalar multiple of the identity. In fact, when we apply the grade involution to this equation, we find that $1 - a\mathbf{e}_1\mathbf{e}_2 \cdots \mathbf{e}_n = 0$. Adding both equations, we arrive at the contradiction; therefore, $\mathbf{e}_1\mathbf{e}_2 \cdots \mathbf{e}_n$ is a scalar multiple of the identity. This property in particular holds for the real and complex fields, which are our main interests. ✓

The Clifford algebras that are *not* universal are restricted by the following theorem:

Theorem 3.5 ▶ *If the Clifford algebra $C\ell(\mathbb{R}^{p,q})$, generated by some orthonormal basis $\{\mathbf{e}_1, \ldots, \mathbf{e}_n\}$, is not universal, then $n = p + q$ is odd. Moreover, if the Clifford algebra is real and not universal, then $p - q - 1$ is an integer multiple of 4.*

Proof: The first part of this theorem comes from the proof of theorem 3.4 (Riesz, 1993). The second part follows from the condition that $\eta^2 = 1$, where $\eta = \mathbf{e}_1\mathbf{e}_2 \cdots \mathbf{e}_n$. Then,

$$\eta^2 = \mathbf{e}_{12\ldots n}\mathbf{e}_{12\ldots n} = (-1)^{n(n-1)/2}\mathbf{e}_{n\ldots 21}\mathbf{e}_{12\ldots n} = (-1)^{n(n-1)/2}(-1)^q 1\,.$$

Since $\eta^2 = 1$, it follows that

$$\frac{1}{2}n(n-1) + q = 2k, \qquad k \in \mathbb{N}\,.$$

As n is odd, by hypothesis, $n = 2m + 1, m \in \mathbb{N}$, and

$$(2m+1)(2m) + 2q = 4k \Rightarrow 2m + 2q = 4(k - m^2).$$

However, as $2m = n - 1 = p + q - 1$, this equation reads

$$\begin{aligned} p + 3q - 1 = 4(k - m^2) &\Rightarrow p + 3q - 1 - 4q = 4(k - m^2) - 4q \\ &\Rightarrow p - q - 1 = 4(k - m^2 - q), \end{aligned}$$

which proves the theorem. ✓

3.4 From the Grassmann Algebra to the Clifford Algebra

The Annihilation and the Creation Operators

For what will be discussed in this section, the exterior algebra would suffices. However, if we endow the exterior algebra with a bilinear form, we get the Grassmann algebra structure; therefore, we will situate our discussion in the context of the Grassmann algebra. Let us then consider the Grassmann algebra $\mathcal{G}(V)$ and define the annihilation and the creation operators.[1]

[1]The terms 'creation' and 'annihilation' are derived from the second quantisation formalism in quantum field theory. Their use here is motivated by the algebraic relations that such operators satisfy, as these relations are the similar to the ones in the second quantisation formalism.

The Operators $\mathbf{E}$ and $\mathbf{E}^\dagger$

Let us consider $\mathbf{v} \in V \subset \bigwedge(V)$, and $A \in \bigwedge(V)$. In the Grassmann algebra $\mathcal{G}(V)$, it is possible to perform the exterior product between the vector $\mathbf{v}$ and the multivector A, either as $\mathbf{v} \wedge A$ or as $A \wedge \mathbf{v}$. Let us at first take the exterior product $\mathbf{v} \wedge A$. Since the result of this product is a multivector, we can interpret the exterior product as an operation of $\bigwedge(V)$ on $\bigwedge(V)$, that is, an element of the endomorphism space of $\bigwedge(V)$; this space is denoted from this point on by $\mathrm{End}(\bigwedge(V))$. Let us define $\mathbf{E}(\mathbf{v})$ as

$$\mathbf{E}(\mathbf{v})(A) = \mathbf{v} \wedge A. \tag{3.63}$$

The operator $\mathbf{E}$ is thus an object such that $\mathbf{E} : V \to \mathrm{End}(\bigwedge(V))$, and $\mathbf{E}(\mathbf{v}) : \bigwedge(V) \to \bigwedge(V)$. This operator $\mathbf{E}$ is called the *creation operator*. Moreover, the exterior product between $\mathbf{v}$ and A can be taken in the reverse order, namely $A \wedge \mathbf{v}$. Hence, we define another operator, denoted by $\mathbf{E}^\dagger$, such that $\mathbf{E}^\dagger : V \to \mathrm{End}(\bigwedge(V))$. Define $\mathbf{E}^\dagger(\mathbf{v})$ as

$$\mathbf{E}^\dagger(\mathbf{v})(A) = A \wedge \mathbf{v}. \tag{3.64}$$

There is a close relationship between the operators $\mathbf{E}$ and $\mathbf{E}^\dagger$. Let us first consider the case where A is a k-vector $A_{[k]}$. In this case,[2] eqn (2.15) yields

$$\mathbf{v} \wedge A_{[k]} = (-1)^k A_{[k]} \wedge \mathbf{v},$$

which implies that

$$\mathbf{v} \wedge A = (\# A) \wedge \mathbf{v}. \tag{3.65}$$

Hence, it follows that

$$\begin{aligned} \mathbf{E}(\mathbf{v}) &= \mathbf{E}^\dagger(\mathbf{v})\#, \\ \mathbf{E}^\dagger(\mathbf{v}) &= \mathbf{E}(\mathbf{v})\#. \end{aligned} \tag{3.66}$$

The Operators $\mathbf{I}$ and $\mathbf{I}^\dagger$

Other operations defined on the space of multivectors are the left and the right contractions by a covector. Using those operations, the operators $\mathbf{I}$ and $\mathbf{I}^\dagger$ can be defined. Given $\alpha \in V^* \subset \bigwedge(V^*)$, and $A \in \bigwedge(V)$, the operator $\mathbf{I} : V^* \to \mathrm{End}(\bigwedge(V))$, called the *annihilation operator*, is defined by

$$\mathbf{I}(\alpha)(A) = \alpha \rfloor A. \tag{3.67}$$

In addition, the operator $\mathbf{I}^\dagger : V^* \to \mathrm{End}(\bigwedge(V))$ is defined by

$$\mathbf{I}^\dagger(\alpha)(A) = A \lfloor \alpha. \tag{3.68}$$

The relationship between those operators follows from eqn (2.48):

$$\begin{aligned} \mathbf{I}(\alpha) &= -\mathbf{I}^\dagger(\alpha)\#, \\ \mathbf{I}^\dagger(\alpha) &= -\mathbf{I}(\alpha)\#. \end{aligned} \tag{3.69}$$

[2] For the content of this section, the notation $\# A_{[k]}$ is more appropriate than $\widehat{A}_{[k]}$.

Commutation Relations

The anti-commutativity of the exterior product between two linearly independent covectors implies that creation operators anti-commute:

$$\boxed{\mathbf{E}(\mathbf{v})\mathbf{E}(\mathbf{u}) + \mathbf{E}(\mathbf{u})\mathbf{E}(\mathbf{v}) = 0} \tag{3.70}$$

for all $\mathbf{v}, \mathbf{u} \in V$. Similarly,

$$\mathbf{E}^{\dagger}(\mathbf{v})\mathbf{E}^{\dagger}(\mathbf{u}) + \mathbf{E}^{\dagger}(\mathbf{u})\mathbf{E}^{\dagger}(\mathbf{v}) = 0. \tag{3.71}$$

From eqn (2.50), the commutation relation between annihilation operators is given by

$$\boxed{\mathbf{I}(\alpha)\mathbf{I}(\beta) + \mathbf{I}(\beta)\mathbf{I}(\alpha) = 0}. \tag{3.72}$$

In full compliance with eqn (2.51), we have

$$\mathbf{I}^{\dagger}(\alpha)\mathbf{I}^{\dagger}(\beta) + \mathbf{I}^{\dagger}(\beta)\mathbf{I}^{\dagger}(\alpha) = 0. \tag{3.73}$$

Finally, there is a commutation relation between creation and annihilation operators. In this case,

$$\boxed{\mathbf{I}(\alpha)\mathbf{E}(\mathbf{v}) + \mathbf{E}(\mathbf{v})\mathbf{I}(\alpha) = \alpha(\mathbf{v})}, \tag{3.74}$$

which follows from eqn (2.41), and

$$\mathbf{I}^{\dagger}(\alpha)\mathbf{E}^{\dagger}(\mathbf{v}) + \mathbf{E}^{\dagger}(\mathbf{v})\mathbf{I}^{\dagger}(\alpha) = \alpha(\mathbf{v}), \tag{3.75}$$

which also follows from eqn (2.46).

The Clifford Algebras $\mathcal{C}\ell(V, +g)$ and $\mathcal{C}\ell(V, -g)$

Let us suppose that the vector space V is endowed with a symmetric correlation $\flat : V \to V^*$ (or, equivalently, a symmetric bilinear form $g : V \times V \to \mathbb{R}$). In the previous section the operators $\mathbf{E}$ and $\mathbf{I}$ were defined in such a way that $\mathbf{E} : V \to \mathrm{End}(\bigwedge(V))$ and $\mathbf{I} : V^* \to \mathrm{End}(\bigwedge(V))$. Since $\flat : V \to V^*$, the composition of $\mathbf{I} \circ \flat$ can now be introduced. It is immediately obvious that $\mathbf{I} \circ \flat : V \to \mathrm{End}(\bigwedge(V))$.

Given the operators $\mathbf{E} : V \to \mathrm{End}(\bigwedge(V))$, and $\mathbf{I} \circ \flat : V \to \mathrm{End}(\bigwedge(V))$, we can then consider the sum of those operators. Let us define $\gamma_+ : V \to \mathrm{End}(\bigwedge(V))$ as

$$\boxed{\gamma_+ = \mathbf{E} + \mathbf{I} \circ \flat}, \tag{3.76}$$

and $\gamma_- : V \to \mathrm{End}(\bigwedge(V))$ as

$$\boxed{\gamma_- = \mathbf{E} - \mathbf{I} \circ \flat}. \tag{3.77}$$

Using these definitions, we can assert the following theorem:

Theorem 3.6 ▶ *The mappings $\gamma_\pm$ are Clifford mappings. The quantities $\gamma_+(\mathbf{v})$ $(\mathbf{v} \in V)$ satisfy*

$$\boxed{\gamma_+(\mathbf{v})\gamma_+(\mathbf{u}) + \gamma_+(\mathbf{u})\gamma_+(\mathbf{v}) = 2g(\mathbf{v}, \mathbf{u})}, \tag{3.78}$$

generating the Clifford algebra $\mathcal{C}\ell(V, +g)$. On the other hand, the quantities $\gamma_-(\mathbf{v})$ $(\mathbf{v} \in V)$ satisfy

$$\boxed{\gamma_-(\mathbf{v})\gamma_-(\mathbf{u}) + \gamma_-(\mathbf{u})\gamma_-(\mathbf{v}) = -2g(\mathbf{v}, \mathbf{u})}, \tag{3.79}$$

generating the Clifford algebra $\mathcal{C}\ell(V, -g)$.

In eqns (3.78) and (3.79), we omit (in order to simplify the notation) the presence of an operator $\mathbf{1}$ in the right-hand side of these operations. Formally, we should have written $g(\mathbf{v}, \mathbf{u})\mathbf{1}$, where $\mathbf{1}A = A$, $\forall A \in \bigwedge(V)$.

Proof: Only the definitions of $\gamma_\pm$ and the commutation relations in eqns (3.70), (3.72), and (3.74) need to be employed. First,

$$\begin{aligned}
\gamma_\pm(\mathbf{v})\gamma_\pm(\mathbf{u}) &= [\mathbf{E}(\mathbf{v}) \pm \mathbf{I}(\mathbf{v}_\flat)][\mathbf{E}(\mathbf{u}) \pm \mathbf{I}(\mathbf{u}_\flat)] \\
&= \mathbf{E}(\mathbf{v})\mathbf{E}(\mathbf{u}) \pm \mathbf{E}(\mathbf{v})\mathbf{I}(\mathbf{u}_\flat) \pm \mathbf{I}(\mathbf{v}_\flat)\mathbf{E}(\mathbf{u}) + \mathbf{I}(\mathbf{v}_\flat)\mathbf{I}(\mathbf{u}_\flat).
\end{aligned}$$

By using the commutation relations, we can see that

$$\begin{aligned}
&\gamma_\pm(\mathbf{v})\gamma_\pm(\mathbf{u}) + \gamma_\pm(\mathbf{u})\gamma_\pm(\mathbf{v}) \\
&\quad = [\mathbf{E}(\mathbf{v})\mathbf{E}(\mathbf{u}) + \mathbf{E}(\mathbf{u})\mathbf{E}(\mathbf{v})] \pm [\mathbf{E}(\mathbf{v})\mathbf{I}(\mathbf{u}_\flat) + \mathbf{I}(\mathbf{u}_\flat)\mathbf{E}(\mathbf{v})] \\
&\qquad \pm [\mathbf{I}(\mathbf{v}_\flat)\mathbf{E}(\mathbf{u}) + \mathbf{E}(\mathbf{u})\mathbf{I}(\mathbf{v}_\flat)] + [\mathbf{I}(\mathbf{v}_\flat)\mathbf{I}(\mathbf{u}_\flat) + \mathbf{I}(\mathbf{u}_\flat)\mathbf{I}(\mathbf{v}_\flat)] \\
&\quad = \pm\mathbf{v}_\flat(\mathbf{u}) \pm \mathbf{u}_\flat(\mathbf{v}) = \pm 2g(\mathbf{v}, \mathbf{u}),
\end{aligned}$$

which shows eqns (3.78) and (3.79). In the last step, the relation $\mathbf{v}_\flat(\mathbf{u}) = g(\mathbf{v}, \mathbf{u}) = g(\mathbf{u}, \mathbf{v}) = \mathbf{u}_\flat(\mathbf{v})$ was used. In the book by Gilbert and Murray (1991), it is shown that the Clifford algebra constructed in this way is indeed universal. ✓

Sufficiency of the Algebra $\mathcal{C}\ell(V, +g)$

The Clifford algebras $\mathcal{C}\ell(V, +g)$ and $\mathcal{C}\ell(V, -g)$ have been already defined; however, we shall now show that it is *not* necessary to take into account both algebras.

The Clifford Mappings $\gamma_\pm^\dagger$

The operator $\mathbf{E}(\mathbf{v})$ consists of the left exterior multiplication by the vector $\mathbf{v}$, whereas the operator $\mathbf{E}^\dagger(\mathbf{v})$ consists of the right exterior multiplication by the vector $\mathbf{v}$. In the same way, the operator $\mathbf{I}(\mathbf{v}_\flat)$ is the left contraction by the covector $\mathbf{v}_\flat$, and the operator $\mathbf{I}^\dagger(\mathbf{v}_\flat)$ consists of the right contraction by the same vector. Since we defined $\gamma_\pm$ using the operations acting on the *left*, it is also possible to define $\gamma_\pm^\dagger$ employing the same operations but now using the right action.

Let us thus define $\gamma^\dagger_\pm : V \to \mathrm{End}(\bigwedge(V))$ as

$$\gamma^\dagger_\pm = \mathbf{E}^\dagger \pm \mathbf{I}^\dagger \circ \flat. \tag{3.80}$$

By using the commutation relations in eqns (3.71), (3.73), and (3.75), we obtain

$$\gamma^\dagger_\pm(\mathbf{v})\gamma^\dagger_\pm(\mathbf{u}) + \gamma^\dagger_\pm(\mathbf{u})\gamma^\dagger_\pm(\mathbf{v}) = \pm 2g(\mathbf{v}, \mathbf{u}). \tag{3.81}$$

This result shows that the quantities $\{\gamma^\dagger_\pm(\mathbf{v}) \mid \mathbf{v} \in V\}$ satisfy the same relations as $\{\gamma_\pm(\mathbf{v}) \mid \mathbf{v} \in V\}$ do.

Let us now denote by $\mathcal{C}\ell^\dagger(V, \pm g)$ the algebras generated by $\{\gamma^\dagger_\pm(\mathbf{v}) \mid \mathbf{v} \in V\}$. Since the generators $\{\gamma^\dagger_\pm(\mathbf{v})\}$ satisfy the same commutation relations as $\{\gamma_\pm(\mathbf{v})\}$, therefore $\mathcal{C}\ell^\dagger(V, +g)$ and $\mathcal{C}\ell(V, g)$ are isomorphic, as they are both universal.

The Relationship between $\gamma_\pm$ and $\gamma^\dagger_\pm$

Two important results established by the creation and the annihilation operators were summarised in eqns (3.66) and (3.69). Using these equations,[3] we find that

$$\gamma^\dagger_\pm = \mathbf{E}\# \pm (-\mathbf{I}\#) \circ \flat = \mathbf{E}\# \mp (\mathbf{I} \circ \flat)\#, \tag{3.82}$$

namely,

$$\boxed{\begin{aligned} \gamma^\dagger_\pm &= \gamma_\mp\#, \\ \gamma_\mp &= \gamma^\dagger_\pm\# \end{aligned}}. \tag{3.83}$$

This result shows that the algebra generated by γ_- is isomorphic to the algebra generated by $\gamma^\dagger_+$.[4] Likewise, the algebra generated by γ_+ is isomorphic to the algebra generated by $\gamma^\dagger_-$. Explicitly,

$$\mathcal{C}\ell^\dagger(V, +g) \simeq \mathcal{C}\ell(V, -g), \qquad \mathcal{C}\ell^\dagger(V, -g) \simeq \mathcal{C}\ell(V, +g). \tag{3.84}$$

We can thus conclude that it is not necessary to consider both of the Clifford algebras $\mathcal{C}\ell(V, +g)$ and $\mathcal{C}\ell(V, -g)$. In each of these algebras, the generators satisfy distinct commutation relations – namely, the relationship given by eqns (3.78) and (3.79), respectively. Therefore, it is sufficient to consider just one of them, for instance, $\mathcal{C}\ell(V, +g)$, and to take into account the left and the right actions of the generators.

Consider now the quantities $\gamma_+(\mathbf{v})$ and $\gamma^\dagger_+(\mathbf{u})$. It is immediately obvious that the product between these quantities commutes, namely,

$$\gamma_+(\mathbf{v})\gamma^\dagger_+(\mathbf{u}) - \gamma^\dagger_+(\mathbf{u})\gamma_+(\mathbf{v}) = 0. \tag{3.85}$$

This result is equivalent to

$$\gamma_+(\mathbf{v})\gamma_-(\mathbf{u}) + \gamma_-(\mathbf{u})\gamma_+(\mathbf{v}) = 0. \tag{3.86}$$

[3] Care must be taken not to confuse the symbols $\#$ and $\sharp$ in eqn (3.82).

[4] The presence of the grade involution $\#$ in eqn (3.83) is not relevant for this type of argument.

Equation (3.85) is particularly important, since it allows us to write the product which acts on both sides *and* satisfies the property of associativity. Indeed, writing

$$A\gamma_+(\mathbf{v}) = \gamma_+^\dagger(\mathbf{v})(A), \qquad \forall \mathbf{v} \in V\,, \forall A \in \bigwedge(V) \tag{3.87}$$

yields

$$(\gamma_+(\mathbf{v})A)\gamma_+(\mathbf{u}) = \gamma_+(\mathbf{v})(A\gamma_+(\mathbf{u})) = \gamma_+(\mathbf{v})A\gamma_+(\mathbf{u}). \tag{3.88}$$

To summarise, it is enough to examine the algebra generated by $\{\gamma_+(\mathbf{v}) \mid \mathbf{v} \in V\}$, via both left and right multiplications, since the same quantity satisfies the same commutation relation, no matter whether the left or the right actions are considered. Since we are just taking $\mathcal{C}\ell(V,+g) = \mathcal{C}\ell(V,g)$ into account, we can omit the index '+' in the generators $\{\gamma(\mathbf{v}) \mid \mathbf{v} \in V\}$.

Chevalley was the first to propose the definition $\mathcal{C}\ell(V,g) = \mathsf{T}(V)/\mathcal{I}_C$ in the context of the mapping $\gamma_+ : V \to \mathrm{End}(\bigwedge(V))$ defined in eqn (3.76). By the universal property of the algebra $\mathsf{T}(V)$, the mapping γ_+ extends to an algebra homomorphism $\mathsf{T} \to \mathrm{End}(\bigwedge(V))$, which can also be denoted by γ_+, vanishing on $\mathcal{I}_C$. Consequently, it factorises through $\mathcal{C}\ell(V,g)$, as $\mathsf{T}(V) \to \mathcal{C}\ell(V,g) \to \mathrm{End}(\bigwedge(V))$. It suffices to remember eqn (3.17), which asserts that $\gamma_+(\mathbf{v}_p\otimes\cdots\otimes\mathbf{v}_2\otimes\mathbf{v}_1)(1)$, where $\mathbf{v}_1, \mathbf{v}_2, \ldots, \mathbf{v}_p \in V$, is the sum of $\mathbf{v}_p\wedge\cdots\wedge\mathbf{v}_2\wedge\mathbf{v}_1$ and some other elements of $\bigwedge(V)$ of degree less than p, which we proved in example 3.3. This result gives a surjective mapping $\mathsf{T}(V) \to \bigwedge(V)$, which factorises through $\mathcal{C}\ell(V,g)$, whence $\dim(\mathcal{C}\ell(V,g)) \geq \dim(\bigwedge(V))$. Since we proved just after eqn (3.6) that the maximal dimension of a Clifford algebra is less than $\dim(\bigwedge(V))$, it is clear that $\dim(\mathcal{C}\ell(V,g)) = \dim(\bigwedge(V))$.

3.5 Grassmann Algebra versus Clifford Algebra

The Relationship between Clifford and Grassmann Products

The *Clifford product* (*or geometric product*) between two multivectors A and B will be denoted by $\gamma(A)\gamma(B)$.[5] In order to realise the meaning of this product, we start by considering the case where there are two vectors $\mathbf{v}$ and $\mathbf{u}$ originating the Clifford product $\gamma(\mathbf{v})\gamma(\mathbf{u})$, where $\gamma : V \to \mathcal{C}\ell(V,g)$. The associativity of Clifford algebras can be expressed by the equation

$$\gamma(\mathbf{v})\gamma(\mathbf{u}) = \gamma[\gamma(\mathbf{v})(\mathbf{u})]. \tag{3.89}$$

Using now the definition of γ, we obtain

$$\gamma(\mathbf{v})(\mathbf{u}) = \mathbf{v}\wedge\mathbf{u} + g(\mathbf{v},\mathbf{u}), \tag{3.90}$$

and then

$$\boxed{\gamma(\mathbf{v})\gamma(\mathbf{u}) = \gamma(\mathbf{v}\wedge\mathbf{u}) + g(\mathbf{v},\mathbf{u})}, \tag{3.91}$$

where we omit the presence of the operator $\mathbf{1}$, that is, we use $\gamma(g(\mathbf{v},\mathbf{u})) = g(\mathbf{v},\mathbf{u})\mathbf{1}$, in order to simplify the notation.

[5] In this section, the explicit use of the Clifford mapping γ in the formulæ involving the product in the Clifford algebra is intended to help (at least we believe it will!) in comprehending the relationship between the Clifford and the Grassmann algebras.

This equation is well known, although by another notation: it is eqn (3.9). This equation shows that the Clifford product can be thought as being composed of two parts. The first part, corresponding to the term $\gamma(\mathbf{v}\wedge\mathbf{u})$, is the image in $\mathcal{C}\ell(V,g)$ of the exterior product $\mathbf{v}\wedge\mathbf{u}$.[6] The second part is the image in $\mathcal{C}\ell(V,g)$ of the contraction $\mathbf{v}_\flat\rfloor\mathbf{u}$, which equals in this case the scalar product $g(\mathbf{v},\mathbf{u})$.

Equation (3.91) is useful for understanding the meaning regarding the Clifford product between a vector and a bivector, namely, the product $\gamma(\mathbf{v})\gamma(\mathbf{u}\wedge\mathbf{w})$. Using eqn (3.91), we can write

$$\gamma(\mathbf{v})\gamma(\mathbf{u}\wedge\mathbf{w}) = \gamma(\mathbf{v})\gamma(\mathbf{u})\gamma(\mathbf{w}) - g(\mathbf{u},\mathbf{w})\gamma(\mathbf{v}). \tag{3.92}$$

On the right-hand side of eqn (3.92), the first term can be calculated by using again eqn (3.89). It reads

$$\begin{aligned}\gamma(\mathbf{v})\gamma(\mathbf{u})\gamma(\mathbf{w}) &= \gamma[\gamma(\mathbf{v})[\gamma(\mathbf{u})(\mathbf{w})]] = \gamma[\gamma(\mathbf{v})(\mathbf{u}\wedge\mathbf{w}+g(\mathbf{u},\mathbf{w}))]\\ &= \gamma(\mathbf{v}\wedge\mathbf{u}\wedge\mathbf{w}+\mathbf{v}_\flat\rfloor(\mathbf{u}\wedge\mathbf{w})+g(\mathbf{u},\mathbf{w})\gamma(\mathbf{v}))\\ &= \gamma(\mathbf{v}\wedge\mathbf{u}\wedge\mathbf{w})+g(\mathbf{v},\mathbf{u})\gamma(\mathbf{w})-g(\mathbf{v},\mathbf{w})\gamma(\mathbf{u})+g(\mathbf{u},\mathbf{w})\gamma(\mathbf{v}).\end{aligned}$$

Using now this result in eqn (3.92), we can obtain

$$\gamma(\mathbf{v})\gamma(\mathbf{u}\wedge\mathbf{w}) = \gamma(\mathbf{v}\wedge\mathbf{u}\wedge\mathbf{w}) + \gamma(\mathbf{v}_\flat\rfloor(\mathbf{u}\wedge\mathbf{w})), \tag{3.93}$$

where

$$\gamma(\mathbf{v}_\flat\rfloor(\mathbf{u}\wedge\mathbf{w})) = g(\mathbf{v},\mathbf{u})\gamma(\mathbf{w}) - g(\mathbf{v},\mathbf{w})\gamma(\mathbf{u}). \tag{3.94}$$

Equation (3.93) shows that, as in the case involving two vectors, the Clifford product between a 1-vector and a 2-vector can be split in two parts: one corresponding to the image of the exterior product and the other part corresponding to the image of the contraction.

A similar reasoning can be repeated for the case involving a 1-vector and a 3-vector, a 4-vector, or in general any p-vector. The result is

$$\gamma(\mathbf{v})\gamma(A_{[k]}) = \gamma(\mathbf{v}\wedge A_{[k]}) + \gamma(\mathbf{v}_\flat\rfloor A_{[k]}), \tag{3.95}$$

where $A_{[k]} \in \bigwedge_k(V)$. Moreover, since the mapping γ is linear, we can write for an arbitrary multivector A the expression

$$\boxed{\gamma(\mathbf{v})\gamma(A) = \gamma(\mathbf{v}\wedge A) + \gamma(\mathbf{v}_\flat\rfloor A)}. \tag{3.96}$$

Another result that follows from a completely analogous procedure is

$$\boxed{\gamma(A)\gamma(\mathbf{v}) = \gamma(A\wedge\mathbf{v}) + \gamma(A\lfloor\mathbf{v}_\flat)}, \tag{3.97}$$

providing the meaning to the product $\gamma(A)\gamma(B)$, in particular when $A=\mathbf{v}$ (vector), and $B=\mathbf{u}\wedge\mathbf{w}$ (bivector). Equations (3.96) and (3.97) are equivalent to eqns (3.21) and (3.22).

[6]We must realise that it does *not* make sense to write something like $\gamma(\alpha)\wedge\gamma(\beta)$. In $\mathcal{C}\ell(V,g)$, there exists only one product – the Clifford product – denoted by juxtaposition most of the time here. In fact, it is well known that the exterior product $\wedge$ only makes sense in the exterior algebra or in the Grassmann algebra. However, the identification between $\bigwedge(V)$ and $\mathcal{C}\ell(V,g)$, coming from the bijection $\bigwedge(V)\to\mathcal{C}\ell(V,g)$ induced by the quotient mapping $\pi: \mathsf{T}(V)\to\mathcal{C}\ell(V,g)$, further provides the identification between $A\in\bigwedge(V)$ and $\pi(A)\in\mathcal{C}\ell(V,g)$. Hence, we can write the image of the exterior product in $\mathcal{C}\ell(V,g)$ by using the Clifford product, as we shall show soon.

It is worthwhile emphasising that the structure of these equations – namely, the decomposition of the geometric product into a part involving the exterior product and a part involving the contraction – holds only when one of the elements in the product is a vector. Let us analyse the following example. The geometric product involving two 2-vectors can be calculated with the aid of eqn (3.91), yielding

$$\begin{aligned}\gamma(\mathbf{v}_1 \wedge \mathbf{v}_2)\gamma(\mathbf{u}_1 \wedge \mathbf{u}_2) &= \gamma(\mathbf{v}_1 \wedge \mathbf{v}_2 \wedge \mathbf{u}_1 \wedge \mathbf{u}_2) + g(\mathbf{v}_1, \mathbf{u}_2)\gamma(\mathbf{v}_2 \wedge \mathbf{u}_1)\\ &- g(\mathbf{v}_1, \mathbf{u}_1)\gamma(\mathbf{v}_2 \wedge \mathbf{u}_2) + g(\mathbf{v}_2, \mathbf{u}_1)\gamma(\mathbf{v}_1 \wedge \mathbf{v}_2) - g(\mathbf{v}_2, \mathbf{u}_2)\gamma(\mathbf{v}_1 \wedge \mathbf{u}_1)\\ &+ g(\mathbf{v}_2, \mathbf{u}_1)g(\mathbf{v}_1, \mathbf{u}_2) - g(\mathbf{v}_2, \mathbf{u}_2)g(\mathbf{v}_1, \mathbf{u}_1).\end{aligned} \tag{3.98}$$

Hence, the Clifford product involving two 2-vectors *does not* consist simply of the sum of a term involving the exterior product and another one involving the contraction. Those two parts are indeed present, where the exterior product is expressed in the first term on the right-hand side of the equation and the contraction in the two last terms on the right-hand side:

$$\begin{aligned}(\mathbf{v}_{1\flat} \wedge \mathbf{v}_{2\flat})\rfloor(\mathbf{u}_1 \wedge \mathbf{u}_2) &= \mathbf{v}_{1\flat}\rfloor(\mathbf{v}_{2\flat}\rfloor(\mathbf{u}_1 \wedge \mathbf{u}_2))\\ &= g(\mathbf{v}_2, \mathbf{u}_1)g(\mathbf{v}_1, \mathbf{u}_2) - g(\mathbf{v}_2, \mathbf{u}_2)g(\mathbf{v}_1, \mathbf{u}_1).\end{aligned} \tag{3.99}$$

Such terms contribute to 4-vector and 0-vector (scalar) parts, and the other terms contribute with a 2-vector part. In general, the Clifford product between a p-vector and a q-vector splits into the sum of a $(p+q)$-vector, a $(p+q-2)$-vector, a $(p+q-4)$-vector, ..., up to a $|p-q|$-vector part, according to eqn (3.47).

It is worthwhile emphasising a peculiarity of the Clifford product between two 2-vectors. From eqn (3.98), we can observe that

$$\begin{aligned}\gamma(\mathbf{v}_1 \wedge \mathbf{v}_2)\gamma(\mathbf{u}_1 \wedge \mathbf{u}_2) - \gamma(\mathbf{u}_1 \wedge \mathbf{u}_2)\gamma(\mathbf{v}_1 \wedge \mathbf{v}_2) &= 2[g(\mathbf{v}_1, \mathbf{u}_2)\gamma(\mathbf{v}_2 \wedge \mathbf{u}_1)\\ - g(\mathbf{v}_1, \mathbf{u}_1)\gamma(\mathbf{v}_2 \wedge \mathbf{u}_2) + g(\mathbf{v}_2, \mathbf{u}_1)\gamma(\mathbf{v}_1 \wedge \mathbf{v}_2) &- g(\mathbf{v}_2, \mathbf{u}_2)\gamma(\mathbf{v}_1 \wedge \mathbf{u}_1)].\end{aligned} \tag{3.100}$$

Here it is opportune to define the *commutator* $[\gamma(A), \gamma(B)]$ as

$$[\gamma(A), \gamma(B)] = \gamma(A)\gamma(B) - \gamma(B)\gamma(A). \tag{3.101}$$

Equation (3.100) shows that the commutator of two 2-vectors is a 2-vector. The Clifford product of two 2-vectors yields

$$\begin{aligned}\gamma(\mathbf{v}_1 \wedge \mathbf{v}_2)\gamma(\mathbf{u}_1 \wedge \mathbf{u}_2) &= \gamma(\mathbf{v}_1 \wedge \mathbf{v}_2 \wedge \mathbf{u}_1 \wedge \mathbf{u}_2)\\ &+ \frac{1}{2}[\gamma(\mathbf{v}_1 \wedge \mathbf{v}_2), \gamma(\mathbf{u}_1 \wedge \mathbf{u}_2)] + (\mathbf{v}_{1\flat} \wedge \mathbf{v}_{2\flat})\rfloor(\mathbf{u}_1 \wedge \mathbf{u}_2).\end{aligned} \tag{3.102}$$

The Exterior Product in Clifford Algebra

Since it was previously demonstrated that the Clifford product contains a part that is the image, in $C\ell(V, g)$, of the exterior product in $\bigwedge(V)$ or $\mathcal{G}(V)$, it is expected that the exterior product (namely, the image of the exterior product) can be expressed by the geometric product. Indeed, eqn (3.97) reads

$$\gamma(A)\gamma(\mathbf{v}) = \gamma(\mathbf{v} \wedge \#A) - \gamma(\mathbf{v}_\flat \rfloor \#A),$$

where eqns (2.48) and (3.65) were used. Hence, it is equivalent to

$$\gamma(\#A)\gamma(\mathbf{v}) = \gamma(\mathbf{v} \wedge A) - \gamma(\mathbf{v}_\flat \rfloor A). \tag{3.103}$$

When we take this equation together with eqn (3.96), we obtain

$$\boxed{\gamma(\mathbf{v} \wedge A) = \frac{1}{2}[\gamma(\mathbf{v})\gamma(A) + \gamma(\#A)\gamma(\mathbf{v})]}. \tag{3.104}$$

This equation is actually eqn (3.26) in a different notation.

Contractions in Clifford Algebra

The expression for the left contraction is straightforward:

$$\boxed{\gamma(\mathbf{v}_\flat \rfloor A) = \frac{1}{2}[\gamma(\mathbf{v})\gamma(A) - \gamma(\#A)\gamma(\mathbf{v})]}. \tag{3.105}$$

It is also presented in another notation by eqn (3.27).

The expression for the right contraction is also immediately obvious:

$$\gamma(A \lfloor \mathbf{v}_\flat) = \frac{1}{2}[\gamma(A)\gamma(\mathbf{v}) - \gamma(\mathbf{v})\gamma(\#A)]. \tag{3.106}$$

Grassmann Algebra in Clifford Algebra and Vice Versa

The last results in the previous section show that the Grassmann algebra $\mathcal{G}(V)$ operations can be completely reproduced by operations in the Clifford algebra $\mathcal{C}\ell(V, g)$, and vice versa.[7] This fact, however, *does not* mean that these algebras are equivalent – or, in a technical sense, *isomorphic*. Specifically, the Grassmann and Clifford algebras *are isomorphic from the point of view of vector spaces but are not algebraically isomorphic*.

The Clifford product can be represented in the Grassmann algebra as a function of eqn (3.90). When $\curlyvee$ is defined as

$$\alpha \curlyvee A = \gamma(\alpha)A, \tag{3.107}$$

it immediately follows that

$$\alpha \curlyvee \beta + \beta \curlyvee \alpha = 2g(\alpha, \beta). \tag{3.108}$$

It is important to emphasise that the product $\curlyvee$ is associated with operations in the Grassmann algebra. In other words, the underlying algebra concerning the product $\curlyvee$ is

[7] We must remember here the distinction between the exterior and the Grassmann algebras. Their equivalence just holds in the context of the Grassmann algebra – the exterior algebra endowed with the respective generalisation of the bilinear functional in V – not the exterior algebra.

the Grassmann algebra, not the Clifford algebra. Hence, the product $\curlyvee$ provides a representation of the geometric product of the Grassmann algebra. Although eqns (3.108) and (3.18) are similar, the products $\curlyvee$ and $\diamond$ involved in those equations are different. As algebras endowed with the products $\curlyvee$ and $\diamond$, they are clearly isomorphic but, since $\curlyvee$ is defined in eqn (3.107) by an operator acting on the Grassmann algebra $\mathcal{G}(V)$, the product $\diamond$ is the image of the tensor product in the quotient algebra $\mathsf{T}(V)/\mathcal{I}_C$.

The isomorphism of vector spaces $\mathcal{C}\ell(V,g) \simeq \mathcal{G}(V)$ is provided by the mapping $\lambda : \mathcal{C}\ell(V,g) \to \mathcal{G}(V)$ defined as

$$\lambda(\gamma(\mathbf{v})\gamma(A)) = \mathbf{v} \curlyvee \lambda(\gamma(A)) = \mathbf{v} \wedge \lambda(\gamma(A)) + \mathbf{v}_\flat \rfloor \lambda(\gamma(A)), \tag{3.109}$$

where

$$\lambda(1) = 1. \tag{3.110}$$

Then eqn (3.110) implies that

$$\lambda(\gamma(\mathbf{v})) = \lambda(\gamma(\mathbf{v})\gamma(1)) = \mathbf{v} \curlyvee \lambda(1) = \mathbf{v},$$

and subsequently that

$$\lambda(\gamma(\mathbf{v})\gamma(\mathbf{u})) = \mathbf{v} \curlyvee \mathbf{u} = \mathbf{v} \wedge \mathbf{u} + g(\mathbf{v}, \mathbf{u}).$$

The mapping λ *is not* an algebraic isomorphism, since it *does not satisfy* $\lambda[\gamma(\Psi)\gamma(\Phi)] = [\lambda(\gamma(\Psi))] \wedge [\lambda(\gamma(\Phi))]$.

To summarise, the exterior product and the contractions of the Grassmann algebra can be 'simulated' in the Clifford algebra, and the product of the Clifford algebra can be emulated in the Grassmann algebra. Thus, any calculation in either one of those structures can be reproduced in the other.

Nevertheless, sometimes it is advantageous to work in one or the other. One of the great advantages of the Clifford algebra as compared to the Grassmann algebra is that, via the Clifford product, we can define the *inverse* of a multivector, which we cannot do via the exterior product. Given a multivector A, there exists an inverse if there exists another multivector B such that, when it is left or right multiplied by A, the result is unity. In some cases, it is possible to accomplish it in the Clifford algebra; one of the simplest examples is that concerning a non-isotropic vector $\mathbf{v}$ ($g(\mathbf{v}, \mathbf{v}) \neq 0$). Clearly, the inverse of $\mathbf{v}$ *does not* exist in the Grassmann algebra, since $\mathbf{v} \wedge \mathbf{v} = 0$. Thus, for any multivector B, it follows that $B \wedge \mathbf{v} = \pm \mathbf{v} \wedge B \neq 1$. In contrast, in the Clifford algebra, there always exists an inverse of $\mathbf{v}$. Indeed,

$$\gamma(\mathbf{v})\gamma(\mathbf{v}) = \gamma(\mathbf{v} \wedge \mathbf{v}) + g(\mathbf{v}, \mathbf{v}) = g(\mathbf{v}, \mathbf{v}),$$

and, since $g(\mathbf{v}, \mathbf{v}) \neq 0$, we can define

$$\gamma(\mathbf{v})^{-1} = \gamma(\mathbf{v})/g(\mathbf{v}, \mathbf{v}).$$

In addition, $\gamma(\mathbf{v})^{-1}\gamma(\mathbf{v}) = \gamma(\mathbf{v})\gamma(\mathbf{v})^{-1} = 1$ holds.

From the computational point of view, it is more straightforward to regard a structure where the inverse does exist than to take a framework where it does not. Doing so can sometimes reduce many pages of calculations to just few lines.

However, the Grassmann algebra has some advantages over the Clifford algebra. In particular, the exterior product permits us to define the *multivector structure* in a *natural* manner which is *independent* of a vector space basis. This advantage comes from the fact that the exterior algebra is the most basic and general structure when we take into account the Clifford and Grassmann algebras, since it does not demand the existence of a correlation. Indeed, in this text, we used the exterior product to define the multivector structure of p-vectors.

As for the question of whether it is possible to define a multivector structure by using the Clifford product, it is straightforward to do so when an orthogonal basis is taken into account. Indeed, if $\mathbf{v}$ and $\mathbf{u}$ are orthogonal, that is, $g(\mathbf{v}, \mathbf{u}) = 0$, then

$$\gamma(\mathbf{v})\gamma(\mathbf{u}) = \gamma(\mathbf{v} \wedge \mathbf{u}).$$

Hence, the geometric product can be used to define a multivector structure since, in this case, $\gamma(\mathbf{v})\gamma(\mathbf{u})$ is a 2-vector. In this case, the Clifford product of p orthogonal vectors is a p-vector.

Obviously, this construction does depend upon the orthogonality between vectors. If $\mathbf{v}$ and $\mathbf{u}$ are not orthogonal, then the geometric product between them is not (only) a 2-vector. Indeed, in order to establish the multivector structure, we must use the geometric product given by eqn (3.104) as the expression for the exterior product. However, this is just a way to disguise using the exterior product! To summarise, in order to define the multivector structure in a basis-independent way, it is best to use the exterior product.

3.6 Notation

Notations seem to satisfy a kind of 'uncertainty principle': the more precise they are, the less operational, and vice versa. Thus, it is important to fully understand the notation and know how to use it appropriately. Let us establish in this section a notation that is adapted for use in calculations.

Let $\mathfrak{B} = \{\mathbf{e}_i\}$ be a basis of V and let $\mathfrak{B}^* = \{\mathbf{e}^i\}$ be its associated dual basis (where $i = 1, \ldots, n$). Let g be a symmetric bilinear functional defined by $g_{ij} = g(\mathbf{e}_i, \mathbf{e}_j) = g_{ji}$, and $g^{ij} = g^{-1}(\mathbf{e}^i, \mathbf{e}^j) = g^{ji}$. It follows that $\flat(\mathbf{e}_i) = \mathbf{e}_{i\flat} = g_{ij}\mathbf{e}^j$, and $\sharp(\mathbf{e}^i) = \mathbf{e}^{i\sharp} = g^{ij}\mathbf{e}_j$. Then,

$$\mathbf{v}_\flat = v^i\mathbf{e}_{i\flat} = v^i g_{ij}\mathbf{e}^j = v_i\mathbf{e}^i,$$

where $v_i = g_{ij}v^j$.

The operators $\mathbf{E}$ and $\mathbf{I}$ are linear. Thus, the action of one of them on elements of a basis of the vector space V suffices to completely determine the action on V. Let us denote

$$\mathbf{E}_i = \mathbf{E}(\mathbf{e}_i), \qquad \mathbf{I}^i = \mathbf{I}(\mathbf{e}^i). \tag{3.111}$$

Hence, $\mathbf{I} \circ \flat(\mathbf{e}_i) = g_{ij}\mathbf{I}^j$ can be written as

$$\gamma_i = \mathbf{E}_i + g_{ij}\mathbf{I}^j, \tag{3.112}$$

where

$$\gamma_i = \gamma(\mathbf{e}_i). \tag{3.113}$$

The quantities $\{\gamma_i\}$ $(i = 1, \ldots, n)$ are the generators of the Clifford algebra $\mathcal{C}\ell(V, g)$ and thus satisfy

$$\boxed{\gamma_i\gamma_j + \gamma_j\gamma_i = 2g_{ij}}. \tag{3.114}$$

In addition,

$$\boxed{\gamma(\mathbf{v}) = v^i\gamma_i}. \tag{3.115}$$

Using an analogous reasoning, we write

$$\gamma(\alpha) = \alpha_i\gamma^i, \tag{3.116}$$

where $\gamma(\mathbf{e}^i) = g^{ij}\gamma(\mathbf{e}_j)$, that is,

$$\boxed{\gamma^i = g^{ij}\gamma_j}, \tag{3.117}$$

which satisfies

$$\boxed{\gamma^i\gamma^j + \gamma^j\gamma^i = 2g^{ij}}. \tag{3.118}$$

Finally, it is common to conceal the mapping γ in the expressions given in eqns (3.115) and (3.116) by directly writing

$$\mathbf{v} = v^i\gamma_i, \qquad \alpha = \alpha_i\gamma^i. \tag{3.119}$$

Sometimes we need to go further and actually omit the mapping. In that case, we can express $\mathbf{v} = v^i\mathbf{e}_i$, where

$$\boxed{\mathbf{e}_i\mathbf{e}_j + \mathbf{e}_j\mathbf{e}_i = 2g_{ij}}. \tag{3.120}$$

The choice of one or the other notation clearly depends on the context.

3.7 Additional Readings

There are many texts that we can suggest for additional reading on the content of this chapter. Besides the references already provided throughout the text, a clear and concise introduction to Clifford algebras, with elements of the theory of Dirac operators and Clifford analysis, has been provided by Garling (2011); a set of introductory lectures about Clifford algebras and analysis, as well as applications in physics and engineering can be found in Abłamowicz and Sobczyk (2004). Further mathematical developments of Clifford algebras are described in the work by Helmstetter and Micali (2008). With respect to applications, there are many, mainly in physics but also in different areas of engineering. An introduction to Clifford algebras, written for physicists, including applications in mechanics, electromagnetism, quantum mechanics, and gravitation, has been provided by Doran and Lasenby (2003). Moreover, a series of papers by Hiley (2011) and by Binz *et al.* (2013) deal with applications in quantum mechanics. Another great book about Clifford algebras, with an emphasis on differential geometry, and on applications in physics, has been written by Benn and Tucker (1987). Many different applications of Clifford algebras in physics, engineering and computer science have been discussed, for example, in the work by Baylis (1996), Sommer, and Dorst and Lasenby (2011). General properties of Clifford algebras can be also found in, for example, the work by Gallier (1997) and Todorov (2011). For a Clifford bundle approach and in particular for the interplay among the Dirac, the Einstein, and the Maxwell equations, see the work by Rodrigues and de Oliveira (2007).

3.8 Exercises

(1) Let W be a vector subspace of $\bigwedge(\mathbb{R}^{5,0})$, given by $W = \mathbb{R} \oplus \mathbb{R}^{5,0} \oplus \bigwedge_2(\mathbb{R}^{5,0})$ and for which the dimension is $1 + 5 + 5(5-1)/2 = 2^5/2 = 2^4$. Define a product $*$ in W as (Lounesto, 2001*a*)

$$\begin{aligned} a * b &= \langle ab(1 + \mathbf{e}_1\mathbf{e}_2\mathbf{e}_3\mathbf{e}_4\mathbf{e}_5)\rangle_{0\oplus1\oplus2} \\ &= \langle ab(1 + \mathbf{e}_1\mathbf{e}_2\mathbf{e}_3\mathbf{e}_4\mathbf{e}_5)\rangle_0 + \langle ab(1 + \mathbf{e}_1\mathbf{e}_2\mathbf{e}_3\mathbf{e}_4\mathbf{e}_5)\rangle_1 + \langle ab(1 + \mathbf{e}_1\mathbf{e}_2\mathbf{e}_3\mathbf{e}_4\mathbf{e}_5)\rangle_2, \end{aligned}$$

where $\{\mathbf{e}_i\}$ $(i = 1, 2, 3, 4, 5)$ is an orthonormal basis of $\mathbb{R}^{5,0}$, and ab is the usual Clifford algebra product $C\ell(\mathbb{R}^5, g)$.
(a) Show that the algebra $\mathcal{A} = (W, *)$ is associative. (b) Show that $\mathbf{v} * \mathbf{v} = g(\mathbf{v}, \mathbf{v})$ for $\mathbf{v} \in \mathbb{R}^{5,0}$. The algebra $\mathcal{A}$ is thus an example of a Clifford algebra that is *not universal*. (c) Show that $\mathbf{e}_1 * \mathbf{e}_2 * \mathbf{e}_3 * \mathbf{e}_4 * \mathbf{e}_5 = 1$. (d) Show that the generators of $\mathcal{A}$ can be represented by the following matrices:

$$\mathbf{e}_1 = \begin{pmatrix} 0 & -i \\ i & 0 \end{pmatrix}, \ \mathbf{e}_2 = \begin{pmatrix} 0 & -j \\ j & 0 \end{pmatrix}, \ \mathbf{e}_3 = \begin{pmatrix} 0 & -k \\ k & 0 \end{pmatrix}, \ \mathbf{e}_4 = \begin{pmatrix} 1 & 0 \\ 0 & -1 \end{pmatrix}, \ \mathbf{e}_5 = \begin{pmatrix} 0 & 1 \\ 1 & 0 \end{pmatrix},$$

where i, j, k denote the quaternionic units.

(2) In $\bigwedge(\mathbb{R}^2) = \mathbb{R} \oplus \mathbb{R}^2 \oplus \bigwedge_2(\mathbb{R}^2)$, define a product $\odot$ as

$$\begin{aligned} &\mathbf{e}_1 \odot \mathbf{e}_1 = 1, && \mathbf{e}_1 \odot \mathbf{e}_2 = \mathbf{e}_1 \wedge \mathbf{e}_2 + b, \\ &\mathbf{e}_1 \odot (\mathbf{e}_1 \wedge \mathbf{e}_2) = \mathbf{e}_2 - b\mathbf{e}_1, && \mathbf{e}_2 \odot \mathbf{e}_1 = -(\mathbf{e}_1 \wedge \mathbf{e}_2) - b, \\ &\mathbf{e}_2 \odot \mathbf{e}_2 = 1, && \mathbf{e}_2 \odot (\mathbf{e}_1 \wedge \mathbf{e}_2) = -\mathbf{e}_1 - b\mathbf{e}_2, \\ &(\mathbf{e}_1 \wedge \mathbf{e}_2) \odot \mathbf{e}_1 = -\mathbf{e}_2 - b\mathbf{e}_1, && (\mathbf{e}_1 \wedge \mathbf{e}_2) \odot \mathbf{e}_2 = \mathbf{e}_1 - b\mathbf{e}_2, \\ &(\mathbf{e}_1 \wedge \mathbf{e}_2) \odot (\mathbf{e}_1 \wedge \mathbf{e}_2) = -1 - b^2 - 2b(\mathbf{e}_1 \wedge \mathbf{e}_2). \end{aligned}$$

(a) Show that the algebra $(\bigwedge(\mathbb{R}^2), \odot)$ is a Clifford algebra for the quadratic space $\mathbb{R}^2$. (b) Show that this algebra is isomorphic to the Clifford algebra $C\ell(\mathbb{R}^2)$. (c) Show that the generators of this algebra can be represented by

$$\mathbf{e}_1 = \begin{pmatrix} 1 & 0 \\ 0 & -1 \end{pmatrix}, \quad \mathbf{e}_2 = \begin{pmatrix} 0 & 1 \\ 1 & 0 \end{pmatrix}, \quad \mathbf{e}_1 \wedge \mathbf{e}_2 = \begin{pmatrix} -b & 1 \\ -1 & -b \end{pmatrix}.$$

(d) Although the Clifford algebras $(\bigwedge(\mathbb{R}^2), \odot)$ and $C\ell(\mathbb{R}^2)$ are isomorphic, they are clearly different Clifford algebras. Which is the relationship between these algebras? (Hint: how could you define a Clifford algebra for a vector space equipped with a bilinear form that is not necessarily symmetric?)

(3) Let $\{\mathbf{v}_1, \ldots, \mathbf{v}_n\}$ be an arbitrary basis of (V, g) and let $\{\theta^1, \ldots, \theta^n\}$ be its associated dual basis. Prove that the product $\psi\phi$ in $C\ell(V, g)$ between the multivectors ψ and ϕ can be written as

$$\psi\phi = \sum_{p=0}^{n} \frac{(-1)^{p(p-1)/2}}{p!} g_{i_1j_1} \cdots g_{i_pj_p} \left[\#^p(\theta^{i_1} \wedge \cdots \wedge \theta^{i_p}) \lrcorner \psi\right] \wedge \left[(\theta^{j_1} \wedge \cdots \wedge \theta^{j_p}) \lrcorner \phi\right],$$

where the sum convention is assumed in the indices $i_1, \ldots, i_p$ and $j_1, \ldots, j_p$. The Clifford algebra represented by the product defined by the right-hand side in this equation is sometimes called Kähler–Atiyah algebra.

(4) Show that, if $n = \dim \mathbb{R}^{p,q}$ is *even*, then $\mathrm{Cen}(\mathcal{C}\ell_{p,q}) = \bigwedge_0(\mathbb{R}^{p,q})$. In addition, prove that, if n is odd, then $\mathrm{Cen}(\mathcal{C}\ell_{p,q}) = \bigwedge_0(\mathbb{R}^{p,q}) \oplus \bigwedge_n(\mathbb{R}^{p,q})$.

(5) Let $\eta = \mathbf{e}_1\mathbf{e}_2\cdots\mathbf{e}_n$ be a unit n-vector of the Clifford algebra $\mathcal{C}\ell(V,g)$, where $V = \mathbb{R}^{p,q}$ is a n-dimensional vector space and $\{\mathbf{e}_1,\ldots,\mathbf{e}_n\}$ is an orthonormal basis of V, namely

$$\begin{aligned} g(\mathbf{e}_i,\mathbf{e}_i) &= 1, \quad i = 1,\ldots,p,\\ g(\mathbf{e}_i,\mathbf{e}_i) &= -1, \quad i = p+1,\ldots,p+q = n,\\ g(\mathbf{e}_i,\mathbf{e}_j) &= 0, \quad i \neq j. \end{aligned}$$

(a) Show that

$$\begin{aligned} \eta^2 &= 1, \quad \text{if } p-q = 0,1 \mod 4,\\ \eta^2 &= -1, \quad \text{if } p-q = 2,3 \mod 4. \end{aligned}$$

This results also holds, of course, for $\eta = -\mathbf{e}_1\mathbf{e}_2\cdots\mathbf{e}_n$, in such a way that it does not depend on the orientation of V. (b) Show that, for $p-q = 0,1 \mod 4$, we can write $\mathcal{C}\ell(V,g) = {}_+\mathcal{C}\ell(V,g) \oplus {}_-\mathcal{C}\ell(V,g)$, where

$${}_\pm\mathcal{C}\ell(V,g) = \{\psi \in \mathcal{C}\ell(V,g) \mid \psi = \pm\psi\eta\}.$$

(c) Show that, if $p-q = 1 \mod 4$, then ${}_\pm\mathcal{C}\ell(V,g)$ are *bilateral ideals* of $\mathcal{C}\ell(V,g)$, that is, that

$${}_\pm\mathcal{C}\ell(V,g)\,\mathcal{C}\ell(V,g) \subset {}_\pm\mathcal{C}\ell(V,g) \quad \text{and} \quad \mathcal{C}\ell(V,g)\,{}_\pm\mathcal{C}\ell(V,g) \subset {}_\pm\mathcal{C}\ell(V,g)\,.$$

(6) Let $\{\mathbf{e}_i\}$ $(i = 1,\ldots,n)$ be an orthogonal basis for the quadratic space (V,g). Use multi-index notation, that is, write $\{\mathbf{e}_I\}$ where

$$I = \{1,\ldots,n,12,\ldots,1n,123,\ldots,\ldots,12\cdots n\}$$

and

$$\mathbf{e}_{i_1\cdots i_k} = \mathbf{e}_{i_1}\cdots\mathbf{e}_{i_k} = \mathbf{e}_1\wedge\cdots\wedge\mathbf{e}_{i_k}$$

such that $\{\mathbf{e}_I\}$ is an orthonormal basis with respect to the extension G of g for $\bigwedge(V)$ (eqns (2.97)–(2.100)). (a) Prove eqn (3.48). (b) Show that any element $\psi \in \mathcal{C}\ell(V,g)$ can be written as

$$\psi = \sum_I \langle\psi\widetilde{\mathbf{e}}^I\rangle_0\mathbf{e}_I,$$

where $\{\mathbf{e}^I\}$ is the dual basis of $\{\mathbf{e}_I\}$.

(7) Let $\{\mathbf{v}_1,\mathbf{v}_2,\ldots,\mathbf{v}_n\}$ be a basis for the quadratic space (V,g) and let $\mathcal{C}\ell(V,g)$ be the associated Clifford algebra. The multivectors $\phi_0,\phi_1,\phi_2,\ldots,\phi_n$ are defined as

$$\phi_0 = 1, \quad \phi_1 = \mathbf{v}_1, \quad \phi_2 = \langle\mathbf{v}_1\mathbf{v}_2\rangle_2,\ldots, \qquad \phi_n = \langle\mathbf{v}_1\mathbf{v}_2\cdots\mathbf{v}_n\rangle_n.$$

Show that the vectors

$$\mathbf{e}_i = \widetilde{\phi}_{i-1}\phi_i, \qquad i = 1,\ldots,n,$$

are orthogonal and that the process of showing that they are orthogonal exactly corresponds to the Gram–Schmidt procedure.

4 Classification and Representation of the Clifford Algebras

In this chapter, the classification and representation of Clifford algebras are introduced and discussed. We denote by $\mathcal{M}(n, \mathbb{K})$ the algebra of $n \times n$ matrices with entries in $\mathbb{K} = \mathbb{R}, \mathbb{C}, \mathbb{H}$, where $\mathbb{H}$ denotes the set of quaternions. Some important theorems regarding the structure of Clifford algebras are presented; these are used later on for the classification and construction of the representations of the Clifford algebras. In addition, procedures to explicitly build Clifford algebras matrix representations are presented.

4.1 Theorems on the Structure of Clifford Algebras

In this section, some important and essential theorems regarding the structure and classification of Clifford algebras are introduced and investigated. As well as being important in themselves, these theorems are useful in the classification of the Clifford algebras as matrix algebras.

Let us start by defining the alternating tensor product of two graded algebras, $\mathcal{A}$ and $\mathcal{B}$. The *alternating tensor product* $\mathcal{A}\hat{\otimes}\mathcal{B}$ between those algebras is defined as the algebra generated by the product $a\hat{\otimes}b$, $a \in \mathcal{A}$, $b \in \mathcal{B}$, with the product being defined by

$$(a_1\hat{\otimes}b_1)(a_2\hat{\otimes}b_2) = (-1)^{\deg(b_1)\deg(a_2)} a_1a_2\hat{\otimes}b_1b_2. \tag{4.1}$$

This definition applies to homogeneous elements and can be extended by linearity. The tensor product $\hat{\otimes}$ is the usual tensor product and the hat notation reminds us that the product must take into account the grading of the elements involved.

The theorem that is key for Clifford algebra mod 8 periodicity shall be proved in what follows. However, first we shall enunciate a lemma that is important for proving it. Details on the proof of the lemma can be found, for example, in the book by Helmstetter and Micali (2008).

Lemma 4.1 ▶ *Let $\mathcal{A}$ and $\mathcal{B}$ be graded algebras, and let $f : \mathcal{A} \to \mathcal{C}$, and $f' : \mathcal{B} \to \mathcal{C}$, be algebra homomorphisms such that $\forall a \in \mathcal{A}$ and $\forall b \in \mathcal{B}$,*

$$f(a)f'(b) = (-1)^{\deg(a)\deg(b)} f'(b)f(a).$$

There exists a unique algebra morphism $\mathring{f} : \mathcal{A}\hat{\otimes}\mathcal{B} \to \mathcal{C}$ such that $f(a) = \mathring{f}(a \otimes 1_{\mathcal{B}})$, and $f'(b) = \mathring{f}(1_{\mathcal{A}} \otimes b)$.

An Introduction to Clifford Algebras and Spinors. First Edition. Jayme Vaz, Jr. and Roldão da Rocha, Jr.

Now we can enunciate and prove the prominent theorem that endows us with the technique necessary to introduce the mod 8 periodicity of Clifford algebras, as well as other structure theorems:

Theorem 4.1 ▶ *Let (V, g) and (V', g') be two quadratic spaces and let $C\ell(V, g)$ and $C\ell(V', g')$ be their respective Clifford algebras. Then*

$$C\ell(V \oplus V', g \oplus g') \simeq C\ell(V, g)\hat{\otimes} C\ell(V', g'), \tag{4.2}$$

where $\simeq$ denotes an isomorphism between Clifford algebras, and $V \oplus V'$ stands for the orthogonal direct sum of V and V'.

Proof: Let $g \oplus g'$ be the symmetric bilinear form defined in $V \oplus V'$, considered to be the orthogonal direct sum of V and V'. Therefore

$$(g \oplus g')(\mathbf{v} + \mathbf{v}', \mathbf{u} + \mathbf{u}') = g(\mathbf{v}, \mathbf{u}) + g'(\mathbf{v}', \mathbf{u}'), \quad \forall \mathbf{v}, \mathbf{u} \in V, \ \forall \mathbf{v}', \mathbf{u}' \in V'.$$

The mapping $\Gamma : V \oplus V' \to C\ell(V, g)\hat{\otimes} C\ell(V', g')$ can be defined as

$$\Gamma(\mathbf{v} + \mathbf{v}') = \mathbf{v}\hat{\otimes}1_{\mathbf{v}'} + 1_{\mathbf{v}}\hat{\otimes}\mathbf{v}',$$

where it is implicit that $\mathbf{v} = \gamma(\mathbf{v})$, and $\mathbf{v}' = \gamma'(\mathbf{v}')$. In addition, since $\gamma : V \to C\ell(V, g)$, and $\gamma\,' : V' \to C\ell(V', g')$, are Clifford mappings, then the mapping Γ is also a Clifford mapping. Indeed,

$$\begin{aligned}
\left(\Gamma(\mathbf{v} + \mathbf{v}')\right)^2 &= (\mathbf{v}\hat{\otimes}1_{\mathbf{v}'} + 1_{\mathbf{v}}\hat{\otimes}\mathbf{v}')(\mathbf{v}\hat{\otimes}1_{\mathbf{v}'} + 1_{\mathbf{v}}\hat{\otimes}\mathbf{v}') \\
&= (-1)^{0\cdot 1}(\mathbf{v})^2\hat{\otimes}1_{\mathbf{v}'} + (-1)^{0\cdot 0}\mathbf{v}\hat{\otimes}\mathbf{v}' \\
&\quad + (-1)^{1\cdot 1}\mathbf{v}\hat{\otimes}\mathbf{v}' + (-1)^{0\cdot 1}1_{\mathbf{v}}\hat{\otimes}(\mathbf{v}')^2 \\
&= g(\mathbf{v}, \mathbf{v})\hat{\otimes}1_{\mathbf{v}'} + 1_{\mathbf{v}}\hat{\otimes}g'(\mathbf{v}', \mathbf{v}') \\
&= \left(g(\mathbf{v}, \mathbf{v}) + g'(\mathbf{v}', \mathbf{v}')\right)1_{\mathbf{v}}\hat{\otimes}1_{\mathbf{v}'} \\
&= (g \oplus g')(\mathbf{v} + \mathbf{v}', \mathbf{v} + \mathbf{v}')1_{\mathbf{v}}\hat{\otimes}1_{\mathbf{v}'}.
\end{aligned}$$

Moreover, for finite-dimensional Clifford algebras, we have $\dim C\ell(V, g)\hat{\otimes} C\ell(V', g') = 2^{\dim V}2^{\dim V'} = 2^{\dim(V \oplus V')}$. On the other hand, let $C\ell(V \oplus V', g \oplus g')$ be the Clifford algebra associated to the quadratic space $(V \oplus V', g \oplus g')$. From the universality theorem, there exists a homomorphism

$$\phi : C\ell(V \oplus V', g \oplus g') \to C\ell(V, g)\hat{\otimes} C\ell(V', g'). \tag{4.3}$$

As $\dim C\ell(V \oplus V', g \oplus g') = 2^{\dim(V \oplus V')} = \dim C\ell(V, g)\hat{\otimes} C\ell(V', g')$, it follows that ϕ is surjective.

Now, the canonical injections $\iota : (V, g) \to (V \oplus V', g \oplus g')$, and $\iota' : (V', g') \to (V \oplus V', g \oplus g')$, respectively induce the homomorphisms $f : C\ell(V, g) \to C\ell(V \oplus V', g \oplus g')$, and $f' : C\ell(V', g') \to C\ell(V \oplus V', g \oplus g')$. Since the vectors $\mathbf{v} + 0$, and $0 + \mathbf{v}'$, are orthogonal in $(V \oplus V', g \oplus g')$, therefore $f(\gamma(\mathbf{v}))$ and $f'(\gamma'(\mathbf{v}))$ anti-commute. Consequently, f and f' satisfy the property required in lemma 4.1 to define an algebra homomorphism $\mathring{f} : C\ell(V, g)\hat{\otimes} C\ell(V', g') \to C\ell(V \oplus V', g \oplus g')$. The action of the application $\phi : C\ell(V \oplus V') \to C\ell(V, g)\hat{\otimes} C\ell(V', g')$, which maps every $\Gamma(\mathbf{v} + \mathbf{v}')$ to $\gamma(\mathbf{v}) \otimes 1_{\mathbf{v}'} + 1_{\mathbf{v}} \otimes \gamma'(\mathbf{v}')$ on the generators $\gamma(\mathbf{v}) \otimes 1_{\mathbf{v}'}$, and $1_{\mathbf{v}} \otimes \gamma'(\mathbf{v}')$, shows that ϕ and $\mathring{f}$ are inverse homomorphisms; therefore, ϕ is an isomorphism. ✓

Up to this point, only real Clifford algebras have been examined, namely, Clifford algebras for a quadratic space over the reals $\mathbb{R}$. Let us now consider the relationship between real and complex Clifford algebras. However, before that, let us define the *complexification* of a vector space. Let V be a real vector space of dimension $\dim V = n$. Its complexification $V_{\mathbb{C}}$ is the space of the elements $\mathbf{v} + i\mathbf{u}$, where $\mathbf{v}, \mathbf{u} \in V$, and i denotes the imaginary unit. The space $V_{\mathbb{C}}$ is a vector space with addition and multiplication by the complex scalar $(a + ib)$, defined by

$$(\mathbf{v}_1 + i\mathbf{u}_1) + (\mathbf{v}_2 + i\mathbf{u}_2) = (\mathbf{v}_1 + \mathbf{v}_2) + i(\mathbf{u}_1 + \mathbf{u}_2),$$
$$(a + ib)(\mathbf{v} + i\mathbf{u}) = (a\mathbf{v} - b\mathbf{u}) + i(b\mathbf{v} + a\mathbf{u}).$$

The dimension of $V_{\mathbb{C}}$ is $\dim_{\mathbb{C}} V_{\mathbb{C}} = n$ over $\mathbb{C}$, and $\dim_{\mathbb{R}} V_{\mathbb{C}} = 2n$ over $\mathbb{R}$. We can see that

$$V_{\mathbb{C}} = \mathbb{C} \otimes V.$$

Given a symmetric bilinear form g endowing V, its extension $g_{\mathbb{C}}$ to $V_{\mathbb{C}}$ is defined as

$$g_{\mathbb{C}}(\mathbf{v}_1 + i\mathbf{u}_1, \mathbf{v}_2 + i\mathbf{u}_2) = g(\mathbf{v}_1, \mathbf{v}_2) - g(\mathbf{u}_1, \mathbf{u}_2) + i\big(g(\mathbf{v}_1, \mathbf{u}_2) + g(\mathbf{u}_1, \mathbf{v}_2)\big).$$

The next theorem concerns Clifford algebras over complex vector spaces.

Theorem 4.2 ▶ *Let (V, g) be a quadratic space over $\mathbb{R}$ and let $\mathcal{C}\ell(V, g)$ be its associated real Clifford algebra. Consider the complex Clifford algebra $\mathcal{C}\ell(V_{\mathbb{C}}, g_{\mathbb{C}})$ for the complexified quadratic space $(V_{\mathbb{C}}, g_{\mathbb{C}})$. Then,*

$$\mathcal{C}\ell(V_{\mathbb{C}}, g_{\mathbb{C}}) \simeq \mathcal{C}\ell_{\mathbb{C}}(V, g), \tag{4.4}$$

where $\mathcal{C}\ell_{\mathbb{C}}(V, g) = \mathbb{C} \otimes \mathcal{C}\ell(V, g)$ denotes the complexification of $\mathcal{C}\ell(V, g)$.

Proof: Let us first observe that the complexification of $\mathcal{C}\ell_{\mathbb{C}}(V, g) = \mathbb{C} \otimes \mathcal{C}\ell(V, g)$ is an algebra endowed with a product given by

$$(a \otimes A)(b \otimes B) = ab \otimes AB, \qquad \forall a, b \in \mathbb{C}, \ \forall\, A, B \in \mathcal{C}\ell(V, g).$$

Since $\dim_{\mathbb{R}} \mathcal{C}\ell(V, g) = 2^{\dim V}$, its dimension $\mathcal{C}\ell_{\mathbb{C}}(V, g)$ over $\mathbb{R}$ is given by $\dim_{\mathbb{R}} \mathcal{C}\ell_{\mathbb{C}}(V, g) = 2 \dim_{\mathbb{R}} \mathcal{C}\ell(V, g) = 2 \cdot 2^{\dim V}$. If γ denotes a Clifford mapping $\gamma : V \to \mathcal{C}\ell(V, g)$, we can define a mapping $\Gamma : V_{\mathbb{C}} \to \mathcal{C}\ell_{\mathbb{C}}(V, g)$ as a linear mapping (in $\mathbb{C}$)

$$\Gamma = 1 \otimes \gamma,$$

where 1 denotes the identity of $V_{\mathbb{C}}$. Hence, for $a \otimes \mathbf{v} \in \mathbb{C} \otimes V = V_{\mathbb{C}}$, it follows that

$$\Gamma(a \otimes \mathbf{v}) = a \otimes \mathbf{v},$$

where it is implicit that $\mathbf{v} = \gamma(\mathbf{v})$. Such a mapping Γ is a Clifford mapping. In fact,

$$\big(\Gamma(1 \otimes \mathbf{v})\big)^2 = (1 \otimes \mathbf{v})(1 \otimes \mathbf{v}) = 1 \otimes g(\mathbf{v}, \mathbf{v}).$$

On the other hand, if $\mathcal{C}\ell(V_{\mathbb{C}}, g_{\mathbb{C}})$ is a Clifford algebra for $(V_{\mathbb{C}}, g_{\mathbb{C}})$, it satisfies the universal property and there exists a homomorphism $\phi : \mathcal{C}\ell(V_{\mathbb{C}}, g_{\mathbb{C}}) \to \mathcal{C}\ell_{\mathbb{C}}(V, g)$. The dimension of $\mathcal{C}\ell(V_{\mathbb{C}}, g_{\mathbb{C}})$ is $\dim_{\mathbb{C}} \mathcal{C}\ell(V_{\mathbb{C}}, g_{\mathbb{C}}) = 2^{\dim V}$ or, equivalently, $\dim_{\mathbb{R}} \mathcal{C}\ell(V_{\mathbb{C}}, g_{\mathbb{C}}) = 2 \cdot 2^{\dim V}$. Since $\dim \mathcal{C}\ell(V_{\mathbb{C}}, g_{\mathbb{C}}) = \dim \mathcal{C}\ell_{\mathbb{C}}(V, g)$, therefore ϕ is an isomorphism. ✓

This result shows us that, in order to describe the Clifford algebra structure, it suffices to understand the *real* Clifford algebra structure. The complex Clifford algebra structure follows from the real case when we use this last theorem. In other words, once the real Clifford algebras structure is known, the complex Clifford algebras are immediately obtained via complexification, as shown in eqn (4.4). Let us then introduce some theorems involving real Clifford algebras.

Theorem 4.3 ▶ *Let $\mathcal{C}\ell_{p,q}$ be a Clifford algebra associated with the quadratic space $\mathbb{R}^{p,q}$. Then, the following isomorphisms hold:*

$$\boxed{\begin{aligned} \mathcal{C}\ell_{p+1,q+1} &\simeq \mathcal{C}\ell_{1,1} \otimes \mathcal{C}\ell_{p,q} \\ \mathcal{C}\ell_{q+2,p} &\simeq \mathcal{C}\ell_{2,0} \otimes \mathcal{C}\ell_{p,q} \\ \mathcal{C}\ell_{q,p+2} &\simeq \mathcal{C}\ell_{0,2} \otimes \mathcal{C}\ell_{p,q} \end{aligned}} \tag{4.5}$$

where either $p > 0$ or $q > 0$, and $\otimes$ denotes the usual tensor product and not the alternating tensor product.

Proof: Let U be a two-dimensional space endowed with a symmetric bilinear form g_U such that, with respect to an orthonormal basis $\{\mathbf{f}_1, \mathbf{f}_2\}$, we can write, for $\mathbf{u} = u^1\mathbf{f}_1 + u^2\mathbf{f}_2 \in U$,

$$g_U = \lambda_1(u^1)^2 + \lambda_2(u^2)^2, \qquad \lambda_1 = \pm 1, \lambda_2 = \pm 1,$$

where the values of λ_1 and λ_2 are chosen according to whether $U = \mathbb{R}^{2,0}$; $\mathbb{R}^{1,1}$; or $\mathbb{R}^{0,2}$. For the quadratic space $\mathbb{R}^{p,q}$, the notation used in eqn (3.35) is adopted. Now, the linear mapping $\Gamma : \mathbb{R}^{p,q} \oplus U \to \mathcal{C}\ell(U, g_U) \otimes \mathcal{C}\ell_{p,q}$ can be defined as

$$\Gamma(\mathbf{v} + \mathbf{u}) = \mathbf{f}_1\mathbf{f}_2 \otimes \mathbf{v} + \mathbf{u} \otimes 1, \qquad \forall\, \mathbf{u} \in U, \forall\, \mathbf{v} \in \mathbb{R}^{p,q},$$

where $\mathbf{f}_1 = \rho(\mathbf{f}_1)$; $\mathbf{f}_2 = \rho(\mathbf{f}_2)$; $\mathbf{u} = \rho(\mathbf{u})$; $\mathbf{v} = \gamma(\mathbf{v})$; and where ρ and γ are the Clifford mappings $\rho : U \to \mathcal{C}\ell(U, g_U)$, and $\gamma : \mathbb{R}^{p,q} \to \mathcal{C}\ell_{p,q}$. Let us show that Γ is also a Clifford mapping:

$$\begin{aligned} \big(\Gamma(\mathbf{v} + \mathbf{u})\big)^2 &= (\mathbf{f}_1\mathbf{f}_2 \otimes \mathbf{v} + \mathbf{u} \otimes 1)(\mathbf{f}_1\mathbf{f}_2 \otimes \mathbf{v} + \mathbf{u} \otimes 1) \\ &= (\mathbf{f}_1\mathbf{f}_2)^2 \otimes \mathbf{v}^2 + (\mathbf{u}\mathbf{f}_1\mathbf{f}_2 + \mathbf{f}_1\mathbf{f}_2\mathbf{u}) \otimes \mathbf{v} + \mathbf{u}^2 \otimes 1. \end{aligned}$$

The term $\mathbf{u}\mathbf{f}_1\mathbf{f}_2 + \mathbf{f}_1\mathbf{f}_2\mathbf{u}$ can be computed if we remember that $\{\mathbf{f}_1, \mathbf{f}_2\}$ is orthogonal – that is, that $\mathbf{f}_1\mathbf{f}_2 = \mathbf{f}_1 \wedge \mathbf{f}_2 = -\mathbf{f}_2\mathbf{f}_1$ – and use eqn (3.26). It then yields

$$\mathbf{u}\mathbf{f}_1\mathbf{f}_2 + \mathbf{f}_1\mathbf{f}_2\mathbf{u} = \mathbf{u}(\mathbf{f}_1 \wedge \mathbf{f}_2) + (\mathbf{f}_1 \wedge \mathbf{f}_2)\mathbf{u} = 2\mathbf{u} \wedge \mathbf{f}_1 \wedge \mathbf{f}_2 = 0,$$

since $\mathbf{u} \in U$ is written as $\mathbf{u} = a\mathbf{f}_1 + b\mathbf{f}_2$. Using this result and that

$$(\mathbf{f}_1\mathbf{f}_2)^2 = -(\mathbf{f}_1)^2(\mathbf{f}_2)^2 = -\lambda_1\lambda_2,$$

for $\big(\Gamma(\mathbf{v} + \mathbf{u})\big)^2$, we obtain

$$\begin{aligned}(\Gamma(\mathbf{v}+\mathbf{u}))^2 &= -\lambda_1\lambda_2 \otimes g(\mathbf{v},\mathbf{v}) + g_V(\mathbf{u},\mathbf{u}) \otimes 1 \\ &= \big[\lambda_1(u^1)^2 + \lambda_2(u^2)^2 \\ &\quad - \lambda_1\lambda_2[(v^1)^2 + \cdots + (v^p)^2 - (v^{p+1})^2 - \cdots - (v^n)^2]\big]1 \otimes 1.\end{aligned}$$

The mapping Γ is thus a Clifford mapping $\Gamma : \mathbb{R}^{p,q} \oplus U \to \mathcal{C}\ell(W, g_W)$, where W is a $(n+2)$-dimensional space endowed with a symmetric bilinear functional g_W given by

$$\begin{aligned}g_W(\mathbf{w},\mathbf{w}) &= \lambda_1(u^1)^2 + \lambda_2(u^2)^2 \\ &\quad - \lambda_1\lambda_2\big[(v^1)^2 + \cdots + (v^p)^2 - (v^{p+1})^2 - \cdots - (v^n)^2\big],\end{aligned}$$

where $\mathbf{w} = u^1\mathbf{f}_1 + u^2\mathbf{f}_2 + v^i\mathbf{e}_i$. Hence, three possibilities arise: (i) if $U = \mathbb{R}^{2,0}$, then $W = \mathbb{R}^{q+2,p}$; (ii) if $U = \mathbb{R}^{1,1}$, then $W = \mathbb{R}^{p+1,q+1}$; and (iii) if $U = \mathbb{R}^{0,2}$, then $W = \mathbb{R}^{p,q+2}$. The isomorphisms follow from the Clifford algebra universality. In the last paragraph of section 3.4, we proved that dim $\mathcal{C}\ell(V,g) \geq$ dim $\bigwedge(V) = 2^{\dim V} = 2^n$. However, we also proved in the beginning of chapter 3 that, from the general definition, dim $\mathcal{C}\ell(V,g) \leq 2^n$. Hence, dim $\mathcal{C}\ell(V,g) = 2^n$, so the Clifford algebra is universal. ✓

By combining these isomorphisms, a plethora of other ones can be obtained. For instance:

$$\mathcal{C}\ell_{p,p} \simeq \otimes^p \mathcal{C}\ell_{1,1}\,, \tag{4.6}$$

which follows from repeated applications of the first isomorphism in eqn (4.5). Similarly, we have

$$\mathcal{C}\ell_{p,q} \simeq \mathcal{C}\ell_{p,p} \otimes \mathcal{C}\ell_{0,q-p} \quad (q > p), \quad \mathcal{C}\ell_{p,q} \simeq \mathcal{C}\ell_{q,q} \otimes \mathcal{C}\ell_{p-q,0} \quad (p > q). \tag{4.7}$$

Some particular cases of great interest are given by the expressions

$$\mathcal{C}\ell_{0,2} \otimes \mathcal{C}\ell_{2,0} \simeq \mathcal{C}\ell_{0,4}, \qquad \mathcal{C}\ell_{0,4} \otimes \mathcal{C}\ell_{4,0} \simeq \mathcal{C}\ell_{0,8}, \tag{4.8}$$

and

$$\mathcal{C}\ell_{2,2} \simeq \mathcal{C}\ell_{0,2} \otimes \mathcal{C}\ell_{0,2} \simeq \mathcal{C}\ell_{1,1} \otimes \mathcal{C}\ell_{1,1}\,. \tag{4.9}$$

From these isomorphisms, it follows in addition that

$$\mathcal{C}\ell_{0,4} \otimes \mathcal{C}\ell_{p,q} = \mathcal{C}\ell_{p,q+4}, \qquad \mathcal{C}\ell_{0,8} \otimes \mathcal{C}\ell_{p,q} = \mathcal{C}\ell_{p,q+8}\,. \tag{4.10}$$

These isomorphisms are due to Cartan (1908). An isomorphism which does not follow from this analysis but which has outstanding importance is

$$\mathcal{C}\ell_{2,0} \simeq \mathcal{C}\ell_{1,1}, \tag{4.11}$$

which can be explicitly constructed. In fact, the elements of $\mathcal{C}\ell_{2,0}$ are written as

$$a_0 + a_1\mathbf{e}_1 + a_2\mathbf{e}_2 + a_{12}\mathbf{e}_1\mathbf{e}_2 \in \mathcal{C}\ell_{2,0},$$

where $(\mathbf{e}_1)^2 = 1$, and $(\mathbf{e}_2)^2 = 1$. On the other hand, the elements of $\mathcal{C}\ell_{1,1}$ are expressed as

$$b_0 + b_1\mathbf{f}_1 + b_2\mathbf{f}_2 + b_{12}\mathbf{f}_1\mathbf{f}_2 \in \mathcal{C}\ell_{1,1},$$

where $(\mathbf{f}_1)^2 = 1$ and $(\mathbf{f}_2)^2 = -1$. It is straightforward to realise that the linear mapping $\phi : \mathcal{C}\ell_{2,0} \to \mathcal{C}\ell_{1,1}$ defined by

$$\phi(1) = 1, \quad \phi(\mathbf{e}_1) = \mathbf{f}_1, \quad \phi(\mathbf{e}_2) = \mathbf{f}_1\mathbf{f}_2, \quad \phi(\mathbf{e}_1\mathbf{e}_2) = \mathbf{f}_2,$$

is an isomorphism. Combining this last isomorphism with the others, we can see that

$$\mathcal{C}\ell_{p+1,q} \simeq \mathcal{C}\ell_{q+1,p}\,. \tag{4.12}$$

Other combinations arising from these isomorphisms can be made, although it is not important to do so now. However, these results indicate that, from the Clifford algebras

$$\boxed{\mathcal{C}\ell_{1,0}, \quad \mathcal{C}\ell_{0,1}, \quad \mathcal{C}\ell_{0,2}, \quad \mathcal{C}\ell_{1,1} \simeq \mathcal{C}\ell_{2,0}}\,, \tag{4.13}$$

all other Clifford algebras in arbitrary finite dimensions can be constructed by using these isomorphisms, thus providing a method for classifying Clifford algebras.

Finally, the following result is of paramount importance:

Theorem 4.4 ► *Let $\mathcal{C}\ell_{p,q}$ be the Clifford algebra associated with the quadratic space $\mathbb{R}^{p,q}$ and let $\mathcal{C}\ell^+_{p,q}$ be its even subalgebra (eqn (3.46)). Then,*

$$\boxed{\mathcal{C}\ell^+_{p,q} \simeq \mathcal{C}\ell_{q,p-1} \simeq \mathcal{C}\ell_{p,q-1} \simeq \mathcal{C}\ell^+_{q,p}}\,. \tag{4.14}$$

Proof: Let $\{\mathbf{e}_i, \mathbf{f}_k\}$ $(i = 1, \ldots, p,\ k = 1, \ldots, q)$ be an orthonormal basis of V such that $\mathcal{C}\ell_{p,q}$ is generated by 1 and $\{\mathbf{e}_i, \mathbf{f}_k\}$ (a Clifford mapping again implied), where $(\mathbf{e}_i)^2 = 1$; $(\mathbf{f}_k)^2 = -1$; $\mathbf{e}_i\mathbf{e}_j + \mathbf{e}_j\mathbf{e}_i = 0$ $(i \neq j)$; $\mathbf{f}_k\mathbf{f}_l + \mathbf{f}_l\mathbf{f}_k = 0$ $(k \neq l)$; and $\mathbf{e}_i\mathbf{f}_k + \mathbf{f}_k\mathbf{e}_i = 0$ $(i, j = 1, \ldots, p,\ k, l = 1, \ldots, q)$. The space $\bigwedge_2(\mathbb{R}^{p,q})$ consists of the elements $\{\mathbf{e}_i\mathbf{e}_j (i \neq j), \mathbf{f}_k\mathbf{f}_l (k \neq l), \mathbf{e}_i\mathbf{f}_k\}$. Not all of these quantities generate the even subalgebra $\mathcal{C}\ell^+_{p,q}$. However, there is redundancy: for example, all the 2-vectors $\{\mathbf{f}_k\mathbf{f}_l\}$ $(k \neq l)$ can be written in terms of the 2-vectors of type $\{\mathbf{e}_i\mathbf{f}_k\}$, since $(\mathbf{e}_i\mathbf{f}_k)(\mathbf{e}_i\mathbf{f}_l) = -(\mathbf{e}_i)^2\mathbf{f}_k\mathbf{f}_l = -\mathbf{f}_k\mathbf{f}_l$ $(k \neq l)$. Choosing an arbitrary vector, for instance $\mathbf{e}_1$, it is clear that the set $\{\mathbf{e}_1\mathbf{e}_m, \mathbf{e}_1\mathbf{f}_k\}$ $(m = 2, \ldots, p,\ k = 1, \ldots, q)$ generates the space $\bigwedge_2(\mathbb{R}^{p,q})$ and therefore generates the even subalgebra $\mathcal{C}\ell^+_{p,q}$. Let us write such generators of $\mathcal{C}\ell^+_{p,q}$ as $\xi_a = \mathbf{e}_1\mathbf{e}_{a+1}$ for $a = 1, \ldots, p-1$ and as $\zeta_b = \mathbf{e}_1\mathbf{f}_b$ for $b = 1, \ldots, q$. It is then straightforward to check that $(\xi_a)^2 = -(\mathbf{e}_1)^2(\mathbf{e}_{a+1})^2 = -1$; $(\zeta_b)^2 = -(\mathbf{e}_1)^2(\mathbf{f}_b)^2 = 1$; $\xi_a\xi_c + \xi_c\xi_a = 0$ $(a \neq c)$; $\zeta_b\zeta_d + \zeta_d\zeta_b = 0$ $(b \neq d)$; and $\xi_a\zeta_b + \zeta_b\xi_a = 0$. The quantities $\{\zeta_b, \xi_a\}$ $(b = 1, \ldots, q,\ a = 1, \ldots, p-1)$ are then the generators of a Clifford algebra associated with a quadratic space $\mathbb{R}^{q,p-1}$, namely, $\mathcal{C}\ell^+_{p,q} \simeq \mathcal{C}\ell_{q,p-1}$. The other isomorphisms naturally follow from the isomorphism in eqn (4.12). ✓

Previously, the grade involution, the reversion, and the conjugation were defined in the context of the Clifford algebras. Their notations are now redefined: the reversion by α_1, and the conjugation by α_{-1}, so that we can unify the two operations in the notation α_ϵ $(\epsilon = \pm 1)$. Taking this notation into account, we now assert a generalisation of the periodicity theorem (Maks, 1989):

Theorem 4.5 ▶

$$\boxed{(\mathcal{C}\ell_{p+1,q+1}, \alpha_\epsilon) \simeq (\mathcal{C}\ell_{p,q}, \alpha_{-\epsilon}) \otimes (\mathcal{C}\ell_{1,1}, \alpha_\epsilon)}. \tag{4.15}$$

Proof: The sets $\{\mathbf{e}_i\}$ and $\{\mathbf{f}_j\}$ are generators for the algebras $\mathcal{C}\ell_{p,q}$ and $\mathcal{C}\ell_{1,1}$, respectively. Take $\{\mathbf{e}_i \otimes \mathbf{f}_1\mathbf{f}_2, 1 \otimes \mathbf{f}_j\}$ as a set of generators for $\mathcal{C}\ell_{p+1,q+1} \simeq \mathcal{C}\ell_{p,q} \otimes \mathcal{C}\ell_{1,1}$. Consider now a mapping $(\alpha_{-\epsilon} \otimes \alpha_\epsilon)(\mathbf{e}_i \otimes \mathbf{f}_1\mathbf{f}_2) = \alpha_{-\epsilon}(\mathbf{e}_i) \otimes \alpha_\epsilon(\mathbf{f}_1\mathbf{f}_2) = \epsilon(\mathbf{e}_i \otimes \mathbf{f}_1\mathbf{f}_2)$. In addition, we have $(\alpha_{-\epsilon} \otimes \alpha_\epsilon)(1 \otimes \mathbf{f}_j) = \alpha_{-\epsilon}(1) \otimes \alpha_\epsilon(\mathbf{f}_j) = \epsilon(1 \otimes \mathbf{f}_j)$. Hence, the generators are multiplied by ϵ; thus the theorem is proved. ✓

This theorem is very useful for describing of the conformal transformations. For example, in the case of Minkowski spacetime, as it uses only the Dirac algebra $\mathbb{C}\otimes\mathcal{C}\ell_{1,3}$, conformal transformations can be represented by 2×2 matrices with entries in $\mathcal{C}\ell_{3,0}$. Thus, operations which act on the Clifford algebra elements of high dimensions can be led to operations on algebras of low dimensions.

4.2 The Classification of Clifford Algebras

Representations

Definition 4.1 ▶ *Let $\mathcal{A}$ be a real algebra and let V be a vector space over $\mathbb{K} = \mathbb{R}, \mathbb{C}, \mathbb{H}$. A linear mapping $\rho : \mathcal{A} \to \mathrm{End}_\mathbb{K}(V)$ satisfying $\rho(1_\mathcal{A}) = 1_V$, and $\rho(ab) = \rho(a)\rho(b)$, $\forall a, b \in \mathcal{A}$, is called a $\mathbb{K}$-representation of $\mathcal{A}$. The vector space V is called the representation space (or carrier space) of $\mathcal{A}$.*

The two representations $\rho_1 : \mathcal{A} \to \mathrm{End}_\mathbb{K}(V_1)$, and $\rho_2 : \mathcal{A} \to \mathrm{End}_\mathbb{K}(V_2)$, are *equivalent* if there exists a $\mathbb{K}$-isomorphism $\phi : V_1 \to V_2$ satisfying $\rho_2(a) = \phi \circ \rho_1(a) \circ \phi^{-1}$, $\forall a \in \mathcal{A}$.

A representation is said to be *faithful* if $\ker \rho = \{0\}$. A representation is *irreducible* or *simple* if the only invariant subspaces of $\rho(a)$, $\forall a \in \mathcal{A}$, are V and $\{0\}$. It is said to be *reducible* or *semisimple* if $V = V_1 \oplus V_2$, where V_1 and V_2 are invariant subspaces under the action of $\rho(a)$, $\forall a \in \mathcal{A}$.

Example 4.1 Let us consider the algebra $\mathbb{C}$ of complex numbers. As an algebra over $\mathbb{C}$, it has two representations: $\rho(a + ib) = a + ib$, and $\bar{\rho}(a + ib) = a - ib$. These two $\mathbb{C}$-representations *are not* equivalent. Indeed, there does not exist any linear mapping $\phi_z : \mathbb{C} \to \mathbb{C}; (a + ib) \mapsto \phi_z(a + ib) = z(a + ib)$, where $z = x + iy$, such that $\bar{\rho}(a + ib) = z\rho(a + ib)z^{-1}$. On the other hand, every $\mathbb{C}$-representation (and also every $\mathbb{H}$-representation) is an $\mathbb{R}$-representation. We can define two real representations $\sigma : \mathbb{C} \to \mathcal{M}(2, \mathbb{R})$ and $\bar{\sigma} : \mathbb{C} \to \mathcal{M}(2, \mathbb{R})$ as

$$\sigma(a + ib) = \begin{pmatrix} a & b \\ -b & a \end{pmatrix}, \qquad \bar{\sigma}(a + ib) = \begin{pmatrix} a & -b \\ b & a \end{pmatrix},$$

respectively. These two representations *are* equivalent, namely, there exists an isomorphism $\phi : \mathbb{R}^2 \to \mathbb{R}^2$ such that $\bar{\sigma}(a + ib) = \phi\sigma(a + ib)\phi^{-1}$. For instance,

$$\phi = \frac{1}{\sqrt{2}} \begin{pmatrix} 1 & 1 \\ 1 & -1 \end{pmatrix} = \phi^{-1}.$$

Those $\mathbb{R}$-representations are irreducible. An example of a reducible $\mathbb{R}$-representation is

$$\xi(a + ib) = \begin{pmatrix} a & b & 0 & 0 \\ -b & a & 0 & 0 \\ 0 & 0 & a & -b \\ 0 & 0 & b & a \end{pmatrix}.$$

The Clifford Algebra $\mathcal{C}\ell_{0,1}$

A Clifford algebra associated with the quadratic space $\mathbb{R}^{0,1}$ was introduced in example 3.2. If $\mathbf{e}$ is an unit vector such that $g(\mathbf{e},\mathbf{e}) = -1$, an arbitrary element in $\mathcal{C}\ell_{0,1}$ reads

$$\psi = a + b\mathbf{e} \in \mathcal{C}\ell_{0,1}, \tag{4.16}$$

where $\mathbf{e}^2 = -1$. This algebra is isomorphic to the algebra of complex numbers $\mathbb{C}$, namely, the set of pairs (a,b), $a,b \in \mathbb{R}$, with multiplication given by

$$(a,b)(c,d) = (ac - bd, ad + bc).$$

The isomorphism is provided by $\rho : \mathcal{C}\ell_{0,1} \to \mathbb{C}$ such that $\rho(1) = (1,0)$ and $\rho(\mathbf{e}) = (0,1) = i$. Therefore,

$$\mathcal{C}\ell_{0,1} \simeq \mathbb{C}. \tag{4.17}$$

The Clifford Algebra $\mathcal{C}\ell_{1,0}$

Let us consider the quadratic space $\mathbb{R}^{1,0}$. Taking the unit vector $\mathbf{e}$, $g(\mathbf{e},\mathbf{e}) = 1$, an arbitrary element of $\mathcal{C}\ell_{1,0}$ can be written as

$$\psi = a + b\mathbf{e} \in \mathcal{C}\ell_{1,0}\,, \tag{4.18}$$

where now $\mathbf{e}^2 = 1$. The difference between this case and the previous one is that here $\mathbf{e}^2 = 1$, before, whereas, $\mathbf{e}^2 = -1$. In order to make this difference explicit, let us consider the pair of numbers (a,b), $a,b \in \mathbb{R}$, with multiplication defined by

$$(a,b)(c,d) = (ac + bd, ad + bc). \tag{4.19}$$

This set might at first sight seem like $\mathbb{C}$, but the difference of sign at the term bd in the right-hand side of the equation has drastic consequences. In particular, this set is *not* a field, although it is a ring. It is also not a division ring, since $(1,1)(1,-1) = (0,0)$. Nonetheless, this set is relevant, as it has some interesting applications in physics – for example, in the theory of relativity (Fjelstad, 1986; Baylis, 1998; da Rocha and Vaz, 2006). Let us denote this set by $\mathbb{D}$, whose elements have distinct denominations: double numbers, perplex numbers, duplex numbers, or Lorentz numbers.

The multiplication defined in eqn (4.19) is appropriate for comparing $\mathbb{D}$ to $\mathbb{C}$, but is not suitable for the classification of Clifford algebras. Instead, let us consider the pairs (a,b), $a,b \in \mathbb{R}$, with multiplication $*$ defined by

$$(a,b) * (c,d) = (ac, bd). \tag{4.20}$$

This algebra consists of the direct sum $\mathbb{R} \oplus \mathbb{R}$ and is isomorphic to the algebra of 2×2 diagonal matrices. This isomorphism is given by

$$\phi(a,b) = \begin{pmatrix} a & 0 \\ 0 & b \end{pmatrix}. \tag{4.21}$$

The algebras $\mathbb{D}$ and $\mathbb{R} \oplus \mathbb{R}$ are isomorphic. The isomorphism $\varphi : \mathbb{D} \to \mathbb{R} \oplus \mathbb{R}$ reads

$$\varphi(a,b) = (a + b, a - b). \tag{4.22}$$

It is trivial to verify that indeed $\varphi((a,b)(c,d)) = \varphi(a,b) * \varphi(c,d)$.

Now going back to the Clifford algebra $\mathcal{C}\ell_{1,0}$, this algebra is clearly isomorphic to $\mathbb{D}$: the isomorphism is $\rho(1) = (1,0)$, and $\rho(\mathbf{e}) = (0,1)$. Hence, $\mathcal{C}\ell_{1,0}$ is isomorphic to $\mathbb{R} \oplus \mathbb{R}$:

$$\mathcal{C}\ell_{1,0} \simeq \mathbb{R} \oplus \mathbb{R}. \tag{4.23}$$

The Clifford algebra $\mathcal{C}\ell_{0,2}$

Let us consider the quadratic space $\mathbb{R}^{0,2}$ and an orthonormal basis $\{\mathbf{e}_1, \mathbf{e}_2\}$,

$$g(\mathbf{e}_1, \mathbf{e}_1) = g(\mathbf{e}_2, \mathbf{e}_2) = -1, \qquad g(\mathbf{e}_1, \mathbf{e}_2) = g(\mathbf{e}_2, \mathbf{e}_1) = 0.$$

An arbitrary element of $\mathcal{C}\ell_{0,2}$ reads

$$\psi = a + b\mathbf{e}_1 + c\mathbf{e}_2 + d\mathbf{e}_1\mathbf{e}_2 \in \mathcal{C}\ell_{0,2}\,, \tag{4.24}$$

where $a, b, c, d \in \mathbb{R}$, and

$$(\mathbf{e}_1)^2 = (\mathbf{e}_2)^2 = -1, \qquad \mathbf{e}_1\mathbf{e}_2 + \mathbf{e}_2\mathbf{e}_1 = 0. \tag{4.25}$$

These relations imply that $(\mathbf{e}_1\mathbf{e}_2)^2 = -1$.

It is clear to see that $\mathcal{C}\ell_{0,2}$ is isomorphic to the quaternion algebra $\mathbb{H}$. This isomorphism is given by

$$\rho(1) = 1, \quad \rho(\mathbf{e}_1) = i, \quad \rho(\mathbf{e}_2) = j, \quad \rho(\mathbf{e}_1\mathbf{e}_2) = k, \tag{4.26}$$

where i, j, and k are the quaternion units $i^2 = j^2 = k^2 = -1$; $ij = -ji = k$; $jk = -kj = i$; and $ki = -ik = j$. Hence, it explicitly follows that

$$\mathcal{C}\ell_{0,2} \simeq \mathbb{H}. \tag{4.27}$$

The Clifford algebra $\mathcal{C}\ell_{2,0} \simeq \mathcal{C}\ell_{1,1}$

The Clifford algebras associated with the quadratic spaces $\mathbb{R}^{2,0}$ and $\mathbb{R}^{1,1}$ have been shown to be isomorphic: $\mathcal{C}\ell_{2,0} \simeq \mathcal{C}\ell_{1,1}$. Hence, it suffices to consider just one of these spaces, for instance, $\mathbb{R}^{2,0}$. Given an orthonormal basis $\{\mathbf{e}_1, \mathbf{e}_2\}$,

$$g(\mathbf{e}_1, \mathbf{e}_1) = g(\mathbf{e}_2, \mathbf{e}_2) = 1, \qquad g(\mathbf{e}_1, \mathbf{e}_2) = g(\mathbf{e}_2, \mathbf{e}_1) = 0.$$

An arbitrary element of $\mathcal{C}\ell_{2,0}$ can be written as

$$\psi = a + b\mathbf{e}_1 + c\mathbf{e}_2 + d\mathbf{e}_1\mathbf{e}_2 \in \mathcal{C}\ell_{0,2}\,, \tag{4.28}$$

where $a, b, c, d \in \mathbb{R}$, and

$$(\mathbf{e}_1)^2 = (\mathbf{e}_2)^2 = 1, \qquad \mathbf{e}_1\mathbf{e}_2 + \mathbf{e}_2\mathbf{e}_1 = 0. \tag{4.29}$$

These relations further imply that

$$(\mathbf{e}_1\mathbf{e}_2)^2 = -1. \tag{4.30}$$

Let $\mathcal{M}(2,\mathbb{R})$ be the algebra of the real 2×2 matrices. The set

$$\left\{\begin{pmatrix}1&0\\0&1\end{pmatrix},\begin{pmatrix}1&0\\0&-1\end{pmatrix},\begin{pmatrix}0&1\\1&0\end{pmatrix},\begin{pmatrix}0&1\\-1&0\end{pmatrix}\right\}$$

generates $\mathcal{M}(2,\mathbb{R})$ and, furthermore,

$$\begin{pmatrix}1&0\\0&-1\end{pmatrix}^2=\begin{pmatrix}1&0\\0&1\end{pmatrix}=\begin{pmatrix}0&1\\1&0\end{pmatrix}^2,\quad \begin{pmatrix}0&1\\-1&0\end{pmatrix}^2=\begin{pmatrix}-1&0\\0&-1\end{pmatrix}. \tag{4.31}$$

Comparing eqns (4.29) and (4.30) to (4.31), ρ can be defined as the linear mapping

$$\begin{aligned}\rho(1)&=\begin{pmatrix}1&0\\0&1\end{pmatrix}, &\quad \rho(\mathbf{e}_2)&=\begin{pmatrix}0&1\\1&0\end{pmatrix},\\ \rho(\mathbf{e}_1)&=\begin{pmatrix}1&0\\0&-1\end{pmatrix}, &\quad \rho(\mathbf{e}_1\mathbf{e}_2)&=\begin{pmatrix}0&1\\-1&0\end{pmatrix}.\end{aligned} \tag{4.32}$$

As ρ is an isomorphism,

$$\mathcal{C}\ell_{2,0}\simeq\mathcal{C}\ell_{1,1}\simeq\mathcal{M}(2,\mathbb{R}). \tag{4.33}$$

Classifying Arbitrary Clifford Algebras

Once the isomorphisms (i) $\mathcal{C}\ell_{0,1}\simeq\mathbb{C}$, (ii) $\mathcal{C}\ell_{1,0}\simeq\mathbb{R}\oplus\mathbb{R}$, (iii) $\mathcal{C}\ell_{0,2}\simeq\mathbb{H}$, and (iv) $\mathcal{C}\ell_{2,0}\simeq\mathcal{C}\ell_{1,1}\simeq\mathcal{M}(2,\mathbb{R})$ have been established, we can, by using the isomorphisms in the previous section, proceed to the classification of arbitrary Clifford algebras. For instance, by using eqn (4.6) and

$$\mathcal{M}(m,\mathbb{R})\otimes\mathcal{M}(n,\mathbb{R})\simeq\mathcal{M}(mn,\mathbb{R}), \tag{4.34}$$

we obtain

$$\mathcal{C}\ell_{p,p}\simeq\mathcal{M}(2^p,\mathbb{R}). \tag{4.35}$$

This result and eqn (4.9) allow us to conclude that

$$\mathbb{H}\otimes\mathbb{H}\simeq\mathcal{M}(4,\mathbb{R}). \tag{4.36}$$

In addition, eqn (4.8) can be used to obtain

$$\mathcal{C}\ell_{0,4}\simeq\mathbb{H}\otimes\mathcal{M}(2,\mathbb{R})\simeq\mathcal{M}(2,\mathbb{H})\simeq\mathcal{C}\ell_{4,0}, \tag{4.37}$$

and

$$\begin{aligned}\mathcal{C}\ell_{0,8}&\simeq\mathcal{M}(2,\mathbb{H})\otimes\mathcal{M}(2,\mathbb{H})\simeq\mathcal{M}(2,\mathbb{R})\otimes\mathbb{H}\otimes\mathbb{H}\otimes\mathcal{M}(2,\mathbb{R})\\ &\simeq\mathcal{M}(2,\mathbb{R})\otimes\mathcal{M}(4,\mathbb{R})\otimes\mathcal{M}(2,\mathbb{R})\simeq\mathcal{M}(16,\mathbb{R}).\end{aligned} \tag{4.38}$$

This result, together with eqn (4.10), implies that

$$\mathcal{C}\ell_{p,q+8}\simeq\mathcal{C}\ell_{p,q}\otimes\mathcal{M}(16,\mathbb{R}). \tag{4.39}$$

Equation (4.39) has an important consequence: we only need to explicitly obtain the classification of the Clifford algebras up to $\dim V=p+q=8$ since, for dimensions higher than that, the isomorphism $\mathcal{C}\ell_{p,q+8} \simeq \mathcal{C}\ell_{p,q} \otimes \mathcal{M}(16,\mathbb{R})$ (called *the periodicity theorem*) can be used to immediately obtain any other isomorphism. In this way, by using all the previous results in this section, we obtain

$$
\begin{aligned}
&\mathcal{C}\ell_{0,0} \simeq \mathbb{R},\\[1ex]
&\mathcal{C}\ell_{0,1} \simeq \mathbb{C},\\
&\mathcal{C}\ell_{1,0} \simeq \mathbb{R} \oplus \mathbb{R},\\[1ex]
&\mathcal{C}\ell_{0,2} \simeq \mathbb{H},\\
&\mathcal{C}\ell_{2,0} \simeq \mathcal{M}(2,\mathbb{R}),\\
&\mathcal{C}\ell_{1,1} \simeq \mathcal{C}\ell_{2,0} \simeq \mathcal{M}(2,\mathbb{R}),\\[1ex]
&\mathcal{C}\ell_{0,3} \simeq \mathcal{C}\ell_{0,2} \otimes \mathcal{C}\ell_{1,0} \simeq \mathbb{H} \oplus \mathbb{H},\\
&\mathcal{C}\ell_{3,0} \simeq \mathcal{C}\ell_{2,0} \otimes \mathcal{C}\ell_{0,1} \simeq \mathcal{M}(2,\mathbb{C}),\\
&\mathcal{C}\ell_{1,2} \simeq \mathcal{C}\ell_{1,1} \otimes \mathcal{C}\ell_{0,1} \simeq \mathcal{M}(2,\mathbb{R}) \otimes \mathbb{C} \simeq \mathcal{M}(2,\mathbb{C}),\\
&\mathcal{C}\ell_{2,1} \simeq \mathcal{C}\ell_{1,1} \otimes \mathcal{C}\ell_{1,0} \simeq \mathcal{M}(2,\mathbb{R}) \otimes \mathbb{R} \oplus \mathbb{R} \simeq \mathcal{M}(2,\mathbb{R} \oplus \mathbb{R}),\\[1ex]
&\mathcal{C}\ell_{0,4} \simeq \mathcal{C}\ell_{0,2} \otimes \mathcal{C}\ell_{2,0} \simeq \mathbb{H} \otimes \mathcal{M}(2,\mathbb{R}) \simeq \mathcal{M}(2,\mathbb{H}),\\
&\mathcal{C}\ell_{4,0} \simeq \mathcal{C}\ell_{2,0} \otimes \mathcal{C}\ell_{0,2} \simeq \mathcal{M}(2,\mathbb{R}) \otimes \mathbb{H} \simeq \mathcal{M}(2,\mathbb{H}),\\
&\mathcal{C}\ell_{1,3} \simeq \mathcal{C}\ell_{1,1} \otimes \mathcal{C}\ell_{0,2} \simeq \mathcal{M}(2,\mathbb{R}) \otimes \mathbb{H} \simeq \mathcal{M}(2,\mathbb{H}),\\
&\mathcal{C}\ell_{3,1} \simeq \mathcal{C}\ell_{1,1} \otimes \mathcal{C}\ell_{2,0} \simeq \mathcal{M}(2,\mathbb{R}) \otimes \mathcal{M}(2,\mathbb{R}) \simeq \mathcal{M}(4,\mathbb{R}),\\
&\mathcal{C}\ell_{2,2} \simeq \mathcal{C}\ell_{1,1} \otimes \mathcal{C}\ell_{1,1} \simeq \mathcal{M}(2,\mathbb{R}) \otimes \mathcal{M}(2,\mathbb{R}) \simeq \mathcal{M}(4,\mathbb{R}),\\[1ex]
&\mathcal{C}\ell_{5,0} \simeq \mathcal{C}\ell_{2,0} \otimes \mathcal{C}\ell_{0,3} \simeq \mathcal{M}(2,\mathbb{R}) \otimes \mathbb{H} \oplus \mathbb{H} \simeq \mathcal{M}(2,\mathbb{H} \oplus \mathbb{H}),\\
&\mathcal{C}\ell_{0,5} \simeq \mathcal{C}\ell_{0,2} \otimes \mathcal{C}\ell_{3,0} \simeq \mathbb{H} \otimes \mathcal{M}(2,\mathbb{C}) \simeq \mathbb{H} \otimes \mathbb{C} \otimes \mathcal{M}(2,\mathbb{R})\\
&\qquad \simeq \mathcal{M}(2,\mathbb{C}) \otimes \mathcal{M}(2,\mathbb{R}) \simeq \mathbb{C} \otimes \mathcal{M}(2,\mathbb{R}) \otimes \mathcal{M}(2,\mathbb{R})\\
&\qquad \simeq \mathbb{C} \otimes \mathcal{M}(4,\mathbb{R}) \simeq \mathcal{M}(4,\mathbb{C}),\\
&\mathcal{C}\ell_{4,1} \simeq \mathcal{C}\ell_{1,1} \otimes \mathcal{C}\ell_{3,0} \simeq \mathcal{M}(2,\mathbb{R}) \otimes \mathcal{M}(2,\mathbb{C}) \simeq \mathcal{M}(4,\mathbb{C}),\\
&\mathcal{C}\ell_{1,4} \simeq \mathcal{C}\ell_{1,1} \otimes \mathcal{C}\ell_{0,3} \simeq \mathcal{M}(2,\mathbb{R}) \otimes (\mathbb{H} \oplus \mathbb{H}) \simeq \mathcal{M}(2,\mathbb{H} \oplus \mathbb{H}),\\
&\mathcal{C}\ell_{3,2} \simeq \mathcal{C}\ell_{1,1} \otimes \mathcal{C}\ell_{1,1} \otimes \mathcal{C}\ell_{1,0} \simeq \mathcal{M}(2,\mathbb{R}) \otimes \mathcal{M}(2,\mathbb{R}) \otimes (\mathbb{R} \oplus \mathbb{R})\\
&\qquad \simeq \mathcal{M}(4,\mathbb{R} \oplus \mathbb{R}),\\
&\mathcal{C}\ell_{2,3} \simeq \mathcal{C}\ell_{1,1} \otimes \mathcal{C}\ell_{1,1} \otimes \mathcal{C}\ell_{0,1} \simeq \mathcal{M}(2,\mathbb{R}) \otimes \mathcal{M}(2,\mathbb{R}) \otimes \mathbb{C} \simeq \mathcal{M}(4,\mathbb{C}),\\[1ex]
&\mathcal{C}\ell_{6,0} \simeq \mathcal{C}\ell_{2,0} \otimes \mathcal{C}\ell_{0,4} \simeq \mathcal{C}\ell_{2,0} \otimes \mathcal{C}\ell_{0,2} \otimes \mathcal{C}\ell_{2,0}\\
&\qquad \simeq \mathcal{M}(2,\mathbb{R}) \otimes \mathbb{H} \otimes \mathcal{M}(2,\mathbb{R}) \simeq \mathcal{M}(4,\mathbb{H}),\\
&\mathcal{C}\ell_{0,6} \simeq \mathcal{C}\ell_{0,2} \otimes \mathcal{C}\ell_{4,0} \simeq \mathcal{C}\ell_{0,2} \otimes \mathcal{C}\ell_{2,0} \otimes \mathcal{C}\ell_{0,2} \simeq \mathbb{H} \otimes \mathcal{M}(2,\mathbb{R}) \otimes \mathbb{H}\\
&\qquad \simeq \mathcal{M}(2,\mathbb{H}) \otimes \mathbb{H} \simeq \mathcal{M}(2,\mathbb{R}) \otimes \mathbb{H} \otimes \mathbb{H} \simeq \mathcal{M}(2,\mathbb{R}) \otimes \mathcal{M}(4,\mathbb{R})\\
&\qquad \simeq \mathcal{M}(8,\mathbb{R}),\\
&\mathcal{C}\ell_{5,1} \simeq \mathcal{C}\ell_{1,1} \otimes \mathcal{C}\ell_{4,0} \simeq \mathcal{M}(2,\mathbb{R}) \otimes \mathcal{M}(2,\mathbb{H}) \simeq \mathcal{M}(4,\mathbb{H}),\\
&\mathcal{C}\ell_{1,5} \simeq \mathcal{C}\ell_{1,1} \otimes \mathcal{C}\ell_{0,4} \simeq \mathcal{M}(2,\mathbb{R}) \otimes \mathcal{M}(2,\mathbb{H}) \simeq \mathcal{M}(4,\mathbb{H}),
\end{aligned}
$$

$$
\begin{aligned}
\mathcal{C}\ell_{4,2} &\simeq \mathcal{C}\ell_{1,1} \otimes \mathcal{C}\ell_{1,1} \otimes \mathcal{C}\ell_{2,0} \simeq \mathcal{M}(2,\mathbb{R}) \otimes \mathcal{M}(2,\mathbb{R}) \otimes \mathcal{M}(2,\mathbb{R}) \simeq \mathcal{M}(8,\mathbb{R}),\\
\mathcal{C}\ell_{2,4} &\simeq \mathcal{C}\ell_{1,1} \otimes \mathcal{C}\ell_{1,1} \otimes \mathcal{C}\ell_{0,2} \simeq \mathcal{M}(2,\mathbb{R}) \otimes \mathcal{M}(2,\mathbb{R}) \otimes \mathbb{H} \simeq \mathcal{M}(4,\mathbb{H}),\\
\mathcal{C}\ell_{3,3} &\simeq \mathcal{C}\ell_{1,1} \otimes \mathcal{C}\ell_{1,1} \otimes \mathcal{C}\ell_{1,1} \simeq \mathcal{M}(2,\mathbb{R}) \otimes \mathcal{M}(2,\mathbb{R}) \otimes \mathcal{M}(2,\mathbb{R}) \simeq \mathcal{M}(8,\mathbb{R}),
\end{aligned}
$$

$$
\begin{aligned}
\mathcal{C}\ell_{7,0} &\simeq \mathcal{C}\ell_{2,0} \otimes \mathcal{C}\ell_{0,5} \simeq \mathcal{M}(2,\mathbb{R}) \otimes \mathcal{M}(4,\mathbb{C}) \simeq \mathcal{M}(8,\mathbb{C}),\\
\mathcal{C}\ell_{0,7} &\simeq \mathcal{C}\ell_{0,2} \otimes \mathcal{C}\ell_{5,0} \simeq \mathbb{H} \otimes \mathcal{M}(2,\mathbb{H} \oplus \mathbb{H})\\
&\simeq \begin{pmatrix} \mathcal{M}(4,\mathbb{R}) & \mathcal{M}(4,\mathbb{R}) \\ \mathcal{M}(4,\mathbb{R}) & \mathcal{M}(4,\mathbb{R}) \end{pmatrix} \oplus \begin{pmatrix} \mathcal{M}(4,\mathbb{R}) & \mathcal{M}(4,\mathbb{R}) \\ \mathcal{M}(4,\mathbb{R}) & \mathcal{M}(4,\mathbb{R}) \end{pmatrix}\\
&\simeq \mathcal{M}(8,\mathbb{R}) \oplus \mathcal{M}(8,\mathbb{R}) \simeq \mathcal{M}(8,\mathbb{R} \oplus \mathbb{R}),\\
\mathcal{C}\ell_{6,1} &\simeq \mathcal{C}\ell_{1,1} \otimes \mathcal{C}\ell_{5,0} \simeq \mathcal{M}(2,\mathbb{R}) \otimes \mathcal{M}(2,\mathbb{H} \oplus \mathbb{H}) \simeq \mathcal{M}(4,\mathbb{H} \oplus \mathbb{H}),\\
\mathcal{C}\ell_{1,6} &\simeq \mathcal{C}\ell_{1,1} \otimes \mathcal{C}\ell_{0,5} \simeq \mathcal{M}(2,\mathbb{R}) \otimes \mathcal{M}(4,\mathbb{C}) \simeq \mathcal{M}(8,\mathbb{C}),\\
\mathcal{C}\ell_{5,2} &\simeq \mathcal{C}\ell_{1,1} \otimes \mathcal{C}\ell_{1,1} \otimes \mathcal{C}\ell_{3,0} \simeq \mathcal{M}(2,\mathbb{R}) \otimes \mathcal{M}(2,\mathbb{R}) \otimes \mathcal{M}(2,\mathbb{C}) \simeq \mathcal{M}(8,\mathbb{C}),\\
\mathcal{C}\ell_{2,5} &\simeq \mathcal{C}\ell_{1,1} \otimes \mathcal{C}\ell_{1,1} \otimes \mathcal{C}\ell_{0,3} \simeq \mathcal{M}(2,\mathbb{R}) \otimes \mathcal{M}(2,\mathbb{R}) \otimes (\mathbb{H} \oplus \mathbb{H}) \simeq \mathcal{M}(4,\mathbb{H} \oplus \mathbb{H}),\\
\mathcal{C}\ell_{4,3} &\simeq \mathcal{C}\ell_{1,1} \otimes \mathcal{C}\ell_{1,1} \otimes \mathcal{C}\ell_{1,1} \otimes \mathcal{C}\ell_{1,0}\\
&\simeq \mathcal{M}(2,\mathbb{R}) \otimes \mathcal{M}(2,\mathbb{R}) \otimes \mathcal{M}(2,\mathbb{R}) \otimes (\mathbb{R} \oplus \mathbb{R}) \simeq \mathcal{M}(8,\mathbb{R} \oplus \mathbb{R}),\\
\mathcal{C}\ell_{3,4} &\simeq \mathcal{C}\ell_{1,1} \otimes \mathcal{C}\ell_{1,1} \otimes \mathcal{C}\ell_{1,1} \otimes \mathcal{C}\ell_{0,1}\\
&\simeq \mathcal{M}(2,\mathbb{R}) \otimes \mathcal{M}(2,\mathbb{R}) \otimes \mathcal{M}(2,\mathbb{R}) \otimes \mathbb{C} \simeq \mathcal{M}(8,\mathbb{C}),
\end{aligned}
$$

where the following isomorphisms hold:

$$
\begin{aligned}
&\mathcal{C}\ell_{2,0} \simeq \mathcal{C}\ell_{1,1} \simeq \mathcal{M}(2,\mathbb{R}),\\
&\mathcal{C}\ell_{3,0} \simeq \mathcal{C}\ell_{1,2} \simeq \mathcal{M}(2,\mathbb{C}),\\
&\mathcal{C}\ell_{0,4} \simeq \mathcal{C}\ell_{4,0} \simeq \mathcal{C}\ell_{1,3} \simeq \mathcal{M}(2,\mathbb{H}),\\
&\mathcal{C}\ell_{3,1} \simeq \mathcal{C}\ell_{2,2} \simeq \mathcal{M}(4,\mathbb{R}),\\
&\mathcal{C}\ell_{5,0} \simeq \mathcal{C}\ell_{1,4} \simeq \mathcal{M}(2,\mathbb{H} \oplus \mathbb{H}),\\
&\mathcal{C}\ell_{0,5} \simeq \mathcal{C}\ell_{4,1} \simeq \mathcal{C}\ell_{2,3} \simeq \mathcal{M}(4,\mathbb{C}),\\
&\mathcal{C}\ell_{6,0} \simeq \mathcal{C}\ell_{5,1} \simeq \mathcal{C}\ell_{1,5} \simeq \mathcal{C}\ell_{2,4} \simeq \mathcal{M}(4,\mathbb{H}),\\
&\mathcal{C}\ell_{0,6} \simeq \mathcal{C}\ell_{4,2} \simeq \mathcal{C}\ell_{3,3} \simeq \mathcal{M}(8,\mathbb{R}),\\
&\mathcal{C}\ell_{7,0} \simeq \mathcal{C}\ell_{1,6} \simeq \mathcal{C}\ell_{5,2} \simeq \mathcal{C}\ell_{3,4} \simeq \mathcal{M}(8,\mathbb{C}),\\
&\mathcal{C}\ell_{0,7} \simeq \mathcal{C}\ell_{4,3} \simeq \mathcal{M}(8,\mathbb{R} \oplus \mathbb{R}),\\
&\mathcal{C}\ell_{6,1} \simeq \mathcal{C}\ell_{2,5} \simeq \mathcal{M}(4,\mathbb{H} \oplus \mathbb{H}).
\end{aligned}
$$

Furthermore, by supposing that $p > q$ and taking $p - q = 8k + r$ with $r < 8$, we can use eqns (4.5) and (4.39) to obtain:

$$
\begin{aligned}
\mathcal{C}\ell_{p,q} &\simeq \mathcal{C}\ell_{q,q} \otimes \mathcal{C}\ell_{p-q,0} \simeq \mathcal{C}\ell_{q,q} \otimes \mathcal{C}\ell_{8k+r,0}\\
&\simeq \mathcal{C}\ell_{q,q} \otimes \mathcal{C}\ell_{8(k-1)+r+6+2,0}\\
&\simeq \mathcal{C}\ell_{q,q} \otimes \mathcal{C}\ell_{2,0} \otimes \mathcal{C}\ell_{0,8(k-1)+r+6}\\
&\simeq \mathcal{C}\ell_{q,q} \otimes \mathcal{C}\ell_{2,0} \otimes \mathcal{C}\ell_{0,8(k-1)} \otimes \mathcal{C}\ell_{0,r+6}\\
&\simeq \mathcal{M}(2^q,\mathbb{R}) \otimes \mathcal{M}(2,\mathbb{R}) \otimes \mathcal{M}(16^{k-1},\mathbb{R}) \otimes \mathcal{C}\ell_{0,r+6}
\end{aligned}
$$

$$\simeq \mathcal{M}(2^{q+4(k-1)+1}, \mathbb{R}) \otimes \mathcal{C}\ell_{0,r+6}$$
$$\simeq \mathcal{M}(2^{q+4k-3}, \mathbb{R}) \otimes \mathcal{C}\ell_{0,r+6}.$$

Since

$$\mathcal{C}\ell_{0,r+6} \simeq \mathcal{C}\ell_{0,2} \otimes \mathcal{C}\ell_{2,0} \otimes \mathcal{C}\ell_{0,2} \otimes \mathcal{C}\ell_{r,0}$$
$$\simeq \mathcal{M}(4, \mathbb{R}) \otimes \mathcal{M}(2, \mathbb{R}) \otimes \mathcal{C}\ell_{r,0} \simeq \mathcal{M}(8, \mathbb{R}) \otimes \mathcal{C}\ell_{r,0}\,,$$

it reads

$$\mathcal{C}\ell_{p,q} \simeq \mathcal{M}(2^{q+4k-3}, \mathbb{R}) \otimes \mathcal{M}(8, \mathbb{R}) \otimes \mathcal{C}\ell_{r,0} \simeq \mathcal{M}(2^{q+4k}, \mathbb{R}) \otimes \mathcal{C}\ell_{r,0}.$$

Now, if $q > p$, and $q - p = 8k + r$, we obtain

$$\mathcal{C}\ell_{p,q} \simeq \mathcal{C}\ell_{p,p} \otimes \mathcal{C}\ell_{0,q-p} = \mathcal{C}\ell_{p,p} \otimes \mathcal{C}\ell_{0,8k+r} \simeq \mathcal{C}\ell_{p,p} \otimes \mathcal{C}\ell_{0,8k} \otimes \mathcal{C}\ell_{0,r}$$
$$\simeq \mathcal{M}(2^p, \mathbb{R}) \otimes \mathcal{M}(2^{4k}, \mathbb{R}) \otimes \mathcal{C}\ell_{0,r} \simeq \mathcal{M}(2^{p+4k}, \mathbb{R}) \otimes \mathcal{C}\ell_{0,r}\,.$$

Hence, we can see that the Clifford algebra is determined by $r = p - q \mod 8$. The possibilities are analysed in what follows:

- $r = 0$, and $p \geq q \quad (p - q \mod 8 = 0)$:

$$\mathcal{C}\ell_{p,q} \simeq \mathcal{M}(2^{q+4k}, \mathbb{R}) \otimes \mathcal{C}\ell_{0,0} \simeq \mathcal{M}(2^{q+4k}, \mathbb{R})$$

- $r = 0$, and $p < q \quad (p - q \mod 8 = 0)$:

$$\mathcal{C}\ell_{p,q} \simeq \mathcal{M}(2^{p+4k}, \mathbb{R}) \otimes \mathcal{C}\ell_{0,0} \simeq \mathcal{M}(2^{p+4k}, \mathbb{R})$$

- $r = 1$, and $p > q \quad (p - q \mod 8 = 1)$:

$$\mathcal{C}\ell_{p,q} \simeq \mathcal{M}(2^{q+4k}, \mathbb{R}) \otimes \mathcal{C}\ell_{1,0} \simeq \mathcal{M}(2^{q+4k}, \mathbb{R}) \otimes (\mathbb{R} \oplus \mathbb{R})$$
$$\simeq \mathcal{M}(2^{q+4k}, \mathbb{R}) \oplus \mathcal{M}(2^{q+4k}, \mathbb{R})$$

- $r = 1$, and $p < q \quad (p - q \mod 8 = 7)$:

$$\mathcal{C}\ell_{p,q} \simeq \mathcal{M}(2^{p+4k}, \mathbb{R}) \otimes \mathcal{C}\ell_{0,1} \simeq \mathcal{M}(2^{p+4k}, \mathbb{R}) \otimes \mathbb{C} \simeq \mathcal{M}(2^{p+4k}, \mathbb{C})$$

  In these cases, $q + 4k = \left[\frac{2q+8k+r}{2}\right] = \left[\frac{p+q}{2}\right] = \left[\frac{n}{2}\right]$, where $[s]$ denotes the integer part of s.

- $r = 2$, and $p > q \quad (p - q \mod 8 = 2)$:

$$\mathcal{C}\ell_{p,q} \simeq \mathcal{M}(2^{q+4k}, \mathbb{R}) \otimes \mathcal{C}\ell_{2,0} \simeq \mathcal{M}(2^{q+4k}, \mathbb{R}) \otimes \mathcal{M}(2, \mathbb{R}) \simeq \mathcal{M}(2^{q+4k+1}, \mathbb{R})$$

- $r = 2$, and $p < q \quad (p - q \mod 8 = 6)$:

$$\mathcal{C}\ell_{p,q} \simeq \mathcal{M}(2^{p+4k}, \mathbb{R}) \otimes \mathcal{C}\ell_{0,2} \simeq \mathcal{M}(2^{q+4k}, \mathbb{R}) \otimes \mathbb{H} \simeq \mathcal{M}(2^{q+4k}, \mathbb{H})$$

- $r = 3$, and $p > q \quad (p - q \mod 8 = 3)$:

$$\mathcal{C}\ell_{p,q} \simeq \mathcal{M}(2^{q+4k}, \mathbb{R}) \otimes \mathcal{C}\ell_{3,0} \simeq \mathcal{M}(2^{q+4k}, \mathbb{R}) \otimes \mathcal{M}(2, \mathbb{C}) \simeq \mathcal{M}(2^{q+4k+1}, \mathbb{C})$$

- $r = 3$, and $p < q$ $\quad (p - q \mod 8 = 5)$:

$$\begin{aligned} \mathcal{C}\ell_{p,q} &\simeq \mathcal{M}(2^{q+4k}, \mathbb{R}) \otimes \mathcal{C}\ell_{0,3} \simeq \mathcal{M}(2^{q+4k}, \mathbb{R}) \otimes (\mathbb{H} \oplus \mathbb{H}) \\ &\simeq \mathcal{M}(2^{q+4k}, \mathbb{H}) \oplus \mathcal{M}(2^{q+4k}, \mathbb{H}) \end{aligned}$$

In the cases where $r = 2, 3$, if $p > q$ then $q + 4k + 1 = \left[\frac{2q+8k+r}{2}\right] = \left[\frac{n}{2}\right]$; and, if $p < q$, then $p + 4k = \left[\frac{2p+8k+r}{2}\right] - 1 = \left[\frac{n}{2}\right] - 1$

- $r = 4$, and $p > q$ $\quad (p - q \mod 8 = 4)$:

$$\mathcal{C}\ell_{p,q} \simeq \mathcal{M}(2^{q+4k}, \mathbb{R}) \otimes \mathcal{C}\ell_{4,0} \simeq \mathcal{M}(2^{q+4k}, \mathbb{R}) \otimes \mathcal{M}(2, \mathbb{H}) \simeq \mathcal{M}(2^{q+4k+1}, \mathbb{H})$$

- $r = 4$, and $p < q$ $\quad (p - q \mod 8 = 4)$:

$$\mathcal{C}\ell_{p,q} \simeq \mathcal{M}(2^{p+4k}, \mathbb{R}) \otimes \mathcal{C}\ell_{0,4} \simeq \mathcal{M}(2^{q+4k}, \mathbb{R}) \otimes \mathcal{M}(2, \mathbb{H}) \simeq \mathcal{M}(2^{q+4k+1}, \mathbb{H})$$

- $r = 5$, and $p > q$ $\quad (p - q \mod 8 = 5)$:

$$\begin{aligned} \mathcal{C}\ell_{p,q} &\simeq \mathcal{M}(2^{q+4k}, \mathbb{R}) \otimes \mathcal{C}\ell_{5,0} \simeq \mathcal{M}(2^{q+4k}, \mathbb{R}) \otimes \mathcal{M}(2, \mathbb{R}) \otimes (\mathbb{H} \oplus \mathbb{H}) \\ &\simeq \mathcal{M}(2^{q+4k}, \mathbb{R}) \otimes (\mathcal{M}(2, \mathbb{H}) \oplus \mathcal{M}(2, \mathbb{H})) \\ &\simeq \mathcal{M}(2^{q+4k+1}, \mathbb{H}) \oplus \mathcal{M}(2^{q+4k+1}, \mathbb{H}) \end{aligned}$$

- $r = 5$, and $p < q$ $\quad (p - q \mod 8 = 3)$:

$$\begin{aligned} \mathcal{C}\ell_{p,q} &\simeq \mathcal{M}(2^{q+4k}, \mathbb{R}) \otimes \mathcal{C}\ell_{0,5} \simeq \mathcal{M}(2^{q+4k}, \mathbb{R}) \otimes \mathbb{H} \otimes \mathcal{M}(2, \mathbb{C}) \\ &\simeq \mathcal{M}(2^{q+4k+2}, \mathbb{C}). \end{aligned}$$

In the cases where $r = 4, 5$, if $p > q$ or $r = 4$, then $q + 4k + 1 = \left[\frac{n}{2}\right] - 1$, whilst, if $p < q$ and $r = 5$, then $q + 4k + 2 = \left[\frac{n}{2}\right]$.

- $r = 6$, and $p > q$ $\quad (p - q \mod 8 = 6)$:

$$\mathcal{C}\ell_{p,q} \simeq \mathcal{M}(2^{q+4k}, \mathbb{R}) \otimes \mathcal{C}\ell_{6,0} \simeq \mathcal{M}(2^{q+4k}, \mathbb{R}) \otimes \mathcal{M}(4, \mathbb{H}) \simeq \mathcal{M}(2^{q+4k+2}, \mathbb{H})$$

- $r = 6$, and $p < q$ $\quad (p - q \mod 8 = 2)$:

$$\mathcal{C}\ell_{p,q} \simeq \mathcal{M}(2^{p+4k}, \mathbb{R}) \otimes \mathcal{C}\ell_{0,6} \simeq \mathcal{M}(2^{q+4k}, \mathbb{R}) \otimes \mathcal{M}(8, \mathbb{R}) \simeq \mathcal{M}(2^{q+4k+3}, \mathbb{R})$$

- $r = 7$, and $p > q$ $\quad (p - q \mod 8 = 7)$:

$$\begin{aligned} \mathcal{C}\ell_{p,q} &\simeq \mathcal{M}(2^{q+4k}, \mathbb{R}) \otimes \mathcal{C}\ell_{7,0} \simeq \mathcal{M}(2^{q+4k}, \mathbb{R}) \otimes \mathcal{M}(4, \mathbb{R}) \otimes \mathbb{H} \otimes \mathbb{C} \\ &\simeq \mathcal{M}(2^{q+4k}, \mathbb{R}) \otimes \mathcal{M}(8, \mathbb{C}) \simeq \mathcal{M}(2^{q+4k+3}, \mathbb{C}) \end{aligned}$$

- $r = 7$, and $p < q$ $\quad (p - q \mod 8 = 1)$:

$$\begin{aligned} \mathcal{C}\ell_{p,q} &\simeq \mathcal{M}(2^{q+4k}, \mathbb{R}) \otimes \mathcal{C}\ell_{0,7} \\ &\simeq \mathcal{M}(2^{q+4k}, \mathbb{R}) \otimes \mathbb{H} \otimes \mathbb{H} \otimes \mathcal{M}(2, \mathbb{R}) \otimes (\mathbb{R} \oplus \mathbb{R}) \\ &\simeq \mathcal{M}(2^{q+4k}, \mathbb{R}) \otimes \mathcal{M}(8, \mathbb{R}) \otimes (\mathbb{R} \oplus \mathbb{R}) \\ &\simeq \mathcal{M}(2^{q+4k+3}, \mathbb{R}) \oplus \mathcal{M}(2^{q+4k+3}, \mathbb{R}) \end{aligned}$$

In the cases where $r = 6$ or 7, if $p < q$, then $q + 4k + 3 = \left[\frac{n}{2}\right]$; and, if $p > q$ and $r = 6$, then $q + 4k + 2 = \left[\frac{n}{2}\right] - 1$.

We can organise the algebras of dimension $n < 8$ according to $p - q$, obtaining:

$$
\begin{array}{llllllll}
p-q=0: & \mathcal{C}\ell_{0,0}, & \mathcal{C}\ell_{1,1}, & \mathcal{C}\ell_{2,2}, & \mathcal{C}\ell_{3,3} & & & \mathcal{M}(2^{[n/2]},\mathbb{R}) \\
p-q=1: & \mathcal{C}\ell_{1,0}, & \mathcal{C}\ell_{2,1}, & \mathcal{C}\ell_{3,2}, & \mathcal{C}\ell_{4,3}, & \mathcal{C}\ell_{0,7} & & \mathcal{M}(2^{[n/2]},\mathbb{R})^{\oplus^2} \\
p-q=2: & \mathcal{C}\ell_{2,0}, & \mathcal{C}\ell_{3,1}, & \mathcal{C}\ell_{4,2}, & \mathcal{C}\ell_{0,6} & & & \mathcal{M}(2^{[n/2]},\mathbb{R}) \\
p-q=3: & \mathcal{C}\ell_{3,0}, & \mathcal{C}\ell_{4,1}, & \mathcal{C}\ell_{5,2}, & \mathcal{C}\ell_{0,5}, & \mathcal{C}\ell_{1,6} & & \mathcal{M}(2^{[n/2]},\mathbb{C}) \\
p-q=4: & \mathcal{C}\ell_{4,0}, & \mathcal{C}\ell_{5,1}, & \mathcal{C}\ell_{0,4}, & \mathcal{C}\ell_{1,5} & & & \mathcal{M}(2^{[n/2]-1},\mathbb{H}) \\
p-q=5: & \mathcal{C}\ell_{5,0}, & \mathcal{C}\ell_{6,1}, & \mathcal{C}\ell_{0,3}, & \mathcal{C}\ell_{1,4}, & \mathcal{C}\ell_{2,5} & & \mathcal{M}(2^{[n/2]-1},\mathbb{R})^{\oplus^2} \\
p-q=6: & \mathcal{C}\ell_{6,0}, & \mathcal{C}\ell_{0,2}, & \mathcal{C}\ell_{1,3}, & \mathcal{C}\ell_{2,4} & & & \mathcal{M}(2^{[n/2]-1},\mathbb{H}) \\
p-q=7: & \mathcal{C}\ell_{7,0}, & \mathcal{C}\ell_{0,1}, & \mathcal{C}\ell_{1,2}, & \mathcal{C}\ell_{2,3}, & \mathcal{C}\ell_{3,4} & & \mathcal{M}(2^{[n/2]},\mathbb{C})
\end{array}
$$

where $\mathcal{A}^{\oplus^2}$ denotes $\mathcal{A} \oplus \mathcal{A}$. Table 4.1 shows the Clifford algebra classification obtained from these isomorphisms.

Let us provide some examples of how this table can be used. First, take the Clifford algebra $\mathcal{C}\ell_{3,0}$, where $p - q = 3$ and which is isomorphic to $\mathcal{M}(2^{[n/2]}, \mathbb{C})$. Since $n = p+q = 3$, it follows that $[n/2] = [3/2] = 1$, and the related matrix algebra is $\mathcal{M}(2^1, \mathbb{C})$. Hence, $\mathcal{C}\ell_{3,0} \simeq \mathcal{M}(2, \mathbb{C})$. Let us now consider the algebra $\mathcal{C}\ell_{0,2}$. In this case $p - q = -2 = 6 \mod 8$ and the corresponding matrix algebra is $\mathcal{M}(2^{[n/2]-1}, \mathbb{H})$. Since $n = p + q = 2$, and $[n/2] = [2/2] = 1$, therefore $\mathcal{M}(2^{1-1}, \mathbb{H}) = \mathcal{M}(1, \mathbb{H}) = \mathbb{H}$, namely, $\mathcal{C}\ell_{0,2} \simeq \mathbb{H}$, as seen previously. The same reasoning leads, for instance, to the conclusion that $\mathcal{C}\ell_{0,3} \simeq \mathbb{H} \oplus \mathbb{H}$; $\mathcal{C}\ell_{1,3} \simeq \mathcal{M}(2, \mathbb{H})$; $\mathcal{C}\ell_{3,1} \simeq \mathcal{M}(4, \mathbb{R})$; $\mathcal{C}\ell_{4,1} \simeq \mathcal{M}(4, \mathbb{C})$; and so on.

With respect to the complex case, the classification can be obtained from $\mathcal{C}\ell(V_\mathbb{C}, g_\mathbb{C}) \simeq \mathbb{C} \otimes \mathcal{C}\ell(V, g)$. The complex Clifford algebra depends only on the parity of $n = p + q$. We therefore denote $\mathbb{C} \otimes \mathcal{C}\ell_{p,q} = \mathcal{C}\ell_\mathbb{C}(n)$. If n is even, $p - q = 0, 2, 4, 6$:

Table 4.1 Real Clifford Algebra Classification, Where $p + q = n$, and $[n/2]$ Denotes the Integer Part of $n/2$

$p-q$ mod 8	0	1	2	3
$\mathcal{C}\ell_{p,q}$	$\mathcal{M}(2^{[n/2]},\mathbb{R})$	$\mathcal{M}(2^{[n/2]},\mathbb{R})$ $\oplus$ $\mathcal{M}(2^{[n/2]},\mathbb{R})$	$\mathcal{M}(2^{[n/2]},\mathbb{R})$	$\mathcal{M}(2^{[n/2]},\mathbb{C})$
$p-q$ mod 8	4	5	6	7
$\mathcal{C}\ell_{p,q}$	$\mathcal{M}(2^{[n/2]-1},\mathbb{H})$	$\mathcal{M}(2^{[n/2]-1},\mathbb{H})$ $\oplus$ $\mathcal{M}(2^{[n/2]-1},\mathbb{H})$	$\mathcal{M}(2^{[n/2]-1},\mathbb{H})$	$\mathcal{M}(2^{[n/2]},\mathbb{C})$

$$\begin{array}{ll} p-q=0, & \mathbb{C}\otimes\mathcal{M}(2^n,\mathbb{R})\simeq\mathcal{M}(2^n,\mathbb{C}),\\ p-q=2, & \mathbb{C}\otimes\mathcal{M}(2^n,\mathbb{R})\simeq\mathcal{M}(2^n,\mathbb{C}),\\ p-q=4, & \mathbb{C}\otimes\mathcal{M}(2^{n-1},\mathbb{H})\simeq\mathcal{M}(2^n,\mathbb{C}),\\ p-q=6, & \mathbb{C}\otimes\mathcal{M}(2^{n-1},\mathbb{H})\simeq\mathcal{M}(2^n,\mathbb{C}). \end{array}$$

If n is odd, $p-q=1,3,5,7$:

$$\begin{array}{ll} p-q=1, & \mathbb{C}\otimes(\mathcal{M}(2^n,\mathbb{R})\oplus\mathcal{M}(2^n,\mathbb{R}))\simeq\mathcal{M}(2^n,\mathbb{C})\oplus\mathcal{M}(2^n,\mathbb{C}),\\ p-q=3, & \mathbb{C}\otimes\mathcal{M}(2^n,\mathbb{C})\simeq\mathcal{M}(2^n,\mathbb{C})\oplus\mathcal{M}(2^n,\mathbb{C}),\\ p-q=5, & \mathbb{C}\otimes(\mathcal{M}(2^{n-1},\mathbb{H})\oplus\mathcal{M}(2^{n-1},\mathbb{H}))\simeq\mathcal{M}(2^n,\mathbb{C})\oplus\mathcal{M}(2^n,\mathbb{C}),\\ p-q=7, & \mathbb{C}\otimes\mathcal{M}(2^n,\mathbb{C})\simeq\mathcal{M}(2^n,\mathbb{C})\oplus\mathcal{M}(2^n,\mathbb{C}). \end{array}$$

Therefore, for the complex Clifford algebras $\mathcal{C}\ell_{\mathbb{C}}(n)$ it follows the classification given in table 4.2.

Although this table provides the isomorphisms between Clifford and matrix algebras (or, in some cases, the direct sums of some algebras), it does not explicitly say how to write such isomorphisms. In other words, the table shows that $\mathcal{C}\ell_{2,0}\simeq\mathcal{M}(2,\mathbb{R})$ but does not supply a way to explicitly obtain the isomorphism given in eqn (4.32). In those cases involving, for instance, low-dimensional algebras like $\mathcal{C}\ell_{2,0}$, it is not that complicated to find an isomorphism; however, as the number of dimensions of the space increases, the procedure becomes non-trivial. The main aim here is to find and exhibit a *matrix representation* for a Clifford algebra. It is not only important from the 'theoretical' point of view but also from the 'practical' viewpoint. From the theoretical point of view, the concept of the *spinor* is associated with such representations (we will discuss spinors in chapter 6). From the practical point of view, it is often better (computationally) to use matrices than to consider the abstract Clifford algebra generators directly. Later on, we will discuss how to explicitly obtain a representation of a Clifford algebra.

4.3 Idempotents and Representations

The Regular Representation

Let $\mathcal{A}$ be an algebra, where there is involved naturally an underlying vector space structure. We can further consider the set of endomorphisms $\mathrm{End}(\mathcal{A})$ in order to construct the representations of the algebra $\mathcal{A}$. We can define a representation $L:\mathcal{A}\to\mathrm{End}(\mathcal{A})$ as

$$L(a)b=ab,\qquad \forall\, b\in\mathcal{A}. \tag{4.40}$$

Table 4.2 Classification of the Complex Clifford Algebras

n even	$\mathcal{C}\ell_{\mathbb{C}}(2k)=\mathcal{M}(2^k,\mathbb{C})$
n odd	$\mathcal{C}\ell_{\mathbb{C}}(2k+1)=\mathcal{M}(2^k,\mathbb{C})\oplus\mathcal{M}(2^k,\mathbb{C})$

It is straightforward to see that L is indeed a representation: $L(1) = 1$, and $L(ab) = L(a)L(b)$. We denominate L a *regular representation*. A regular representation is faithful. Indeed, if $L(a)c = L(b)c$, then $L(a-b)c = (a-b)c = 0$, and when $c = 1$, it follows that $a = b$. In other words, $\ker L = \{0\}$. We can also define $R : \mathcal{A} \to \text{End}(\mathcal{A})$ as

$$R(a)b = ba, \qquad \forall\, b \in \mathcal{A}. \tag{4.41}$$

In this case, $R(1) = 1$ but $R(ab) = R(b)R(a)$, namely, R is a representation of the opposed algebra $\mathcal{A}^{\text{op}}$, where $\mathcal{A}^{\text{op}}$ is an algebra with multiplication opposed with respect to $\mathcal{A}$, namely, $m_{\mathcal{A}^{\text{op}}}(a,b) = m_{\mathcal{A}}(b,a) = ba$. Therefore, $R(\mathcal{A}) = \mathcal{A}^{\text{op}}$. The sets $L(\mathcal{A})$ and $R(\mathcal{A})$ are subalgebras of $\text{End}(\mathcal{A})$ that commute $L(a)R(b)c = L(a)cb = acb = (L(a)c)b = R(b)L(a)c$, since $\mathcal{A}$ is associative.

Now let us suppose that there exist subspaces $\mathcal{B}_1$ and $\mathcal{B}_2$ of $\mathcal{A}$ that are invariant with respect to the regular representation, that is, $L(a)(\mathcal{B}_1) \subset \mathcal{B}_1$, and $L(a)(\mathcal{B}_2) \subset \mathcal{B}_2$, $\forall\, a \in \mathcal{A}$. Hence, it is possible to express $L = L_1 \oplus L_2$, where $L_1 : \mathcal{A} \to \text{End}(\mathcal{B}_1)$, and $L_2 : \mathcal{A} \to \text{End}(\mathcal{B}_2)$. If there exist other invariant subspaces L_1 and L_2, the same procedure is employed until a space S and a representation $\mathcal{L} : \mathcal{A} \to \text{End}(S)$ are obtained, such that the unique invariant subspaces are S and $\{0\}$. This representation is then irreducible. The space S satisfies $\mathcal{L}(a)(S) \subset S$, $\forall\, a \in \mathcal{A}$, that is, $\forall\, a \in \mathcal{A}$ and $\forall\, x \in S$, it follows that $ax \in S$. By definition, S is a *left ideal* with respect to the algebra $\mathcal{A}$ (see chapter 2, section 2.5), and indeed a *minimal* left ideal, where by minimal it is meant that S does not contain any non-trivial subideal (the unique subideals of S are either S or $\{0\}$).

To summarise: *the representation space associated with an irreducible regular representation is a left minimal ideal of the algebra.*

Idempotents

An element $f \in \mathcal{A}$ is an *idempotent* if $f^2 = f$. If the algebra $\mathcal{A}$ is a division algebra,[1] then the unique idempotent is the identity 1. Indeed, if $f^2 = f$, where $f \neq 0$ and $\mathcal{A}$ is a division algebra, then $f = f^{-1}f^2 = f^{-1}f = 1$. However, most Clifford algebras are not division algebras.

Two idempotents, f_1 and f_2, are called *orthogonal* if $f_1 f_2 = f_2 f_1 = 0$. An idempotent f is said to be *primitive* if it cannot be written as the sum of other two orthogonal idempotents, that is, $f \neq f_1 + f_2$, where $(f_1)^2 = f_1$; $(f_2)^2 = f_2$; and $f_1 f_2 = f_2 f_1 = 0$ (Lounesto and Wene, 1987).

Simple Algebras

Looking at table 4.1 we can see that every Clifford algebra can be expressed either by $\mathbb{K} \otimes \mathcal{M}(N, \mathbb{R})$ or by $[\mathbb{K} \otimes \mathcal{M}(N, \mathbb{R})] \oplus [\mathbb{K} \otimes \mathcal{M}(N, \mathbb{R})]$, for $\mathbb{K} = \mathbb{R}, \mathbb{C}, \mathbb{H}$, and for some

[1] A division algebra is defined as an algebra where every non-zero element has an inverse. Equivalently, an algebra $\mathcal{A}$ is a division algebra when, if $ab = 0$ (or if $ba = 0$), $\forall\, b \in \mathcal{A}$, then $a = 0$. A theorem attributed to Frobenius asserts that the unique, real, associative finite-dimensional algebras are $\mathbb{R}$, $\mathbb{C}$, and $\mathbb{H}$ (Schafer, 1954). The octonion algebra (or Cayley algebra) $\mathbb{O}$ is also a division algebra, but it is not associative. For the Clifford algebraic formulation of octonions see Lounesto (2001*a*, de Andrade and Toppan (1999, da Rocha and Vaz (2007); an advanced discussion of this topics, see the articles by da Rocha *et al.* (2012, da Rocha and Traesel (2012, da Rocha and Vaz Jr (2006).

$N = 0, 1, \ldots$. In addition, an algebra that can be represented by $\mathbb{K} \otimes \mathcal{M}(N, \mathbb{R})$ is a simple algebra. An algebra $\mathcal{A}$ is said to be *simple* if the unique bilateral ideals of $\mathcal{A}$ are $\mathcal{A}$ and $\{0\}$. In order to check that a matrix algebra $\mathcal{A} = \mathbb{K} \otimes \mathcal{M}(N, \mathbb{R})$ is a simple algebra, let us consider an element x in an ideal $I \in \mathcal{A}$. According to exercise 4 in chapter 1, the matrices $\{E_{AB}\}$ defined by $(E_{AB})_{CD} = \delta_{AC}\delta_{BD}$ and satisfying $E_{AB}E_{CD} = \delta_{BC}E_{AD}$ $(A, B, C, D = 1, \ldots, N)$ form a basis for the space of matrices $\mathcal{M}(N, \mathbb{R})$. Since $I \subset \mathcal{A} = \mathbb{K} \otimes \mathcal{M}(N, \mathbb{R})$, therefore $x \in I$ can be written as $x = \sum_{AB} x_{AB}E_{AB}$, where $x_{AB} \in \mathbb{K}$. If $x \neq 0$, then at least one of the components x_{AB} is null. Now we can see that

$$\begin{aligned}\sum_C E_{CA}xE_{BC} &= \sum_{CDE} x_{DE}E_{CA}E_{DE}E_{BC} \\ &= \sum_{CDE} x_{DE}\delta_{AD}\delta_{BE}E_{CC} = x_{AB}\sum_C E_{CC} = x_{AB}1,\end{aligned}$$

where $1 = \sum_C E_{CC}$ denotes the identity matrix. Since $x_{AB} \in \mathbb{K}$ – which is a division algebra – then there exists x_{AB}^{-1} and

$$\sum_C x_{AB}^{-1}E_{CA}xE_{BC} = 1.$$

This last equation means that $1 \in \mathcal{A}x\mathcal{A}$. However, $\mathcal{A}x\mathcal{A} \in \mathcal{A}I\mathcal{A} \subset I$, and then $1 \in I$. Taking $x = 1$, it follows that $\mathcal{A}\mathcal{A} = \mathcal{A} \subset I$. Since $\mathcal{A} \subset I$ and $I \subset \mathcal{A}$, therefore $I = \mathcal{A}$. The other possibility is the case where $x \neq 0$ and therefore $I = \{0\}$. Hence, an algebra $\mathcal{A} = \mathbb{K} \otimes \mathcal{M}(N, \mathbb{R})$ is a simple algebra. It is possible to show – indeed, it will be shown in the next section – that the reciprocal holds, namely, if an algebra is simple, then it can be written (not uniquely) as $\mathbb{K} \otimes \mathcal{M}(N, \mathbb{R})$.

Therefore, *every Clifford algebra is either a simple algebra or the direct sum of simple algebras*. An algebra that is the sum of simple algebras is said to be a *semisimple* algebra.

Idempotents and Simple Algebras

Consider a set of N primitive idempotents $\{f_1, \ldots, f_N\}$ which are mutually orthogonal: $f_Af_B = \delta_{AB}f_A$. Let us denote by $\mathcal{A}_{AB}$ the set $\mathcal{A}_{AB} = f_A\mathcal{A}f_B$, $(A, B = 1, \ldots, N)$. For those sets, we have

$$\mathcal{A}_{AB}\mathcal{A}_{CD} = f_A\mathcal{A}f_Bf_C\mathcal{A}f_D = \delta_{BC}f_A\mathcal{A}f_B\mathcal{A}f_D.$$

In addition, the set $\mathcal{A}f_B\mathcal{A}$ is a bilateral ideal of $\mathcal{A}$. Since the algebra $\mathcal{A}$ is simple, therefore $\mathcal{A}f_B\mathcal{A}$ equals either $\mathcal{A}$ or $\{0\}$. Since $f_B \neq 0$, therefore $\mathcal{A}f_B\mathcal{A} = \mathcal{A}$ and thus

$$\mathcal{A}_{AB}\mathcal{A}_{CD} = \delta_{BC}\mathcal{A}_{AD}. \tag{4.42}$$

On the other hand, there exists idempotents $f_A \in \mathcal{A}_{AA}$ $(A = 1, \ldots, N)$. Let us choose one of those idempotents, for example, f_1. Because of eqn (4.42), it follows that $f_1 \in$

$\mathcal{A}_{11} = \mathcal{A}_{1A}\mathcal{A}_{A1}$ for any value of A. Hence, there exist $\boldsymbol{\mathcal{E}}_{1A} \in \mathcal{A}_{1A}$, and $\boldsymbol{\mathcal{E}}_{A1} \in \mathcal{A}_{A1}$, such that $f_1 = \boldsymbol{\mathcal{E}}_{1A}\boldsymbol{\mathcal{E}}_{A1}$. Now, choose $\boldsymbol{\mathcal{E}}_{1A}$ and $\boldsymbol{\mathcal{E}}_{A1}$ such that

$$f_B\boldsymbol{\mathcal{E}}_{A1} = \delta_{AB}\boldsymbol{\mathcal{E}}_{A1}, \qquad \boldsymbol{\mathcal{E}}_{1A}f_B = \delta_{AB}\boldsymbol{\mathcal{E}}_{1A}. \tag{4.43}$$

Let us define the quantities $\boldsymbol{\mathcal{E}}_{AB}$ as

$$\boldsymbol{\mathcal{E}}_{AB} = \boldsymbol{\mathcal{E}}_{A1}\boldsymbol{\mathcal{E}}_{1B}, \tag{4.44}$$

which yields

$$f_C\boldsymbol{\mathcal{E}}_{AB} = \delta_{AC}\boldsymbol{\mathcal{E}}_{AB}, \qquad \boldsymbol{\mathcal{E}}_{AB}f_C = \delta_{BC}\boldsymbol{\mathcal{E}}_{AB}. \tag{4.45}$$

Now let us take into account the product $\boldsymbol{\mathcal{E}}_{AB}\boldsymbol{\mathcal{E}}_{CD}$. With these definitions it reads

$$\begin{aligned}\boldsymbol{\mathcal{E}}_{AB}\boldsymbol{\mathcal{E}}_{CD} &= \boldsymbol{\mathcal{E}}_{A1}\boldsymbol{\mathcal{E}}_{1B}\boldsymbol{\mathcal{E}}_{C1}\boldsymbol{\mathcal{E}}_{1D} = \delta_{BC}\boldsymbol{\mathcal{E}}_{A1}\boldsymbol{\mathcal{E}}_{1B}\boldsymbol{\mathcal{E}}_{B1}\boldsymbol{\mathcal{E}}_{1D}\\ &= \delta_{BC}\boldsymbol{\mathcal{E}}_{A1}f_1\boldsymbol{\mathcal{E}}_{1D} = \delta_{BC}\boldsymbol{\mathcal{E}}_{A1}\boldsymbol{\mathcal{E}}_{1D},\end{aligned}$$

namely,

$$\boldsymbol{\mathcal{E}}_{AB}\boldsymbol{\mathcal{E}}_{CD} = \delta_{BC}\boldsymbol{\mathcal{E}}_{AD}. \tag{4.46}$$

This equation shows that the quantities $\{\boldsymbol{\mathcal{E}}_{AB}\}$ $(A, B = 1, \ldots, N)$ comprise a *basis for the space of* $N \times N$ *matrices*. In addition, it shows that $\boldsymbol{\mathcal{E}}_{AA}$ is an idempotent. The identity $1_{\mathcal{A}}$ de $\mathcal{A}$ is given by

$$1_{\mathcal{A}} = \sum_A \boldsymbol{\mathcal{E}}_{AA}, \tag{4.47}$$

since $1_{\mathcal{A}}\boldsymbol{\mathcal{E}}_{BC} = \sum_A \boldsymbol{\mathcal{E}}_{AA}\boldsymbol{\mathcal{E}}_{BC} = \sum_A \delta_{AB}\boldsymbol{\mathcal{E}}_{AC} = \boldsymbol{\mathcal{E}}_{BC}$.

The set $\mathcal{A}_{AA} = f_A\mathcal{A}f_A$ is an ideal of $\mathcal{A}$. For $x_A \in \mathcal{A}_{AA}$, we have $x_Af_A = f_Ax_A = x_A$, that is, the idempotent f_A is the *unit* of $\mathcal{A}_{AA}$. This idempotent f_A is the unique idempotent of $\mathcal{A}_{AA}$ if and only if f_A is a primitive. This assertion follows from the following: (i) if f_A is not a primitive, then $f_A = g_A + h_A$, where $h_Ag_A = g_Ah_A = 0$; therefore, $f_Ag_A = g_Af_A = g_A$, and $f_Ah_A = h_Af_A = h_A$, namely, $g_A, h_A \in \mathcal{A}_{AA}$, which negates the hypothesis that f_A is the unique idempotent of $\mathcal{A}_{AA}$; (ii) if there exists another idempotent $g_A \in \mathcal{A}_{AA}$ besides f_A, then $f_A - g_A$ is another idempotent, since, if $g_A \in \mathcal{A}_{AA}$, then $g_Af_A = f_Ag_A = g_A$. This result implies that $(f_A - g_A)(f_A - g_A) = f_A - f_Ag_A - g_Af_A + g_A = f_A - g_A$. In addition, the idempotents g_A and $f_A - g_A$ are orthogonal: $g_A(f_A - g_A) = g_A - g_A = 0$, and $(f_A - g_A)g_A = g_A - g_A = 0$; since $f_A = (f_A - g_A) + g_A$, the idempotent f_A is not primitive. Moreover, since $\boldsymbol{\mathcal{E}}_{AA} \subset f_A\mathcal{A}f_A$, therefore the idempotent f_A is primitive, yielding

$$\boldsymbol{\mathcal{E}}_{AA} = f_A. \tag{4.48}$$

Moreover, $\mathcal{A}_{AA} = f_A\mathcal{A}f_A$ is a division algebra with identity f_A, and the algebras $\mathcal{A}_{AA}$ $(A = 1, \ldots, N)$ are isomorphic. Let us first show that $f_A\mathcal{A}f_A$ is a division algebra. Let $I_A = \mathcal{A}f_A$ be a left ideal of $\mathcal{A}$. Since f_A is primitive, this ideal is minimal; thus, the unique primitive subideals are I_A and $\{0\}$. Let now J_A be a (non-null) left ideal of $\mathcal{A}_{AA}$. Obviously, $J_A \subset \mathcal{A}_{AA}$, and $\mathcal{A}J_A \subset \mathcal{A}f_A\mathcal{A}f_A = f_A\mathcal{A}I_A \subset I_A$, that is, $\mathcal{A}J_A \subset I_A$. However, since I_A is minimal and J_A is non-null, the unique possibility is that $\mathcal{A}J_A = I_A$. On the other hand, $\mathcal{A}_{AA} = f_A\mathcal{A}f_A = f_AI_A = f_A\mathcal{A}J_Af_A\mathcal{A}f_AJ_A \subset J_A$. Since

$J_A \subset \mathcal{A}_{AA}$ and we just proved that $\mathcal{A}_{AA} \subset J_A$, it follows that $J_A = \mathcal{A}_{AA}$. We then conclude that the unique left ideal of $\mathcal{A}_{AA}$ is either $\mathcal{A}_{AA}$ itself or $\{0\}$. Consider now a non-null element $z \in \mathcal{A}_{AA}$. The set $\mathcal{A}_{AA}z$ is a left ideal of $\mathcal{A}_{AA}$. However, since $\mathcal{A}_{AA}$ does not contain non-trivial subideals, and $z \neq 0$, therefore $\mathcal{A}_{AA} = \mathcal{A}_{AA}z$. This result means that there exists $w, w' \in \mathcal{A}_{AA}$ such that $wz = w'$. Hence, there exists a non-null $z' \in \mathcal{A}_{AA}$ such that $z'z = f_A$, since f_A is the identity in $\mathcal{A}_{AA}$. Similarly, there exists $z'' \in \mathcal{A}_{AA}$ such that $z''z' = f_A$. Now $z'' = z''z'z = z$, namely, $z'z = zz' = f_A$; thus, $\mathcal{A}_{AA}$ is a *unital division algebra* f_A. In order to show that $\mathcal{A}_{AA} \simeq \mathcal{A}_{BB}$, given $x_A \in \mathcal{A}_{AA}$, and $x_B \in \mathcal{A}_{BB}$, we define $\phi_{AB} : \mathcal{A}_{AA} \to \mathcal{A}_{BB}$ as

$$\phi_{AB}(x_A) = \boldsymbol{\mathcal{E}}_{BA} x_A \boldsymbol{\mathcal{E}}_{AB}. \tag{4.49}$$

This mapping is linear and by using eqn (4.46) and given the fact that $f_A = \boldsymbol{\mathcal{E}}_{AA}$ is the identity in $\mathcal{A}_{AA}$, we can see that it satisfies

$$\begin{aligned}\phi_{AB}(x_A y_A) &= \boldsymbol{\mathcal{E}}_{BA} x_A y_A \boldsymbol{\mathcal{E}}_{AB} = \boldsymbol{\mathcal{E}}_{BA} x_A f_A y_A \boldsymbol{\mathcal{E}}_{AB} \\ &= \boldsymbol{\mathcal{E}}_{BA} x_A \boldsymbol{\mathcal{E}}_{AB} \boldsymbol{\mathcal{E}}_{BA} y_A \boldsymbol{\mathcal{E}}_{AB} = \phi_{AB}(x_A)\phi_{AB}(y_A).\end{aligned}$$

Moreover, $\phi_{AB}^{-1} = \phi_{BA}$. Therefore, the division algebras $\mathcal{A}_{AA}$ $(A = 1, \ldots, N)$ are isomorphic.

Given, for instance, $x_1 \in \mathcal{A}_{11}$, and $x_A = \boldsymbol{\mathcal{E}}_{A1} x_1 \boldsymbol{\mathcal{E}}_{1A}$, we define $x \in \mathbb{K}$ as

$$x = \sum_A x_A = \sum_A \boldsymbol{\mathcal{E}}_{A1} x_1 \boldsymbol{\mathcal{E}}_{1A}. \tag{4.50}$$

This expression defines a linear mapping $\mathcal{A}_{11} \to \mathbb{K}$ whose inverse is provided by $\boldsymbol{\mathcal{E}}_{11} x \boldsymbol{\mathcal{E}}_{11} = x_1$. Hence, it implies that $\mathbb{K} \simeq \mathcal{A}_{11} \simeq \mathcal{A}_{AA}$ $(A = 1, \ldots, N)$. Moreover, $x \in \mathbb{K}$ commutes with any matrix in the set generated by $\{\boldsymbol{\mathcal{E}}_{AB}\}$. Indeed,

$$\begin{aligned}x\boldsymbol{\mathcal{E}}_{AB} &= \sum_C \boldsymbol{\mathcal{E}}_{C1} x_1 \boldsymbol{\mathcal{E}}_{1C} \boldsymbol{\mathcal{E}}_{AB} = \boldsymbol{\mathcal{E}}_{A1} x_1 \boldsymbol{\mathcal{E}}_{1A} \boldsymbol{\mathcal{E}}_{AB} = \boldsymbol{\mathcal{E}}_{A1} x_1 \boldsymbol{\mathcal{E}}_{1B} \\ &= \boldsymbol{\mathcal{E}}_{AB} \boldsymbol{\mathcal{E}}_{B1} x_1 \boldsymbol{\mathcal{E}}_{1B} = \sum_C \boldsymbol{\mathcal{E}}_{AB} \boldsymbol{\mathcal{E}}_{C1} x_1 \boldsymbol{\mathcal{E}}_{1C} = \boldsymbol{\mathcal{E}}_{AB} x.\end{aligned}$$

Consider now $x \in \mathcal{A}$; since $1_{\mathcal{A}} = \sum_A f_A = \sum_A \boldsymbol{\mathcal{E}}_{AA}$, it is then possible to write

$$x = \sum_A \boldsymbol{\mathcal{E}}_{AA} x \sum_B \boldsymbol{\mathcal{E}}_{BB} = \sum_{AB} \boldsymbol{\mathcal{E}}_{AA} x \boldsymbol{\mathcal{E}}_{BA} \boldsymbol{\mathcal{E}}_{AB} = \sum_{ABC} \boldsymbol{\mathcal{E}}_{CA} x \boldsymbol{\mathcal{E}}_{BC} \boldsymbol{\mathcal{E}}_{AB}, \tag{4.51}$$

and subsequently

$$x = \sum_{ABC} (x_C)_{AB} \boldsymbol{\mathcal{E}}_{AB}, \tag{4.52}$$

where we define

$$(x_C)_{AB} = \boldsymbol{\mathcal{E}}_{CA} x \boldsymbol{\mathcal{E}}_{BC}. \tag{4.53}$$

Since $(x_C)_{AB} = f_C (x_C)_{AB} f_C$, it follows that $(x_C)_{AB} \in \mathcal{A}_{CC}$. As expected, eqn (4.49) holds, that is,

$$(x_D)_{AB} = \boldsymbol{\mathcal{E}}_{DC} (x_C)_{AB} \boldsymbol{\mathcal{E}}_{CD}. \tag{4.54}$$

According to eqn (4.50), we define

$$x_{AB} = \sum_C (x_C)_{AB}, \tag{4.55}$$

namely

$$\boxed{x_{AB} = \sum_C \boldsymbol{\mathcal{E}}_{CA} x \boldsymbol{\mathcal{E}}_{BC}} \tag{4.56}$$

Finally, we can write eqn (4.52) as

$$\boxed{x = \sum_{AB} x_{AB} \boldsymbol{\mathcal{E}}_{AB}} \tag{4.57}$$

where $x_{AB} \in \mathbb{K}$, and $\{\boldsymbol{\mathcal{E}}_{AB}\}$ forms a basis for the space of matrices $N \times N$. This result shows that, if $\mathcal{A}$ is simple, then $\mathcal{A}$ can be written as $\mathbb{K} \otimes \mathcal{M}(N, \mathbb{R})$. The element x can be represented by the matrix with coefficients x_{AB}, accordingly. In fact, there is a representation ρ given by

$$\rho(x) = \begin{pmatrix} x_{11} & x_{12} & \dots & x_{1N} \\ x_{21} & x_{22} & \dots & x_{2N} \\ \vdots & \vdots & \ddots & \vdots \\ x_{N1} & x_{N2} & \dots & x_{NN} \end{pmatrix} \tag{4.58}$$

Clearly, the expression (4.56) is not unique. In fact, any invertible element $u \in \mathcal{A}$ can be used to define another basis of $\mathcal{M}(N, \mathbb{R})$ by

$$\boldsymbol{\mathcal{E}}'_{AB} = u \boldsymbol{\mathcal{E}}_{AB} u^{-1}, \tag{4.59}$$

and so yields

$$x = \sum_{AB} x'_{AB} \boldsymbol{\mathcal{E}}'_{AB}, \tag{4.60}$$

where $x'_{AB} = \sum_C \boldsymbol{\mathcal{E}}'_{CA} a \boldsymbol{\mathcal{E}}'_{BC} = \sum_C u \boldsymbol{\mathcal{E}}_{CA} u^{-1} a u \boldsymbol{\mathcal{E}}_{BC} u^{-1} = u(u^{-1} a u)_{AB} u^{-1}$.

The representation space here is a minimal left ideal $\mathcal{A}f_1$, because of the choice of the idempotent f_1 when the quantities $\boldsymbol{\mathcal{E}}_{AB} = \boldsymbol{\mathcal{E}}_{A1} \boldsymbol{\mathcal{E}}_{1B}$ $(A, B = 1, \dots, N)$ are defined via eqn (4.44). If an idempotent f_C were chosen, the representation space would clearly be $\mathcal{A}f_C$. The set of objects $\{\boldsymbol{\mathcal{E}}_{AB}\}$ acts on $\boldsymbol{\mathcal{E}}_{A1}$ $(A = 1, \dots, N)$, and $\{\boldsymbol{\mathcal{E}}_{A1}\}$ is a basis for the ideal $\mathcal{A}f_1$. Indeed, for $x \in \mathcal{A}$, we can write $x f_1 = \sum_{AB} x_{AB} \boldsymbol{\mathcal{E}}_{AB} \boldsymbol{\mathcal{E}}_{11} = \sum_A x_{A1} \boldsymbol{\mathcal{E}}_{A1}$, which shows that $\{\boldsymbol{\mathcal{E}}_{A1}\}$ generates $\mathcal{A}_1$. It is clear that the set $\{\boldsymbol{\mathcal{E}}_{A1}\}$ is linearly independent, so there is a basis for the minimal left ideal $\mathcal{A}f_1$. Since $\mathcal{A} \simeq \mathbb{K} \otimes \mathcal{M}(N, \mathbb{R})$, therefore $\mathcal{A}f_1 \simeq \mathbb{K} \otimes \mathbb{R}^N$, a result which follows from $\boldsymbol{\mathcal{E}}_{A1} \simeq \mathbf{e}_A$, where $\mathbf{e}_A$ $(A = 1, \dots, N)$ is a basis of $\mathbb{R}^N$. The isomorphism $\mathcal{A}f_1 \simeq \mathbb{K} \otimes \mathbb{R}^N$, although obvious, has leading and substantial consequences, which shall be explored in chapter 6.

4.4 Clifford Algebra Representations

The Clifford algebras are already known to be either isomorphic to simple algebras or isomorphic to the direct sum of simple algebras (namely, a semisimple algebra). As a result of the discussion in section 4.3, we now have a method to obtain matrix representations of Clifford algebras. Let us summarise the steps of this procedure:

(1) Choose a set of N primitive idempotents f_A $(A = 1, \ldots, N)$ of $C\ell_{p,q}$ such that $\sum_A f_A = 1$, with one of them being a primitive idempotent, for instance, f_1.

(2) Choose elements $\{\mathcal{E}_{A1}\}$ and $\{\mathcal{E}_{1A}\}$ $(A = 1, \ldots, N)$ such that $f_1 = \mathcal{E}_{1A}\mathcal{E}_{A1}$ and that eqn (4.43) holds, namely, $f_B\mathcal{E}_{A1} = \delta_{AB}\mathcal{E}_{A1}$, and $\mathcal{E}_{1A}f_B = \delta_{AB}\mathcal{E}_{1A}$.

It is possible to switch these steps with other, equivalent ones:

(1) Choose a primitive idempotent f_1 of $C\ell_{p,q}$.

(2) Find a basis of the ideal $C\ell_{p,q}\, f_1$, denoted by $\{\mathcal{E}_{A1}\}$, and the associated dual basis $\{\mathcal{E}_{1A}\}$ which satisfies $\mathcal{E}_{1A}\mathcal{E}_{B1} = \delta_{AB} f_1$.

The third step, whichever of the first two steps is chosen, is

(3) Define a basis for $\mathcal{M}(N, \mathbb{R})$ as $\mathcal{E}_{AB} = \mathcal{E}_{A1}\mathcal{E}_{1B}$. If $\{\gamma_i = \gamma(\mathbf{e}_i)\}$ $(i = 1, \ldots, n)$ are the generators of $C\ell_{p,q}$ $(p + q = n)$, its matrix representation is given by eqn (4.56):

$$(\gamma_i)_{AB} = \sum_C \mathcal{E}_{CA}\gamma_i\mathcal{E}_{BC}. \tag{4.61}$$

The scalars are isomorphic to the set $f_1\, C\ell_{p,q}\, f_1$, where f_1 is the identity. The representation space is isomorphic to the minimal left ideal $C\ell_{p,q}\, f_1$.

Example 4.2 Let us consider the Clifford algebra $C\ell_{2,0}$ with $\{\mathbf{e}_1, \mathbf{e}_2\}$ being an orthonormal basis of $\mathbb{R}^{2,0}$ satisfying $(\mathbf{e}_1)^2 = (\mathbf{e}_2)^2 = 1$, and $\mathbf{e}_1\mathbf{e}_2 + \mathbf{e}_2\mathbf{e}_1 = 0$. Obviously,

$$f_\pm = \frac{1}{2}(1 \pm \mathbf{e}_1)$$

are primitive idempotents of $C\ell_{2,0}$ and satisfy $1 = f_+ + f_-$. Other idempotents are $g_\pm = (1/2)(1 \pm \mathbf{e}_2)$, as well as suitable linear combinations of $f_\pm$ and $g_\pm$.

Let us begin with the first procedure. The idempotents are $f_1 = f_+$, and $f_2 = f_-$, and let us choose f_1. Now let us choose elements $\{\mathcal{E}_{A1}\}$ and $\{\mathcal{E}_{1A}\}$ $(A = 1, 2)$ such that

$$f_1 = \mathcal{E}_{1A}\mathcal{E}_{A1}$$

and

$$f_B\mathcal{E}_{A1} = \delta_{AB}\mathcal{E}_{A1}, \qquad \mathcal{E}_{1A}f_B = \delta_{AB}\mathcal{E}_{1A},$$

for $A, B = 1, 2$. Since $\{\mathcal{E}_{A1}\}$ and $\{\mathcal{E}_{1A}\}$ are elements of $C\ell_{2,0}$, they can be written as $\phi = a + b\mathbf{e}_1 + c\mathbf{e}_2 + d\mathbf{e}_1\mathbf{e}_2$. Let us calculate the products $f_1\phi$, $f_2\phi$, ϕf_1, and ϕf_2:

$$f_1\phi = (a + b)f_1 + (c + d)f_1\mathbf{e}_2, \qquad f_2\phi = (a - b)f_2 + (c - d)f_2\mathbf{e}_2,$$
$$\phi f_1 = (a + b)f_1 + (c - d)f_2\mathbf{e}_2, \qquad \phi f_2 = (a - b)f_2 + (c + d)f_1\mathbf{e}_2.$$

These equations imply that $f_1\phi f_1 = (a + b)f_1 = a' f_1$, $(a' \in \mathbb{R})$, in such a way that $f_1\phi f_1 \simeq \mathbb{R}$.

Using these results, it can be straightforwardly seen that the conditions $f_2\mathcal{E}_{11} = \mathcal{E}_{11}f_2 = 0$, and $f_1\mathcal{E}_{11} = \mathcal{E}_{11}f_1 = \mathcal{E}_{11}$ hold, and hence $\mathcal{E}_{11} = a(1 + \mathbf{e}_1)$. The conditions $f_1\mathcal{E}_{21} = \mathcal{E}_{21}f_2 = 0$ and $f_2\mathcal{E}_{21} = \mathcal{E}_{21}f_1 = \mathcal{E}_{21}$ are valid for $\mathcal{E}_{21} = c(1 - \mathbf{e}_1)\mathbf{e}_2$. Finally, the conditions $f_2\mathcal{E}_{12} = \mathcal{E}_{12}f_1 = 0$, and $f_1\mathcal{E}_{12} = \mathcal{E}_{12}f_2 = \mathcal{E}_{12}$ hold for $\mathcal{E}_{12} = c'(1 - \mathbf{e}_1)\mathbf{e}_2$. The condition $f_1 = \mathcal{E}_{11}\mathcal{E}_{11}$ holds if $a = 1/2$ in $\mathcal{E}_{11}$. On the other hand, the condition $f_1 = \mathcal{E}_{12}\mathcal{E}_{21}$ implies that $cc' = 1/4$. This is the same condition that is obtained with $f_2 = \mathcal{E}_{22} = \mathcal{E}_{21}\mathcal{E}_{12}$. The simplest solution is obviously $c = c' = 1/2$. Furthermore, we can express

$$\mathcal{E}_{11} = \frac{1}{2}(1+\mathbf{e}_1), \qquad \mathcal{E}_{12} = \frac{1}{2}(1+\mathbf{e}_1)\mathbf{e}_2,$$
$$\mathcal{E}_{21} = \frac{1}{2}(1-\mathbf{e}_1)\mathbf{e}_2, \qquad \mathcal{E}_{22} = \frac{1}{2}(1-\mathbf{e}_1).$$

In order to find a matrix representation of an element ϕ, let us calculate the matrix components $\phi_{AB} = \sum_C \mathcal{E}_{CA}\phi\mathcal{E}_{BC} = \mathcal{E}_{1A}\phi\mathcal{E}_{B1} + \mathcal{E}_{2A}\phi\mathcal{E}_{B2}$. This calculation is certainly a lot of work. A straightforward manner is to use directly the equations for $\mathcal{E}_{11}$, $\mathcal{E}_{12}$, $\mathcal{E}_{21}$, and $\mathbf{e}_{22}$. From the first and last equations, it can be seen that

$$1 = \mathcal{E}_{11} + \mathcal{E}_{22}, \qquad \mathbf{e}_1 = \mathcal{E}_{11} - \mathcal{E}_{22}.$$

From the second and third ones, we can see that

$$\mathbf{e}_2 = \mathcal{E}_{12} + \mathcal{E}_{21}, \qquad \mathbf{e}_1\mathbf{e}_2 = \mathcal{E}_{12} - \mathcal{E}_{21}.$$

It is immediately clear that

$$\rho(1) = \begin{pmatrix} 1 & 0 \\ 0 & 1 \end{pmatrix}, \qquad \rho(\mathbf{e}_1) = \begin{pmatrix} 1 & 0 \\ 0 & -1 \end{pmatrix},$$
$$\rho(\mathbf{e}_2) = \begin{pmatrix} 0 & 1 \\ 1 & 0 \end{pmatrix}, \qquad \rho(\mathbf{e}_1\mathbf{e}_2) = \begin{pmatrix} 0 & 1 \\ -1 & 0 \end{pmatrix}.$$

This representation is precisely that described in the discussion of the isomorphism $\mathcal{C}\ell_{2,0} = \mathcal{M}(2,\mathbb{R})$.

Let us consider the second procedure. Since $\phi f_1 = (a+b)(1/2)(1+\mathbf{e}_1) + (c-d)(1/2)(1-\mathbf{e}_1)\mathbf{e}_2$, it follows that the ideal $\mathcal{C}\ell_{2,0}\, f_1$ is given by

$$\mathcal{C}\ell_{2,0}\, f_1 = \{a'\frac{1}{2}(1+\mathbf{e}_1) + b'\frac{1}{2}(1-\mathbf{e}_1)\mathbf{e}_2 \mid a', b' \in \mathbb{R}\}.$$

It is clear that $\mathfrak{B} = \{(1/2)(1+\mathbf{e}_1), (1/2)(1-\mathbf{e}_1)\mathbf{e}_2\} = \{\mathcal{E}_{11}, \mathcal{E}_{21}\}$ is a basis for the ideal $\mathcal{C}\ell_{2,0}\, f_1$. The dual basis can be straightforwardly obtained, and the action of the dual basis elements $\{\mathcal{E}_{1A}\}$ can be written in this case by the multiplication, as $\mathcal{E}_{1A}(\mathcal{E}_{B1}) = \delta_{AB} f_1 = \mathcal{E}_{1A}\mathcal{E}_{B1}$. The dual basis is $\mathfrak{B}^* = \{(1/2)(1+\mathbf{e}_1), (1/2)(1+\mathbf{e}_1)\mathbf{e}_2\} = \{\mathcal{E}_{11}, \mathcal{E}_{12}\}$. The objects $\mathcal{E}_{AB}$ $(A, B = 1, 2)$ are exactly the ones already obtained and, therefore, the representation is completely derived.

Example 4.3 Another important example involves the Clifford algebra $\mathcal{C}\ell_{3,0}$. If $\{\mathbf{e}_1, \mathbf{e}_2, \mathbf{e}_3\}$ denotes an orthonormal basis of $\mathbb{R}^{3,0}$, we have $(\mathbf{e}_i)^2 = 1$ $(i = 1, 2, 3)$, and $\mathbf{e}_i\mathbf{e}_j + \mathbf{e}_j\mathbf{e}_i = 0$ $(i \neq j)$, and an arbitrary element of $\mathcal{C}\ell_{3,0}$ is given by

$$\phi = a_0 + a_1\mathbf{e}_1 + a_2\mathbf{e}_2 + a_3\mathbf{e}_3 + a_{12}\mathbf{e}_1\mathbf{e}_2 + a_{13}\mathbf{e}_1\mathbf{e}_3 + a_{23}\mathbf{e}_2\mathbf{e}_3 + a_{123}\mathbf{e}_1\mathbf{e}_2\mathbf{e}_{33}.$$

This algebra has an interesting property: the pseudoscalar (or 3-vector) $\mathbf{e}_1\mathbf{e}_2\mathbf{e}_3$ commutes with any element of $\mathcal{C}\ell_{3,0}$. Indeed, it is immediate to see that $\mathbf{e}_1\mathbf{e}_2\mathbf{e}_3\mathbf{e}_i = \mathbf{e}_i\mathbf{e}_1\mathbf{e}_2\mathbf{e}_3$ for $i = 1, 2, 3$, so that if $\mathbf{e}_1\mathbf{e}_2\mathbf{e}_3$ commutes with $\{\mathbf{e}_1, \mathbf{e}_2, \mathbf{e}_3\}$ which are generators of the algebra, then it commutes with any element in this algebra. The centre of $\mathcal{C}\ell_{3,0}$ is therefore $\mathrm{Cen}(\mathcal{C}\ell_{3,0}) = \bigwedge_0(\mathbb{R}^{3,0}) \oplus \bigwedge_3(\mathbb{R}^{3,0})$. On the other hand, the algebra $\mathcal{C}\ell_{3,0}$ presents a property that is not shared by any Clifford algebra. Indeed, calculating $(\mathbf{e}_1\mathbf{e}_2\mathbf{e}_3)^2$, we find that

$$(\mathbf{e}_1\mathbf{e}_2\mathbf{e}_3)^2 = -1.$$

The centre of $\mathcal{C}\ell_{3,0}$ is therefore such that $\mathrm{Cen}(\mathcal{C}\ell_{3,0}) \simeq \mathbb{C}$. This result suggests the following notation:

$$\mathbf{I} = \mathbf{e}_1\mathbf{e}_2\mathbf{e}_3.$$

A set of primitive idempotents of $\mathcal{C}\ell_{3,0}$ is

$$f_1 = \frac{1}{2}(1+\mathbf{e}_3), \qquad f_2 = \frac{1}{2}(1-\mathbf{e}_3).$$

The choice of the vector $\mathbf{e}_3$ is arbitrary. The vector $\mathbf{e}_1$ could be chosen as well as $\mathbf{e}_2$ or, in the general case, any vector $\mathbf{v}$ such that $\mathbf{v}^2 = 1$ can be placed, instead. Let us calculate $f_1\phi$, $f_2\phi$, ϕf_1, and ϕf_2. In order to do so, some preliminaries are required. First,

$$\mathbf{e}_3 f_1 = f_1\mathbf{e}_3 = f_1, \qquad \mathbf{e}_3 f_2 = f_2\mathbf{e}_3 = -f_2.$$

It then follows that

$$\mathbf{e}_1\mathbf{e}_2 f_1 = f_1\mathbf{e}_1\mathbf{e}_2 = \mathbf{I}f_1, \qquad \mathbf{e}_1\mathbf{e}_2 f_2 = f_2\mathbf{e}_1\mathbf{e}_2 = -\mathbf{I}f_2.$$

In addition,

$$\mathbf{e}_2 f_1 = f_2\mathbf{e}_2 = \mathbf{e}_1\mathbf{I}f_1, \qquad \mathbf{e}_2 f_2 = f_1\mathbf{e}_2 = -\mathbf{e}_1\mathbf{I}f_2.$$

By using those results and some other manipulations, we find that

$$\begin{aligned}
f_1\phi &= [(a_0+a_3)+(a_{12}+a_{123})\mathbf{I}]f_1 + [(a_1-a_{13})+(-a_2+a_{23})\mathbf{I}]f_1\mathbf{e}_1,\\
f_2\phi &= [(a_0-a_3)+(-a_{12}+a_{123})\mathbf{I}]f_2 + [(a_1+a_{13})+(a_2+a_{23})\mathbf{I}]f_2\mathbf{e}_1,\\
\phi f_1 &= [(a_0+a_3)+(a_{12}+a_{123})\mathbf{I}]f_1 + [(a_1+a_{13})+(a_2+a_{23})\mathbf{I}]\mathbf{e}_1 f_1,\\
\phi f_2 &= [(a_0-a_3)+(-a_{12}+a_{123})\mathbf{I}]f_2 + [(a_1-a_{13})+(-a_2+a_{23})\mathbf{I}]\mathbf{e}_1 f_2.
\end{aligned}$$

Furthermore, we can observe that

$$f_1\phi f_1 = [(a_0+a_3)+(a_{12}+a_{123})\mathbf{I}]f_1 \simeq \mathbb{C}.$$

The element 1 is represented by the identity matrix I, and the pseudoscalar $\mathbf{I}$ is represented by the matrix iI, where i denotes the imaginary unit.

Given these results, the conditions $f_2\mathcal{E}_{11} = \mathcal{E}_{11}f_2 = 0$ and $f_1\mathcal{E}_{11} = \mathcal{E}_{11}f_1 = \mathcal{E}_{11}$ hold for $\mathcal{E}_{11} = (a+b\mathbf{I})(1+\mathbf{e}_3)$. In addition, $\mathcal{E}_{11}\mathcal{E}_{11} = \mathcal{E}_{11}$, in such a way that $4(a^2-b^2) = 2a$ and $2ab = b$, the solution of which is $a = 1/2$, and $b = 0$. Therefore, $\mathcal{E}_{11}$ is given by

$$\mathcal{E}_{11} = \frac{1}{2}(1+\mathbf{e}_3).$$

The conditions $f_1\mathcal{E}_{21} = \mathcal{E}_{21}f_2 = 0$, and $f_2\mathcal{E}_{21} = \mathcal{E}_{21}f_1 = \mathcal{E}_{21}$, hold for $\mathcal{E}_{21} = (a'\mathbf{e}_1 + b'\mathbf{e}_2)(1+\mathbf{e}_3)$, inasmuch as the conditions $f_2\mathcal{E}_{12} = \mathcal{E}_{12}f_1 = 0$, and $f_1\mathcal{E}_{12} = \mathcal{E}_{12}f_2 = \mathcal{E}_{12}$, hold for $\mathcal{E}_{12} = (a''\mathbf{e}_1 + b''\mathbf{e}_2)(1-\mathbf{e}_3)$. The condition $\mathbf{e}_{12}\mathcal{E}_{21} = \mathcal{E}_{11}$ implies that $a'b'' = a''b'$, and $a'a'' + b'b'' = 1/4$. One possible solution is $b' = b'' = 0$, and $a' = a'' = 1/2$. Then,

$$\mathcal{E}_{12} = \mathbf{e}_1\frac{1}{2}(1-\mathbf{e}_3), \qquad \mathcal{E}_{21} = \mathbf{e}_1\frac{1}{2}(1+\mathbf{e}_3).$$

When $\mathcal{E}_{22} = \mathcal{E}_{21}\mathcal{E}_{12}$, it reads

$$\mathcal{E}_{22} = \frac{1}{2}(1-\mathbf{e}_3).$$

The equations for $\mathcal{E}_{AB}$ $(A, B = 1, 2)$ imply that

$$\begin{aligned}
1 &= \mathcal{E}_{11} + \mathcal{E}_{22}, \qquad & \mathbf{e}_3 &= \mathcal{E}_{11} - \mathcal{E}_{22},\\
\mathbf{e}_1 &= \mathcal{E}_{12} + \mathcal{E}_{21}, \qquad & \mathbf{e}_1\mathbf{e}_3 &= \mathcal{E}_{21} - \mathcal{E}_{12}.
\end{aligned}$$

For the remaining elements, we need just to take the representation of $\mathbf{I} = \mathbf{e}_1\mathbf{e}_2\mathbf{e}_3$ as being the imaginary unit i, namely, $\mathbf{I}\mathcal{E}_{AB} = i\mathcal{E}_{AB}$. In addition to these equations, we can express

$$\begin{aligned}
\mathbf{e}_1\mathbf{e}_2\mathbf{e}_3 &= \mathbf{I}1 = i\mathcal{E}_{11} + i\mathcal{E}_{22}, \qquad & \mathbf{e}_1\mathbf{e}_2 &= \mathbf{I}\mathbf{e}_3 = i\mathcal{E}_{11} - i\mathcal{E}_{22},\\
\mathbf{e}_2\mathbf{e}_3 &= \mathbf{I}\mathbf{e}_1 = i\mathcal{E}_{12} + i\mathcal{E}_{21}, \qquad & \mathbf{e}_2 &= \mathbf{I}\mathbf{e}_1\mathbf{e}_3 = i\mathcal{E}_{21} - i\mathcal{E}_{12}.
\end{aligned}$$

A matrix representation of the vectors $\mathbf{e}_1$, $\mathbf{e}_2$, and $\mathbf{e}_3$ is given by

$$\rho(\mathbf{e}_1) = \begin{pmatrix} 0 & 1\\ 1 & 0\end{pmatrix}, \qquad \rho(\mathbf{e}_2) = \begin{pmatrix} 0 & -i\\ i & 0\end{pmatrix}, \qquad \rho(\mathbf{e}_3) = \begin{pmatrix} 1 & 0\\ 0 & -1\end{pmatrix},$$

which are the Pauli matrices $\sigma_i = \rho(\mathbf{e}_i)$ $(i = 1, 2, 3)$ already seen in example 3.3.

Using the other procedure described in the text, we must regard the ideal $\mathcal{C}\ell_{3,0}\, f_1$, whose elements are expressed by

$$\mathcal{C}\ell_{3,0}\, f_1 = \{(a+b\mathbf{I})f_1 + (c+d\mathbf{I})\mathbf{e}_1 f_1 \mid a, b, c, d \in \mathbb{R}\}.$$

A basis for this ideal is provided by

$$\mathfrak{B} = \{f_1, \mathbf{e}_1 f_1\} = \{\mathcal{E}_{11}, \mathcal{E}_{21}\}.$$

Since $\mathbf{e}_1 f_1 = f_2\mathbf{e}_1$, and $f_1 f_2 = f_2 f_1 = 0$, and $\mathcal{E}_{1A}(\mathcal{E}_{B1}) = \mathcal{E}_{1A}\mathcal{E}_{B1} = \delta_{AB} f_1$, it follows that the dual basis is

$$\mathfrak{B}^* = \{f_1, f_1\mathbf{e}_1\} = \{\mathcal{E}_{11}, \mathcal{E}_{12}\},$$

and therefore $\mathcal{E}_{22} = \mathcal{E}_{21}\mathcal{E}_{12} = \mathbf{e}_1 f_1 f_1 \mathbf{e}_1 = \mathbf{e}_1 f_1 \mathbf{e}_1 = (\mathbf{e}_1)^2 f_2 = f_2$. The representation is derived in the same way as the previous procedure was.

In these methods for obtaining a representation of a Clifford algebra, the starting point is a primitive idempotent that generates – by right multiplication – a minimal left ideal $C\ell_{p,q}\, f$, which is the space of representation of a simple Clifford algebra. In order to close this topic, we shall establish a result that makes it possible to obtain this idempotent.

In what follows, the multi-index notation shall be used. Let $\{\mathbf{e}_i\}$ $(i = 1, \ldots, n = p + q)$ be an orthonormal basis of $\mathbb{R}^{p,q}$. An element $\mathbf{e}_{\mu_1} \cdots \mathbf{e}_{\mu_k}$ of $C\ell_{p,q}$ is denoted by

$$\mathbf{e}_{\mu_1 \cdots \mu_k} = \mathbf{e}_{\mu_1} \cdots \mathbf{e}_{\mu_k}. \tag{4.62}$$

and the set of multi-indices by I.

Theorem 4.6 ▶ *Let $C\ell_{p,q}$ be the Clifford algebra associated with $\mathbb{R}^{p,q}$ and let $\{\mathbf{e}_i\}$ $(i = 1, \ldots, n)$ be an orthonormal basis of this quadratic space. A primitive idempotent of $C\ell_{p,q}$ is given by*

$$f = \frac{1}{2}(1 + \mathbf{e}_{I_1}) \cdots \frac{1}{2}(1 + \mathbf{e}_{I_k}), \tag{4.63}$$

where $\{\mathbf{e}_{I_1}, \ldots, \mathbf{e}_{I_k}\}$ is a set of elements which are in $C\ell_{p,q}$, which commute, and such that $(\mathbf{e}_{I_\alpha})^2 = 1$ for $\alpha = 1, \ldots, k$. This idempotent then generates a group of order 2^k, where $k = q - r_{q-p}$, and r_j are the Radon–Hurwitz numbers defined by table 4.3 with the recurrence relation $r_{j+8} = r_j + 4$.

Table 4.3 The Radon–Hurwitz Numbers

j	0	1	2	3	4	5	6	7
r_j	0	1	2	2	3	3	3	3

Proof: Let us take an element $\mathbf{e}_{I_1}$ of $C\ell_{p,q}$ such that $(\mathbf{e}_{I_1})^2 = 1$. Hence, $f_1^{\pm} = (1/2)(1 \pm \mathbf{e}_{I_1})$ are orthogonal idempotents: in fact, $f_1^+ + f_1^- = 1$, and $f_1^+ f_1^- = f_1^- f_1^+ = 0$. The Clifford algebra $C\ell_{p,q}$ can be therefore decomposed into the sum of two ideals $C\ell_{p,q}\, f_1^+ \oplus C\ell_{p,q}\, f_1^-$, and the dimension of each one of these ideals is $2^n/2 = 2^{n-1}$. Let now $\mathbf{e}_{I_2}$ be another element of $C\ell_{p,q}$ such that $(\mathbf{e}_{I_2})^2 = 1$. In this way, $f_2^{\pm} = (1/2)(1 \pm \mathbf{e}_{I_2})$ are idempotents satisfying $f_2^+ + f_2^- = 1$, and $f_2^+ f_2^- = f_2^- f_2^+ = 0$. If $\mathbf{e}_{I_1}$ and $\mathbf{e}_{I_2}$ commute, four mutually orthogonal idempotents can be constructed: $f_1^+ f_2^+$, $f_1^- f_2^-$, $f_1^- f_2^+$, and $f_1^+ f_2^-$, the sum of which is the identity. The algebra $C\ell_{p,q}$ can be thus decomposed into the sum of four ideals, each one with dimension 2^{n-2}. Continuing this procedure, a set of 2^k idempotents

$$\frac{1}{2}(1 \pm \mathbf{e}_{I_1}) \cdots \frac{1}{2}(1 \pm \mathbf{e}_{I_k})$$

can be obtained, and we can decompose $C\ell_{p,q}$ in the sum of 2^k ideals, each one with dimension 2^{n-k}. Such reasoning can be again used until 2^{n-k} equals the dimension of the irreducible representation space of $C\ell_{p,q}$, a value that can be obtained from table 4.1. With a little work, it can be concluded that k is given by this formula. ✓

Example 4.4 Let us consider the Clifford algebra $\mathcal{C}\ell_{0,7}$. According to theorem 4.6, $k = 7 - r_7 = 7 - 3 = 4$. We must look for a set of four elements, that is, $\mathbf{e}_{I_1}$, $\mathbf{e}_{I_2}$, $\mathbf{e}_{I_3}$, and $\mathbf{e}_{I_4}$, which are of $\mathcal{C}\ell_{0,7}$, which commute, and the square of which equals 1. Since, for all vectors in $\mathcal{C}\ell_{0,7}$, we have $(\mathbf{e}_i)^2 = -1$ $(i = 1, \ldots, 7)$, the elements that we search for are not 1-vectors and, furthermore, cannot be 2-vectors, since $(\mathbf{e}_i\mathbf{e}_j)^2 = -(\mathbf{e}_i)^2(\mathbf{e}_j)^2 = -1$ $(i \neq j)$. Meanwhile, the four elements can be 3-vectors, since $(\mathbf{e}_i\mathbf{e}_j\mathbf{e}_k)^2 = -(\mathbf{e}_i)^2(\mathbf{e}_j)^2(\mathbf{e}_k)^2 = 1$ $(i \neq j \neq k)$. Let us choose one of those elements as $\mathbf{e}_{I_1} = \mathbf{e}_{123} = \mathbf{e}_1\mathbf{e}_2\mathbf{e}_3$. Moreover, 3-vectors, having only one of their indices equal, commute. Hence, we can identify, for instance, $\mathbf{e}_{I_2} = \mathbf{e}_{145}$. Moreover, $\mathbf{e}_{I_3} = \mathbf{e}_{167}$, and the fourth element might be $\mathbf{e}_{I_4} = \mathbf{e}_{347}$, for example. Therefore, the idempotent

$$f = \frac{1}{2}(1 + \mathbf{e}_1\mathbf{e}_2\mathbf{e}_3)\frac{1}{2}(1 + \mathbf{e}_1\mathbf{e}_4\mathbf{e}_5)\frac{1}{2}(1 + \mathbf{e}_1\mathbf{e}_6\mathbf{e}_7)\frac{1}{2}(1 + \mathbf{e}_3\mathbf{e}_4\mathbf{e}_7)$$

is primitive in $\mathcal{C}\ell_{0,7}$.

Example 4.5 Let us consider $\mathcal{C}\ell_{1,3}(\mathbb{C}) \simeq \mathcal{M}(4, \mathbb{C})$ and let $\{\mathbf{e}_0, \mathbf{e}_1, \mathbf{e}_2, \mathbf{e}_3\}$ be an orthonormal basis of $\mathbb{R}^{1,3}$. Four primitive idempotents f_1, f_2, f_3, and f_4 must be obtained such that $1 = f_1 + f_2 + f_3 + f_4$. By theorem 4.6, we know that two elements, $\mathbf{e}_{I_1}$, $\mathbf{e}_{I_2}$, which are of $\mathcal{C}\ell_{1,3}(\mathbb{C})$, which commute, and for which $(\mathbf{e}_{I_1})^2 = (\mathbf{e}_{I_2})^2 = 1$, must be obtained in such a way that such four primitive idempotents are $(1/2)(1 \pm \mathbf{e}_{I_1})(1/2)(1 \pm \mathbf{e}_{I_2})$. For the *standard representation* of the so-called *Dirac gamma matrices*, these elements are $\mathbf{e}_{I_1} = \mathbf{e}_0$, and $\mathbf{e}_{I_2} = i\mathbf{e}_1\mathbf{e}_2$. Hence,

$$f_1 = \frac{1}{2}(1 + \mathbf{e}_0)\frac{1}{2}(1 + i\mathbf{e}_1\mathbf{e}_2), \qquad f_2 = \frac{1}{2}(1 + \mathbf{e}_0)\frac{1}{2}(1 - i\mathbf{e}_1\mathbf{e}_2),$$
$$f_3 = \frac{1}{2}(1 - \mathbf{e}_0)\frac{1}{2}(1 + i\mathbf{e}_1\mathbf{e}_2), \qquad f_4 = \frac{1}{2}(1 - \mathbf{e}_0)\frac{1}{2}(1 - i\mathbf{e}_1\mathbf{e}_2). \tag{4.64}$$

These four primitive idempotents are similar. Indeed,

$$\mathbf{e}_{13}f_1(\mathbf{e}_{13})^{-1} = f_2, \qquad \mathbf{e}_{30}f_1(\mathbf{e}_{30})^{-1} = f_3, \qquad \mathbf{e}_{10}f_1(\mathbf{e}_{10})^{-1} = f_4. \tag{4.65}$$

Thus, $\mathbf{e}_{13}f_1 \subset f_2\mathcal{C}\ell_{1,3}(\mathbb{C})f_1$; $\mathbf{e}_{30}f_1 \subset f_3\mathcal{C}\ell_{1,3}(\mathbb{C})f_1$; and $\mathbf{e}_{10}f_1 \subset f_4\mathcal{C}\ell_{1,3}(\mathbb{C})f_1$; so

$$\mathcal{E}_{11} = f_1, \quad \mathcal{E}_{21} = -\mathbf{e}_{13}f_1, \quad \mathcal{E}_{31} = \mathbf{e}_{30}f_1, \quad \mathcal{E}_{41} = \mathbf{e}_{10}f_1. \tag{4.66}$$

With the restriction that $\mathcal{E}_{1j} \subset f_1\mathcal{C}\ell_{1,3}(\mathbb{C})f_j$, and $\mathcal{E}_{1j}\mathcal{E}_{j1} = f_1$, this result implies that

$$\mathcal{E}_{11} = f_1, \quad \mathcal{E}_{12} = \mathbf{e}_{13}f_2, \quad \mathcal{E}_{13} = \mathbf{e}_{30}f_3, \quad \mathcal{E}_{14} = \mathbf{e}_{10}f_4. \tag{4.67}$$

All the $\mathcal{E}_{ij}$ are shown in table 4.4. From these identities, the matrix representations of $\mathbf{e}_\mu$, denoted by $\gamma_\mu = \gamma(\mathbf{e}_\mu)$, can be immediately obtained:

(i) $\mathbf{e}_0 = f_1 + f_2 - f_3 - f_4 = \mathcal{E}_{11} + \mathcal{E}_{22} - \mathcal{E}_{33} - \mathcal{E}_{44}$. Hence,

$$\gamma(\mathbf{e}_0) = \gamma_0 = \begin{pmatrix} 1 & 0 & 0 & 0 \\ 0 & 1 & 0 & 0 \\ 0 & 0 & -1 & 0 \\ 0 & 0 & 0 & -1 \end{pmatrix}. \tag{4.68}$$

(ii) $\mathbf{e}_{10} = \mathbf{e}_{10}f_1 + \mathbf{e}_{10}f_2 + \mathbf{e}_{10}f_3 + \mathbf{e}_{10}f_4 = \mathcal{E}_{41} + \mathcal{E}_{32} + \mathcal{E}_{23} + \mathcal{E}_{14}$. Therefore,

$$\gamma(\mathbf{e}_{10}) = \gamma_{10} = \begin{pmatrix} 0 & 0 & 0 & 1 \\ 0 & 0 & 1 & 0 \\ 0 & 1 & 0 & 0 \\ 1 & 0 & 0 & 0 \end{pmatrix}, \quad \gamma(\mathbf{e}_1) = \gamma_1 = \gamma_{10}\gamma_0 = \begin{pmatrix} 0 & 0 & 0 & -1 \\ 0 & 0 & -1 & 0 \\ 0 & 1 & 0 & 0 \\ 1 & 0 & 0 & 0 \end{pmatrix} = \begin{pmatrix} 0 & -\sigma_1 \\ \sigma_1 & 0 \end{pmatrix}. \tag{4.69}$$

Table 4.4 The Elements $\mathcal{E}_{ij}$ for $i, j = 1, 2, 3, 4$.

$\mathcal{E}_{ij}$	$\mathcal{E}_{i1}$	$\mathcal{E}_{i2}$	$\mathcal{E}_{i3}$	$\mathcal{E}_{i4}$
$\mathcal{E}_{1j}$	f_1	$\mathbf{e}_{13}f_2$	$\mathbf{e}_{30}f_3$	$\mathbf{e}_{10}f_4$
$\mathcal{E}_{2j}$	$-\mathbf{e}_{13}f_1$	f_2	$\mathbf{e}_{10}f_3$	$-\mathbf{e}_{30}f_4$
$\mathcal{E}_{3j}$	$\mathbf{e}_{30}f_1$	$\mathbf{e}_{10}f_2$	f_3	$\mathbf{e}_{13}f_4$
$\mathcal{E}_{4j}$	$\mathbf{e}_{10}f_1$	$\mathbf{e}_{03}f_2$	$-\mathbf{e}_{13}f_3$	f_4

(iii) $\mathbf{e}_{30} = \mathbf{e}_{30}f_1 + \mathbf{e}_{30}f_2 + \mathbf{e}_{30}f_3 + \mathbf{e}_{30}f_4 = \mathcal{E}_{31} - \mathcal{E}_{42} + \mathcal{E}_{13} - \mathcal{E}_{24}$. This result implies that

$$\gamma(\mathbf{e}_{30}) = \gamma_{30} = \begin{pmatrix} 0 & 0 & 1 & 0 \\ 0 & 0 & 0 & -1 \\ 1 & 0 & 0 & 0 \\ 0 & -1 & 0 & 0 \end{pmatrix}, \quad \text{and} \quad \gamma_3 = -\gamma_0\gamma_{03} = \begin{pmatrix} 0 & -\sigma_3 \\ \sigma_3 & 0 \end{pmatrix}. \tag{4.70}$$

(iv) $f_1+f_3-f_2-f_4 = i\mathbf{e}_1\mathbf{e}_2 \Longrightarrow \mathbf{e}_2 = i\mathbf{e}_0(\mathbf{e}_{01}f_1+\mathbf{e}_{01}f_3-\mathbf{e}_{01}f_2-\mathbf{e}_{01}f_4) = i\mathbf{e}_0(\mathcal{E}_{14}+\mathcal{E}_{32}-\mathcal{E}_{23}-\mathcal{E}_{41})$. It follows that

$$\gamma_2 = i\begin{pmatrix} 1 & 0 & 0 & 0 \\ 0 & 1 & 0 & 0 \\ 0 & 0 & -1 & 0 \\ 0 & 0 & 0 & -1 \end{pmatrix}\begin{pmatrix} 0 & 0 & 0 & 1 \\ 0 & 0 & -1 & 0 \\ 0 & 1 & 0 & 0 \\ -1 & 0 & 0 & 0 \end{pmatrix} = \begin{pmatrix} 0 & -\sigma_2 \\ \sigma_2 & 0 \end{pmatrix}. \tag{4.71}$$

The Dirac matrices standard representation is consequently given by

$$\gamma_0 = \begin{pmatrix} I & 0 \\ 0 & -I \end{pmatrix}, \quad \gamma(\mathbf{e}_k) = \gamma_k = \begin{pmatrix} 0 & -\sigma_k \\ \sigma_k & 0 \end{pmatrix} \tag{4.72}$$

Example 4.6 In view of the isomorphism $\mathcal{C}\ell_{4,1} \simeq \mathbb{C}\otimes\mathcal{C}\ell_{1,3}$ and the matrix representation of $\mathbb{C}\otimes\mathcal{C}\ell_{1,3} \simeq \mathcal{C}\ell_{1,3}(\mathbb{C})$ obtained in example 4.5, we can construct a matrix representation of $\mathcal{C}\ell_{4,1}$. But in order to implement this representation, we need to explicitly construct such an isomorphism. Let us denote the generators of $\mathcal{C}\ell_{4,1}$ by E_A $(A = 0,1,2,3,4)$ such that $E_1^2 = E_2^2 = E_3^2 = E_4^2 = -E_0^2 = 1$. An isomorphism $\rho_1 : \mathcal{C}\ell_{4,1} \to \mathbb{C}\otimes\mathcal{C}\ell_{1,3}$ can be defined by

$$\rho_1(E_\mu) = -i\star\gamma_\mu = -i\gamma_\mu\gamma_{0123} \quad (\mu = 0,1,2,3), \quad \rho_1(E_4) = -i\star 1 = -i\gamma_{0123}.$$

If $\varrho : \mathcal{C}\ell_{1,3}(\mathbb{C}) \to \mathcal{M}(4,\mathbb{C})$ is the standard representation from example 4.5, we have a representation $\varrho_1 = \varrho\circ\rho_1 : \mathcal{C}\ell_{4,1} \to \mathcal{M}(4,\mathbb{C})$ such that, if $Z \in \mathcal{C}\ell_{4,1}$ is an arbitrary element of the form

$$Z = H + H^A E_A + H^{AB}E_{AB} + H^{ABC}E_{ABC} + H^{ABCD}E_{ABCD} + H^{01234}E_{01234},$$

its matrix representation $\varrho_1(Z)$ reads

$$\varrho_1(Z) = \begin{pmatrix} z_{11} & z_{12} & z_{13} & z_{14} \\ z_{21} & z_{22} & z_{23} & z_{24} \\ z_{31} & z_{32} & z_{33} & z_{34} \\ z_{41} & z_{42} & z_{43} & z_{44} \end{pmatrix} = \begin{pmatrix} \phi_1 & \phi_2 \\ \phi_3 & \phi_4 \end{pmatrix},$$

where

$$\begin{aligned}
z_{11} &= (H + H^{04} + H^{034} - H^3) + i(H^{01234} - H^{123} + H^{12} + H^{0124}),\\
z_{12} &= (-H^{13} - H^{0134} + H^{014} - H^1) + i(-H^{024} + H^2 + H^{23} + H^{0234}),\\
z_{13} &= (H^{03} - H^{34} + H^4 + H^0) + i(H^{124} + H^{012} + H^{0123} - H^{1234}),\\
z_{14} &= (H^{01} - H^{14} + H^{134} - H^{013}) + i(H^{234} + H^{023} - H^{02} + H^{24}),\\
z_{21} &= (H^{13} + H^{0134} + H^{014} - H^1) + i(H^{024} - H^2 + H^{23} + H^{0234}),\\
z_{22} &= (H + H^{04} - H^{034} + H^3) + i(H^{01234} - H^{123} - H^{12} - H^{0124}),\\
z_{14} &= (H^{01} - H^{14} + H^{134} + H^{013}) + i(H^{234} + H^{023} + H^{02} - H^{24}),\\
z_{24} &= (-H^{03} + H^{34} + H^4 + H^0) + i(-H^{124} - H^{012} + H^{0123} - H^{1234}),\\
z_{31} &= (H^{03} + H^{34} + H^4 - H^0) + i(H^{124} - H^{012} + H^{0123} + H^{1234}),\\
z_{32} &= (H^{01} + H^{14} - H^{134} + H^{013}) + i(H^{234} - H^{023} - H^{02} - H^{24}),\\
z_{33} &= (H - H^{04} + H^{034} + H^3) + i(H^{01234} + H^{123} + H^{12} - H^{0124}),\\
z_{34} &= (-H^{13} + H^{0134} + H^{014} + H^1) + i(-H^{024} - H^2 + H^{23} - H^{0234}),\\
z_{41} &= (H^{01} + H^{14} + H^{134} - H^{013}) + i(H^{234} - H^{023} + H^{02} + H^{24}),\\
z_{42} &= (-H^{03} - H^{34} + H^4 - H^0) + i(-H^{124} + H^{012} + H^{0123} + H^{1234}),\\
z_{43} &= (H^{13} - H^{0134} + H^{014} + H^1) + i(H^{024} + H^2 + H^{23} - H^{0234}),\\
z_{44} &= (H - H^{04} - H^{034} - H^3) + i(H^{01234} + H^{123} - H^{12} + H^{0124}).
\end{aligned}$$

The matrix representations of the conjugation $\bar{Z}$, the reversion $\tilde{Z}$, and the graded involution $\hat{Z}$ are given, respectively, by

$$\varrho_1(\bar{Z}) = \begin{pmatrix} \phi_4^\dagger & -\phi_2^\dagger \\ -\phi_3^\dagger & \phi_1^\dagger \end{pmatrix}, \quad \varrho_1(\tilde{Z}) = \begin{pmatrix} \mathrm{adj}(\phi_4) & \mathrm{adj}(\phi_2) \\ \mathrm{adj}(\phi_3) & \mathrm{adj}(\phi_1) \end{pmatrix},$$

and

$$\varrho_1(\hat{Z}) = \begin{pmatrix} \mathrm{cof}(\phi_1^*) & -\mathrm{cof}(\phi_2^*) \\ -\mathrm{cof}(\phi_3^*) & \mathrm{cof}(\phi_4^*) \end{pmatrix}, \qquad \text{with} \quad \mathrm{cof}\begin{pmatrix} a & b \\ c & d \end{pmatrix} = \begin{pmatrix} d & -c \\ -b & a \end{pmatrix},$$

where $\dagger$ denotes Hermitian conjugation, and $\mathrm{cof}(\phi) = \mathrm{adj}(\phi^\mathsf{T})$.

Another isomorphism, $\rho_2 : \mathcal{C}\ell_{4,1} \to \mathbb{C} \otimes \mathcal{C}\ell_{1,3}$, is given by

$$\rho_2(E_\mu) = -i\gamma_\mu \quad (\mu = 0, 1, 2, 3), \quad \rho_2(E_4) = -i\gamma_{0123}.$$

Now we have the matrix representation $\varrho_2 = \varrho \circ \rho_2$ such that

$$\varrho_2(Z) = \begin{pmatrix} z'_{11} & z'_{12} & z'_{13} & z'_{14} \\ z'_{21} & z'_{22} & z'_{23} & z'_{24} \\ z'_{31} & z'_{32} & z'_{33} & z'_{34} \\ z'_{41} & z'_{42} & z'_{43} & z'_{44} \end{pmatrix} = \begin{pmatrix} \phi'_1 & \phi'_2 \\ \phi'_3 & \phi'_4 \end{pmatrix},$$

where

$$\begin{aligned}
z'_{11} &= (H - H^{1234} + H^{034} + H^{012}) + i(H^{01234} - H^0 + H^{12} + H^{34}), \\
z'_{12} &= (-H^{13} + H^{24} + H^{014} + H^{023}) + i(-H^{024} + H^{013} + H^{23} + H^{14}), \\
z'_{13} &= (H^{03} - H^{0124} + H^4 - H^{123}) + i(H^{124} + H^3 + H^{0123} - H^{04}), \\
z'_{14} &= (H^{01} + H^{0234} - H^{134} + H^2) + i(H^{234} + H^1 - H^{02} + H^{0134}), \\
z'_{21} &= (H^{13} - H^{24} + H^{014} + H^{023}) + i(H^{024} - H^{013} + H^{23} + H^{14}), \\
z'_{22} &= (H - H^{1234} - H^{034} - H^{012}) + i(H^{01234} - H^0 - H^{12} - H^{34}), \\
z'_{23} &= (H^{01} + H^{0234} + H^{134} - H^2) + i(H^{234} + H^1 + H^{02} - H^{0134}), \\
z'_{24} &= (-H^{03} - H^{0124} + H^4 + H^{123}) + i(-H^{124} - H^3 + H^{0123} + H^{04}), \\
z'_{31} &= (H^{03} - H^{0124} + H^4 + H^{123}) + i(H^{124} - H^3 + H^{0123} + H^{04}), \\
z'_{32} &= (H^{01} - H^{0234} - H^{134} - H^2) + i(H^{234} - H^1 - H^{02} - H^{0134}), \\
z'_{33} &= (H + H^{1234} + H^{034} - H^{012}) + i(H^{01234} + H^0 + H^{12} - H^{34}), \\
z'_{34} &= (-H^{13} - H^{24} + H^{014} - H^{023}) + i(-H^{024} - H^{013} + H^{23} - H^{14}), \\
z'_{41} &= (H^{01} - H^{0234} + H^{134} + H^2) + i(H^{234} - H^1 + H^{02} + H^{0134}), \\
z'_{42} &= (-H^{03} + H^{0124} + H^4 + H^{123}) + i(-H^{124} + H^3 + H^{0123} + H^{04}), \\
z'_{43} &= (H^{13} + H^{24} + H^{014} - H^{023}) + i(H^{024} + H^{013} + H^{23} - H^{14}), \\
z'_{44} &= (H + H^{1234} - H^{034} + H^{012}) + i(H^{01234} + H^0 - H^{12} + H^{34}).
\end{aligned}$$

The matrix representation of the conjugation $\bar{Z}$, the reversion $\tilde{Z}$, and the graded involution $\hat{Z}$ are given, respectively, by

$$\varrho_2(\bar{Z}) = \begin{pmatrix} \phi_1'^\dagger & -\phi_3'^\dagger \\ -\phi_2'^\dagger & \phi_4'^\dagger, \end{pmatrix}, \quad \varrho_2(\tilde{Z}) = \begin{pmatrix} \mathrm{adj}(\phi'_1) & \mathrm{adj}(\phi'_3) \\ \mathrm{adj}(\phi'_2) & \mathrm{adj}(\phi'_4) \end{pmatrix},$$

and

$$\varrho_2(\hat{Z}) = \begin{pmatrix} \mathrm{cof}(\phi_1'^*) & -\mathrm{cof}(\phi_2'^*) \\ -\mathrm{cof}(\phi_3'^*) & \mathrm{cof}(\phi_4'^*) \end{pmatrix}.$$

Now a few remarks are deserved. First, notice that the matrix representations $\varrho_1(Z)$ and $\varrho_2(Z)$ are different, although we have used the same standard representation of the gamma matrices. Second, $\varrho_1(\bar{Z})$ and $\varrho_2(\bar{Z})$ are obtained from $\varrho_1(Z)$ and $\varrho_2(Z)$, respectively, by different matrix operations. The same is true for $\varrho_1(\tilde{Z})$ and $\varrho_2(\tilde{Z})$. However, $\varrho_1(\hat{Z})$ and $\varrho_2(\hat{Z})$ are obtained from $\varrho_1(Z)$ and $\varrho_2(Z)$, respectively, by the same matrix operations. The first fact follows from ρ_1 and ρ_2 being different isomorphisms; the images of $Z \in \mathcal{C}\ell_{4,1}$ in $\mathbb{C} \otimes \mathcal{C}\ell_{1,3}$ are different, and it is not difficult to see that

$$\rho_1(Z) = \mathsf{T}_+[\rho_2(Z)], \quad \rho_2(Z) = \mathsf{T}_-[\rho_1(Z)],$$

where

$$\mathsf{T}_{\pm}[\psi] = \frac{1}{2}(\psi + \hat{\psi}) \pm \frac{1}{2}(\psi - \hat{\psi})\gamma_{0123}.$$

The second follows from the fact that there are different images of the anti-automorphisms but the same one for the graded involution, that is

$$\begin{aligned}
\rho_1(\overline{\mathcal{C}\ell_{4,1}}) &\simeq \mathbb{C}^* \otimes \overline{\mathcal{C}\ell_{1,3}}, & \rho_2(\overline{\mathcal{C}\ell_{4,1}}) &\simeq \mathbb{C}^* \otimes \widetilde{\mathcal{C}\ell_{1,3}},\\
\rho_1(\widetilde{\mathcal{C}\ell_{4,1}}) &\simeq \mathbb{C} \otimes \overline{\mathcal{C}\ell_{1,3}}, & \rho_2(\widetilde{\mathcal{C}\ell_{4,1}}) &\simeq \mathbb{C} \otimes \widetilde{\mathcal{C}\ell_{1,3}},\\
\rho_1(\widehat{\mathcal{C}\ell_{4,1}}) &\simeq \mathbb{C}^* \otimes \mathcal{C}\ell_{1,3}, & \rho_2(\widehat{\mathcal{C}\ell_{4,1}}) &\simeq \mathbb{C}^* \otimes \mathcal{C}\ell_{1,3},
\end{aligned}$$

Notice that $\mathsf{T}_{\pm}[\widetilde{\mathcal{C}\ell_{1,3}}] = \overline{\mathcal{C}\ell_{1,3}}$, and $\mathsf{T}_{\pm}[\overline{\mathcal{C}\ell_{1,3}}] = \widetilde{\mathcal{C}\ell_{1,3}}$. It is also opportune to observe that the complex conjugation in $\mathbb{C}^* \otimes \mathcal{C}\ell_{1,3}$ is not the same as the complex conjugation in $\mathcal{M}(4,\mathbb{C})$, as will be seen in what follows. Moreover, the graded involution in $\mathcal{C}\ell_{1,3}$ is the image of the automorphism

$$Z^{\triangle} = E_4 Z E_4,$$

that is,

$$\rho_1(\mathcal{C}\ell_{4,1}^{\triangle}) \simeq \mathbb{C} \otimes \widehat{\mathcal{C}\ell_{1,3}} \qquad \rho_2(\mathcal{C}\ell_{4,1}^{\triangle}) \simeq \mathbb{C} \otimes \widehat{\mathcal{C}\ell_{1,3}}.$$

Their matrix representations are, respectively,

$$\varrho_1(Z) = \begin{pmatrix} \phi_4 & \phi_3 \\ \phi_2 & \phi_1 \end{pmatrix}, \qquad \varrho_2(Z) = \begin{pmatrix} \phi_4' & \phi_3' \\ \phi_2' & \phi_1' \end{pmatrix}.$$

Another interesting isomorphism is $\rho_3 : \mathcal{C}\ell_{4,1} \to \mathbb{C} \otimes \mathcal{C}\ell_{1,3}$, which will be useful when discussing twistors from the Clifford algebraic point of view (section 6.15), is

$$\rho_3(E_0) = i\gamma_0, \quad \rho_3(E_i) = \gamma_{i0}\ (i = 1, 2, 3), \quad \rho_3(E_4) = -\gamma_{123}.$$

The Hermitian Conjugation

The Hermitian conjugation is defined in the matrix algebra as a composition of the complex conjugation and the transposition operations. Let us discuss the features related to each of these operations in this context.

It is well-known that complex Clifford algebras are isomorphic either to the complex algebra of the matrices or to the sum of two such algebras. In addition, the operation of complex conjugation necessarily means that the complex conjugate of the matrices components is taken into account. Consider an algebra $\mathcal{A} \simeq \mathbb{C} \otimes \mathcal{M}(r, \mathbb{R})$ with an involutive automorphism which is denoted by * and which induces a non-trivial automorphism at the centre of the algebra, and let $\mathcal{B}$ be a real subalgebra of $\mathcal{A}$ – namely, $a \in \mathcal{B}$ if and only if $a = a^*$ (Benn and Tucker, 1987). Since any element $a \in \mathcal{A}$ can be expressed as the sum of real and imaginary parts, it follows that $\mathcal{A} = \mathbb{C} \otimes \mathcal{B}$, and therefore $\mathbb{C} \otimes \mathcal{B} \simeq \mathbb{C} \otimes \mathcal{M}(r, \mathbb{R})$. In order for these statements to be true, the algebra $\mathcal{B}$ must be simple; and, since only real algebras are isomorphic to the tensor product of matrix algebras by the algebras $\mathbb{R}, \mathbb{C}$ or $\mathbb{H}$, either $\mathcal{B} \simeq \mathcal{M}(r, \mathbb{R})$, or $\mathcal{B} \simeq \mathbb{H} \otimes \mathcal{M}(\frac{r}{2}, \mathbb{R})$. Now let us define in $\mathcal{A}$ another involutive automorphism * which leaves elements of $\mathcal{M}(r, \mathbb{R})$ invariant, and which conjugates the elements in the centre. Choose the bases $\{\mathcal{E}_{ij}\}$ and $\{\mathcal{E}^*_{ij}\}$ for $\mathcal{M}(r, \mathbb{C})$, so that $\mathcal{E}^*_{ij} = m\mathcal{E}_{ij}m^{-1}$ for some $m \in \mathcal{A}$. Hence, if

$$a = \sum_{i,j=1}^{r} a_{ij}\mathcal{E}_{ij}, \tag{4.73}$$

then

$$a^* = \sum_{i,j=1}^{r} a_{ij}^* m\mathcal{E}_{ij}m^{-1} = m\left(\sum_{i,j=1}^{r} a_{ij}^*\mathcal{E}_{ij}\right)m^{-1}, \tag{4.74}$$

which implies that

$$a^* = ma^\star m^{-1}. \tag{4.75}$$

Since * and * are involutions, eqn (4.75) asserts that $m^*m = \rho$, where ρ is an element of the centre of the algebra. The involutions * and * induce the same automorphism in the centre, and $(m^*m)^* = (m^*m)^\star = m^{-1}(m^*m)m$, namely, $mm^* = m^*m$, and ρ is real. Thus, we can redefine m up to a scale factor, and $m^*m = \pm 1$, or, equivalently, $m^\star m = \pm 1$. In what follows, we shall discuss how the sign chosen in the relation $m^*m = \pm 1$ corresponds to the classification of the real algebra $\mathcal{B}$ (Benn and Tucker, 1987).

Theorem 4.7 ▶ *If $\mathcal{A} \simeq \mathbb{C}\otimes\mathcal{B}$, where * and * are automorphisms that conjugate the centre and leave $\mathcal{B}$ and $\mathcal{M}(r,\mathbb{R})$ invariant, then $a^* = ma^\star m^{-1}$, where $mm^* = 1$ if and only if $\mathcal{B} \simeq \mathcal{M}(r,\mathbb{R})$, and $mm^* = -1$ if and only if $\mathcal{B} \simeq \mathbb{H}\otimes\mathcal{M}(\frac{r}{2},\mathbb{R})$.*

Proof: Since there are two mutually exclusive possibilities for $\mathcal{B}$, and similarly for m, if we prove that $mm^* = 1 \Leftrightarrow \mathcal{B} \simeq \mathcal{M}(r,\mathbb{R})$, it immediately follows that $mm^* = -1 \Leftrightarrow \mathbb{H}\otimes\mathcal{M}(\frac{r}{2},\mathbb{R})$. Suppose first that $\mathcal{B} \simeq \mathcal{M}(r,\mathbb{R})$. Using $\mathbf{b}_{ij}$ and $\mathcal{E}_{ij}$ to denote bases of $\mathcal{B}$ and $\mathcal{M}(r,\mathbb{R})$ respectively, then we obtain $\mathcal{E}_{ij} = s\mathbf{b}_{ij}s^{-1}$, for some $s \in \mathcal{A}$. In addition, by eqn (4.73), it follows that

$$\begin{aligned} a^* &= \sum_{i,j=1}^{r} a_{ij}^*(s\mathbf{b}_{ij}s^{-1})^* = \sum_{i,j=1}^{r} a_{ij}^* s^*\mathbf{b}_{ij}s^{*-1} \\ &= \sum_{i,j=1}^{r} a_{ij}^* s^*s^{-1}\mathcal{E}_{ij}ss^{*-1} = s^*s^{-1}a^\star(s^*s^{-1})^{-1}. \end{aligned} \tag{4.76}$$

Therefore, $m = s^*s^{-1}$ can be chosen, so $m^* = m^{-1}$. The reciprocal can be shown when a linear mapping – complex conjugation – is introduced into $\mathcal{A}$ by $a^c = a^*m = ma^\star$, and therefore ^{c} preserves the columns of $\mathcal{M}(r,\mathbb{R})$. When $m^* = m^{-1}$, the mapping ^{c} is involutive, and for any $a \in \mathcal{A}$, we write $a = \frac{1}{2}(a + a^c) + \frac{1}{2}(a - a^c)$. In particular, the left minimal ideals of $\mathcal{A}$ – which are columns of $\mathcal{M}(r,\mathbb{R})$ with entries in $\mathbb{C}$ – can be decomposed into eigenspaces of ^{c}; since the dimension (over the real field) of a left minimal ideal of $\mathcal{A}$ equals $2r$, the associated eigenspaces have dimension r. Let ψ be an element of such eigenspaces. If $a \in \mathcal{B}$, it follows that $a\psi$ is in a left minimal ideal of $\mathcal{A}$ and, since $(a\psi)^c = a^*\psi^c = a\psi^c$, it is indeed an autospace carrying representations of $\mathcal{B}$. Thus, if $m^* = m^{-1}$, then irreducible representations of $\mathcal{A}$ induce reducible representations in $\mathcal{B}$. However, either $\mathcal{B} \simeq \mathcal{M}(r,\mathbb{R})$ – in which case, its irreducible representations are r-dimensional – or $\mathcal{B} \simeq \mathbb{H}\otimes\mathcal{M}(\frac{r}{2},\mathbb{R})$, with irreducible representations of dimension $2r$. Thus, $m^* = m^{-1}$ implies that $\mathcal{B} \simeq \mathcal{M}(r,\mathbb{R})$. ✓

Let us now consider the transposition. Define

$$A_k = \mathbf{e}_{i_1 \cdots i_k} = \mathbf{e}_{i_1} \cdots \mathbf{e}_{i_k}. \tag{4.77}$$

The transposition of A_k, denoted by A_k^T, is defined by

$$A_k^\mathsf{T} = \begin{cases} A_k, & \text{if} \quad A_k^2 = 1, \\ -A_k, & \text{if} \quad A_k^2 = -1, \end{cases} \tag{4.78}$$

and satisfies

$$(AB)^\mathsf{T} = B^\mathsf{T} A^\mathsf{T}, \tag{4.79}$$

for A, B arbitrary multivectors.

An interesting question to be posed is whether it is possible to write the transposition as an inner automorphism, that is, if we can write A^T as UAU^{-1}, $U\tilde{A}U^{-1}$, or even as $U\bar{A}U^{-1}$. By the property in eqn (4.79), the first possibility is excluded.

If $p = 0$ or $q = 0$, the transposition can be written in terms of the Clifford conjugation and the reversion, respectively. Indeed, since

$$\tilde{A}_k = \mathbf{e}_{i_k} \cdots \mathbf{e}_{i_1} = (-1)^{k(k-1)/2} A_k, \tag{4.80}$$

it follows that

$$A_k^2 = (-1)^{k(k-1)/2} \mathbf{e}_{i_1}^2 \cdots \mathbf{e}_{i_k}^2. \tag{4.81}$$

If $p = 0$, then $\mathbf{e}_i^2 = -1$ and therefore

$$A_k^2 = (-1)^{k(k-1)/2} (-1)^k, \tag{4.82}$$

or, equivalently,

$$A_k^\mathsf{T} = (-1)^{k(k-1)/2} (-1)^k A_k = \bar{A}_k. \tag{4.83}$$

On the other hand, if $q = 0$, then $\mathbf{e}_i^2 = 1$, yielding

$$A_k^2 = (-1)^{k(k-1)/2}, \tag{4.84}$$

that is,

$$A_k^\mathsf{T} = (-1)^{k(k-1)/2} A_k = \tilde{A}_k. \tag{4.85}$$

Now the cases where $p \neq 0$ and $q \neq 0$ are analysed. First, let us suppose that the transposition is given by $A^\mathsf{T} = U\tilde{A}U^{-1}$. For a basis $\{\mathbf{e}_i\}$ of $\mathcal{C}\ell_{p,q}$, in this case $\mathbf{e}_i^\mathsf{T} = \mathbf{e}_i$ for $i = 1, \ldots, p$, and $\mathbf{e}_i^\mathsf{T} = -\mathbf{e}_i$ for $i = p+1, \ldots, p+q$, in such a way that, if $A^\mathsf{T} = U\tilde{A}U^{-1}$, then U has to commute with $\mathbf{e}_i$ for $i = 1, \ldots, p$ and must anti-commute with $\mathbf{e}_i$ for $i = p+1, \ldots, p+q$. If p is *odd*, there exists U satisfying the properties given by $U = \mathbf{e}_{1 \cdots p} = \mathbf{e}_1 \cdots \mathbf{e}_p$. If q is *even*, then there also exists U given by $U = \mathbf{e}_{p+1 \cdots p+q} = \mathbf{e}_{p+1} \cdots \mathbf{e}_{p+q}$. If p is odd *and* q even, then a linear combination of these elements also satisfies the properties. On the other hand, if p is even *and* q is odd, there exists U such that $A^\mathsf{T} = U\tilde{A}U^{-1}$.

In addition, there exists another possibility, namely, $A^\mathsf{T} = U\bar{A}U^{-1}$. In this case, U must anti-commute with $\mathbf{e}_i$ for $i = 1, \ldots, p$ and has to commute with $\mathbf{e}_i$ for $i =$

$p+1,\dots,p+q$. Therefore, if p is *even*, there exists U for which such properties hold, given by $U=\mathbf{e}_{1\dots p}=\mathbf{e}_1\cdots\mathbf{e}_p$. If q is *odd*, there in addition exists U given by $U=\mathbf{e}_{p+1\dots p+q}=\mathbf{e}_{p+1}\cdots\mathbf{e}_{p+q}$. Finally, if p is even *and* q is odd, a linear combination of these elements also satisfies the properties.

To summarise:

$$p \text{ is odd} \Longrightarrow A^{\mathsf{T}}=U\tilde{A}U^{-1},\quad U=\mathbf{e}_1\cdots\mathbf{e}_p, \tag{4.86}$$

$$p \text{ is even} \Longrightarrow A^{\mathsf{T}}=U\bar{A}U^{-1},\quad U=\mathbf{e}_1\cdots\mathbf{e}_p, \tag{4.87}$$

$$q \text{ is odd} \Longrightarrow A^{\mathsf{T}}=U\bar{A}U^{-1},\quad U=\mathbf{e}_{p+1}\cdots\mathbf{e}_{p+q}, \tag{4.88}$$

$$p \text{ is even} \Longrightarrow A^{\mathsf{T}}=U\tilde{A}U^{-1},\quad U=\mathbf{e}_{p+1}\cdots\mathbf{e}_{p+q}. \tag{4.89}$$

Now the Hermitian conjugation, denoted by $\dagger$, is defined by

$$A^{\dagger}=(A^*)^{\mathsf{T}}=(A^{\mathsf{T}})^*, \tag{4.90}$$

where $*$ is the complex conjugation in the Clifford algebra (when it is the case), and the transposition is expressed according to eqns (4.86–4.89).

Example 4.7 In example 4.5, we obtained a matrix representation of the Clifford algebra $\mathbb{C}\otimes\mathcal{C}\ell_{1,3}\simeq\mathcal{C}\ell_{1,3}(\mathbb{C})\simeq\mathcal{M}(4,\mathbb{C})$. What was discussed there will be illustrated here using this algebra as an example. Consider an arbitrary multivector

$$\begin{aligned}
A=&(\alpha+i\beta)+(\alpha_0+i\beta_0)\mathbf{e}_0+(\alpha_1+i\beta_1)\mathbf{e}_1+(\alpha_2+i\beta_2)\mathbf{e}_2+(\alpha_3+i\beta_3)\mathbf{e}_3\\
&+(\alpha_{01}+i\beta_{01})\mathbf{e}_0\mathbf{e}_1+(\alpha_{02}+i\beta_{02})\mathbf{e}_0\mathbf{e}_2+(\alpha_{03}+i\beta_{03})\mathbf{e}_0\mathbf{e}_3\\
&+(\alpha_{12}+i\beta_{12})\mathbf{e}_1\mathbf{e}_2+(\alpha_{13}+i\beta_{13})\mathbf{e}_1\mathbf{e}_3+(\alpha_{23}+i\beta_{23})\mathbf{e}_2\mathbf{e}_3\\
&+(\alpha_{012}+i\beta_{012})\mathbf{e}_0\mathbf{e}_1\mathbf{e}_2+(\alpha_{013}+i\beta_{013})\mathbf{e}_0\mathbf{e}_1\mathbf{e}_3+(\alpha_{023}+i\beta_{023})\mathbf{e}_0\mathbf{e}_2\mathbf{e}_3\\
&+(\alpha_{123}+i\beta_{123})\mathbf{e}_1\mathbf{e}_2\mathbf{e}_3+(\alpha_{0123}+i\beta_{0123})\mathbf{e}_0\mathbf{e}_1\mathbf{e}_2\mathbf{e}_3.
\end{aligned} \tag{4.91}$$

Using this matrix representation, this multivector is represented by the matrix $[A]$, whose components A_{ij} are

$$\begin{aligned}
A_{11}&=(\alpha+\alpha_0+\beta_{12}+\beta_{012})+i(\beta+\beta_0-\alpha_{12}-\alpha_{012}),\\
A_{12}&=(\alpha_{13}+\alpha_{013}+\beta_{23}+\beta_{023})+i(\beta_{13}+\beta_{013}-\alpha_{23}-\alpha_{023}),\\
A_{13}&=(-\alpha_3-\alpha_{03}-\beta_{123}-\beta_{0123})+i(-\beta_3-\beta_{03}+\alpha_{123}+\alpha_{0123}),\\
A_{14}&=(-\alpha_1-\alpha_{01}-\beta_2-\beta_{02})+i(-\beta_1-\beta_{01}+\alpha_2+\alpha_{02}),\\
A_{21}&=(-\alpha_{13}-\alpha_{013}+\beta_{23}+\beta_{023})+i(-\beta_{13}-\beta_{013}-\alpha_{23}-\alpha_{023}),\\
A_{22}&=(\alpha+\alpha_0-\beta_{12}-\beta_{012})+i(\beta+\beta_0+\alpha_{12}+\alpha_{012}),\\
A_{23}&=(-\alpha_1-\alpha_{01}+\beta_2+\beta_{02})+i(-\beta_1-\beta_{01}-\alpha_2-\alpha_{02}),\\
A_{24}&=(\alpha_3+\alpha_{03}-\beta_{123}-\beta_{0123})+i(\beta_3+\beta_{03}+\alpha_{123}+\alpha_{0123}),\\
A_{31}&=(\alpha_3-\alpha_{03}+\beta_{123}-\beta_{0123})+i(\beta_3-\beta_{03}-\alpha_{123}+\alpha_{0123}),\\
A_{32}&=(\alpha_1-\alpha_{01}+\beta_2-\beta_{02})+i(\beta_1-\beta_{01}-\alpha_2+\alpha_{02}),\\
A_{33}&=(\alpha-\alpha_0+\beta_{12}-\beta_{012})+i(\beta-\beta_0-\alpha_{12}+\alpha_{012}),\\
A_{34}&=(\alpha_{13}-\alpha_{013}+\beta_{23}-\beta_{023})+i(\beta_{13}-\beta_{013}-\alpha_{23}+\alpha_{023}),\\
A_{41}&=(\alpha_1-\alpha_{01}-\beta_2+\beta_{02})+i(\beta_1-\beta_{01}+\alpha_2-\alpha_{02}),\\
A_{42}&=(-\alpha_3+\alpha_{03}+\beta_{123}-\beta_{0123})+i(-\beta_3+\beta_{03}-\alpha_{123}+\alpha_{0123}),\\
A_{43}&=(-\alpha_{13}+\alpha_{013}+\beta_{23}-\beta_{023})+i(-\beta_{13}+\beta_{013}-\alpha_{23}+\alpha_{023}),\\
A_{44}&=(\alpha-\alpha_0-\beta_{12}+\beta_{012})+i(\beta-\beta_0+\alpha_{12}-\alpha_{012}).
\end{aligned} \tag{4.92}$$

The complex conjugation is now considered in $\mathbb{C}\otimes\mathcal{C}\ell_{1,3}\simeq\mathcal{C}\ell_{1,3}(\mathbb{C})$. The multivector A^* is obtained when we take $i\mapsto -i$ in eqn (4.91). Then we can obtain a matrix representation $[A^*]$ of the multivector A^* according to these expressions. Let us denote the matrix components $[A^*]$ by B_{ij}. It follows that

$$\begin{aligned} B_{11} &= A^*_{22}, & B_{31} &= -A^*_{42},\\ B_{12} &= -A^*_{21}, & B_{32} &= A^*_{41},\\ B_{13} &= -A^*_{24}, & B_{33} &= A^*_{44},\\ B_{14} &= A^*_{23}, & B_{34} &= -A^*_{43},\\ B_{21} &= -A^*_{12}, & B_{41} &= A^*_{32},\\ B_{22} &= A^*_{11}, & B_{42} &= -A^*_{31},\\ B_{23} &= A^*_{14}, & B_{42} &= -A^*_{34},\\ B_{24} &= -A^*_{13}, & B_{44} &= A^*_{33}. \end{aligned} \tag{4.93}$$

If we denote by $[A]^*$ the matrix obtained from the complex conjugation A^*_{ij} and evaluated on its components, it is clear to see that

$$[A^*] \neq [A]^*, \tag{4.94}$$

as already discussed. By theorem 4.7 and since $\mathcal{B} = \mathcal{C}\ell_{1,3} \simeq \mathcal{M}(2,\mathbb{H})$, we know that there must exist $m \in \mathcal{C}\ell_{1,3}$ satisfying $mm^* = -1$ such that $[mA^*m^{-1}] = [A]^*$. Furthermore, $m = \mathbf{e}_{013} = \mathbf{e}_0\mathbf{e}_1\mathbf{e}_3$; or, equivalently,

$$[\mathbf{e}_{013}A^*\mathbf{e}_{013}^{-1}] = [A]^* \iff [A^*] = \gamma_{013}[A]^*\gamma_{013}^{-1}. \tag{4.95}$$

Regarding the Hermitian conjugation, as $p = 1$, it follows that $U = \mathbf{e}_0$, that is,

$$A^\dagger = \mathbf{e}_0\tilde{A}^*\mathbf{e}_0. \tag{4.96}$$

By accomplishing this operation on the multivector A and then taking the matrix representation, we clearly see that

$$[A^\dagger] = [\mathbf{e}_0\tilde{A}^*\mathbf{e}_0] = [A]^\dagger, \tag{4.97}$$

where $[A]^\dagger$ denotes the matrix which is Hermitian conjugated and associated with $[A]$. Consequently, it follows that

$$[\tilde{A}^*] = \gamma_0[A]^\dagger\gamma_0 = [\bar{A}], \tag{4.98}$$

which is the expression for the Dirac adjoint $[\bar{A}]$. Using the expression for the complex conjugation, we then obtain

$$[\tilde{A}] = \gamma_{13}[A]^\mathsf{T}\gamma_{13}^{-1}, \tag{4.99}$$

where $[A]^\mathsf{T}$ is the matrix transposed with respect to $[A]$. Finally, we can prove that

$$[\hat{A}] = \gamma_{0123}[A]\gamma_{0123}^{-1}. \tag{4.100}$$

4.5 Additional Readings

Some texts with a focus on Clifford algebras (and applications) in terms of their matrix representations are those by Charlier, Bérard, Charlier, and Fristot (1992) and Snygg (1997, 2010). In the first two works, one can find many applications in physics – for example, in relativity, electromagnetism, quantum mechanics, and son on – whereas the third contains applications in differential geometry. Garling (2011) uses Pauli spin matrices to represent the angular momentum of particles with spin 1/2 to construct the Dirac equation and expresses Maxwell's equations as a single equation using the Dirac operator.

4.6 Exercises

(1) Consider the Clifford algebra $\mathcal{C}\ell_{1,3} \simeq \mathcal{M}(2,\mathbb{H})$ and let $\{\mathbf{e}_\mu\}$ $(\mu = 0,1,2,3)$ be an orthonormal basis of $\mathbb{R}^{1,3}$ such that $\mathbf{e}_0^2 = 1$, $\mathbf{e}_i^2 = -1$, $(i = 1,2,3)$. Using the idempotent $f = (1/2)(1+\mathbf{e}_0)$, obtain the following matrix representation for the generators of this algebra:

$$\rho(\mathbf{e}_0) = \begin{pmatrix} 1 & 0 \\ 0 & -1 \end{pmatrix}, \quad \rho(\mathbf{e}_1) = \begin{pmatrix} 0 & \boldsymbol{i} \\ \boldsymbol{i} & 0 \end{pmatrix},$$
$$\rho(\mathbf{e}_2) = \begin{pmatrix} 0 & \boldsymbol{j} \\ \boldsymbol{j} & 0 \end{pmatrix}, \quad \rho(\mathbf{e}_3) = \begin{pmatrix} 0 & \boldsymbol{k} \\ \boldsymbol{k} & 0 \end{pmatrix},$$

where $\boldsymbol{i}, \boldsymbol{j}, \boldsymbol{k}$ are the quaternionic units.

(2) Consider the complexification $\mathbb{C} \otimes \mathcal{C}\ell_{1,3}$ of the Clifford algebra $\mathcal{C}\ell_{1,3}$ regarded in the previous exercise. The standard representation of the Dirac matrices is such that $\mathbf{e}_{I_1} = \mathbf{e}_0$, and $\mathbf{e}_{I_2} = i\mathbf{e}_1\mathbf{e}_2$. Using

$$\mathbf{e}_{I_1} = \mathbf{e}_5 = \mathbf{e}_{0123}, \qquad \mathbf{e}_{I_2} = i\mathbf{e}_1\mathbf{e}_2,$$

obtain the so-called *Weyl representation* or *chiral representation* of the gamma matrices, namely,

$$\gamma_0 = \begin{pmatrix} 0 & I \\ -I & 0 \end{pmatrix}, \quad \gamma_1 = \begin{pmatrix} 0 & -\sigma_1 \\ \sigma_1 & 0 \end{pmatrix},$$
$$\gamma_2 = \begin{pmatrix} 0 & -\sigma_2 \\ \sigma_2 & 0 \end{pmatrix}, \quad \gamma_3 = \begin{pmatrix} 0 & -\sigma_3 \\ \sigma_3 & 0 \end{pmatrix},$$

where I is the identity matrix 2×2, and σ_i denotes the Pauli matrices.

(3) Consider the Clifford algebra $\mathcal{C}\ell_{3,1} \simeq \mathcal{M}(4,\mathbb{R})$. Using

$$\mathbf{e}_{I_1} = \mathbf{e}_1 \qquad \mathbf{e}_{I_2} = \mathbf{e}_0\mathbf{e}_2,$$

obtain the *Majorana representation* of the gamma matrices:

$$\gamma_0 = \begin{pmatrix} 0 & -i\sigma_2 \\ -i\sigma_2 & 0 \end{pmatrix}, \quad \gamma_1 = \begin{pmatrix} I & 0 \\ 0 & -I \end{pmatrix},$$
$$\gamma_2 = \begin{pmatrix} 0 & -I \\ I & 0 \end{pmatrix}, \quad \gamma_3 = \begin{pmatrix} 0 & \sigma_3 \\ \sigma_3 & 0 \end{pmatrix}.$$

5 Clifford Algebras, and Associated Groups

In this chapter, we study the groups that can be defined within a Clifford algebra. These groups, which deserve special attention, are the Clifford–Lipschitz, the Pin group, and Spin group. The Lie algebras associated with those groups are hence constructed and implemented, together with some of their applications. Conformal transformations and the standard twistors of Penrose (Penrose, 1967; Penrose and Rindler, 1984) are introduced from the algebraic point of view (Klotz, 1974; Keller, 1997; da Rocha and Vaz Jr, 2007) in a straightforward formulation, as geometric multivectors.

5.1 Orthogonal Transformations and the Cartan–Dieudonné Theorem

Orthogonal Transformations

Let g be a symmetric bilinear form endowing the vector space V. A linear mapping $T : V \to V$ is said to be an *isometry*, or *orthogonal transformation*, if

$$g(T(\mathbf{v}), T(\mathbf{u})) = g(\mathbf{v}, \mathbf{u}), \qquad \forall\, \mathbf{v}, \mathbf{u} \in V. \tag{5.1}$$

Defining T_i^j by $T(\mathbf{e}_i) = T_i^j \mathbf{e}_j$ and letting $g_{ij} = g(\mathbf{e}_i, \mathbf{e}_j)$, we can consequently write eqn (5.1) as $T_i^k g_{kl} T_j^l = g_{ij}$. When matrices are used, this equation is equivalent to $T^\intercal G T = G$, where now T denotes a matrix with entries $\{T_i^j\}$, and $T^\intercal$ denotes its associated transposed matrix. Since $\det(AB) = \det A \det B$, and $\det A = \det A^\intercal$, it immediately follows that

$$\left(\det T\right)^2 = 1. \tag{5.2}$$

The orthogonal transformations for which $\det T = 1$ are called *rotations*, and those for which $\det T = -1$ are *reflections*. The set of isometries forms a group called the *orthogonal group* and is denoted by $\mathrm{O}(p,q)$ for $V = \mathbb{R}^{p,q}$. The subgroup of $\mathrm{O}(p,q)$ formed only by rotations is called the *special orthogonal group* and is denoted by $\mathrm{SO}(p,q)$ – in general, the use of the notation S indicates that the group is restricted to the case where $\det T = 1$.

An Introduction to Clifford Algebras and Spinors. First Edition. Jayme Vaz, Jr. and Roldão da Rocha, Jr.

The Components of the Orthogonal Group

Given a group G, define a *path* in this group as a continuous mapping $\chi : [0,1] \to G$. A subset G' of G is said to be *connected* if, for any elements $g_0, g_1 \in G'$, there exists a path $\chi(t)$ linking these elements, namely, $\chi(0) = g_0$ and $\chi(1) = g_1$. A connected subset that is not contained in any other connected subset is called a *component* of the group G.

The orthogonal groups $\mathrm{O}(n,0)$ and $\mathrm{O}(0,n)$ have two components. Indeed, given orthogonal transformations T_0 and T_1 such that, for instance, $\det T_0 = 1$ and $\det T_1 = -1$, there is no continuous path that links such transformations. Between these two components, the one satisfying $\det T = 1$ is either the subgroup $\mathrm{SO}(n,0)$ or the subgroup $\mathrm{SO}(0,n)$.

The orthogonal groups $\mathrm{O}(p,q)$, where $p \neq 0$ or $q \neq 0$, have *four* components. In fact, let us consider an orthonormal basis $\{\mathbf{e}_1, \dots, \mathbf{e}_{p+q}\}$ of $\mathbb{R}^{p,q}$ in such a way that, in terms of this basis, the symmetric bilinear functional g can be represented by the matrix

$$G = \begin{pmatrix} 1_p & 0 \\ 0 & -1_q \end{pmatrix}, \tag{5.3}$$

where 1_p and 1_q denote the identity matrices of order p and q, respectively. Representing T by the matrix

$$T = \begin{pmatrix} A_p & B_{p,q} \\ C_{q,p} & D_q \end{pmatrix}, \tag{5.4}$$

where A_p is the matrix $p \times p$, and $B_{p,q}$ is the matrix $p \times q$, the condition $T^\intercal G T = G$ implies that

$$\begin{aligned} A_p^\intercal A_p - C_{p,q}^\intercal C_{q,p} &= 1_p, \\ D_q^\intercal D_q - B_{q,p}^\intercal B_{p,q} &= 1_q, \\ A_p^\intercal B_{p,q} &= C_{p,q}^\intercal D_q, \end{aligned} \tag{5.5}$$

where $B_{q,p}^\intercal = (B_{p,q})^\intercal$, and $C_{p,q}^\intercal = (C_{q,p})^\intercal$.

The matrices A_p and D_q satisfy $\det A_p \neq 0$, and $\det D_q \neq 0$. Let us consider the case for the matrix A_p – the case involving D_q is completely analogous. The matrix A_p must satisfy $A_p^\intercal A_p = 1_p + C_{p,q}^\intercal C_{q,p}$. This result implies that

$$(\det A_p)^2 = \det(1_p + C_{p,q}^\intercal C_{q,p}). \tag{5.6}$$

Let us suppose that $\det(1_p + C_{p,q}^\intercal C_{q,p}) = 0$. In this case, the equation $(1_p + C_{p,q}^\intercal C_{q,p})X = 0$ has a non-trivial solution X. However, $X = -C_{p,q}^\intercal C_{q,p} X$, and therefore

$$X^\intercal X = -X^\intercal C_{p,q}^\intercal C_{q,p} X = -(C_{q,p}X)^\intercal (C_{q,p}X). \tag{5.7}$$

On the right-hand side, $X^\intercal X = (X_1)^2 + \cdots + (X_n)^2 > 0$, where X_i represents the components of X, since $X \neq 0$. Similarly, the left-hand side yields $-(C_{q,p}X)^\intercal (C_{q,p}X) \leq 0$, where the equality corresponds to the possibility $C_{q,p}X = 0$. Then we obtain a contradiction; thus we must conclude that $\det(1_p + C_{p,q}^\intercal C_{q,p}) \neq 0$.

In addition, this result implies that

$$(\det A_p)^2 \neq 0, \qquad (\det D_q)^2 \neq 0. \tag{5.8}$$

It is possible to divide the orthogonal transformations $T \in \mathrm{O}(p,q)$ into four classes:

$$\begin{aligned}
&\text{(i) } \mathrm{O}_+^{\uparrow}(p,q): \ \det A_p > 0, \det D_q > 0,\\
&\text{(ii) } \mathrm{O}_-^{\uparrow}(p,q): \ \det A_p > 0, \det D_q < 0,\\
&\text{(iii) } \mathrm{O}_+^{\downarrow}(p,q): \ \det A_p < 0, \det D_q > 0,\\
&\text{(iv) } \mathrm{O}_-^{\downarrow}(p,q): \ \det A_p < 0, \det D_q < 0.
\end{aligned} \tag{5.9}$$

It is possible to show (it is left as an exercise) that, if $\det A_p \det D_q > 0$, then $\det T > 0$. However, if $\det A_p \det D_q < 0$, then $\det T < 0$. In addition, the following sets can be shown to be subsets of $\mathrm{O}(p,q)$:

$$\begin{aligned}
&\text{(i) } \mathrm{O}_+^{\uparrow}(p,q),\\
&\text{(ii) } \mathrm{O}^{\uparrow}(p,q) = \mathrm{O}_+^{\uparrow}(p,q) \cup \mathrm{O}_-^{\uparrow}(p,q),\\
&\text{(iii) } \mathrm{O}_+(p,q) = \mathrm{O}_+^{\uparrow}(p,q) \cup \mathrm{O}_+^{\downarrow}(p,q),\\
&\text{(iv) } \mathrm{O}_+^{\uparrow}(p,q) \cup \mathrm{O}_-^{\downarrow}(p,q).
\end{aligned} \tag{5.10}$$

We can moreover verify that

$$\mathrm{SO}_+(p,q) = \mathrm{SO}^{\uparrow}(p,q) = \mathrm{SO}_+^{\uparrow}(p,q) = \mathrm{O}_+(p,q) \cap \mathrm{O}^{\uparrow}(p,q). \tag{5.11}$$

Example 5.1 Let us consider the orthogonal group $\mathrm{O}(1,1)$. A matrix

$$T = \begin{pmatrix} a & b \\ c & d \end{pmatrix} \in \mathrm{O}(1,1)$$

must satisfy

$$\begin{pmatrix} a & c \\ b & d \end{pmatrix}\begin{pmatrix} 1 & 0 \\ 0 & -1 \end{pmatrix}\begin{pmatrix} a & b \\ c & d \end{pmatrix} = \begin{pmatrix} 1 & 0 \\ 0 & -1 \end{pmatrix},$$

so

$$a^2 - c^2 = 1, \qquad d^2 - b^2 = 1, \qquad ab = cd.$$

The two first equations hold when

$$\begin{aligned}
a &= \pm\cosh\alpha, \quad c = \sinh\alpha,\\
d &= \pm\cosh\beta, \quad b = \sinh\beta,
\end{aligned}$$

and the last one provides the relation between α and β. It reads as follows: (i) if $a = \cosh\alpha$, and $d = \cosh\beta$, then $\alpha = \beta$; (ii) if $a = -\cosh\alpha$, and $d = \cosh\beta$, then $\alpha = -\beta$; (iii) if $a = \cosh\alpha$, and $d = -\cosh\beta$, it implies that $\alpha = -\beta$; and (iv) if $a = -\cosh\alpha$, and $d = -\cosh\beta$, then $\alpha = \beta$. Four possibilities then exist:

$$T_1 = \begin{pmatrix} \cosh\alpha & \sinh\alpha \\ \sinh\alpha & \cosh\alpha \end{pmatrix}, \qquad T_2 = \begin{pmatrix} \cosh\alpha & -\sinh\alpha \\ \sinh\alpha & -\cosh\alpha \end{pmatrix},$$

$$T_3 = \begin{pmatrix} -\cosh\alpha & -\sinh\alpha \\ \sinh\alpha & \cosh\alpha \end{pmatrix}, \qquad T_4 = \begin{pmatrix} -\cosh\alpha & \sinh\alpha \\ \sinh\alpha & \cosh\alpha \end{pmatrix}.$$

It is clear that there is no path that links any two matrices T_i and T_j, for $i \neq j$. According to the notation used, we have

$$T_1 \in \mathrm{O}_+^{\uparrow}(1,1), \qquad T_2 \in \mathrm{O}_-^{\uparrow}(1,1),$$
$$T_3 \in \mathrm{O}_+^{\downarrow}(1,1), \qquad T_4 \in \mathrm{O}_-^{\downarrow}(1,1).$$

Orthogonal Symmetries and Reflections

Let us write V as $V = U \oplus U^\perp$, where U is a non-isotropic subspace of V, that is, $g(\mathbf{u},\mathbf{u}) \neq 0, \forall\, \mathbf{u} \in V$. We define the *orthogonal symmetry* S_U with respect to U as

$$S_U(\mathbf{v}_\parallel + \mathbf{v}_\perp) = -\mathbf{v}_\parallel + \mathbf{v}_\perp, \qquad \forall\, \mathbf{v}_\parallel \in U, \mathbf{v}_\perp \in U^\perp. \tag{5.12}$$

It is clear that $S_U = -\,\mathrm{id}_U + \mathrm{id}_{U^\perp}$, where id denotes the identity operator. Since the matrix representation of id is the identity matrix, the determinant $\det S_U$ of the matrix that represents S_U is obviously

$$\det S_U = (-1)^{\dim U}. \tag{5.13}$$

Let us choose a vector $\mathbf{u} \in V$ such that $g(\mathbf{u},\mathbf{u}) \neq 0$ and take U as the subspace generated by $\mathbf{u}$, namely, $U = \{a\mathbf{u} \mid a \in \mathbb{R}\}$. We can write a vector $\mathbf{v} \in V$ as $\mathbf{v} = \mathbf{v}_\parallel + \mathbf{v}_\perp$, where $g(\mathbf{v}_\perp, \mathbf{u}) = 0$. It is immediately obvious that the expression

$$\mathbf{v}_\perp = \mathbf{v} - \frac{g(\mathbf{v},\mathbf{u})}{g(\mathbf{u},\mathbf{u})}\mathbf{u} \tag{5.14}$$

satisfies $g(\mathbf{v}_\perp, \mathbf{u}) = 0$; therefore, we can split $\mathbf{v} = \mathbf{v}_\parallel + \mathbf{v}_\perp$, where $\mathbf{v}_\perp$ is provided by eqn (5.14), and

$$\mathbf{v}_\parallel = \frac{g(\mathbf{v},\mathbf{u})}{g(\mathbf{u},\mathbf{u})}\mathbf{u}. \tag{5.15}$$

The symmetry $S_\mathbf{u}$ is thus given by

$$\begin{aligned} S_\mathbf{u}(\mathbf{v}) &= S_\mathbf{u}(\mathbf{v}_\parallel + \mathbf{v}_\perp) = -\mathbf{v}_\parallel + \mathbf{v}_\perp \\ &= -\frac{g(\mathbf{v},\mathbf{u})}{g(\mathbf{u},\mathbf{u})}\mathbf{u} + \mathbf{v} - \frac{g(\mathbf{v},\mathbf{u})}{g(\mathbf{u},\mathbf{u})}\mathbf{u}, \end{aligned} \tag{5.16}$$

that is,

$$S_\mathbf{u} = \mathbf{v} - 2\frac{g(\mathbf{v},\mathbf{u})}{g(\mathbf{u},\mathbf{u})}\mathbf{u}. \tag{5.17}$$

The symmetry $S_\mathbf{u}$ is a *reflection* – indeed, from eqn (5.13) it follows that $\det S_\mathbf{u} = (-1)^1 = -1$. This reflection occurs on the hyperplane orthogonal to the vector $\mathbf{u}$.

Two arbitrary non-isotropic vectors $\mathbf{v}$ and $\mathbf{u}$ with the same norm – $g(\mathbf{v},\mathbf{v}) = g(\mathbf{u},\mathbf{u}) \neq 0$ – can be related by at most two reflections. Indeed, let us first suppose the case where $g(\mathbf{v}-\mathbf{u}, \mathbf{v}-\mathbf{u}) \neq 0$. Hence,

$$\begin{aligned} S_{\mathbf{v}-\mathbf{u}}(\mathbf{v}) &= \mathbf{v} - 2\frac{g(\mathbf{v},\mathbf{v}-\mathbf{u})}{g(\mathbf{v}-\mathbf{u},\mathbf{v}-\mathbf{u})}(\mathbf{v}-\mathbf{u}) \\ &= \mathbf{v} - 2\frac{g(\mathbf{v},\mathbf{v}) - g(\mathbf{v},\mathbf{u})}{g(\mathbf{v},\mathbf{v}) + g(\mathbf{u},\mathbf{u}) - 2g(\mathbf{v},\mathbf{u})}(\mathbf{v}-\mathbf{u}) \\ &= \mathbf{v} - \frac{g(\mathbf{v},\mathbf{v}) - g(\mathbf{v},\mathbf{u})}{g(\mathbf{v},\mathbf{v}) - g(\mathbf{v},\mathbf{u})}(\mathbf{v}-\mathbf{u}) \\ &= \mathbf{v} - (\mathbf{v}-\mathbf{u}) = \mathbf{u}. \end{aligned} \tag{5.18}$$

In the case where $g(\mathbf{v}-\mathbf{u},\mathbf{v}-\mathbf{u})=0$, it follows that $g(\mathbf{v},\mathbf{v})=g(\mathbf{u},\mathbf{u})=g(\mathbf{v},\mathbf{u})\neq 0$, and $g(\mathbf{v}+\mathbf{u},\mathbf{v}+\mathbf{u})\neq 0$. Then,

$$\begin{aligned} S_{\mathbf{v}+\mathbf{u}}(\mathbf{v}) &= \mathbf{v} - 2\frac{g(\mathbf{v},\mathbf{v}+\mathbf{u})}{g(\mathbf{v}+\mathbf{u},\mathbf{v}+\mathbf{u})}(\mathbf{v}+\mathbf{u}) \\ &= \mathbf{v} - \frac{g(\mathbf{v},\mathbf{v})+g(\mathbf{v},\mathbf{u})}{g(\mathbf{v},\mathbf{v})+g(\mathbf{v},\mathbf{u})}(\mathbf{v}+\mathbf{u}) \\ &= \mathbf{v} - (\mathbf{v}+\mathbf{u}) = -\mathbf{u}, \end{aligned} \tag{5.19}$$

implying that

$$S_{\mathbf{u}}S_{\mathbf{v}+\mathbf{u}}(\mathbf{v}) = S_{\mathbf{u}}(-\mathbf{u}) = -\mathbf{u} - 2\frac{g(-\mathbf{u},\mathbf{u})}{g(\mathbf{u},\mathbf{u})}\mathbf{u} = \mathbf{u}. \tag{5.20}$$

The Cartan–Dieudonné Theorem

We have just proved that two arbitrary non-isotropic vectors with the same norm – $g(\mathbf{v},\mathbf{v})=g(\mathbf{u},\mathbf{u})\neq 0$ – can be related by at most two reflections. This result is a particular case of the Cartan–Dieudonné theorem, presented in what follows as its 'weak version':

Theorem 5.1 ▶ *Any orthogonal transformation T in a finite dimensional vector space V can be expressed as the product of symmetries (reflections) with respect to non-isotropic hyperplanes.*

Proof: The first assertion can be demonstrated by using finite induction, since it holds for $n=1$. Let us then assume that the assertion holds for $\dim V=n$ and show that, in this case, it holds for $\dim V=n+1$. Let $\mathbf{v}\in V$ be such that $g(\mathbf{v},\mathbf{v})\neq 0$; $\dim V=n+1$; and $U=\operatorname{span}\{\mathbf{v}\}$. The vector subspace $U^\perp$ is n-dimensional. If we denote by T an orthogonal transformation, then, by definition, $g(T(\mathbf{v}),T(\mathbf{v}))=g(\mathbf{v},\mathbf{v})$. Moreover, from eqns (5.18) and (5.20), $T(\mathbf{v})$ and $\mathbf{v}$ are related by at most two symmetries (reflections) S; thus, $S(T(\mathbf{v}))=\mathbf{v}$. Since the space U is $S\circ T$-invariant, the orthogonal complement $U^\perp$ is also $S\circ T$ invariant; therefore, $S\circ T$ is an orthogonal transformation. Since $\dim U^\perp=n$, supposing that the first assertion holds for dimension n, we can conclude that $S\circ T$ is the product Σ of a finite number of symmetries and therefore $T=S^{-1}\circ\Sigma$, that is, the orthogonal transformation T in a space V such that $\dim V=n+1$ is the product of a finite number of symmetries. ✓

Observation ☞ The 'strong version' of the Cartan–Dieudonné theorem asserts that, *if $\dim V=n$, then T ($T\neq \mathrm{id}$) can be expressed as the product of at most n symmetries.* The proof of this version is not as straightforward as that for the weak version. Thus, since it will not be necessary to use the strong version in this book, we shall not go into it in detail.

Now let us focus on eqn (5.17). In $\mathcal{C}\ell_{p,q}$ we have $\mathbf{vu}+\mathbf{uv}=2g(\mathbf{v},\mathbf{u})$, and $\mathbf{u}^2=g(\mathbf{u},\mathbf{u})$. The object $\mathbf{u}/g(\mathbf{u},\mathbf{u})$ can be interpreted as the inverse of the element $\mathbf{u}$:

$$\mathbf{u}^{-1} = \frac{\mathbf{u}}{g(\mathbf{u},\mathbf{u})} = \frac{\mathbf{u}}{\mathbf{u}^2}, \tag{5.21}$$

where $\mathbf{u}^{-1}\mathbf{u} = \mathbf{u}\mathbf{u}^{-1} = 1$. Hence, eqn (5.17) can be written as

$$S_{\mathbf{u}}(\mathbf{v}) = \mathbf{v} - \frac{(\mathbf{v}\mathbf{u} + \mathbf{u}\mathbf{v})}{g(\mathbf{u}, \mathbf{u})}\mathbf{u} = \mathbf{v} - (\mathbf{v}\mathbf{u} + \mathbf{u}\mathbf{v})\mathbf{u}^{-1} = \mathbf{v} - \mathbf{v} - \mathbf{u}\mathbf{v}\mathbf{u}^{-1}, \tag{5.22}$$

namely,

$$\boxed{S_{\mathbf{u}}(\mathbf{v}) = -\mathbf{u}\mathbf{v}\mathbf{u}^{-1} = \hat{\mathbf{u}}\mathbf{v}\mathbf{u}^{-1}}. \tag{5.23}$$

This equation has prominent usefulness in what follows.

5.2 The Clifford–Lipschitz Group

Group of The Invertible Elements

Various groups can be defined in a Clifford algebra $\mathcal{C}\ell_{p,q}$. The largest one is the *group of invertible* (or *regular*) *elements* $\mathcal{C}\ell^*_{p,q}$,

$$\mathcal{C}\ell^*_{p,q} = \{a \in \mathcal{C}\ell_{p,q} \mid \exists a^{-1}\}. \tag{5.24}$$

The Clifford–Lipschitz Group

A subgroup of $\mathcal{C}\ell^*_{p,q}$ of great interest is the *Clifford–Lipschitz group* $\Gamma_{p,q}$, defined by

$$\boxed{\Gamma_{p,q} = \{a \in \mathcal{C}\ell^*_{p,q} \mid a\mathbf{v}a^{-1} \in V, \forall \mathbf{v} \in V = \mathbb{R}^{p,q}\}}. \tag{5.25}$$

It is straightforward to see that this set presents a group structure – in general, a non-abelian structure.

Adjoint Representation

A representation ρ of the Clifford algebra $\mathcal{C}\ell_{p,q}$ obviously defines a representation of the Clifford–Lipschitz group $\Gamma_{p,q}$. Other representations of this group can also be defined; a particular representation is called the *adjoint representation*, or *vector representation* $\sigma : \Gamma_{p,q} \to \text{Aut}(\mathcal{C}\ell_{p,q})$, which is defined by,[1]

$$\sigma(a)(x) = axa^{-1}. \tag{5.26}$$

Here, $\sigma(a)$ is an element of the group of the automorphisms $\text{Aut}(\mathcal{C}\ell_{p,q})$ of $\mathcal{C}\ell_{p,q}$.[2]

[1]Another notation commonly employed is $\sigma = \text{Ad}$.

[2]An endomorphism is a homomorphism of a set X on X. If this homomorphism is also an isomorphism, it is said to be an automorphism. The set of all automorphisms is the group of the automorphisms.

Consider $\mathbf{v}, \mathbf{u} \in V = \mathbb{R}^{p,q}$, such that

$$\mathbf{v}\mathbf{u} + \mathbf{u}\mathbf{v} = 2g(\mathbf{v}, \mathbf{u}). \tag{5.27}$$

Let us consider the vectors $\sigma(a)\mathbf{v}$ and $\sigma(a)\mathbf{u}$. We have

$$\begin{aligned} 2g(\sigma(a)\mathbf{v}, \sigma(a)\mathbf{u}) &= \sigma(a)\mathbf{v}\sigma(a)\mathbf{u} + \sigma(a)\mathbf{u}\sigma(a)\mathbf{v} \\ &= a\mathbf{v}a^{-1}a\mathbf{u}a^{-1} + a\mathbf{u}a^{-1}a\mathbf{v}a^{-1} \\ &= a\mathbf{v}\mathbf{u}a^{-1} + a\mathbf{u}\mathbf{v}a^{-1} = 2ag(\mathbf{v}, \mathbf{u})a^{-1} \\ &= 2g(\mathbf{v}, \mathbf{u}). \end{aligned} \tag{5.28}$$

Since $\sigma(a)$ satisfies

$$g(\sigma(a)\mathbf{v}, \sigma(a)\mathbf{u}) = g(\mathbf{v}, \mathbf{u}), \tag{5.29}$$

therefore,

$$\sigma(a) \in \mathrm{O}(p, q)\,, \tag{5.30}$$

where $\mathrm{O}(p, q)$ denotes the group of *orthogonal* transformations of $\mathbb{R}^{p,q}$. Hence, σ is both a mapping $\sigma : \Gamma_{p,q} \to \mathrm{O}(p, q)$, and a group homomorphism. In fact,

$$\sigma(ab)(\mathbf{v}) = ab\mathbf{v}(ab)^{-1} = ab\mathbf{v}b^{-1}a^{-1} = \sigma(a)\sigma(b)(\mathbf{v}), \tag{5.31}$$

and $\sigma(ab) = \sigma(a)\sigma(b)$. We will show in what follows that σ is, furthermore, surjective. In some cases, the image of $\sigma(\Gamma_{p,q})$ may not be the whole group $\mathrm{O}(p, q)$ but instead merely a subgroup, depending upon the dimension $n = p + q$ of the vector space. To determine which is the case, $\det \sigma(a)$ must be calculated.

The determinant of a linear transformation T can be usually defined by $T(\mathbf{e}_1) \wedge \cdots \wedge T(\mathbf{e}_n) = (\det T)\mathbf{e}_1 \wedge \cdots \wedge \mathbf{e}_n$, where $\{\mathbf{e}_1, \ldots, \mathbf{e}_n\}$ is a basis of $V = \mathbb{R}^{p,q}$. Hence, $\det \sigma(a)$ is given by

$$\sigma(a)(\mathbf{e}_1) \wedge \cdots \wedge \sigma(a)(\mathbf{e}_n) = \big(\det \sigma(a)\big)\mathbf{e}_1 \wedge \cdots \wedge \mathbf{e}_n. \tag{5.32}$$

On the other hand, for vectors $\mathbf{v}$ and $\mathbf{u}$, it follows that

$$\begin{aligned} \sigma(a)(\mathbf{v}) \wedge \sigma(a)(\mathbf{u}) &= (a\mathbf{v}a^{-1}) \wedge (a\mathbf{u}a^{-1}) \\ &= \frac{1}{2}(a\mathbf{v}a^{-1}a\mathbf{u}a^{-1} - a\mathbf{u}a^{-1}a\mathbf{v}a^{-1}) \\ &= a\frac{1}{2}(\mathbf{v}\mathbf{u} - \mathbf{u}\mathbf{v})a^{-1} = a(\mathbf{v} \wedge \mathbf{u})a^{-1}, \end{aligned} \tag{5.33}$$

that is,

$$\sigma(a)(\mathbf{v}) \wedge \sigma(a)(\mathbf{u}) = \sigma(a)(\mathbf{v} \wedge \mathbf{u}), \tag{5.34}$$

and, for eqn (5.32), it follows that

$$a(\mathbf{e}_1 \wedge \cdots \wedge \mathbf{e}_n)a^{-1} = \big(\det \sigma(a)\big)\mathbf{e}_1 \wedge \cdots \wedge \mathbf{e}_n. \tag{5.35}$$

In order to compute the left-hand side of (5.35), let us take a as the homogeneous multivector $a = a_{[k]} \in \bigwedge_k(V)$ and then generalise the result for an arbitrary multivector by linearity. Since $\mathbf{e}_1 \wedge \cdots \wedge \mathbf{e}_n$ is an n-vector (pseudoscalar), it follows that

$$a_{[k]}(\mathbf{e}_1 \wedge \cdots \wedge \mathbf{e}_n) = a_{[k]^\flat} \rfloor (\mathbf{e}_1 \wedge \cdots \wedge \mathbf{e}_n). \tag{5.36}$$

Now by using eqn (2.60), we see that

$$a_{[k]^\flat} \rfloor (\mathbf{e}_1 \wedge \cdots \wedge \mathbf{e}_n) = (-1)^{k(n-1)}(\mathbf{e}_1 \wedge \cdots \wedge \mathbf{e}_n) \lfloor a_{[k]^\flat}, \tag{5.37}$$

which yields

$$a_{[k]}(\mathbf{e}_1 \wedge \cdots \wedge \mathbf{e}_n) = (-1)^{k(n-1)}(\mathbf{e}_1 \wedge \cdots \wedge \mathbf{e}_n) a_{[k]}. \tag{5.38}$$

If n is *odd*, then $n-1$ is even and consequently

$$a_{[k]}(\mathbf{e}_1 \wedge \cdots \wedge \mathbf{e}_n) = (\mathbf{e}_1 \wedge \cdots \wedge \mathbf{e}_n) a_{[k]}, \qquad k = 0, 1, \ldots, n. \tag{5.39}$$

Hence, if n is *odd*, it follows that

$$a(\mathbf{e}_1 \wedge \cdots \wedge \mathbf{e}_n) = (\mathbf{e}_1 \wedge \cdots \wedge \mathbf{e}_n) a \qquad (n \text{ odd}), \tag{5.40}$$

which can be substituted into eqn (5.35), so that

$$(\mathbf{e}_1 \wedge \cdots \wedge \mathbf{e}_n) = \big(\det \sigma(a)\big) \mathbf{e}_1 \wedge \cdots \wedge \mathbf{e}_n, \tag{5.41}$$

and therefore

$$\det \sigma(a) = 1, \qquad (n \text{ odd}). \tag{5.42}$$

On the other hand, if n is *even*, then $n-1$ is odd and, from eqn (5.38), we conclude that

$$a_{[k]}(\mathbf{e}_1 \wedge \cdots \wedge \mathbf{e}_n) = (-1)^k (\mathbf{e}_1 \wedge \cdots \wedge \mathbf{e}_n) a_{[k]} \qquad (n \text{ even}). \tag{5.43}$$

For n even, the expression $\det \sigma(a) = \pm 1$ holds. Specifically, when n is even, we have

$$\begin{aligned} \det \sigma(a) &= 1 \qquad \text{if } a \in \mathcal{C}\ell^+_{p,q}, \\ \det \sigma(a) &= -1 \qquad \text{if } a \in \mathcal{C}\ell^-_{p,q}. \end{aligned} \tag{5.44}$$

Thus, we can conclude that

$$\begin{aligned} \sigma(\Gamma_{p,q}) &= \mathrm{O}(p,q) \qquad (n \text{ even}), \\ \sigma(\Gamma_{p,q}) &= \mathrm{SO}(p,q) \qquad (n \text{ odd}). \end{aligned} \tag{5.45}$$

This result is somewhat disappointing: we have two distinct situations, depending upon the dimension of the vector space V. It is clear that, if there is no way to deal with this situation, we must separately consider each situation. It would be preferable to deal with it independently of the dimension of the space V.

Twisted Adjoint Representation

Instead of considering the representation σ, the Clifford group representation $\widehat{\sigma}$ is defined by

$$\widehat{\sigma}(a)(\mathbf{v}) = \widehat{a}\mathbf{v}a^{-1}. \tag{5.46}$$

The mapping $\widehat{\sigma}$ denotes the *twisted adjoint* (or *vector*) *representation*. Clearly, the Clifford–Lipschitz group could be defined by the condition $\widehat{a}\mathbf{v}a^{-1} \in V$, instead of

$ava^{-1} \in V$ for $\mathbf{v} \in V$. However, the difference is irrelevant if only the group definition is taken into account. Either way, the twisted Clifford–Lipschitz group $\widehat{\Gamma}_{p,q}$ is defined as

$$\widehat{\Gamma}_{p,q} = \{a \in \mathcal{C}\ell_{p,q} \mid \widehat{a}\mathbf{v}a^{-1} \in V, \forall \mathbf{v} \in V = \mathbb{R}^{p,q}\}. \tag{5.47}$$

Obviously, the isomorphism $\widehat{\Gamma}_{p,q} \simeq \Gamma_{p,q}$ holds. In terms of a representation, the presence of the grade involution $\widehat{a}$ drastically modifies the rule of each one of these groups, as we will discuss.

The results obtained for σ can be straightforwardly adapted to $\widehat{\sigma}$. In order to avoid unnecessary complication, let us take $a = a_{[k]} \in \bigwedge_k(V)$ and then generalise the results by linearity. First, the analogue of eqn (5.34) for $\widehat{\sigma}$ is

$$\begin{aligned}\widehat{\sigma}(a_{[k]})(\mathbf{v}_1) \wedge \cdots \wedge \widehat{\sigma}(a_{[k]})(\mathbf{v}_i) &= (-1)^{k(i-1)}\widehat{\sigma}(a_{[k]})(\mathbf{v}_1 \wedge \cdots \wedge \mathbf{v}_i)\\ &= (-1)^{k(i-1)}(-1)^k\sigma(a_{[k]})(\mathbf{v}_1 \wedge \cdots \wedge \mathbf{v}_i)\\ &= (-1)^{ki}\sigma(a_{[k]})(\mathbf{v}_1 \wedge \cdots \wedge \mathbf{v}_k).\end{aligned} \tag{5.48}$$

By using eqn (5.38), we obtain

$$\begin{aligned}\widehat{\sigma}(a_{[k]})(\mathbf{e}_1) \wedge \cdots \wedge \widehat{\sigma}(a_{[k]})(\mathbf{e}_n) &= (-1)^{nk}a_{[k]}(\mathbf{e}_1 \wedge \cdots \wedge \mathbf{e}_n)a_{[k]}^{-1}\\ &= (-1)^{nk}(-1)^{k(n-1)}\mathbf{e}_1 \wedge \cdots \wedge \mathbf{e}_n,\end{aligned} \tag{5.49}$$

which yields

$$\widehat{\sigma}(a_{[k]})(\mathbf{e}_1) \wedge \cdots \wedge \widehat{\sigma}(a_{[k]})(\mathbf{e}_n) = (-1)^k(\mathbf{e}_1 \wedge \cdots \wedge \mathbf{e}_n). \tag{5.50}$$

Equation (5.35) implies that

$$\det \widehat{\sigma}(a_{[k]}) = (-1)^k. \tag{5.51}$$

What should be realised here is that $\det \widehat{\sigma}(a_{[k]})$ *does not* depend on n, which is the case for $\det \sigma(a_{[k]})$. Since $\det \widehat{\sigma}(a_{[k]}) = \pm 1$, therefore

$$\widehat{\sigma}(\Gamma_{p,q}) = \mathrm{O}(p,q). \tag{5.52}$$

However, if we define

$$\Gamma^+_{p,q} = \Gamma_{p,q} \cap \mathcal{C}\ell^+_{p,q}, \tag{5.53}$$

so that $\Gamma^+_{p,q}$ consists of the *even* elements of the Clifford–Lipschitz group, then

$$\widehat{\sigma}(\Gamma^+_{p,q}) = \mathrm{SO}(p,q). \tag{5.54}$$

The mapping $\widehat{\sigma}$ is a surjective homomorphism. We show it for $\widehat{\sigma} : \Gamma_{p,q} \to \mathrm{O}(p,q)$, and we conclude that the same holds for $\widehat{\sigma} : \Gamma^+_{p,q} \to \mathrm{SO}(p,q)$ – and consequently also for σ, as already asserted. Indeed, in section 5.1, it was shown (eqn (5.23)) that a reflection with respect to the orthogonal hyperplane to a given vector $\mathbf{u}$ is given by $S_{\mathbf{u}}(\mathbf{v}) = \widehat{\mathbf{u}}\mathbf{v}\mathbf{u}^{-1}$. Therefore,

$$\widehat{\sigma}(\mathbf{u}) = S_{\mathbf{u}}. \tag{5.55}$$

Clearly $\mathbf{u} \in \Gamma_{p,q}$. Nevertheless, according to the Cartan–Dieudonné theorem, any orthogonal transformation $T \in \mathrm{O}(p,q)$ can be written as the product of a finite number of reflections of type $S_{\mathbf{u}} = \widehat{\sigma}(\mathbf{u})$. It follows that there exists a finite number of vectors $\{\mathbf{u}_1, \ldots, \mathbf{u}_k\}$ such that any orthogonal transformation reads

$$\widehat{\sigma}(\mathbf{u}_1) \cdots \widehat{\sigma}(\mathbf{u}_k) = \widehat{\sigma}(\mathbf{u}_1 \cdots \mathbf{u}_k)\,. \tag{5.56}$$

Hence, $\mathbf{u}_1 \cdots \mathbf{u}_k \in \Gamma_{p,q}$, showing that $\widehat{\sigma}$ is surjective. The same holds for σ and for the restriction to the even elements $\widehat{\sigma} : \Gamma^+_{p,q} \to \mathrm{SO}(p,q)$ as well.

Another equivalent characterisation of the Clifford–Lipschitz group can be obtained by the last result, namely, as the group consisting of the product of all non-null vectors of $\mathcal{C}\ell_{p,q}$:

$$\Gamma_{p,q} = \{a \in \mathcal{C}\ell^*_{p,q} \mid a = \mathbf{v}_1 \cdots \mathbf{v}_k,\ \text{where}\, \mathbf{v}_i \in \mathbb{R}^{p,q}\ \text{and}\ g(\mathbf{v}_i, \mathbf{v}_i) \neq 0\}. \tag{5.57}$$

The decomposition $a = \mathbf{v}_1 \cdots \mathbf{v}_k$ is *not* unique. Moreover, we have

$$\begin{aligned} k &= \text{odd} \iff \text{reflection}, \\ k &= \text{even} \iff \text{rotation}\,. \end{aligned} \tag{5.58}$$

We can further prove that

$$\ker \widehat{\sigma} = \mathbb{R}^*, \tag{5.59}$$

where $\mathbb{R}^* = \mathbb{R}\backslash\{0\}$ (and $\ker \sigma = \mathbb{R}^*$). Indeed, the condition $\widehat{\sigma}(a) = 1$,[3] where $a \in \ker \widehat{\sigma}$, can be equivalently written as $\widehat{\sigma}(a) = \widehat{\sigma}(b)$, where $a, b \in \ker \widehat{\sigma}$; therefore

$$\begin{aligned} \widehat{\sigma}(a)(\mathbf{v}) &= \widehat{\sigma}(b)(\mathbf{v})\,, \\ \hat{a}\mathbf{v}a^{-1} &= \hat{b}\mathbf{v}b^{-1}\,, \\ \widehat{(b^{-1}a)}\mathbf{v} &= \mathbf{v}(b^{-1}a). \end{aligned} \tag{5.60}$$

By writing $b^{-1}a = (b^{-1}a)_+ + (b^{-1}a)_-$, where $(b^{-1}a)_+$ denotes the even part of $b^{-1}a$, and $(b^{-1}a)_-$ denotes its odd part, we can see that

$$(b^{-1}a)_+\mathbf{v} = \mathbf{v}(b^{-1}a)_+, \qquad (b^{-1}a)_-\mathbf{v} = -\mathbf{v}(b^{-1}a)_-. \tag{5.61}$$

The second condition holds only if $(b^{-1}a)_- = 0$, since there is no element of $\mathcal{C}\ell_{p,q}$ that anti-commutes with *all* generators $\mathbf{e}_i$ $(i = 1, \ldots, n)$ of $\mathcal{C}\ell_{p,q}$. The first condition holds for $(b^{-1}a)_+ \in \mathrm{Cen}(\mathcal{C}\ell_{p,q})$. As shown in chapter 4, exercise 1, if n is even, then $\mathrm{Cen}(\mathcal{C}\ell_{p,q}) = \bigwedge_0(\mathbb{R}^{p,q})$; on the other hand, if n is odd, then $\mathrm{Cen}(\mathcal{C}\ell_{p,q}) = \bigwedge_0(\mathbb{R}^{p,q}) \oplus \bigwedge_n(\mathbb{R}^{p,q})$. However, if n is odd, an element of $\bigwedge_n(\mathbb{R}^{p,q})$ is odd and, consequently, the second condition must hold, instead of the first one. Hence, the two conditions are only satisfied when $b^{-1}a \in \bigwedge_0(\mathbb{R}^{p,q}) = \mathbb{R}$. Moreover, $b^{-1}a$ is invertible and consequently non-null. It follows that $b^{-1}a \in \mathbb{R}^*$, and $\ker \widehat{\sigma} = \mathbb{R}^*$.

[3]The definition given in chapter 1 – $\ker f = \{\mathbf{v} \in V \mid f(\mathbf{v}) = 0\}$ for $f : V \to W$ – was provided with respect to a linear mapping that exists between vector spaces V and W and where the *unity* is the null vector 0. Here, the mapping $\widehat{\sigma} : \Gamma_{p,q} \to \mathrm{O}(p,q)$ must be denoted by $\ker \widehat{\sigma} = \{a \in \Gamma_{p,q} \mid \widehat{\sigma}(a) = 1\}$, where 1 is the *unity* in $\mathrm{O}(p,q)$.

Example 5.2 Let us consider the Clifford algebra $\mathcal{C}\ell_{2,0}$, where an arbitrary element is given by

$$\psi = a_0 + a_1\mathbf{e}_1 + a_2\mathbf{e}_2 + a_{12}\mathbf{e}_1\mathbf{e}_2.$$

In addition,

$$\tilde{\psi}\psi = a_0^2 + a_1^2 + a_2^2 + a_{12}^2 + 2(a_0a_1 - a_2a_{12})\mathbf{e}_1 + 2(a_0a_2 - a_1a_{12})\mathbf{e}_2,$$
$$\bar{\psi}\psi = a_0^2 - a_1^2 - a_2^2 + a_{12}^2.$$

The last expression shows that, if

$$\bar{\psi}\psi = a_0^2 - a_1^2 - a_2^2 + a_{12}^2 \neq 0,$$

then ψ^{-1} can be defined as

$$\psi^{-1} = \frac{\bar{\psi}}{\bar{\psi}\psi} = \frac{a_0 - a_1\mathbf{e}_1 - a_2\mathbf{e}_2 - a_{12}\mathbf{e}_1\mathbf{e}_2}{a_0^2 - a_1^2 - a_2^2 + a_{12}^2}.$$

The group of the invertible elements $\mathcal{C}\ell_{2,0}^*$ consists of the elements of $\mathcal{C}\ell_{2,0}$ such that $a_0^2 - a_1^2 - a_2^2 + a_{12}^2 \neq 0$.

In order to determine the Clifford–Lipschitz group, the relation $\psi\mathbf{v}\psi^{-1} \in \mathbb{R}^{2,0}$ must hold for all $\mathbf{v} \in \mathbb{R}^{2,0}$. Since $\psi^{-1} = \bar{\psi}/(\bar{\psi}\psi)$, we just need to consider the similar condition $\psi\mathbf{v}\bar{\psi} \in \mathbb{R}^{2,0}$. By calculating $\psi\mathbf{v}\bar{\psi}$, we obtain

$$\begin{aligned}\psi\mathbf{v}\bar{\psi} =& [v_1(a_0^2 - a_1^2 + a_2^2 - a_{12}^2) + v_2(2a_0a_{12} - 2a_1a_2)]\mathbf{e}_1 \\ &+ [v_1(-2a_0a_{12} - 2a_1a_2) + v_2(a_0^2 + a_1^2 - a_2^2 - a_{12}^2)]\mathbf{e}_2 \\ &+ [v_1(-2a_1a_{12} - 2a_0a_2) + v_2(2a_0a_1 - 2a_2a_{12})]\mathbf{e}_1\mathbf{e}_2.\end{aligned}$$

Hence, in order for $\psi\mathbf{v}\bar{\psi} \in \mathbb{R}^{2,0}$, the equalities

$$a_0a_2 + a_1a_{12} = 0, \qquad a_0a_1 - a_2a_{12} = 0$$

must hold. Here, there are two possibilities: either $a_1 = a_2 = 0$, or $a_0 = a_{12} = 0$. In the first case, elements of $\Gamma_{2,0}$ are written as $\psi = a_0 + a_{12}\mathbf{e}_1\mathbf{e}_2$, with the condition $a_0^2 + a_{12}^2 \neq 0$. In the second case, the elements of $\Gamma_{2,0}$ are $a_1\mathbf{e}_1 + a_2\mathbf{e}_2 = (a_2 + a_1\mathbf{e}_1\mathbf{e}_2)\mathbf{e}_2$, with the condition $a_1^2 + a_2^2 \neq 0$.

In order to understand what is being accomplished here, we use the matrix representation of $\mathcal{C}\ell_{2,0}$, as studied in chapter 4. The element $\psi \in \mathcal{C}\ell_{2,0}$ can be represented by the matrix $\Psi = \rho(\psi) \in \mathcal{M}(2,\mathbb{R})$ given by

$$\Psi = \begin{pmatrix} a_0 + a_1 & a_2 + a_{12} \\ a_2 - a_{12} & a_0 - a_1 \end{pmatrix}.$$

The condition $\bar{\psi}\psi \neq 0$, necessary for the existence of ψ^{-1}, is equivalent to $\det\Psi \neq 0$, where

$$\det\Psi = \bar{\psi}\psi = a_0^2 - a_1^2 - a_2^2 + a_{12}^2.$$

The inverse matrix Ψ^{-1} corresponds to

$$\Psi^{-1} = \frac{1}{\det\Psi}\begin{pmatrix} a_0 - a_1 & -a_2 - a_{12} \\ -a_2 + a_{12} & a_0 + a_1 \end{pmatrix},$$

which is exactly the matrix representation of ψ^{-1}. Notice that

$$\rho(\bar{\psi}) = \begin{pmatrix} a_0 - a_1 & -a_2 - a_{12} \\ -a_2 + a_{12} & a_0 + a_1 \end{pmatrix}.$$

5.3 The Pin Group and the Spin Group

The property $\ker\widehat{\sigma} = \mathbb{R}^*$ can be interpreted as an assertion that the Clifford–Lipschitz group $\Gamma_{p,q}$ is 'bigger than necessary'. It is not necessary to use the whole group $\Gamma_{p,q}$ to describe orthogonal transformations; instead, a subgroup can be used, which can be obtained via a suitable 'normalisation process' can be used.

In chapter 3, we showed that the norm of a multivector – obtained by the 'Grassmannian' extension of g – reads in $\mathcal{C}\ell_{p,q}$ as eqn (3.49)

$$N(a) = \langle\tilde{a}a\rangle_0, \tag{5.62}$$

where $a \in \mathcal{C}\ell_{p,q}$. In addition, another definition for the norm in $\mathcal{C}\ell_{p,q}$ was shown to be

$$N'(a) = \langle\bar{a}a\rangle_0. \tag{5.63}$$

From eqn (3.51), it follows that

$$\left|N'\left(a_{[k]}\right)\right| = \left|N\left(a_{[k]}\right)\right|. \tag{5.64}$$

It is worth emphasising here that, as we proved in section 3.3, this equality does not hold when the multivectors are not simple.

The mappings N and N' are the homomorphisms $N : \Gamma_{p,q} \to \mathbb{R}^*$, and $N' : \Gamma_{p,q} \to \mathbb{R}^*$, as can be shown by eqn (5.57). Let us now consider the norm N, since the case involving N' is completely similar. By denoting $a = \mathbf{v}_1 \cdots \mathbf{v}_k$,

$$\begin{aligned} N(a) &= \langle \mathbf{v}_k \cdots \mathbf{v}_1 \mathbf{v}_1 \cdots \mathbf{v}_k\rangle_0 \\ &= \mathbf{v}_k \cdots \mathbf{v}_1 \mathbf{v}_1 \cdots \mathbf{v}_k = N(\mathbf{v}_1)\ldots N(\mathbf{v}_k). \end{aligned} \tag{5.65}$$

It follows that

$$\begin{aligned} N(ab) &= \langle\widetilde{(ab)}ab\rangle_0 = \langle\tilde{b}\tilde{a}ab\rangle_0 = \langle\tilde{a}ab\tilde{b}\rangle_0 \\ &= \langle \mathbf{v}_k \cdots \mathbf{v}_1 \mathbf{v}_1 \cdots \mathbf{v}_k \mathbf{u}_1 \cdots \mathbf{u}_l \mathbf{u}_l \cdots \mathbf{u}_1\rangle_0 \\ &= N(\mathbf{v}_1)\ldots N(\mathbf{v}_k) N(\mathbf{u}_1)\ldots N(\mathbf{u}_l) \\ &= N(a)N(b). \end{aligned} \tag{5.66}$$

The elements of $\Gamma_{p,q}$ can thus be normalised to obtain subgroups whose kernels are smaller than that for $\Gamma_{p,q}$.

The Pin Group

The group $\mathrm{Pin}(p,q)$ is defined as

$$\boxed{\mathrm{Pin}(p,q) = \{a \in \Gamma_{p,q} \mid N(a) = \pm 1\}}. \tag{5.67}$$

Using this definition, we have

$$\widehat{\sigma} : \mathrm{Pin}(p,q) \to \mathrm{O}(p,q), \tag{5.68}$$

where

$$\ker \widehat{\sigma}\big|_{\mathrm{Pin}(p,q)} = \{\pm 1\} = \mathbb{Z}_2. \tag{5.69}$$

By using the norm N', we can define the group $\mathrm{Pin}\hat{\ }(p,q)$ as

$$\boxed{\mathrm{Pin}\hat{\ }(p,q) = \{a \in \Gamma_{p,q} \mid N'(a) = \pm 1\}}. \tag{5.70}$$

From eqn (5.64), it can be seen that $N'(a) = N(a)$ if $a \in \mathcal{C}\ell^+(p,q)$. Consequently,

$$\mathrm{Pin}(p,q) \cap \mathcal{C}\ell^+(p,q) = \mathrm{Pin}\hat{\ }(p,q) \cap \mathcal{C}\ell^+(p,q). \tag{5.71}$$

The Spin Group

The group $\mathrm{Spin}(p,q)$ is defined as

$$\boxed{\mathrm{Spin}(p,q) = \{a \in \Gamma^+_{p,q} \mid N(a) = \pm 1\}}\,, \tag{5.72}$$

which implies that

$$\widehat{\sigma} : \mathrm{Spin}(p,q) \to \mathrm{SO}(p,q)\,, \tag{5.73}$$

where

$$\ker \widehat{\sigma}\big|_{\mathrm{Spin}(p,q)} = \{\pm 1\} = \mathbb{Z}_2. \tag{5.74}$$

It immediately follows that

$$\mathrm{Spin}(p,q) = \mathrm{Pin}(p,q) \cap \mathcal{C}\ell^+_{p,q}\,. \tag{5.75}$$

The definition of the group $\mathrm{Spin}\hat{\ }(p,q)$ using the norm N' is no different, since, in this case,

$$\mathrm{Spin}\hat{\ }(p,q) = \mathrm{Pin}\hat{\ }(p,q) \cap \mathcal{C}\ell^+_{p,q}\,, \tag{5.76}$$

and, from eqn (5.71), it follows that

$$\mathrm{Spin}(p,q) = \mathrm{Spin}\hat{\ }(p,q). \tag{5.77}$$

The Reduced Pin Group and the Reduced Spin Group

The $\mathrm{Pin}_+(p,q)$ and the $\mathrm{Pin}_+\hat{\ }(p,q)$ subgroups are respectively defined as

$$\boxed{\mathrm{Pin}_+(p,q) = \{a \in \Gamma_{p,q} \mid N(a) = 1\}}\,, \tag{5.78}$$

and

$$\boxed{\mathrm{Pin}\hat{\ }_+(p,q) = \{a \in \Gamma_{p,q} \mid N'(a) = 1\}}\,; \tag{5.79}$$

the subgroup $\mathrm{Spin}_+(p,q)$ is defined as

$$\boxed{\mathrm{Spin}_+(p,q) = \{a \in \Gamma^+_{p,q} \mid N(a) = 1\}}\,. \tag{5.80}$$

From these definitions it is straightforward to see that

$$\mathrm{Spin}_+(p,q) = \mathrm{Pin}_+(p,q) \cap \mathrm{Pin}\hat{\ }_+(p,q). \tag{5.81}$$

Let us look at these groups in more detail. Concerning $\mathbf{v} \in \mathbb{R}^{p,q}$, eqn (5.55) asserts that $\hat{\sigma}(\mathbf{v}) = S_{\mathbf{v}}$ is an orthogonal reflection with respect to the orthogonal hyperplane to $\mathbf{v}$. On the other hand, $N(\mathbf{v}) = \mathbf{v}^2 = g(\mathbf{v},\mathbf{v})$ and, if $N(\mathbf{v}) = 1$, then $g(\mathbf{v},\mathbf{v}) = 1$. Hence, the elements of $\mathrm{Pin}_+(p,q)$ are such that $\hat{\sigma}(a)$ consists of the product of reflections on hyperplanes orthogonal to vectors of type $g(\mathbf{v},\mathbf{v}) = 1$. If $T(a)$ is the matrix of type (5.4) representing this orthogonal transformation, then $\det D_q$ cannot change sign under this transformation. Consequently, according to the notation established in section 5.1, it yields

$$\hat{\sigma}(\text{Pin}_+(p,q)) = \text{O}_+(p,q). \tag{5.82}$$

On the other hand, $N'(\mathbf{v}) = -\mathbf{v}^2 = -g(\mathbf{v},\mathbf{v})$, and $N(\mathbf{v}) = 1$ implies that $g(\mathbf{v},\mathbf{v}) = -1$. Thus, the elements of $\text{Pin}\hat{\ }_+(p,q)$ are such that $\hat{\sigma}(a)$ consists of the product of reflections with respect to orthogonal hyperplanes to vectors $\mathbf{v}$ such that $g(\mathbf{v},\mathbf{v}) = -1$. In this case, the matrix $T(a)$ representing this orthogonal transformation is given by $\det A_p$, which does not change sign. Consequently, according to the notation adopted in section 5.1,

$$\hat{\sigma}(\text{Pin}\hat{\ }_+(p,q)) = \text{O}^{\uparrow}(p,q). \tag{5.83}$$

Equations (5.11) and (5.81) imply that

$$\hat{\sigma}(\text{Spin}_+(p,q)) = \text{SO}_+(p,q). \tag{5.84}$$

The kernel of the mappings (5.82–5.84) is $\mathbb{Z}_2$. In this case, the group $\text{Spin}_+(p,q)$ is said to be the *two-fold covering* of the group $\text{SO}_+(p,q)$. Finally, we can write

$$\boxed{\begin{array}{c} \text{Pin}_+(p,q)/\mathbb{Z}_2 \simeq \text{O}_+(p,q) \\ \text{Pin}\hat{\ }_+(p,q)/\mathbb{Z}_2 \simeq \text{O}^{\uparrow}(p,q) \\ \text{Spin}_+(p,q)/\mathbb{Z}_2 \simeq \text{SO}_+(p,q) \end{array}}\,. \tag{5.85}$$

A very useful result, since it involves cases of great interest – mostly involving important applications in physics – is the following:

Theorem 5.2 ▶ *Let $C\ell_{p,q}$ be the Clifford algebra associated with the quadratic space $\mathbb{R}^{p,q}$ and let $C\ell^+_{p,q}$ be the associated even subalgebra. If $n = p+q \leq 5$, then*

$$\boxed{\text{Spin}_+(p,q) = \{R \in C\ell^+_{p,q} \mid R\tilde{R} = \tilde{R}R = 1\}}\,. \tag{5.86}$$

Proof: Let us first consider the case $n < 5$. Take $x = \hat{R}\mathbf{v}R^{-1}$ for $\mathbf{v} \in V = \mathbb{R}^{p,q}$, and $R \in C\ell^+_{p,q}$, such that $R\tilde{R} = \tilde{R}R = 1$. Since $\hat{R} = R$, and $R^{-1} = \tilde{R}$, therefore $x = R\mathbf{v}\tilde{R}$. In addition, we have $\hat{x} = \hat{R}\hat{\mathbf{v}}\hat{\tilde{R}} = R(-\mathbf{v})\tilde{R} = -x$, and $\tilde{x} = \tilde{\tilde{R}}\tilde{\mathbf{v}}\tilde{R} = R\mathbf{v}\tilde{R} = x$. Only vectors of $C\ell_{p,q}$ can satisfy the conditions $\hat{x} = -x$ and $\tilde{x} = x$, if $n < 5$, and in this case $x = R\mathbf{v}\tilde{R} \in V$. In addition, $N(R) = |\langle\tilde{R}R\rangle_0| = |\tilde{R}R| = 1$, and $R \in \text{Spin}_+(p,q)$. For $n = 5$, the elements of $C\ell_{p,q}$ are, such that $\hat{x} = -x$, and $\tilde{x} = x$, in general a sum of 1-vectors and 5-vectors: $x = \mathbf{u} + p\eta$, where $\eta = \mathbf{e}_1\mathbf{e}_2\mathbf{e}_3\mathbf{e}_4\mathbf{e}_5$. Since η commutes with all elements of $C\ell_{p,q}$, namely, $\eta \in \text{Cen}(C\ell_{p,q})$, therefore $x^2 = \mathbf{u}^2 + p^2\eta^2 + 2p\mathbf{u}\eta$. All the terms x^2, $\mathbf{u}^2$, and $p^2\eta^2$ are scalars, and $\mathbf{u}\eta$ is a 4-vector. Hence, $\mathbf{u}\eta = 0$, and thus either $\mathbf{u} = 0$, or $\eta = 0$. If $\mathbf{u} = 0$, and consequently $R\mathbf{v}R^{-1} = \eta$, on the left-hand side there is a quantity that is not in the centre of $C\ell_{p,q}$, but on the right-hand side there is a quantity that is in the centre of $C\ell_{p,q}$, a result that is absurd. Hence, the unique possibility is $\eta = 0$, and therefore $x = R\mathbf{v}\tilde{R} \in V$ for $n = 5$. Since $N(R) = 1$, it follows that $R \in \mathrm{Spin}_+(p,q)$. This reasoning cannot be generalised for $n = 6$ since, in this case, a 5-vector is not an element of the centre of the algebra. For instance, for $R = (1/\sqrt{2})(\mathbf{e}_1\mathbf{e}_2 + \mathbf{e}_3\mathbf{e}_4\mathbf{e}_5\mathbf{e}_6) \in \mathcal{C}\ell_{p,q}$, it follows that $\tilde{R}R = R\tilde{R} = 1$ but that $R\mathbf{e}_1\tilde{R} = -\mathbf{e}_2\mathbf{e}_3\mathbf{e}_4\mathbf{e}_5\mathbf{e}_6$. ✓

We can actually obtain a slightly more general result than the one in eqn (5.86):

$$\mathrm{Pin}(p,q) = \{R \in \mathcal{C}\ell_{p,q} \mid \hat{R} = \pm R,\ R\tilde{R} = \tilde{R}R = \pm 1\}. \tag{5.87}$$

As the proof of eqn (5.87) is quite similar to that for eqn (5.86), the details of the proof have been omitted.

Example 5.3 Let us consider the Clifford algebra $\mathcal{C}\ell_{1,1}$. According to eqn (5.86), the group $\mathrm{Spin}_+(1,1)$ is defined by the elements $R \in \mathcal{C}\ell^+_{1,1}$ such that $R\tilde{R} = \tilde{R}R = 1$. Hence,

$$\mathrm{Spin}_+(1,1) = \{a + b\mathbf{e}_1\mathbf{e}_2 \mid a, b \in \mathbb{R},\quad a^2 - b^2 = 1\},$$

that is, $\mathrm{Spin}_+(1,1) \simeq \mathbb{R} \oplus \mathbb{R}$. This group has two components, with representatives $R = \cosh\alpha + \sinh\alpha\mathbf{e}_1\mathbf{e}_2$, and $R = -\cosh\alpha + \sinh\alpha\mathbf{e}_1\mathbf{e}_2$, respectively, $-\infty < \alpha < \infty$. Therefore, the two-fold covering $\mathrm{Spin}_+(1,1)$ *is not* connected, although $\mathrm{SO}_+(1,1)$ *is* connected. Indeed, the groups $\mathrm{Spin}_+(p,q)$ with $p + q \geq 2$ are connected, and the exception is precisely $\mathrm{Spin}_+(1,1)$.

Example 5.4 The group Spin(3) is given by

$$\mathrm{Spin}(3) = \{R \in \mathcal{C}\ell^+_{3,0}\,|\tilde{R}R = R\tilde{R} = 1\} = \{R \in \mathcal{C}\ell_{3,0}\,|R = \hat{R},\ \tilde{R}R = R\tilde{R} = 1\}.$$

In terms of the matrix representation of $\mathcal{C}\ell_{3,0}$, we have from section 3.3 that $\tilde{R}R = 1$ is translated for $\mathsf{R} = \rho(R)$ as $\mathsf{R}^\dagger\mathsf{R} = 1$, that is, $\mathsf{R} \in \mathrm{U}(2)$. But the condition $R = \hat{R}$ is translated as $\mathsf{R} = \mathrm{adj}(\mathsf{R}^\dagger)$, and it is straightforward to see that, from both conditions, it follows that $\det\mathsf{R} = 1$. As a consequence,

$$\mathrm{Spin}(3) \simeq \mathrm{SU}(2).$$

The Lie Algebra of the Associated Groups

The groups that have been presented in this section are called *Lie groups*. The *Lie algebra* associated with these groups can be identified as a subspace of the Clifford algebra $\mathcal{C}\ell_{p,q}$, and the product of the Lie algebra – the *Lie bracket* – is given by the commutator $[a,b] = ab - ba$, for a, b in this subspace of $\mathcal{C}\ell_{p,q}$. In fact, it is expected, since a Clifford algebra is isomorphic to a matrix algebra, and therefore the group of invertible elements $\mathcal{C}\ell^*_{p,q}$ is isomorphic to some subgroup in the invertible matrices group, which is itself a Lie group. From the formalism of Lie algebras, the *exponential* mapping maps an element of the Lie algebra to the Lie group component connected to the identity of this group (Meinrenken, 2013). It is thus pertinent to consider the definition of the exponential mapping inside a Clifford algebra. The *exponential* of $a \in \mathcal{C}\ell_{p,q}$ is defined by

$$\exp a = \mathrm{e}^a = \sum_{n=0}^{\infty} \frac{a^n}{n!}. \tag{5.88}$$

Since $\exp(-a) = (\exp a)^{-1}$, the exponential mapping links an element of $\mathcal{C}\ell_{p,q}$ with an element from the group of invertible elements $\mathcal{C}\ell^*_{p,q}$; namely, $\exp : \mathcal{C}\ell_{p,q} \to \mathcal{C}\ell^*_{p,q}$.

The vector space $V = \mathcal{C}\ell_{p,q}$ endowed with a product defined by $[a,b] = ab - ba$ can be identified as the Lie algebra associated with $\mathcal{C}\ell^*_{p,q}$.

The Clifford–Lipschitz group $\Gamma_{p,q}$ is the Lie subgroup of the Lie group $\mathcal{C}\ell^*_{p,q}$. Its associated Lie algebra is a vector subspace of $\mathcal{C}\ell_{p,q}$. Let us suppose that X is an element in the Lie algebra of $\Gamma_{p,q}$, in such a way that $\exp(tX)$ is an element of $\Gamma_{p,q}$, that is,

$$f(t) = \mathrm{Ad}\exp(tX)(\mathbf{v}) = \exp(tX)\mathbf{v}\exp(-tX) \in \mathbb{R}^{p,q}, \quad \forall \mathbf{v} \in \mathbb{R}^{p,q}. \tag{5.89}$$

Let us analyse $f(t)$. It is straightforward to see by finite induction that

$$\frac{\mathrm{d}^n f(t)}{\mathrm{d}t^n} = \exp(tX)\big(\mathrm{ad}(X)\big)^n(\mathbf{v})\exp(-tX), \tag{5.90}$$

where

$$\big(\mathrm{ad}(X)\big)(\mathbf{v}) = [X,\mathbf{v}] = X\mathbf{v} - \mathbf{v}X. \tag{5.91}$$

Constructing a Taylor expansion to $f(t)$ around the point $t = 0$, we can write

$$\begin{aligned} f(t) &= \mathbf{v} + t\big(\mathrm{ad}(X)\big)(\mathbf{v}) + \frac{t^2}{2!}\big(\mathrm{ad}(X)\big)^2(\mathbf{v}) + \cdots \\ &= \sum_{n=0}^{\infty}\left[\frac{t^n}{n!}\big(\mathrm{ad}(X)\big)^n\right](\mathbf{v}) = \big[\exp\mathrm{ad}(tX)\big](\mathbf{v}). \end{aligned} \tag{5.92}$$

This expression is equivalent to

$$\mathrm{Ad}\big(\exp(tX)\big) = \exp\big(\mathrm{ad}(tX)\big), \tag{5.93}$$

which is a classical result in group theory.

Accordingly, $f(t) \in \mathbb{R}^{p,q}$ if and only if

$$\mathrm{ad}(X)(\mathbf{v}) = [X,\mathbf{v}] = X\mathbf{v} - \mathbf{v}X \in \mathbb{R}^{p,q}. \tag{5.94}$$

Let us split X in even and odd parts: $X = X_+ + X_-$, where $X_+ = \hat{X}_+$, and $X_- = -\hat{X}_-$. It is thus possible to express

$$\begin{aligned} X\mathbf{v} &= X\lfloor\mathbf{v} + X\wedge\mathbf{v} = X_+\lfloor\mathbf{v} + X_-\lfloor\mathbf{v} + X_+\wedge\mathbf{v} + X_-\wedge\mathbf{v} \\ &= -\mathbf{v}\rfloor X_+ + \mathbf{v}\rfloor X_- + \mathbf{v}\wedge X_+ - \mathbf{v}\wedge X_-, \end{aligned} \tag{5.95}$$

and therefore

$$X\mathbf{v} - \mathbf{v}X = 2\mathbf{v}\rfloor X_+ - 2\mathbf{v}\wedge X_-. \tag{5.96}$$

If $X_+ \in \bigwedge_b(\mathbb{R}^{p,q})$, then $\mathbf{v}\rfloor X_+ \in \bigwedge_{b-1}(\mathbb{R}^{p,q})$. In order for the expression $\bigwedge_{b-1}(\mathbb{R}^{p,q}) = \bigwedge_1(\mathbb{R}^{p,q})$ to hold, we must have the condition $b = 2$, which means that X_+ is a 2-vector. For the other cases, we must have $\mathbf{v}\rfloor X_+ = 0$, $\forall \mathbf{v} \in \mathbb{R}^{p,q}$. This equation only holds for $X_+ \in \bigwedge_0(\mathbb{R}^{p,q}) = \mathbb{R}$. The most general expression for X_+ is the sum of a *scalar* and a *2-vector*. In addition, $\mathbf{v}\wedge X_- \in \bigwedge_{b+1}(\mathbb{R}^{p,q})$ if $X_- \in \bigwedge_n(\mathbb{R}^{p,q})$. In this case, when $\bigwedge_{b+1}(\mathbb{R}^{p,q}) = \bigwedge_1(\mathbb{R}^{p,q})$ holds, we must have $b = 0$. However, X_- is odd, and $b = 1, 3, 5, \ldots$, which is not possible. Therefore, we must have $\mathbf{v}\wedge X_- = 0, \forall \mathbf{v} \in \mathbb{R}^{p,q}$.

It only happens if X_- is an n-*vector*, and X_- is odd only if n is *odd*; in this situation, an n-vector is an element of the centre of $\mathcal{C}\ell_{p,q}$. Hence, if $n = p + q$ is even, then

$$X \in \bigwedge{}_0(\mathbb{R}^{p,q}) \oplus \bigwedge{}_2(\mathbb{R}^{p,q}) \tag{5.97}$$

and, if $n = p + q$ is odd,

$$X \in \bigwedge{}_0(\mathbb{R}^{p,q}) \oplus \bigwedge{}_2(\mathbb{R}^{p,q}) \oplus \bigwedge{}_n(\mathbb{R}^{p,q}). \tag{5.98}$$

In either case, we can write

$$X \in \mathrm{Cen}(\mathcal{C}\ell_{p,q}) \oplus \bigwedge{}_2(\mathbb{R}^{p,q}). \tag{5.99}$$

In this way, $\exp(tX) \in \Gamma_{p,q}$.

A subgroup that is particularly important is the group $\mathrm{Spin}_+(p,q)$, which is the connected component to the identity of $\mathrm{Spin}(p,q)$. Since $R \in \mathrm{Spin}_+(p,q)$ consists of the even elements, it is equivalent to $\hat{R} = R$. Hence, for $R = \exp(tX)$, we must have $\hat{X} = X$, and therefore X must be of the form $X = a + B$, where $a \in \mathbb{R}$, and $B \in \bigwedge_2(\mathbb{R})$. Moreover, the condition $N(R) = 1$ implies that $1 = \exp(t\tilde{X})\exp(tX) = \exp[t(a-B)]\exp[t(a+B)] = \exp(2ta)$, implying that $a = 0$. Hence, we can conclude that

$$R = \exp(tB) \in \mathrm{Spin}_+(p,q)\,, \quad B \in \bigwedge{}_2(\mathbb{R}^{p,q}). \tag{5.100}$$

The Lie algebra associated with the group $\mathrm{Spin}_+(p,q)$ consists of the vector space defined by the *2-vectors* and endowed with the *commutator*. Moreover, the commutator of two 2-vectors is a 2-vector. In fact, if B and C are 2-vectors, then

$$BC = \langle BC\rangle_0 + \langle BC\rangle_2 + \langle BC\rangle_4 \tag{5.101}$$

and, since $\widetilde{B} = -B$, $\widetilde{C} = -C$,

$$\widetilde{(BC)} = \widetilde{C}\widetilde{B} = CB. \tag{5.102}$$

By applying the reversion in eqn (5.101), we obtain

$$CB = \langle BC\rangle_0 - \langle BC\rangle_2 + \langle BC\rangle_4. \tag{5.103}$$

Now, by subtracting eqn (5.103) from eqn (5.103), we find that

$$BC - CB = [B, C] = 2\langle BC\rangle_2. \tag{5.104}$$

Hence, the commutator of 2-vectors is another 2-vector, and the algebra $\left(\bigwedge_2(\mathbb{R}^{p,q}), [\ ,\ ]\right)$ is indeed the Lie algebra of $\mathrm{Spin}_+(p,q)$. From eqn (5.104), it is often convenient to consider $(1/2)[B,C]$ instead of the commutator $[B,C]$. For another approach to Clifford algebras and Lie theory, the reader may consult, for example, the work by Meinrenken (2013).

Example 5.5 Consider $\mathcal{C}\ell_{3,0}$, the Clifford algebra associated with the Euclidean space $\mathbb{R}^{3,0}$. If $\{\mathbf{e}_1, \mathbf{e}_2, \mathbf{e}_3\}$ is an orthonormal basis of $\mathbb{R}^{3,0}$, the space of the 2-vectors has elements of the form

$$\{B = a\mathbf{e}_1\mathbf{e}_2 + b\mathbf{e}_2\mathbf{e}_3 + c\mathbf{e}_3\mathbf{e}_1 \mid a, b, c \in \mathbb{R}\}.$$

The element $\exp(tX) \in \mathrm{Spin}_+(3,0)$ and indeed it is straightforward to realise that $\mathrm{Spin}_+(3,0) = \mathrm{Spin}(3,0)$. In order to further present the Lie algebra of $\mathrm{Spin}(3,0)$, let us denote by L_i $(i = 1,2,3)$ the objects

$$L_1 = \frac{1}{2}\mathbf{e}_2\mathbf{e}_3, \quad L_2 = \frac{1}{2}\mathbf{e}_1\mathbf{e}_3, \quad L_3 = \frac{1}{2}\mathbf{e}_1\mathbf{e}_2,$$

satisfying

$$[L_1, L_2] = L_3, \quad [L_2, L_3] = L_1, \quad [L_3, L_1] = L_2.$$

These show that the Lie algebra of $\mathrm{Spin}(3,0)$ is isomorphic to the Lie algebra associated with the Lie group $\mathrm{SU}(2)$. Indeed, these groups are isomorphic: $\mathrm{Spin}(3,0) \simeq \mathrm{SU}(2)$.

5.4 Conformal Transformations in Clifford Algebras

The main goal of this section is to describe the conformal transformations in the context of the Clifford algebras. As a main application, we aim to study the conformal transformations associated with the Minkowski spacetime $\mathbb{R}^{1,3}$ and derived from the action of elements of the group $\$\mathrm{pin}_+(2,4)$ on elements of $\mathbb{R} \oplus \mathbb{R}^{4,1}$, as an immediate application of the periodicity theorem. Given a vector space V and its associated real universal Clifford algebra $\mathcal{C}\ell(V,g)$, the subspace $\mathbb{R} \oplus V$ of $\mathcal{C}\ell(V,g)$ is defined to be the paravector space of V (Porteous, 1969; Maks, 1989; Baylis, 1996; da Rocha and Vaz Jr, 2007), denoted by V_π.

Möbius Transformations on the Plane

The algebra $\mathcal{C}\ell_{0,1} \simeq \mathbb{C}$ is appropriate for describing rotations in $\mathbb{R}^2$. Using the periodicity theorem $\mathcal{C}\ell_{p+1,q+1} \simeq \mathcal{C}\ell_{1,1} \otimes \mathcal{C}\ell_{p,q}$, we see that the Lorentz transformations in Minkowski spacetime, generated from the action of elements of the group $\$\mathrm{pin}_+(1,3) = \{s \in \mathcal{C}\ell_{3,0}\,|s\bar{s} = 1\} \simeq \mathrm{SL}(2,\mathbb{C})$, are closely associated with the Möbius transformations in the plane, since

$$\mathcal{C}\ell_{3,0} \simeq \mathcal{C}\ell_{2,0} \otimes \mathcal{C}\ell_{0,1} \simeq \mathcal{C}\ell_{1,1} \otimes \mathcal{C}\ell_{0,1}. \tag{5.105}$$

The conformal transformations in the plane are described by the algebra $\mathcal{C}\ell_{1,1} \simeq \mathcal{M}(2,\mathbb{R})$. Using the isomorphism in (5.105), we can represent a paravector $\mathfrak{a} \in \mathbb{R} \oplus \mathbb{R}^3$ of $\mathcal{C}\ell_{3,0}$ by an element of $\mathcal{M}(2,\mathbb{C})$:

$$\mathfrak{a} = \begin{pmatrix} z & \lambda \\ \mu & \bar{z} \end{pmatrix} \in \mathbb{R} \oplus \mathbb{R}^3,$$

where $z, \lambda, \mu \in \mathbb{C}$. Consider now an element in the group $\$\mathrm{pin}_+(1,3)$. The generalised periodicity theorem asserts that, in this particular case of $\mathbb{R}^3$, the reversion of a matrix which represents an element of $\mathcal{C}\ell_{3,0}$ is accompanied by the conjugation of each one of its entries. It is in fact represented by

$$\widetilde{\begin{pmatrix} a & c \\ b & d \end{pmatrix}} = \begin{pmatrix} \bar{d} & \bar{c} \\ \bar{b} & \bar{a} \end{pmatrix}, \qquad \text{where } a, b, c, d \in \mathbb{C}. \tag{5.106}$$

The rotation of a paravector $\mathfrak{a} \in \mathbb{R} \oplus \mathbb{R}^3$ can be then performed by a transformation $\mathfrak{a} \mapsto \mathfrak{a}' = \eta\mathfrak{a}\tilde{\eta}$ for $\eta \in \$\text{pin}_+(1,3)$. A rotation can be represented by

$$\begin{pmatrix} a & c \\ b & d \end{pmatrix} \begin{pmatrix} z & \lambda \\ \mu & \bar{z} \end{pmatrix} \widetilde{\begin{pmatrix} a & c \\ b & d \end{pmatrix}} = \begin{pmatrix} a & c \\ b & d \end{pmatrix} \begin{pmatrix} z & \lambda \\ \mu & \bar{z} \end{pmatrix} \begin{pmatrix} \bar{d} & \bar{c} \\ \bar{b} & \bar{a} \end{pmatrix}. \tag{5.107}$$

When $\mu = 1$, and $\lambda = z\bar{z}$, it follows that the paravector $\mathfrak{a}$ is mapped into

$$\begin{pmatrix} a & c \\ b & d \end{pmatrix} \begin{pmatrix} z & z\bar{z} \\ 1 & \bar{z} \end{pmatrix} \begin{pmatrix} \bar{d} & \bar{c} \\ \bar{b} & \bar{a} \end{pmatrix} = \omega \begin{pmatrix} z' & z'\bar{z}' \\ 1 & \bar{z}' \end{pmatrix},$$

where $z' := \frac{az+c}{bz+d}$, and $\omega := |bz+d|^2 \in \mathbb{R}$.

The formalism of Möbius transformations in the plane, constructed from the generalised periodicity theorem, leads to the classical framework regarding rotations performed by the group $\mathrm{SL}(2,\mathbb{C})$. It can be further generalised, to describe conformal transformations in Minkowski spacetime, as we shall see.

Example 5.6 Consider the space $\mathbb{R}^{2,0}$. The image of the mapping

$$\begin{aligned} \mathbb{R}^{2,0} &\to \widehat{\mathbb{R}^{0,4}} \\ (x,y) &\mapsto [x, y, 1, x^2+y^2] \end{aligned}$$

lies in the quadric defined by the equation $x^2 + y^2 - \mu\nu = 0$, which is the conformal compactification of $\mathbb{R}^{2,0}$ (Porteous, 1969, 1995), denoted by $\widehat{\mathbb{R}^{0,4}}$. Suppose now that a change of variables is taken into account in order to express the quadric equation as the sum of two squares. An appropriate choice reads $x = x$; $y = y$; $z = \mu - \nu$; and $t = \mu + \nu$. The equation of the quadric is led to $x^2 + y^2 + z^2 - t^2$, and the image lies in the chart $t = 1$, under the mapping $\mathbb{R}^2 \to \mathbb{R}^3$ given by

$$(x,y) \mapsto \left(\frac{2x}{1+x^2+y^2}, \frac{2y}{1+x^2+y^2}, \frac{1-x^2-y^2}{1+x^2+y^2}\right), \tag{5.108}$$

which image is a subset of the sphere S^2 in $\mathbb{R}^3$. Indeed, the image is the sphere devoid of the south pole (0,0,−1).

Conformal Compactification

In this section, we revisit some prominent results from the work by Porteous (1969), Maks (1989), and Crumeyrolle (1990). Given the quadratic space $\mathbb{R}^{p,q}$, consider the injective mapping given by

$$\begin{aligned} \varkappa : \mathbb{R}^{p,q} &\to \mathbb{R}^{p+1,q+1} \\ x &\mapsto \varkappa(x) = (x, x\cdot x, 1) = (x, \lambda, \mu). \end{aligned} \tag{5.109}$$

The image of $\mathbb{R}^{p,q}$ is a subset of the quadric $Q \hookrightarrow \mathbb{R}^{p+1,q+1}$, described by the equation:

$$x \cdot x - \lambda\mu = 0. \tag{5.110}$$

This quadric is a high-dimensional generalisation of the quadric equation and is called the *Klein absolute*. The mapping $\varkappa$ induces an injective mapping from Q into the projective space $\mathbb{RP}^{p+1,q+1}$. In addition, Q is compact and is defined as the conformal compactification $\widehat{\mathbb{R}^{p,q}}$ of $\mathbb{R}^{p,q}$. (Here, the hat denotes the conformal compactification, instead of the grade involution, in order to preserve the usual notation).

It has been shown that $Q \approx \widehat{\mathbb{R}^{p,q}}$ is homeomorphic to $(S^p \times S^q)/\mathbb{Z}_2$ (Porteous, 1969). In the particular case where $p = 0$, and $q = n$, the quadric is homeomorphic to the n-sphere S^n, seen as the compactification of $\mathbb{R}^n$ by the addition of a point at infinity.

There also exists an injective mapping

$$s : \mathbb{R} \oplus \mathbb{R}^3 \to \mathbb{R} \oplus \mathbb{R}^{4,1},$$
$$v \mapsto s(v) = \begin{pmatrix} v & v\bar{v} \\ 1 & \bar{v} \end{pmatrix}. \tag{5.111}$$

The following theorem was introduced by Porteous (1969, 1995); also see Maks (1989):

Theorem 5.3 ▶ (i) *The mapping* $\varkappa : \mathbb{R}^{p,q} \to \mathbb{R}^{p+1,q+1}$; $x \mapsto (x, x \cdot x, 1)$ *is an isometry.* (ii) *The mapping* $\pi : Q \to \mathbb{R}^{p,q}$; $(x, \lambda, \mu) \mapsto x/\mu$ *is defined where* $\lambda \neq 0$ *is conformal.* (iii) *If* $U : \mathbb{R}^{p+1,q+1} \to \mathbb{R}^{p+1,q+1}$ *is an orthogonal mapping, the map* $\Omega = \pi \circ U \circ \varkappa : \mathbb{R}^{p,q} \to \mathbb{R}^{p,q}$ *is conformal.*

The mapping Ω maps conformal spheres onto conformal spheres, which can be either quasi-spheres or hyperplanes. A quasi-sphere is a submanifold of $\mathbb{R}^{p,q}$, defined by the equation $a\, x \cdot x + b \cdot x + c = 0$, where $a, c \in \mathbb{R}$, and $b \in \mathbb{R}^{p,q}$. A quasi-sphere is a sphere when a quadratic form g in $\mathbb{R}^{p,q}$ is positive defined, and $a \neq 0$. Otherwise, a quasi-sphere is a plane when $a = 0$. From assertion (iii) of theorem 5.3, we see that both U and $-U$ induce the very same conformal transformation in $\mathbb{R}^{p,q}$. The *conformal group* is defined as

$$\mathrm{Conf}(p, q) \simeq \mathrm{O}(p + 1, q + 1)/\mathbb{Z}_2, \tag{5.112}$$

where $\mathrm{O}(p + 1, q + 1)$ has four components and, in the Minkowski spacetime case, where $p = 1$, and $q = 3$, the group Conf(1,3) has four components (Porteous, 1995; Anglès, 2008). The component of $\mathrm{Conf}(1, 3)$ connected to the identity, denoted by $\mathrm{Conf}_+(1, 3)$, is known as *the Möbius group* of $\mathbb{R}^{1,3}$. In addition, $\mathrm{SConf}_+(1,3)$ denotes the time-preserving and future-pointing component connected to the identity.

Möbius Transformations in Minkowski Spacetime

Let the group $\$\mathrm{pin}_+(2, 4)$ be defined as

$$\$\mathrm{pin}_+(2, 4) = \{g \in \mathcal{C}\ell_{4,1} \mid g\bar{g} = 1,\; g\mathfrak{b}g^{-1} \subset \mathbb{R} \oplus \mathbb{R}^{4,1}, \forall \mathfrak{b} \in \mathbb{R} \oplus \mathbb{R}^{4,1}\}. \tag{5.113}$$

The matrix $g = \begin{pmatrix} a & c \\ b & d \end{pmatrix}$ is in the group $\$\mathrm{pin}_+(2, 4)$ if, and only if, its entries $a, b, c, d \in \mathcal{C}\ell_{3,0}$ satisfy the following conditions (Maks, 1989):

$$\begin{aligned}
&\text{(i)} && a\bar{a},\; b\bar{b},\; c\bar{c},\; d\bar{d} \in \mathbb{R},\\
&\text{(ii)} && a\bar{b},\; c\bar{d} \;\in \mathbb{R} \oplus \mathbb{R}^3,\\
&\text{(iii)} && av\bar{c} + c\bar{v}\bar{a},\;\; cv\bar{d} + d\bar{v}\bar{c} \in \mathbb{R}, \quad \forall v \in \mathbb{R} \oplus \mathbb{R}^3,\\
&\text{(iv)} && av\bar{d} + c\bar{v}\bar{b} \in \mathbb{R} \oplus \mathbb{R}^3, \quad \forall v \in \mathbb{R} \oplus \mathbb{R}^3,\\
&\text{(v)} && a\tilde{c} = c\tilde{a},\; b\tilde{d} = d\tilde{b},\\
&\text{(vi)} && a\tilde{d} - c\tilde{b} = 1.
\end{aligned} \tag{5.114}$$

Conditions (i), (ii), (iii), and (iv) are equivalent to the condition $\hat{\sigma}(g)(\mathfrak{b}) := g\mathfrak{b}\tilde{g} \in \mathbb{R} \oplus \mathbb{R}^{4,1}$, $\forall \mathfrak{b} \in \mathbb{R} \oplus \mathbb{R}^{4,1}$ (Ahlford, 1986; Fillmore and Springer, 1990) where $\hat{\sigma}$: $\$\text{pin}_+(2,4) \to \text{SO}_+(2,4)$ denotes the twisted adjoint representation. Indeed,

$$\begin{aligned} g\mathfrak{b}\tilde{g} &= \begin{pmatrix} a & c \\ b & d \end{pmatrix} \begin{pmatrix} x & \lambda \\ \mu & \bar{x} \end{pmatrix} \begin{pmatrix} \bar{d} & \bar{c} \\ \bar{b} & \bar{a} \end{pmatrix} \\ &= \begin{pmatrix} ax\bar{d} + \lambda a\bar{b} + \mu c\bar{d} + c\bar{x}\bar{b} & ax\bar{c} + \lambda a\bar{a} + \mu c\bar{c} + b x\bar{a} \\ bx\bar{d} + \lambda d\bar{b} + \mu d\bar{d} + d\bar{x}\bar{b} & bx\bar{c} + \lambda b\bar{a} + \mu d\bar{b} + d\bar{x}\bar{a} \end{pmatrix} \\ &= \begin{pmatrix} w & \lambda' \\ \mu' & \bar{w} \end{pmatrix} \in \mathbb{R} \oplus \mathbb{R}^{4,1}, \end{aligned} \tag{5.115}$$

where the last equality (considering $w \in \mathbb{R} \oplus \mathbb{R}^3$ and $\lambda', \mu' \in \mathbb{R}$) comes from the requirement that $g \in \$\text{pin}_+(2,4)$, that is, $g\mathfrak{b}\tilde{g} \in \mathbb{R} \oplus \mathbb{R}^{4,1}$. If these conditions are required, (i), (ii), (iii), and (iv) follow from the results given by Vahlen (1902) and Ahlford (1986). More details can be seen in the work by Maks (1989) and Fillmore and Springer (1990).

Conditions (v) and (vi) lead to $g\bar{g} = 1$ since, for all $g \in \$\text{pin}_+(2,4)$, we have

$$g\bar{g} = 1 \quad \Leftrightarrow \quad \begin{pmatrix} a\tilde{d} - c\tilde{b} & a\tilde{c} - c\tilde{a} \\ b\tilde{d} - d\tilde{b} & d\tilde{a} - b\tilde{c} \end{pmatrix} = \begin{pmatrix} 1 & 0 \\ 0 & 1 \end{pmatrix}. \tag{5.116}$$

Conformal Transformations

The paravector $\mathfrak{b} \in \mathbb{R} \oplus \mathbb{R}^{4,1} \hookrightarrow \mathcal{C}\ell_{4,1}$ is represented as

$$\begin{pmatrix} x & x\bar{x} \\ 1 & \bar{x} \end{pmatrix} = \begin{pmatrix} x & \lambda \\ \mu & \bar{x} \end{pmatrix}, \tag{5.117}$$

where $x \in \mathbb{R} \oplus \mathbb{R}^3$ is a paravector of $\mathcal{C}\ell_{3,0}$.

Consider an element of the group $\$\text{pin}_+(2,4)$. It is possible to represent it as an element $g \in \mathcal{C}\ell_{4,1} \simeq \mathcal{C}\ell_{1,1} \otimes \mathcal{C}\ell_{3,0}$, that is, $g = \begin{pmatrix} a & c \\ b & d \end{pmatrix}$, where $a, b, c, d \in \mathcal{C}\ell_{3,0}$. The rotation of $\mathfrak{b} \in \mathbb{R} \oplus \mathbb{R}^{4,1} \hookrightarrow \mathcal{C}\ell_{4,1}$ is performed by the use of the twisted adjoint representation $\hat{\sigma} : \$\text{pin}_+(2,4) \Rightarrow \text{SO}_+(2,4)$, defined as

$$\begin{aligned} \hat{\sigma}(g)(\mathfrak{b}) &= g\mathfrak{b}\hat{g}^{-1} \\ &= g\mathfrak{b}\tilde{g}, \qquad g \in \$\text{pin}_+(2,4). \end{aligned} \tag{5.118}$$

Using the matrix representation, the action of $\$\text{pin}_+(2,4)$ is given by

$$\begin{pmatrix} a & c \\ b & d \end{pmatrix} \begin{pmatrix} x & \lambda \\ \mu & \bar{x} \end{pmatrix} \widetilde{\begin{pmatrix} a & c \\ b & d \end{pmatrix}} = \begin{pmatrix} a & c \\ b & d \end{pmatrix} \begin{pmatrix} x & \lambda \\ \mu & \bar{x} \end{pmatrix} \begin{pmatrix} \bar{d} & \bar{c} \\ \bar{b} & \bar{a} \end{pmatrix}. \tag{5.119}$$

By fixing $\mu = 1$, the paravector $\mathfrak{b}$ is mapped on

$$\begin{pmatrix} a & c \\ b & d \end{pmatrix} \begin{pmatrix} x & x\bar{x} \\ 1 & \bar{x} \end{pmatrix} \begin{pmatrix} \bar{d} & \bar{c} \\ \bar{b} & \bar{a} \end{pmatrix} = \Delta \begin{pmatrix} x' & x'\bar{x}' \\ 1 & \bar{x}' \end{pmatrix}, \tag{5.120}$$

where

$$x' := (ax+c)(bx+d)^{-1}, \qquad \Delta := (bx+d)(\overline{bx+d}) \in \mathbb{R}. \tag{5.121}$$

The transformation (5.121) is conformal (Vahlen, 1902; Lawson and Michelson, 1990; Hestenes, 1991).

From the isomorphisms

$$\mathcal{C}\ell_{4,1} \simeq \mathbb{C} \otimes \mathcal{C}\ell_{1,3} \simeq \mathcal{M}(4,\mathbb{C}), \tag{5.122}$$

elements of \$pin$_+$(2,4) are elements of the Dirac algebra $\mathbb{C} \otimes \mathcal{C}\ell_{1,3}$. From eqn (6.192), we denote $x \in \mathbb{R} \oplus \mathbb{R}^3$ a paravector. The conformal mappings are expressed by the action of \$pin$_+$(2,4), via the matrices given in table 5.1 (Vahlen, 1902; Porteous, 1969; Maks, 1989; Hestenes, 1991).

Table 5.1 Representation of the Conformal Mappings

	Explicit Transformation	Matrix of \$pin$_+(2,4)$
Translation	$x \mapsto x + h, \ \ h \in \mathbb{R} \oplus \mathbb{R}^3$	$\begin{pmatrix} 1 & h \\ 0 & 1 \end{pmatrix}$
Dilation	$x \mapsto \rho x, \ \rho \in \mathbb{R}$	$\begin{pmatrix} \sqrt{\rho} & 0 \\ 0 & 1/\sqrt{\rho} \end{pmatrix}$
Rotation	$x \mapsto \mathfrak{g} x \hat{\mathfrak{g}}^{-1}, \ \mathfrak{g} \in \$\mathrm{pin}_+(1,3)$	$\begin{pmatrix} \mathfrak{g} & 0 \\ 0 & \hat{\mathfrak{g}} \end{pmatrix}$
Inversion	$x \mapsto -\overline{x}$	$\begin{pmatrix} 0 & -1 \\ 1 & 0 \end{pmatrix}$
Transvection	$x \mapsto x + x(hx+1)^{-1}, \ h \in \mathbb{R} \oplus \mathbb{R}^3$	$\begin{pmatrix} 1 & 0 \\ h & 1 \end{pmatrix}$

This index-free geometric formulation makes it possible to trivially generalise the conformal mappings of $\mathbb{R}^{1,3}$ to the ones of $\mathbb{R}^{p,q}$, if the periodicity theorem of Clifford algebras is used.

The group SConf$_+$(1,3) is fourfold covered by SU(2,2) (Laufer, 1997), and the identity element $id_{\mathrm{SConf}_+(1,3)}$ of the group SConf$_+(1,3)$ corresponds to the following elements of $\mathrm{SU}(2,2) \simeq$ \$pin$_+$(2,4):

$$\begin{pmatrix} 1_2 & 0 \\ 0 & 1_2 \end{pmatrix}, \quad \begin{pmatrix} -1_2 & 0 \\ 0 & -1_2 \end{pmatrix}, \quad \begin{pmatrix} i_2 & 0 \\ 0 & i_2 \end{pmatrix}, \quad \begin{pmatrix} -i_2 & 0 \\ 0 & -i_2 \end{pmatrix}. \tag{5.123}$$

The element 1_2 denotes the 2×2 identity matrix, and i_2 denotes the matrix diag(i,i).

In this way, elements of \$pin$_+$(2,4) generate the orthochronous Möbius transformations. The isomorphisms

$$\mathrm{Conf}(1,3) \simeq \mathrm{O}(2,4)/\mathbb{Z}_2 \simeq \mathrm{Pin}(2,4)/\{\pm 1, \pm i\} \tag{5.124}$$

are constructed as in the work by Porteous (1969); consequently,

$$\mathrm{SConf}_+(1,3) \simeq \mathrm{SO}_+(2,4)/\mathbb{Z}_2 \simeq \$\mathrm{pin}_+(2,4)/\{\pm 1, \pm i\}. \tag{5.125}$$

The homomorphisms

$$\$\text{pin}_+(2,4) \xrightarrow{2-1} \text{SO}_+(2,4) \xrightarrow{2-1} \text{SConf}_+(1,3) \tag{5.126}$$

are explicitly constructed in the work by Klotz (1974) and Laufer (1997).

The Lie Algebra of the Conformal Group

The Lie algebra of \$pin$_+$(2,4) is generated by $\bigwedge^2(\mathbb{R}^{2,4})$, which has dimension 15. Since dim Conf(1,3) = 15, the relation between these groups is investigated now. In chapter 4, example 4.6, we introduced the isomorphisms

$$E_\mu = -i\gamma_\mu \; (\mu = 0,1,2,3), \quad E_4 = -i\gamma_{0123}, \tag{5.127}$$

and

$$E_A = \varepsilon_A \varepsilon_5, \tag{5.128}$$

where $\{\varepsilon_{\mathring{A}}\}_{\mathring{A}=0}^5$ is basis of $\mathbb{R}^{2,4}$; $\{E_A\}_{A=0}^4$ is basis of $\mathbb{R}^{4,1}$; and $\{\gamma_\mu\}_{\mu=0}^3$ is basis of $\mathbb{R}^{1,3}$. The generators of Conf(1,3), as elements of $\bigwedge_2(\mathbb{R}^{2,4})$, are defined as follows:

$$P_\mu = \frac{i}{2}(\varepsilon_\mu\varepsilon_5 + \varepsilon_\mu\varepsilon_4), \qquad D = -\frac{1}{2}\varepsilon_4\varepsilon_5, \tag{5.129}$$

$$K_\mu = -\frac{i}{2}(\varepsilon_\mu\varepsilon_5 - \varepsilon_\mu\varepsilon_4), \qquad M_{\mu\nu} = \frac{i}{2}\varepsilon_\nu\varepsilon_\mu. \tag{5.130}$$

From the relations in (5.127) and (5.128), the generators of Conf(1,3) are expressed from the $\{\gamma_\mu\} \in \mathcal{C}\ell_{1,3}$ as

$$P_\mu = \frac{1}{2}(\gamma_\mu + i\gamma_\mu\gamma_5), \qquad D = \frac{1}{2}i\gamma_5, \tag{5.131}$$

$$K_\mu = -\frac{1}{2}(\gamma_\mu - i\gamma_\mu\gamma_5), \qquad M_{\mu\nu} = \frac{1}{2}(\gamma_\nu \wedge \gamma_\mu). \tag{5.132}$$

They satisfy the following relations:

$$[P_\mu, P_\nu] = 0, \qquad [K_\mu, K_\nu] = 0, \qquad [M_{\mu\nu}, D] = 0, \tag{5.133}$$

$$[M_{\mu\nu}, P_\lambda] = -(g_{\mu\lambda}P_\nu - g_{\nu\lambda}P_\mu), \tag{5.134}$$

$$[M_{\mu\nu}, K_\lambda] = -(g_{\mu\lambda}K_\nu - g_{\nu\lambda}K_\mu), \tag{5.135}$$

$$[M_{\mu\nu}, M_{\sigma\rho}] = g_{\mu\rho}M_{\nu\sigma} + g_{\nu\sigma}M_{\mu\rho} - g_{\mu\sigma}M_{\nu\rho} - g_{\nu\rho}M_{\mu\sigma}, \tag{5.136}$$

$$[P_\mu, K_\nu] = 2(g_{\mu\nu}D - M_{\mu\nu}), \qquad [P_\mu, D] = P_\mu, \qquad [K_\mu, D] = -K_\mu. \tag{5.137}$$

These commutation relations are invariant under $P_\mu \mapsto -K_\mu$; $K_\mu \mapsto -P_\mu$; and $D \mapsto -D$. In chapter 6, we shall apply what has been developed here.

5.5 Additional Readings

There are some good texts for those who want to go deeply into the material discussed in this chapter but keep connection to the theory of Clifford algebras: the classical groups are regarded in the book by Porteous (1995); a detailed study of the theory of Lie groups and Lie algebras is given by Meinreken (2013); and a detailed study of the conformal group, by Anglès (2008). Problems related to the construction of Clifford algebras and of Spin groups on differentiable manifolds are examined, for example, in the book by Lawson and Michelson (1990).

5.6 Exercises

(1) Consider the spacetime algebra $\mathcal{C}\ell_{3,1} \simeq \mathcal{M}(4,\mathbb{R})$ generated by the basis $\{\mathbf{e}_1, \mathbf{e}_2, \mathbf{e}_3, \mathbf{e}_4\}$ such that $\mathbf{e}_i^2 = 1$, $\mathbf{e}_4^2 = -1$. Show that the matrix

$$M = \frac{1}{2}\begin{pmatrix} 1+\mathbf{e}_1\mathbf{e}_4 & -\mathbf{e}_1+\mathbf{e}_4 \\ \mathbf{e}_1+\mathbf{e}_4 & 1-\mathbf{e}_1\mathbf{e}_4 \end{pmatrix}$$

can be written as the composition of a transvection, a translation, and again a transvection. In addition, show that

$$\frac{1}{\sqrt{2}}\begin{pmatrix} 1 & \mathbf{e}_1 \\ -\mathbf{e}_1 & 1 \end{pmatrix}$$

can be written as the product of a transvection, a dilatation, and a translation.

(2) Show that, in an Euclidean space $\mathbb{R}^n$, the conditions $a\tilde{b}, b\tilde{d}, d\tilde{c}, \tilde{c}a \in \mathbb{R}^n$, and $\bar{a}b, b\bar{d}, \bar{d}c, c\bar{a} \in \mathbb{R}^n$, are equivalent, where $a, b, c, d \in \mathcal{C}\ell_n$.

(3) Define the Clifford and the exterior exponential mappings, respectively, by

$$e^A = 1 + A + \frac{1}{2!}A^2 + \frac{1}{3!}A^3 + \cdots, \qquad A \in \mathcal{C}\ell_{p,q}$$
$$e^{\wedge A} = 1 + A + \frac{1}{2!}A\wedge A + \frac{1}{3!}A\wedge A\wedge A + \cdots.$$

Given $B \in \bigwedge^2(\mathbb{R}^{p,q})$, show that $e^B \in \text{Spin}_+(p,q)$ and that $e^{\wedge B} \in \Gamma_{p,q}$.

(4) Define the groups

$$\text{Pin}_\pm(n,\mathbb{C}) = \{g \in \mathcal{C}\ell(\mathbb{C}^n) \,|\, N(g) = \pm 1,\ \ g\mathbf{v}\hat{g}^{-1} \in \mathbb{C}^n\},$$
$$\text{Spin}_\pm(n,\mathbb{C}) = \{g \in \mathcal{C}\ell^+(\mathbb{C}^n) \,|\, N(g) = \pm 1,\ \ g\mathbf{v}\hat{g}^{-1} \in \mathbb{C}^n\}.$$

Show that those groups are not connected when $n = 1$. In other words, show that $\text{Pin}_+(1,\mathbb{C}) = \{\pm 1, \pm i\mathbf{e}_1\}$ and that $\text{Spin}_+(1,\mathbb{C}) = \{\pm 1\}$.

(5) Show that $\text{Spin}_+(p,q) \simeq \text{Spin}_+(q,p)$ and that

$$\begin{aligned}
&\text{Spin}(2) \simeq \text{U}(1), && \text{Spin}_+(1,1) \simeq \mathbb{R}\oplus\mathbb{R},\\
&\text{Spin}(3) \simeq \text{SU}(2), && \text{Spin}_+(1,2) \simeq \text{SL}(2,\mathbb{R}),\\
&\text{Spin}_+(4) \simeq \text{SU}(2)\times\text{SU}(2), && \text{Spin}_+(1,3) \simeq \text{SL}(2,\mathbb{C}),\\
&\text{Spin}_+(2,2) \simeq \text{SL}(2,\mathbb{R})\times\text{SL}(2,\mathbb{R}), && \text{Spin}(5) \simeq \text{Sp}(2,\mathbb{H}),\\
&\text{Spin}_+(1,4) \simeq \text{Sp}(1,1,\mathbb{H}), && \text{Spin}_+(2,3) \simeq \text{Sp}(4,\mathbb{R}).
\end{aligned}$$

6 Spinors

In this chapter, we introduce and discuss the theory of spinors as well as some of their prominent applications. Three different definitions of spinors and their properties are presented. Pure spinors are also introduced and the triality principle, twistors, and Penrose flagpoles are discussed. A detailed study of the so-called Weyl spinors, which are the basis of the Penrose and Rindler formalism (Penrose, 1967; Penrose and Rindler, 1984; da Rocha and Vaz Jr, 2007), concludes this chapter. The comparison between the Clifford algebraic framework and the van der Waerden notation is left to the appendix.

First, we must assert that there is not a unique definition of spinor in the literature. Although the differences between the definitions are small, they do exist. Perhaps the origin of these differences is the fact that the theory of spinors was developed independently by physicists and mathematicians. To try to clarify this situation, we begin with a brief discussion about these differences, to pave the way for the definitions and the developments to be presented in this chapter.

6.1 The Babel of Spinors

There are essentially three different definitions of spinors, each one emphasising a different point of view. Two of them are well established, whereas the third one is gradually becoming known in the literature. We classify these three definitions as (i) *classical*, (ii) *algebraic*, and (iii) *operatorial*. The 'well-established' definitions are the classical and the algebraic ones. A comparative study of these different definitions in some situations of interest in physics is presented in the article by Figueiredo, de Oliveira, and Rodrigues Jr (1990).

In physics, spinors effectively emerged as a result of the study of quantum mechanics, specifically, from the Pauli theory (1926), about non-relativistic quantum mechanics (Pauli, 1927), and Dirac's (1928) theory regarding relativistic quantum mechanics (Bethe and Salpeter, 1957). The first confusion arises when we naively think that because of these origins, the spinor is necessarily and closely related to the spin of a particle, in the same way that the electron. Spinors had already appeared before in physics, although within the context of the classical mechanics, precisely, in the context of rigid-body dynamics. In a standard classical mechanics textbook like that by Goldstein, Poole, and Safko (2001), we can find a detailed discussion about the so-called Cayley–Klein parameters, which are widely used to describe spatial rotations. These parameters are in fact entries of a 2×2 unitary matrix $A \in \mathrm{SU}(2)$. The group $\mathrm{SU}(2)$ is well known to be the double covering of the special orthogonal group $\mathrm{SO}(3)$. The representation space of the group $\mathrm{SU}(2)$ is obviously $\mathbb{C}^2$. Such elements of $\mathbb{C}^2$ are called spinors – and in our classification they are called classical spinors. However, there is no relationship between this name and the emergence of spinors in classical mechanics. The adjective 'classical' suggests that it was because of the classical approach that these objects appeared initially in physics and mathematics as well. Spinors, seen as elements of $\mathbb{C}^2$, were largely used by Pauli to describe the behaviour of an electron according to quantum mechanics, taking into account the spin of this electron (which is not the case in the Schrödinger theory). In the literature of quantum mechanics, such objects are known as *Pauli spinors*.

The essential fact here is that the group SU(2) is isomorphic to the group Spin(3) = Spin(3, 0). From the isomorphism SU(2) $\simeq$ Spin(3), we conclude that $\mathbb{C}^2$ is the representation space of the group Spin(3). Hence, a Pauli spinor is an element of the representation space of the group Spin(3), which is the spin 1/2 representation of the group of three-dimensional spatial rotations (Goldstein, Poole, and Safko, 2001). Moreover, the group SO(3) is the representation of spin 1 regarding these rotations, with Spin(3) $\simeq$ SU(2), and Spin(3)/$\mathbb{Z}_2 \simeq$ SO(3).

Let us now analyse the relativistic case. Spacetime rotations are described by elements of the group $\mathrm{SO}_+(1,3)$, the so-called orthochronous proper Lorentz group – remembering that $\mathrm{SO}_+(1,3) \simeq \mathrm{SO}^{\uparrow}(1,3) \simeq \mathrm{SO}_+^{\uparrow}(1,3)$ – or, simply, the Lorentz group. A similar role is played here by the group SL(2, $\mathbb{C}$) of 2 × 2 complex matrices and determinant equal to 1. The group SL(2, $\mathbb{C}$) is the double covering of $\mathrm{SO}_+(1,3)$. We have here the isomorphism SL(2, $\mathbb{C}$) $\simeq \mathrm{Spin}_+(1,3)$. Obviously, the representation space of SL(2, $\mathbb{C}$) is $\mathbb{C}^2$. However, here the situation is quite different. Indeed, there are two representations of SL(2, $\mathbb{C}$) that are *not* equivalent: we can define $\rho(A)$ and $\bar{\rho}(A)$ for $A \in \mathrm{SL}(2,\mathbb{C})$ as $\rho(A)(z) = Az$, and $\bar{\rho}(A)(z) = \bar{A}z$, where $\bar{A}$ is the complex conjugate matrix associated with A, and $z \in \mathbb{C}^2$. These two representations should be equivalent if there exists an isomorphism $\phi : \mathbb{C}^2 \to \mathbb{C}^2$ such that $\bar{\rho}(A) = \phi \circ \rho(A) \circ \phi^{-1}$. In other words, those two representations should be equivalent if there exists an invertible 2 × 2 complex matrix T such that $\bar{\rho}(A)T = T\rho(A)$. By explicit computation, we can show that $\bar{\rho}(A)T = T\rho(A)$ does not have solution for $A \in \mathrm{SL}(2,\mathbb{C})$ – on the other hand, $\bar{\rho}(A)T = T\rho(A)$ does have a solution for $A \in \mathrm{SU}(2)$. Hence, there are two inequivalent representations of SL(2, $\mathbb{C}$), denoted by $D^{(1/2,0)}$ and $D^{(0,1/2)}$, respectively. The elements of a space that carries each one of such representations are called *Weyl spinors*. As in the previous case, such spinors are elements of the representation space of a Spin group, that is, $\mathrm{Spin}_+(1,3) \simeq \mathrm{SL}(2,\mathbb{C})$, and they fit into what we call the classical definition of a spinor. Moreover, the so-called *Dirac spinors*, according to the classical definition, are elements of $\mathbb{C}^4$ and carry a reducible representation of $\mathrm{Spin}_+(1,3) \simeq \mathrm{SL}(2,\mathbb{C})$ composed of the sum of two Weyl spinors, each one corresponding to one of the irreducible representations $D^{(1/2,0)}$ and $D^{(0,1/2)}$.

To summarise, from the classical point of view, spinors can be defined as objects which carry an irreducible representation of the Spin group, which is the double covering of the special orthogonal group and therefore the spin 1/2 representation of the group of rotations in a quadratic space.

On the other hand, we already discussed the Spin group in a Clifford algebra. Hence, another definition of spinor can be introduced: the algebraic one. When we discussed the representations of a Clifford algebra (section 4.3), the representation space associated with an irreducible regular representation was shown to be a minimal left ideal related to the Clifford algebra. An *algebraic spinor* is an element of a minimal left ideal in a Clifford algebra. The representation of the Clifford algebra obtained is called a spinor representation.

There is a little variation when the algebraic definition of spinors is regarded. We know that a Clifford algebra is isomorphic either to a simple algebra or to the direct sum which is a semisimple algebra – of two simple algebras. In the latter case, there is a minimal left ideal associated with each simple algebra. Let us denote such ideals by I_1 and I_2. In this case, we have two inequivalent irreducible representations: one associated with the representation space I_1, and the other associated with I_2. According to the definition, elements in these ideals I_1 and I_2 are spinors. Some authors call elements of I_1 and I_2 *semispinors*, reserving the term *spinors* for the elements of $I_1 \oplus I_2$. Hence, the two inequivalent irreducible representations are called semispinor representations. Increasing the amount of confusion, some authors use the term *pinors* for the elements of a minimal left ideal of a Clifford algebra and use the term *spinors* for the elements of a minimal left ideal of the *even subalgebra/* associated with this Clifford algebra.

A spinor representation of a Clifford algebra naturally induces a representation in any subset, by simply restricting to the elements of this subset the left multiplication on the ideal. An irreducible representation of the Clifford algebra induces therefore an irreducible representation of the Clifford–Lipschitz group – the irreducibility arises from the fact that non-isotropic vectors generate the Clifford algebra and the Clifford group. Moreover, it also induces a representation of the even subalgebra related to this Clifford algebra. The question is whether this representation is reducible or irreducible. In what follows, we will see that, if the even subalgebra representation is reducible, then it consists of a sum of two irreducible representations.

Similarly, an irreducible representation of the even subalgebra induces a representation in its subsets. In particular, an irreducible representation of the even subalgebra induces an irreducible representation of the Spin group. In this point, the contact between the classical and algebraic definitions of a spinor is accomplished. The representation space of the Spin group – whose elements are classical spinors – is a minimal left ideal of the even subalgebra. This minimal left ideal of the even subalgebra is not necessarily a minimal ideal of the whole Clifford algebra. If a primitive idempotent of the even subalgebra is also a primitive in the whole Clifford algebra, then the ideals are similar. However, it does not hold in general, and the algebraic and classical definitions do differ.

Another possible definition of spinor, which is denominated *operatorial*, can be introduced from another representation – distinct from the regular representation – of a Clifford algebra. This representation uses the representation space associated with the even subalgebra of a Clifford algebra. Although this definition seems to be distinct from the algebraic and classical definitions, it is equivalent to these in most cases, in particular in the cases of great interest for physical applications.

Because these definitions are so different, we must be careful to clearly state which one which we are using. The best way to do this is to attribute different names to each case: algebraic spinors, classical spinors, and operator spinors. These adjectives *are not standard*, but it seems extremely appropriate to use them in order not to be lost in this Babel of spinors.

6.2 Algebraic Spinors

Definition 6.1 ▶ An element of a minimal left ideal associated with a Clifford algebra $\mathcal{C}\ell(V,g)$ is said to be an *algebraic spinor* if $\mathcal{C}\ell(V,g)$ is a simple algebra, and an *algebraic semispinor* if $\mathcal{C}\ell(V,g)$ is semisimple.

Given this definition, the algebraic spinors can be now identified according to the classification of the Clifford algebras. For a simple Clifford algebra, we have the isomorphism $\mathcal{C}\ell_{p,q} \simeq \mathcal{M}(N,\mathbb{K})$, and a minimal left ideal of $\mathcal{C}\ell_{p,q}$ is isomorphic to $\mathbb{K}^N$, which is used for algebraic spinor classification. In the case of a semisimple Clifford algebra, we have the isomorphism $\mathcal{C}\ell_{p,q} \simeq \mathcal{M}(N,\mathbb{K})\oplus\mathcal{M}(N,\mathbb{K})$. In this case, a minimal left ideal of $\mathcal{C}\ell_{p,q}$ is isomorphic to $\mathbb{K}^N$. The algebraic semispinors classification follows from this isomorphism. The sum of algebraic semispinors is called an algebraic spinor – although, for this case, the ideal is not minimal. For a semisimple Clifford algebra, an algebraic spinor can be classified as $\mathbb{K}^N \oplus \mathbb{K}^N$.

Table 6.1 shows the algebraic spinor classification.

$p-q=0,2 \mod 8$ In this case, $\mathcal{C}\ell_{p,q} \simeq \mathcal{M}(2^{[n/2]},\mathbb{R})$. An algebraic spinor is thus an element of a minimal left ideal isomorphic to $\mathbb{R}^{2^{[n/2]}}$.

$p-q=4,6 \mod 8$ Here, the isomorphism $\mathcal{C}\ell_{p,q} \simeq \mathcal{M}(2^{[n/2]-1},\mathbb{H})$ holds. The space of algebraic spinors is isomorphic to $\mathbb{H}^{2^{[n/2]-1}}$.

$p-q=3,7 \mod 8$ In this case, $\mathcal{C}\ell_{p,q} \simeq \mathcal{M}(2^{[n/2]},\mathbb{C})$. Hence, the space of algebraic spinors is isomorphic to the representation space $\mathbb{C}^{2^{[n/2]}}$. The possibility $p-q = 3,7 \mod 8$ holds only if the dimension $n=p+q$ is odd. In this case, the pseudoscalar (n-vector) η commutes with all elements of $\mathcal{C}\ell_{p,q}$ and satisfies $\eta^2 = -1$, defining a complex structure.

$p-q=5 \mod 8$ For this metric signature, the Clifford algebra $\mathcal{C}\ell_{p,q}$ is semisimple, and $\mathcal{C}\ell_{p,q} \simeq \mathcal{M}(2^{[n/2]-1},\mathbb{H}) \oplus \mathcal{M}(2^{[n/2]-1},\mathbb{H})$. This is a situation where algebraic semispinors exist. The space of algebraic semispinors is isomorphic to $\mathbb{H}^{2^{[n/2]-1}}$, whereas the space of algebraic spinors is isomorphic to $\mathbb{H}^{2^{[n/2]-1}} \oplus \mathbb{H}^{2^{[n/2]-1}}$. For $p-q = 5 \mod 8$, a n-vector η commutes with all elements of $\mathcal{C}\ell_{p,q}$ and satisfies $\eta^2 = 1$. We can thus write (see chapter 3, exercise 4) $\mathcal{C}\ell_{p,q} = {}_+\mathcal{C}\ell_{p,q} \oplus {}_-\mathcal{C}\ell_{p,q}$, where ${}_\pm\mathcal{C}\ell_{p,q} \simeq \mathcal{M}(2^{[n/2]-1},\mathbb{H})$.

$p-q=1 \mod 8$ In this case, the Clifford algebra $\mathcal{C}\ell_{p,q}$ is semisimple, and $\mathcal{C}\ell_{p,q} \simeq \mathcal{M}(2^{[n/2]},\mathbb{R}) \oplus \mathcal{M}(2^{[n/2]},\mathbb{R})$. The space of the algebraic semispinors is isomorphic to $\mathbb{R}^{2^{[n/2]}}$, and the space of the algebraic spinors is isomorphic to $\mathbb{R}^{2^{[n/2]}} \oplus \mathbb{R}^{2^{[n/2]}}$. For $p-q=1 \mod 8$, an n-vector η commutes with all the elements of $\mathcal{C}\ell_{p,q}$ and is such that $\eta^2 = 1$. In this case we can write (see again chapter 3, Exercise 4) $\mathcal{C}\ell_{p,q} = {}_+\mathcal{C}\ell_{p,q} \oplus {}_-\mathcal{C}\ell_{p,q}$, where ${}_\pm\mathcal{C}\ell_{p,q} \simeq \mathcal{M}(2^{[n/2]},\mathbb{R})$.

This analysis is summarised in table 6.1.

For the case involving complex Clifford algebras, the situation is much simpler than that for reals. When $\dim V = n$ is an even number, we have the complex Clifford algebra isomorphism $\mathcal{C}\ell_{\mathbb{C}}(2k) \simeq \mathcal{M}(2^k, \mathbb{C})$. Hence, the space of algebraic spinors is isomorphic to $\mathbb{C}^{2^k}$. For $\dim V = n$ odd, the isomorphism $\mathcal{C}\ell_{\mathbb{C}}(2k+1) \simeq \mathcal{M}(2^k, \mathbb{C}) \oplus \mathcal{M}(2^k, \mathbb{C})$ holds for the complex Clifford algebra. The space of the algebraic semispinors is then isomorphic to $\mathbb{C}^{2^k}$, and the space of algebraic spinors in this case is isomorphic to $\mathbb{C}^{2^k} \oplus \mathbb{C}^{2^k}$ as summarised in table 6.2.

Example 6.1 Some of the most outstanding and important examples in physics regard dimensions 3 and 4. For the quadratic space $\mathbb{R}^{3,0}$, we have the Clifford algebra $\mathcal{C}\ell_{3,0} \simeq \mathcal{M}(2, \mathbb{C})$. The space of algebraic spinors of $\mathcal{C}\ell_{3,0}$ is isomorphic to $\mathbb{C}^2$. For the quadratic space $\mathbb{R}^{0,3}$, we have $\mathcal{C}\ell_{0,3} \simeq \mathbb{H} \oplus \mathbb{H}$. Algebraic semispinors are elements of a space isomorphic to $\mathbb{H}$. For the quadratic space $\mathbb{R}^{1,3}$, we have $\mathcal{C}\ell_{1,3} \simeq \mathcal{M}(2, \mathbb{H})$ and therefore the space of the algebraic spinors is $\mathbb{H}^2$. In the case of $\mathbb{R}^{3,1}$, we have $\mathcal{C}\ell_{3,1} \simeq \mathcal{M}(4, \mathbb{R})$, and the space of the algebraic spinors is provided by $\mathbb{R}^4$. For complex Clifford algebras, we have $\mathcal{C}\ell_{\mathbb{C}}(3) \simeq \mathcal{M}(2, \mathbb{C}) \oplus \mathcal{M}(2, \mathbb{C})$, and the space of algebraic semispinors is $\mathbb{C}^2$. Regarding $\mathcal{C}\ell_{\mathbb{C}}(4) \simeq \mathcal{M}(4, \mathbb{C})$, the algebraic spinors space is given by $\mathbb{C}^4$.

6.3 Classical Spinors

Definition 6.2 ▶ *Consider $\mathbb{R}^{p,q}$ a quadratic space, the Clifford algebra $\mathcal{C}\ell_{p,q}$ associated with this space, and the reduced Spin group $\mathrm{Spin}_+(p,q)$, associated with $\mathcal{C}\ell_{p,q}$. An element of the irreducible representation space of $\mathrm{Spin}_+(p,q)$ is said to be a classical spinor.*

The group $\mathrm{Spin}_+(p,q) = \{a \in \Gamma^+_{p,q} \mid N(a) = 1\}$ is the set of even elements of the Clifford–Lipschitz group. Then, an irreducible representation of $\mathrm{Spin}_+(p,q)$ descends from an irreducible representation of the even subalgebra $\mathcal{C}\ell^+_{p,q}$. On the other hand,

Table 6.1 Algebraic Spinors Classification: The Real Case, Where $p+q=n$, and $[n/2]$ Denotes the Integer Part of $n/2$

$p-q$ mod 8	0	1	2	3
$S^A_{p,q}$	$\mathbb{R}^{2^{[n/2]}}$	$\mathbb{R}^{2^{[n/2]}} \oplus \mathbb{R}^{2^{[n/2]}}$	$\mathbb{R}^{2^{[n/2]}}$	$\mathbb{C}^{2^{[n/2]}}$
$p-q$ mod 8	4	5	6	7
$S^A_{p,q}$	$\mathbb{H}^{2^{[n/2]-1}}$	$\mathbb{H}^{2^{[n/2]-1}} \oplus \mathbb{H}^{2^{[n/2]-1}}$	$\mathbb{H}^{2^{[n/2]-1}}$	$\mathbb{C}^{2^{[n/2]}}$

Table 6.2 Algebraic Spinors: The Complex Case

$n = 2k$	$\mathbb{C}^{2^k}$
$n = 2k+1$	$\mathbb{C}^{2^k} \oplus \mathbb{C}^{2^k}$

when we discussed the classification of the Clifford algebras, an important result was established: $\mathcal{C}\ell^+_{p,q} \simeq \mathcal{C}\ell_{q,p-1} \simeq \mathcal{C}\ell_{p,q-1} \simeq \mathcal{C}\ell^+_{q,p}$. An irreducible representation of $\mathcal{C}\ell^+_{p,q} \simeq \mathcal{C}\ell^+_{q,p}$ is thus obtained from an irreducible representation of $\mathcal{C}\ell_{q,p-1} \simeq \mathcal{C}\ell_{p,q-1}$, as was previously established. In other words, a classical spinor in a quadratic space $\mathbb{R}^{p,q}$ or $\mathbb{R}^{q,p}$ is an algebraic spinor (or an algebraic semispinor) in a quadratic space that is either $\mathbb{R}^{q,p-1}$ or $\mathbb{R}^{p,q-1}$. In order to make our analysis easier, let us use such isomorphisms to construct a classification of the even Clifford algebras, as shown in table 6.3.

We can now describe the classification of classical spinors in what follows.

$\boxed{\mathbf{p-q=1,7 \mod 8}}$ For $p-q=1,7 \mod 8$, it follows that $\mathcal{C}\ell^+_{p,q} \simeq \mathcal{C}\ell_{p,q-1} = \mathcal{C}\ell_{p',q'}$, where $p'-q'=p-q+1=0,2 \mod 8$, namely, $\mathcal{C}\ell^+_{p,q} \simeq \mathcal{M}(2^{[(n-1)/2]},\mathbb{R})$, where $n=p+q$. Hence, a classical spinor is an element of the representation space $\mathbb{R}^{2^{[(n-1)/2]}}$.

$\boxed{\mathbf{p-q=2,6 \mod 8}}$ In this case, $p'-q'=p-q+1=3,7 \mod 8$, yielding $\mathcal{C}\ell^+_{p,q} \simeq \mathcal{M}(2^{[(n-1)/2]},\mathbb{C})$. A classical spinor is therefore an element of $\mathbb{C}^{2^{[(n-1)/2]}}$. In this case, the n-vector η defines a complex structure in the classical spinors space. Hence, there are two inequivalent irreducible representations: one of them with η equals the complex structure induced by i, whereas the other is achieved when η is equal to the complex structure induced by $-i$. The two corresponding classical spinors are conjugate.

$\boxed{\mathbf{p-q=3,5 \mod 8}}$ Here, $p'-q'=4,6 \mod 8$, and then $\mathcal{C}\ell^+_{p,q} \simeq \mathcal{M}(2^{[(n-1)/2]-1},\mathbb{H})$. A classical spinor is an element of $\mathbb{H}^{2^{[(n-1)/2]-1}}$.

$\boxed{\mathbf{p-q=4 \mod 8}}$ We have $p'-q'=5 \mod 8$. The even subalgebra is semisimple, and $\mathcal{C}\ell_{p,q} \simeq \mathcal{M}(2^{[(n-1)/2]-1},\mathbb{H}) \oplus \mathcal{M}(2^{[(n-1)/2]-1},\mathbb{H})$. There are two inequivalent representations of $\mathrm{Spin}_+(p,q)$. In this case, we can write $\mathcal{C}\ell^+_{p,q} = {}_+\mathcal{C}\ell^+_{p,q} \oplus {}_-\mathcal{C}\ell^+_{p,q}$ and denominate as positive classical spinors the elements of the representation space

Table 6.3 Real Even Subalgebra Classification Table, Where $p+q=n$, and $[\kappa]$ Denotes the Integer Part of $\kappa=(n-1)/2$

$p-q$ mod 8	0	1	2	3
$\mathcal{C}\ell^+_{p,q}$	$\mathcal{M}(2^{[\kappa]},\mathbb{R}) \oplus \mathcal{M}(2^{[\kappa]},\mathbb{R})$	$\mathcal{M}(2^{[\kappa]},\mathbb{R})$	$\mathcal{M}(2^{[\kappa]},\mathbb{C})$	$\mathcal{M}(2^{[\kappa]-1},\mathbb{H})$
$p-q$ mod 8	4	5	6	7
$\mathcal{C}\ell^+_{p,q}$	$\mathcal{M}(2^{[\kappa]-1},\mathbb{H}) \oplus \mathcal{M}(2^{[\kappa]-1},\mathbb{H})$	$\mathcal{M}(2^{[\kappa]-1},\mathbb{H})$	$\mathcal{M}(2^{[\kappa]},\mathbb{C})$	$\mathcal{M}(2^{[\kappa]},\mathbb{R})$

of ${}_+\mathcal{C}\ell^+_{p,q}$, and the elements of the representation space of ${}_-\mathcal{C}\ell^+_{p,q}$ as negative classical spinors. A classical spinor, positive or negative, in this case is an element of $\mathbb{H}^{2^{[(n-1)/2]-1}}$.

$p - q = 0 \mod 8$ In this case, $p' - q' = 1 \mod 8$. The even subalgebra $\mathcal{C}\ell^+_{p,q}$ is semisimple, and $\mathcal{C}\ell^+_{p,q} \simeq \mathcal{M}(2^{[(n-1)/2]}, \mathbb{R}) \oplus \mathcal{M}(2^{[(n-1)/2]}, \mathbb{R})$. There are two inequivalent representations of $\mathrm{Spin}_+(p,q)$. As in the previous case, we can write $\mathcal{C}\ell^+_{p,q} = {}_+\mathcal{C}\ell^+_{p,q} \oplus {}_-\mathcal{C}\ell^+_{p,q}$ and denominate as positive classical spinors the elements of the representation space of ${}_+\mathcal{C}\ell^+_{p,q}$, and as negative classical spinors the elements of the representation space ${}_-\mathcal{C}\ell^+_{p,q}$. A positive or negative classical spinor in this case is an element of $\mathbb{R}^{2^{[(n-1)/2]}}$.

This analysis is summarised in table 6.4.

For the case involving complex Clifford algebras, the situation is again straightforward (see Table 6.5). Clearly, $\mathcal{C}\ell^+_{\mathbb{C}}(n) \simeq \mathcal{C}\ell_{\mathbb{C}}(n-1)$. Hence, if $\dim V = n$ is even, it follows that $\mathcal{C}\ell^+_{\mathbb{C}}(2k) = \mathcal{C}\ell_{\mathbb{C}}(2k-1) \simeq \mathcal{M}(2^{k-1}, \mathbb{C}) \oplus \mathcal{M}(2^{k-1}, \mathbb{C})$. There are two inequivalent irreducible representations, and the classical positive or negative spinors are elements of $\mathbb{C}^{2^{k-1}}$. If $\dim V = n$ is odd, we have $\mathcal{C}\ell^+_{\mathbb{C}}(2k+1) \simeq \mathcal{C}\ell_{\mathbb{C}}(2k) \simeq \mathcal{M}(2^k, \mathbb{C})$. The classical spinors are therefore elements of $\mathbb{C}^{2^k}$.

Regarding the space of classical spinors $S_{p,q}$, an idempotent endomorphism R $\in \mathrm{End}(S_{p,q})$ can be defined, which for some dimensions and signatures is usually identified with the volume element η (Lazaroiu, Babalic, and Coman, 2013; Bonora, de Brito,

Table 6.4 Classical Spinors Classification: The Real Case, Where $p + q = n$, and $[n/2]$ Denotes the Integer Part of $n/2$

$p-q$ mod 8	0	1	2	3
$S^C_{p,q}$	$\mathbb{R}^{2^{[(n-1)/2]}} \oplus \mathbb{R}^{2^{[(n-1)/2]}}$	$\mathbb{R}^{2^{[(n-1)/2]}}$	$\mathbb{C}^{2^{[(n-1)/2]}}$	$\mathbb{H}^{2^{[(n-1)/2]-1}}$
$p-q$ mod 8	4	5	6	7
$S^C_{p,q}$	$\mathbb{H}^{2^{[(n-1)/2]-1}} \oplus \mathbb{H}^{2^{[(n-1)/2]-1}}$	$\mathbb{H}^{2^{[(n-1)/2]-1}}$	$\mathbb{C}^{2^{[(n-1)/2]}}$	$\mathbb{R}^{2^{[(n-1)/2]}}$

Table 6.5 Classical Spinors: The Complex Case

$n = 2k$	$\mathbb{C}^{2^{k-1}} \oplus \mathbb{C}^{2^{k-1}}$
$n = 2k+1$	$\mathbb{C}^{2^k}$

and da Rocha, 2015). Spin projectors are then defined by $\Pi_\pm = \frac{1}{2}(I \pm \mathrm{R})$, where I denotes the identity operator on $S_{p,q}$, providing the direct sum $S_{p,q} = S^+_{p,q} \oplus S^-_{p,q}$, where $S^\pm_{p,q} = \Pi_\pm(S_{p,q})$. Elements of $S^\pm_{p,q}$ are called (symplectic) Majorana–Weyl spinors when $p - q = 0 \mod 8$ ($p - q = 4 \mod 8$), whereas elements of $S^+_{p,q}$ are known as (symplectic) Majorana spinors when $p - q = 7 \mod 8$ ($p - q = 6 \mod 8$). More details shall be discussed in section 6.6 and can be verified in the article by de Andrade, Rojas, and Toppan (2001).

6.4 Spinor Operators

Given a Clifford algebra $\mathcal{C}\ell_{p,q}$, the $\mathbb{Z}_2$-grading can be taken into account to use the even subalgebra $\mathcal{C}\ell^+_{p,q}$ as a representation space for the algebra $\mathcal{C}\ell_{p,q}$. In other words, we can define a representation $\rho : \mathcal{C}\ell_{p,q} \to \mathrm{End}(\mathcal{C}\ell^+_{p,q})$, which is called a *graded regular representation*. An arbitrary element $a \in \mathcal{C}\ell_{p,q}$ can be split as $a = a_+ + a_-$, where

$$a_\pm = \frac{1}{2}(a \pm \hat{a}). \tag{6.1}$$

Let us further split the representation ρ as $\rho = \rho_+ + \rho_-$, such that

$$\rho(a) = \rho_+(a_+) + \rho_-(a_-). \tag{6.2}$$

The part ρ_+ is a regular representation of a_+ in $\mathcal{C}\ell^+_{p,q}$, namely,

$$\rho_+(a_+)(\phi) = a_+\phi, \qquad \forall \phi \in \mathcal{C}\ell^+_{p,q}. \tag{6.3}$$

On the other hand, for $a_- \in \mathcal{C}\ell^-_{p,q}$, we have $a_-\phi \in \mathcal{C}\ell^-_{p,q}$, for any $\phi \in \mathcal{C}\ell^+_{p,q}$. In order to define $\rho_-(a_-)$ we must have $\rho_-(a_-)(\phi) \in \mathcal{C}\ell^+_{p,q}$. It is possible if an odd element κ is taken into the following definition

$$\rho_-(a_-)(\phi) = a_-\phi\kappa, \qquad \forall \phi \in \mathcal{C}\ell^+_{p,q}. \tag{6.4}$$

If we choose κ such that

$$\kappa^2 = 1, \qquad \kappa \in \mathcal{C}\ell^-_{p,q}, \tag{6.5}$$

then $\rho = \rho_+ + \rho_-$ is a representation of $\mathcal{C}\ell_{p,q}$.

In order to see that ρ is a representation of $\mathcal{C}\ell_{p,q}$, let us calculate $\rho(ab)$ for $a, b \in \mathcal{C}\ell_{p,q}$:

$$\begin{aligned} \rho(ab) &= \rho(a_+b_+ + a_+b_- + a_-b_+ + a_-b_-) \\ &= \rho_+(a_+b_+) + \rho_-(a_+b_-) + \rho_-(a_-b_+) + \rho_+(a_-b_-). \end{aligned} \tag{6.6}$$

Moreover, each part in the above sum can be further analysed:

$$\begin{aligned} \rho_+(a_+b_+)(\phi) &= a_+b_+\phi = \rho_+(a_+)\rho_+(b_+)(\phi), \\ \rho_-(a_+b_-)(\phi) &= a_+b_-\phi\kappa = \rho_+(a_+)\rho_-(b_-)(\phi), \\ \rho_-(a_-b_+)(\phi) &= a_-b_+\phi\kappa = \rho_-(a_-)\rho_+(b_+)(\phi), \\ \rho_+(a_-b_-)(\phi) &= a_-b_-\phi = a_-b_-\phi\kappa^2 = \rho_-(a_-)\rho_-(b_-)(\phi), \end{aligned} \tag{6.7}$$

which therefore proves that ρ is indeed a representation:

$$\rho(ab) = \rho(a)\rho(b). \tag{6.8}$$

The definition of the irreducible regular representation depends upon the existence of an odd element κ such that $\kappa^2 = 1$. In *two* cases, this element does not exist: when $\mathcal{C}\ell_{0,1}$ and $\mathcal{C}\ell_{0,2}$, which are exactly the cases $\mathcal{C}\ell_{0,1} \simeq \mathbb{C}$ and $\mathcal{C}\ell_{0,2} \simeq \mathbb{H}$. For the other cases, we can find an element κ.

Now we want to know whether the representation ρ is reducible or not. Suppose that there exists elements $\eta, \epsilon \in \mathcal{C}\ell^+_{p,q}$ such that

$$\eta^2 = \pm 1, \qquad \eta\kappa = \kappa\eta, \tag{6.9}$$

$$\epsilon^2 = 1, \qquad \epsilon\kappa = \kappa\epsilon, \qquad \epsilon\eta = -\eta\epsilon. \tag{6.10}$$

In this case, we can write

$$\mathcal{C}\ell^+_{p,q} = {}^+\mathcal{C}\ell^+_{p,q} \oplus {}^-\mathcal{C}\ell^+_{p,q}, \tag{6.11}$$

where

$${}^\pm\mathcal{C}\ell^+_{p,q} = \frac{1}{2}[\mathcal{C}\ell^+_{p,q} \pm \eta\, \mathcal{C}\ell^+_{p,q}\, \eta^{-1}]. \tag{6.12}$$

Hence, for arbitrary elements $\phi_\pm \in {}^\pm\mathcal{C}\ell^+_{p,q}$, we have

$$\eta\phi_\pm = \pm\phi_\pm\eta. \tag{6.13}$$

These subspaces consist of $\mathcal{C}\ell^+_{p,q}$ elements which either *commute* or *anti-commute* with η. Since

$${}^\pm\mathcal{C}\ell^+_{p,q}\; {}^\pm\mathcal{C}\ell^+_{p,q} \subset {}^+\mathcal{C}\ell^+_{p,q}, \qquad {}^\pm\mathcal{C}\ell^+_{p,q}\; {}^\mp\mathcal{C}\ell^+_{p,q} \subset {}^\mp\mathcal{C}\ell^+_{p,q},$$

we can see that only ${}^+\mathcal{C}\ell^+_{p,q}$ is a subalgebra of $\mathcal{C}\ell^+_{p,q}$. In other words, an involution

$$a^\eta = \eta a \eta^{-1} \tag{6.14}$$

can be defined (note that $(a^\eta)^\eta = a$) where an element a such that $a^\eta = a$ is said to be η-even, and $a^\eta = -a$ is said to be η-odd, where ${}^+\mathcal{C}\ell^+_{p,q}$ is an η-even subalgebra of $\mathcal{C}\ell^+_{p,q}$. Note that ϵ is η-odd.

If there exists $\eta, \epsilon \in \mathcal{C}\ell^+_{p,q}$ satisfying these conditions, we can define a representation $\rho : \mathcal{C}\ell_{p,q} \to \mathrm{End}({}^+\mathcal{C}\ell^+_{p,q})$. Given $a \in \mathcal{C}\ell_{p,q}$, we can express $a = {}^+a_+ + {}^-a_+ + {}^+a_- + {}^-a_-$, where

$${}^\pm a = \frac{1}{2}(a \pm \eta a \eta^{-1}), \tag{6.15}$$

and $a_\pm$ are given by eqn (6.1). Now let us write ρ as

$$\rho = {}^+\rho_+ + {}^-\rho_+ + {}^+\rho_- + {}^-\rho_-,$$

in such a way that

$$\rho(a) = {}^+\rho_+({}^+a_+) + {}^-\rho_+({}^-a_+) + {}^+\rho_-({}^+a_-) + {}^-\rho_-({}^-a_-), \tag{6.16}$$

where, for $\phi_+ \in {}^+\mathcal{C}\ell^+_{p,q}$, it follows that

$$
\begin{aligned}
{}^+\rho_+({}^+a_+)(\phi_+) &= {}^+a_+\,\phi_+,\\
{}^-\rho_+({}^-a_+)(\phi_+) &= {}^-a_+\,\phi_+\epsilon,\\
{}^+\rho_-({}^+a_-)(\phi_+) &= {}^+a_-\,\phi_+\kappa,\\
{}^-\rho_-({}^-a_-)(\phi_+) &= {}^-a_-\,\phi_+\kappa\epsilon.
\end{aligned}
\tag{6.17}
$$

We can see that ${}^\pm\rho_\pm({}^\pm a_\pm)(\phi_+) \in {}^+\mathcal{C}\ell^+_{p,q}$ and is in fact a representation, namely, $\rho(ab) = \rho(a)\rho(b)$.

If there still exist other odd elements and η-even elements η' and ϵ' such that $(\eta')^2 = \pm 1$, and $(\epsilon')^2 = 1$, and $\eta'\epsilon' = -\epsilon'\eta'$, and which furthermore commute with κ, η, and ϵ, we can construct a new subalgebra ${}^{++}\mathcal{C}\ell^+_{p,q}$ which is invariant under the action of ρ in a completely similar way.

When even elements satisfying such conditions do not exist, we then arrive at an *irreducible representation*. The space which carries this graded irreducible representation is a subalgebra of $\mathcal{C}\ell_{p,q}$ and is called *spinor algebra*. The spinor algebra is a subalgebra of the even subalgebra, and, in some cases, it can be the even subalgebra itself. Meanwhile, it is worth emphasising that the subalgebras ${}^+\mathcal{C}\ell^+_{p,q}$ or ${}^{++}\mathcal{C}\ell^+_{p,q}$ are not in general Clifford algebras (as we will see with an example in what follows).

Definition 6.3 ▶ *An element of the graded irreducible representation space of $\mathcal{C}\ell_{p,q}$ is said to be a spinor operator.*

Example 6.2 Let us consider the case $\mathcal{C}\ell_{3,0}$. Let $\{\mathbf{e}_1, \mathbf{e}_2, \mathbf{e}_3\}$ be an orthonormal basis in such a way that $(\mathbf{e}_i)^2 = 1$, and $\mathbf{e}_i\mathbf{e}_j = -\mathbf{e}_j\mathbf{e}_i$ $(i \neq j)$. An odd element such that $\kappa^2 = 1$ is, for instance, $\kappa = \mathbf{e}_3$. In this case, there are no even elements (besides the scalar 1) such that $\eta^2 = 1$. The even subalgebra $\mathcal{C}\ell^+_{3,0} \simeq \mathcal{C}\ell_{0,2} \simeq \mathbb{H}$ is therefore an irreducible representation space of $\mathcal{C}\ell_{3,0}$. Spinor operators are elements of $\mathcal{C}\ell^+_{3,0}$, reading $a + a_{12}\mathbf{e}_1\mathbf{e}_2 + a_{13}\mathbf{e}_1\mathbf{e}_3 + a_{23}\mathbf{e}_2\mathbf{e}_3$. These are the so-called Pauli spinor operators.

Example 6.3 Another interesting example, particularly for physical applications, is the algebra $\mathcal{C}\ell_{1,3}$. Let $\{\mathbf{e}_0, \mathbf{e}_1, \mathbf{e}_2, \mathbf{e}_3\}$ be an orthonormal basis which satisfies $(\mathbf{e}_0)^2 = 1$; $(\mathbf{e}_i)^2 = -1$ $(i = 1, 2, 3)$; and $\mathbf{e}_\mu\mathbf{e}_\nu = -\mathbf{e}_\nu\mathbf{e}_\mu$ $(\mu \neq \nu)$. An odd element satisfying $\kappa^2 = 1$ is given by $\kappa = \mathbf{e}_0$. The unique even elements that squared equals 1 are $\mathbf{e}_0\mathbf{e}_i$ $(i = 1, 2, 3)$. However, these elements do not commute with κ. The even subalgebra $\mathcal{C}\ell^+_{1,3} \simeq \mathcal{C}\ell_{3,0}$ is then an irreducible representation space of $\mathcal{C}\ell_{1,3}$, and its elements are called Dirac spinor operators.

Example 6.4 Let us now consider $\mathcal{C}\ell_{4,1}$ with $(\mathbf{e}_0)^2 = -1$; $(\mathbf{e}_a)^2 = 1$ $(a = 1, 2, 3, 4)$; and $\mathbf{e}_\mu\mathbf{e}_\nu = -\mathbf{e}_\nu\mathbf{e}_\mu$ $(\mu \neq \nu)$. We can choose κ as, for instance, $\kappa = \mathbf{e}_4$. Here, there already exist even elements of type η and ϵ commuting with κ; for example $\eta = \mathbf{e}_1\mathbf{e}_2$, $\epsilon = \mathbf{e}_0\mathbf{e}_1$, in such a way that $\eta^2 = -1$; $\epsilon^2 = 1$; $\eta\epsilon = -\epsilon\eta$. We can further verify that there is no other element of this same type. Hence, ${}^+\mathcal{C}\ell^+_{4,1} \simeq \mathcal{C}\ell_{3,0}$ is an irreducible representation space of $\mathcal{C}\ell_{4,1}$.

Example 6.5 Consider the Clifford algebra $\mathcal{C}\ell_{2,1}$. The element κ can be taken as $\kappa = \mathbf{e}_1\mathbf{e}_2\mathbf{e}_3$, since κ is odd, and $\kappa^2 = 1$. In this way, we can take $\mathcal{C}\ell^+_{2,1}$ as being the representation space. This representation is not irreducible. Indeed, there are elements $\epsilon = \mathbf{e}_1\mathbf{e}_3$, and $\eta = \mathbf{e}_2\mathbf{e}_3$, which commute with κ and such that $\eta\epsilon = -\epsilon\eta$, and $\epsilon^2 = 1$, besides the relation $\eta^2 = 1$. Thus, the subalgebra ${}^+\mathcal{C}\ell^+_{2,1}$ carries a graded irreducible representation of $\mathcal{C}\ell_{2,1}$. In addition, we have ${}^+\mathcal{C}\ell^+_{2,1} \simeq \mathbb{R} \oplus \mathbb{R}$. On the other hand, there exists another possible choice for η that also defines an irreducible representation. In fact, let $\acute{\eta} = \mathbf{e}_1\mathbf{e}_2$ be the element that commutes with κ and anti-commutes with ϵ, and $\acute{\eta}^2 = -1$. We have another irreducible representation carried by the subalgebra ${}^0\mathcal{C}\ell^+_{2,1}$, where we used another index to improve the notation. The interesting fact to be mentioned here is that such representations are not equivalent. Indeed, we can see that ${}^0\mathcal{C}\ell^+_{2,1} \simeq \mathbb{C}$.

Example 6.6 Consider now the algebra $\mathcal{C}\ell_{2,2}$. In this case, we can choose $\kappa = \mathbf{e}_1$; $\epsilon = \mathbf{e}_2\mathbf{e}_3$; and $\eta = \mathbf{e}_3\mathbf{e}_4$. The subalgebra ${}^+\mathcal{C}\ell^+_{2,2}$ hence carries an irreducible graded representation of $\mathcal{C}\ell_{2,2}$. The interesting fact here is that ${}^+\mathcal{C}\ell^+_{2,2} \simeq \mathbb{R}\oplus\mathbb{R}\oplus\mathbb{R}\oplus\mathbb{R}$, namely, ${}^+\mathcal{C}\ell^+_{2,2}$ is not a Clifford algebra, an observation which can be straightforwardly verified from the classification table. There is still another possible choice given by κ and ϵ, and where $\acute{\eta} = \mathbf{e}_3\mathbf{e}_4$; this choice leads us to a inequivalent representation in terms of the subalgebra ${}^0\mathcal{C}\ell^+_{2,2} \simeq \mathbb{C}\otimes\mathbb{C} \simeq \mathbb{C}\oplus\mathbb{C}$. This algebra is not a real Clifford algebra, although it is the complex Clifford algebra $\mathcal{C}\ell_{\mathbb{C}}(1) \simeq \mathbb{C}\oplus\mathbb{C}$.

Example 6.7 The graded irreducible representation of a complex Clifford algebra is relatively easy to study. Let us consider, for example, $\mathcal{C}\ell_{2k}(\mathbb{C})$ and let us choose $\kappa = \mathbf{e}_1$; $\epsilon_j = i\mathbf{e}_{2j}\mathbf{e}_{2j+1}$; and $\eta_j = \mathbf{e}_{2j-1}\mathbf{e}_{2j}$ for $j = 1, \dots, k-1$. We then obtain ${}^{+\cdots+}\mathcal{C}\ell^+_{2k}(\mathbb{C}) = \underbrace{\mathbb{C}\oplus\mathbb{C}\oplus\cdots\oplus\mathbb{C}}_{2^k \text{ times}} = \mathbb{C}^{\oplus 2^k}$.

A Generalised Spinor Algebra

A detailed study concerning $\mathbb{Z}_2$-gradings in Clifford algebras can be seen in the article by Mosna, Miralles, and Vaz Jr (2003). Let us denote an arbitrary $\mathbb{Z}_2$-grading of $\mathcal{C}\ell(V,g)$ by $\mathcal{C}\ell^0 \oplus \mathcal{C}\ell^1$ and let α be a vector space isomorphism defined by $\alpha|_{\mathcal{C}\ell^i} = (-1)^i \operatorname{id}_{\mathcal{C}\ell^i}$, where $\operatorname{id}_{\mathcal{C}\ell^i}$ is the identity map on $\mathcal{C}\ell^i$ $(i = 0, 1)$. We say that $\mathcal{C}\ell^0$ and $\mathcal{C}\ell^1$ are the α-even and α-odd parts, respectively, of $\mathcal{C}\ell(V,g)$. For the usual $\mathbb{Z}_2$-grading given by $\mathcal{C}\ell(V,g) = \mathcal{C}\ell^+ \oplus \mathcal{C}\ell^-$, we have $\alpha = \#$.

A grading automorphism α is said to preserve the multivector structure of $\mathcal{C}\ell(V,g)$ if $\alpha(\bigwedge^k(V)) \subset \bigwedge^k(V)$. In the article by Mosna, Miralles, and Vaz Jr (2003), it is proven that, if $\mathcal{C}\ell_{p,q}(\mathbb{R}) = \mathcal{C}\ell^0_{p,q} \oplus \mathcal{C}\ell^1_{p,q}$ is a $\mathbb{Z}_2$-grading preserving the multivector structure of $\bigwedge(\mathbb{R}^{p,q})$, then

$$\mathcal{C}\ell^0_{p,q} \simeq \mathcal{C}\ell_{p_0,q_0} \otimes \mathcal{C}\ell^+_{p-p_0,q-q_0}, \tag{6.18}$$

where $p_0(q_0)$ is the number of α-even elements of an orthonormal basis of $\mathbb{R}^{p,q}$ squaring to $+1(-1)$.

Given $\mathcal{C}\ell^0_{p,q}$, we can use it to define a graded regular representation, as we did with $\mathcal{C}\ell^+_{p,q}$, that is, we write $a = a_0 + a_1$, with $a_i \in \mathcal{C}\ell^i_{p,q}$ $(i = 0, 1)$ and define a representation $\rho = \rho_0 + \rho_1$ such that

$$\rho_0(a_0)(\phi) = a_0\phi, \qquad \rho_1(a_1)(\phi) = a_1\phi\kappa, \tag{6.19}$$

where $\phi \in \mathcal{C}\ell^0_{p,q}$ and $\kappa \in \mathcal{C}\ell^1_{p,q}$ such that $\kappa^2 = 1$. Moreover, if there exist elements $\eta, \epsilon \in \mathcal{C}\ell^0_{p,q}$ satisfying relations as in eqns (6.9) and (6.10), the representation space can be reduced to ${}^0\mathcal{C}\ell^0_{p,q}$, and so on.

The question now is how to actually find the spinor algebra using $\mathcal{C}\ell^0_{p,q}$ as in eqn (6.18); in order to do so, we need to find the α grading automorphism. First, let us denote by $\{\mathbf{e}_i, \mathbf{f}_k\}$ $(i = 1, \dots, p,\ k = 1, \dots, q)$ an orthonormal basis of V in such a way that $\mathcal{C}\ell_{p,q}$ is generated by $\{\mathbf{e}_i, \mathbf{f}_k\}$ (and 1) with $(\mathbf{e}_i)^2 = 1$; $(\mathbf{f}_k)^2 = -1$; and

$\mathbf{e}_i\mathbf{f}_k+\mathbf{f}_k\mathbf{e}_i = 0$. Let us suppose that $\mathcal{C}\ell_{p_0,q_0}$ is generated by $\{\mathbf{e}_1,\ldots,\mathbf{e}_{p_0},\mathbf{f}_1,\ldots,\mathbf{f}_{q_0}\}$ and that $\mathcal{C}\ell_{p-p_0,q-q_0}$ is generated by $\{\mathbf{e}_{p_0+1},\ldots,\mathbf{e}_p,\mathbf{f}_{q_0+1},\ldots,\mathbf{f}_q\}$. Now we remember that, if $V = \text{span}\{\mathbf{v}_1,\ldots,\mathbf{v}_n\}$, then $\mathbf{v}\in V$ if and only if $\mathbf{v}\wedge\Omega_V = 0$, where $\Omega_V = \mathbf{v}_1\wedge\cdots\wedge\mathbf{v}_n$ is the pseudoscalar of V. Let us denote the pseudoscalars of $\mathbb{R}^{p,q}$, $\mathbb{R}^{p_0,q_0}$, and $\mathbb{R}^{p-p_0,q-q_0}$ by I, Ω, and Θ, respectively, that is,

$$\begin{aligned} I &= \mathbf{e}_1\cdots\mathbf{e}_p\mathbf{f}_1\cdots\mathbf{f}_q, \\ \Omega &= \mathbf{e}_1\cdots\mathbf{e}_{p_0}\mathbf{f}_1\cdots\mathbf{f}_{q_0}, \\ \Theta &= \mathbf{e}_{p_0+1}\cdots\mathbf{e}_p\mathbf{f}_{q_0+1}\cdots\mathbf{f}_q, \end{aligned} \tag{6.20}$$

where we used the fact that $\{\mathbf{e}_i,\mathbf{f}_k\}$ is an orthonormal basis. Note that

$$I = (-1)^{(p-p_0)q_0}\Omega\Theta. \tag{6.21}$$

The condition that $\mathbf{v}\in\mathbb{R}^{p_0,q_0}$ is equivalent to $\mathbf{v}\wedge\Omega = 0$ (Mosna, Miralles, and Vaz Jr, 2003), which can be written as

$$\mathbf{v}\Omega + \widehat{\Omega}\mathbf{v} = 0, \tag{6.22}$$

yields

$$\mathbf{v} = -\widehat{\Omega}\mathbf{v}\Omega^{-1} = -\Omega\mathbf{v}\widehat{\Omega}^{-1}. \tag{6.23}$$

This result suggests the definition of α as

$$\alpha(\mathbf{v}) = -\Omega\mathbf{v}\widehat{\Omega}^{-1}, \tag{6.24}$$

in such a way that $\mathbf{v}\in\mathbb{R}^{p_0,q_0}$ if and only if $\alpha(\mathbf{v}) = \mathbf{v}$. We can extend this definition to $\phi\in\mathcal{C}\ell_{p,q}$ in such a way to satisfy $\alpha(\mathbf{vu}) = \alpha(\mathbf{v})\alpha(\mathbf{u})$ as either

$$\alpha(\phi) = \Omega\widehat{\phi}\Omega^{-1}, \quad \text{if} \quad \widehat{\Omega} = \Omega, \tag{6.25}$$

or

$$\alpha(\phi) = \Omega\phi\Omega^{-1}, \quad \text{if} \quad \widehat{\Omega} = -\Omega. \tag{6.26}$$

Hence, if $\phi\in\mathcal{C}\ell_{p_0,q_0}$, then we have $\alpha(\phi) = \phi$.

Similarly, we have that $\mathbf{u}\in\mathbb{R}^{p-p_0,q-q_0}$ if and only if $\mathbf{u}\wedge\Theta = 0$, which can be written as

$$\mathbf{u} = \Theta\widehat{\mathbf{u}}\widehat{\Theta}^{-1}, \tag{6.27}$$

and generalised to $\psi\in\mathcal{C}\ell_{p-p_0,q-q_0}$ as either

$$\psi = \Theta\widehat{\psi}\Theta^{-1}, \quad \text{if} \quad \widehat{\Theta} = \Theta, \tag{6.28}$$

or

$$\psi = \Theta\psi\Theta^{-1}, \quad \text{if} \quad \widehat{\Theta} = -\Theta. \tag{6.29}$$

Let us see now what happens in the four different situations according to whether the pseudoscalars are even or odd elements.

(1) Let us suppose that $\widehat{\Theta} = \Theta$. Then, for $\psi \in \mathcal{C}\ell_{p-p_0,q-q_0}$, we have

$$\psi = \Theta\widehat{\psi}\Theta^{-1} = \Omega^{-1}I\widehat{\psi}I^{-1}\Omega = \Omega I\widehat{\psi}I^{-1}\Omega^{-1}, \tag{6.30}$$

where we used eqn (6.21). We must also distinguish two other cases.

(a) Let us suppose that $\widehat{\Omega} = \Omega$. In this case, I is an even element, and $I\psi I^{-1} = \widehat{\psi}$. Then,

$$\psi = \Omega\psi\Omega^{-1} = \begin{cases} \Omega\widehat{\psi}\Omega^{-1}, & \text{if } \widehat{\psi} = \psi, \\ -\Omega\widehat{\psi}\Omega^{-1}, & \text{if } \widehat{\psi} = -\psi, \end{cases} \tag{6.31}$$

that is,

$$\psi = \begin{cases} \alpha(\psi), & \text{if } \widehat{\psi} = \psi, \\ -\alpha(\psi), & \text{if } \widehat{\psi} = -\psi. \end{cases} \tag{6.32}$$

(b) Let us suppose that $\widehat{\Omega} = -\Omega$. In this case, I is an odd element, and $I\psi I^{-1} = \psi$. Accordingly,

$$\psi = \Omega\widehat{\psi}\Omega^{-1} = \begin{cases} \Omega\psi\Omega^{-1}, & \text{if } \widehat{\psi} = \psi, \\ -\Omega\psi\Omega^{-1}, & \text{if } \widehat{\psi} = -\psi, \end{cases} \tag{6.33}$$

that is

$$\psi = \begin{cases} \alpha(\psi), & \text{if } \widehat{\psi} = \psi, \\ -\alpha(\psi), & \text{if } \widehat{\psi} = -\psi. \end{cases} \tag{6.34}$$

(2) Let us suppose that $\widehat{\Theta} = -\Theta$. Then, for $\psi \in \mathcal{C}\ell_{p-p_0,q-q_0}$ we have

$$\psi = \Theta\psi\Theta^{-1} = \Omega^{-1}I\psi I^{-1}\Omega = \Omega I\psi I^{-1}\Omega^{-1}, \tag{6.35}$$

where we used eqn (6.21). We must also distinguish two other cases.

(a) Let us suppose that $\widehat{\Omega} = \Omega$. In this case, I is an odd element, and $I\psi I^{-1} = \psi$. Then,

$$\psi = \Omega\psi\Omega^{-1} = \begin{cases} \Omega\widehat{\psi}\Omega^{-1}, & \text{if } \widehat{\psi} = \psi, \\ -\Omega\widehat{\psi}\Omega^{-1}, & \text{if } \widehat{\psi} = -\psi, \end{cases} \tag{6.36}$$

that is,

$$\psi = \begin{cases} \alpha(\psi), & \text{if } \widehat{\psi} = \psi, \\ -\alpha(\psi), & \text{if } \widehat{\psi} = -\psi. \end{cases} \tag{6.37}$$

(b) Let us suppose that $\widehat{\Omega} = -\Omega$. In this case, I is an even element, and $I\psi I^{-1} = \widehat{\psi}$. Thus, we have

$$\psi = \Omega\psi\Omega^{-1} = \begin{cases} \Omega\widehat{\psi}\Omega^{-1}, & \text{if } \widehat{\psi} = \psi, \\ -\Omega\widehat{\psi}\Omega^{-1}, & \text{if } \widehat{\psi} = -\psi, \end{cases} \tag{6.38}$$

that is,

$$\psi = \begin{cases} \alpha(\psi), & \text{if } \widehat{\psi} = \psi, \\ -\alpha(\psi), & \text{if } \widehat{\psi} = -\psi. \end{cases} \tag{6.39}$$

Then, from all the four cases, we conclude that

$$\mathcal{C}\ell^{+}_{p-p_0,q-q_0} \in \mathcal{C}\ell^{0}_{p,q}, \qquad \mathcal{C}\ell^{-}_{p-p_0,q-q_0} \in \mathcal{C}\ell^{1}_{p,q}, \tag{6.40}$$

in relation to the α grading automorphism. Therefore, in order to use eqn (6.18) to define a generalised spinor algebra, we define the α grading automorphism using either eqn (6.25) or eqn (6.26), depending upon whether Ω is even or odd, respectively.

Example 6.8 Let us consider $\mathcal{C}\ell_{1,1}$. We have that $\mathcal{C}\ell^{+}_{1,1} \simeq \mathcal{C}\ell_{1,0}$, and the spinor algebra obtained using the usual $\mathbb{Z}_2$-grading is $\mathbb{R}\oplus\mathbb{R}$. However, if we use the α grading automorphism with $\Omega = \mathbf{e}_2$ ($(\mathbf{e}_2)^2 = -1$) in eqn (6.26), that is, $\alpha(\psi) = -\mathbf{e}_2\psi\mathbf{e}_2$, we obtain that $\mathcal{C}\ell^{0}_{1,1} \simeq \mathbb{C}$. In this case, we can choose either $\kappa = \mathbf{e}_1$, or $\kappa = \mathbf{e}_1\mathbf{e}_2$.

Example 6.9 Consider the case $\mathcal{C}\ell_{3,0}$. We saw that $\mathcal{C}\ell^{+}_{3,0} \simeq \mathbb{H}$ is an irreducible representation space of $\mathcal{C}\ell_{3,0}$. Consider now $\Omega = \mathbf{v}$, where $\mathbf{v}$ is a unit vector. Then, $\alpha(\phi) = \mathbf{v}\phi\mathbf{v}$. We choose $\kappa = \mathbf{u}$, where $\mathbf{u}$ is an unit vector orthogonal to $\mathbf{v}$. We have that $\mathcal{C}\ell^{0}_{3,0}$ is generated by $\{1, \mathbf{v}, \mathbf{i}\mathbf{v}, \mathbf{i}\}$, where $\mathbf{i} = \mathbf{e}_1\mathbf{e}_2\mathbf{e}_3$, and $\mathcal{C}\ell^{0}_{3,0} \simeq \mathbb{C} \oplus \mathbb{C}$. We can also choose $\Omega = \mathbf{u}\mathbf{v}$, with $\mathbf{u}$ and $\mathbf{v}$ orthogonal to each other. Then, $\alpha(\psi) = \mathbf{u}\mathbf{v}\widehat{\psi}\mathbf{v}\mathbf{u}$. In this case, we choose $\kappa = \mathbf{i}\mathbf{u}\mathbf{v}$. The spinor algebra $\mathcal{C}\ell^{0}_{3,0}$ in this case is generated by $\{1, \mathbf{u}, \mathbf{v}, \mathbf{u}\mathbf{v}\}$ and $\mathcal{C}\ell^{0}_{3,0} \simeq \mathcal{C}\ell_{2,0} \simeq \mathcal{M}(2,\mathbb{R})$.

Example 6.10 We noticed that $\mathbb{H}\oplus\mathbb{H}$ and $\mathcal{M}(2,\mathbb{C})$ are irreducible representation spaces of $\mathcal{C}\ell_{4,1}$ by using the usual $\mathbb{Z}_2$-grading. If we use the α grading, we can obtain another spinor algebra, that is, $\mathbb{C}\oplus\mathbb{C}\oplus\mathbb{C}\oplus\mathbb{C}$, which is not a Clifford algebra. In fact, we can define $\alpha(\phi) = \mathbf{e}_1\phi\mathbf{e}_1$, and $\kappa = \mathbf{e}_2$. But we can also define another grading as $\alpha'(\phi) = \mathbf{e}_1\mathbf{e}_2\mathbf{e}_3\phi\mathbf{e}_3\mathbf{e}_2\mathbf{e}_1$, with $\kappa' = \mathbf{e}_3\mathbf{e}_5$. Note that $\kappa\kappa' = \kappa'\kappa$. Then, the space ${}^{0}\mathcal{C}\ell^{0}_{4,1}$ of α-even and α'-even elements is generated by $\{1, \mathbf{e}_1, \mathbf{e}_2\mathbf{e}_3, \mathbf{e}_1\mathbf{e}_2\mathbf{e}_3, \mathbf{e}_4\mathbf{e}_5, \mathbf{e}_1\mathbf{e}_4\mathbf{e}_5, \mathbf{e}_2\mathbf{e}_3\mathbf{e}_4\mathbf{e}_5, \mathbf{e}_1\mathbf{e}_2\mathbf{e}_3\mathbf{e}_4\mathbf{e}_5\}$, and it is not difficult to see that ${}^{0}\mathcal{C}\ell^{0}_{4,1} = \mathbb{C}\oplus\mathbb{C}\oplus\mathbb{C}\oplus\mathbb{C}$.

Further applications and development of the α grading (6.24) can be found in the article by da Rocha and Vaz Jr (2006*c*).

6.5 A Comparison of the Different Definitions of Spinors

A question that naturally arises now regards the relationship among the different definitions of spinors. Let us first consider the case of real Clifford algebras. Unfortunately, we do not have a classification table for the spinor operators for real Clifford algebras, as it would make it easy to compare the different definitions. In fact, for spinor operators, we have even cases with different and non-equivalent spinor algebras, and we were not able to design a table that could take into account all possible cases. Thus, we will study the problem for each case up to dimension 5, leaving for the interested readers the analysis for dimensions higher than that. The spinor operators for dimensions 1–5 can be found in table 6.6.[1]

We can easily see that, given $\mathcal{C}\ell_{p,q}$, algebraic spinors, classical spinors, and spinor operators are in general different. However, in most of the cases, all these different

[1]The notation $A \mid B$ in the $S^{O}_{p,q}$ column in table 6.6 means that A uses the usual parity automorphism, and B uses the α grading automorphism.

spinor spaces are isomorphic from the point of view of vector spaces. In fact, given $\mathcal{C}\ell_{p,q}$, we see that, when $p - q = 0, 2, 3, 4 \mod 8$, the dimensions of these spaces are equal. This case includes the distinctive example of the three-dimensional Euclidean space. When $p - q = 1, 5 \mod 8$, if we consider the space of algebraic semispinors instead of the space of algebraic spinors, we see that the dimensions of the spaces are also equal. However, when $p - q = 6, 7 \mod 8$, the dimension of the space of classical spinors is half that of the algebraic spinors and the spinor operators. The space of spinor operators has therefore the same dimension as the space of algebraic spinors or the space of algebraic semispinors, when the latter is defined.

Table 6.6 Algebraic Spinors ($S^A_{p,q}$), Classical Spinor s ($S^C_{p,q}$) and Spinor Operators ($S^O_{p,q}$) for $\mathcal{C}\ell_{p,q}$, with $p+q \leq 5$ (Where $\mathcal{A}^{\oplus^4} = \mathcal{A} \oplus \mathcal{A} \oplus \mathcal{A} \oplus \mathcal{A}$)

	$S^A_{p,q}$	$S^C_{p,q}$	$S^O_{p,q}$
$\mathcal{C}\ell_{1,0}$	$\mathbb{R} \oplus \mathbb{R}$	$\mathbb{R}$	$\mathbb{R}$
$\mathcal{C}\ell_{0,1}$	$\mathbb{C}$	$\mathbb{R}$	$\mathbb{C}$
$\mathcal{C}\ell_{2,0}$	$\mathbb{R}^2$	$\mathbb{C}$	$\mathbb{C} \mid \mathbb{R} \oplus \mathbb{R}$
$\mathcal{C}\ell_{1,1}$	$\mathbb{R}^2$	$\mathbb{R} \oplus \mathbb{R}$	$\mathbb{R} \oplus \mathbb{R} \mid \mathbb{C}$
$\mathcal{C}\ell_{0,2}$	$\mathbb{H}$	$\mathbb{C}$	$\mathbb{H}$
$\mathcal{C}\ell_{3,0}$	$\mathbb{C}^2$	$\mathbb{H}$	$\mathbb{H} \mid \mathbb{C} \oplus \mathbb{C};\ \mathcal{M}(2,\mathbb{R})$
$\mathcal{C}\ell_{2,1}$	$\mathbb{R}^2 \oplus \mathbb{R}^2$	$\mathbb{R}^2$	$\mathbb{C};\ \mathbb{R} \oplus \mathbb{R}$
$\mathcal{C}\ell_{1,2}$	$\mathbb{C}^2$	$\mathbb{R}^2$	$\mathcal{M}(2,\mathbb{R}) \mid \mathbb{C} \oplus \mathbb{C}; \mathbb{H}$
$\mathcal{C}\ell_{0,3}$	$\mathbb{H} \oplus \mathbb{H}$	$\mathbb{H}$	$\mathbb{H}$
$\mathcal{C}\ell_{4,0}$	$\mathbb{H}^2$	$\mathbb{H} \oplus \mathbb{H}$	$\mathbb{H} \oplus \mathbb{H} \mid \mathcal{M}(2,\mathbb{C})$
$\mathcal{C}\ell_{3,1}$	$\mathbb{R}^4$	$\mathbb{C}^2$	$\mathbb{C} \oplus \mathbb{C} \mid \mathbb{R}^{\oplus^4}$
$\mathcal{C}\ell_{2,2}$	$\mathbb{R}^4$	$\mathbb{R}^2 \oplus \mathbb{R}^2$	$\mathbb{R}^{\oplus^4};\ \mathbb{C} \oplus \mathbb{C}$
$\mathcal{C}\ell_{1,3}$	$\mathbb{H}^2$	$\mathbb{C}^2$	$\mathcal{M}(2,\mathbb{C}) \mid \mathbb{H} \oplus \mathbb{H}$
$\mathcal{C}\ell_{0,4}$	$\mathbb{H}^2$	$\mathbb{H} \oplus \mathbb{H}$	$\mathbb{H} \oplus \mathbb{H} \mid \mathcal{M}(2,\mathbb{C})$
$\mathcal{C}\ell_{5,0}$	$\mathbb{H}^2 \oplus \mathbb{H}^2$	$\mathbb{H}^2$	$\mathcal{M}(2,\mathbb{C}) \mid \mathbb{H} \oplus \mathbb{H}$
$\mathcal{C}\ell_{4,1}$	$\mathbb{C}^4$	$\mathbb{H}^2$	$\mathbb{H} \oplus \mathbb{H};\ \mathcal{M}(2,\mathbb{C}) \mid \mathbb{C}^{\oplus^4}$
$\mathcal{C}\ell_{3,2}$	$\mathbb{R}^4 \oplus \mathbb{R}^4$	$\mathbb{R}^4$	$\mathbb{C} \oplus \mathbb{C};\ \mathbb{R}^{\oplus^4}$
$\mathcal{C}\ell_{2,3}$	$\mathbb{C}^4$	$\mathbb{R}^4$	$\mathcal{M}(2,\mathbb{C}) \mid \mathbb{C}^{\oplus^4}$
$\mathcal{C}\ell_{1,4}$	$\mathbb{H}^2 \oplus \mathbb{H}^2$	$\mathbb{H}^2$	$\mathbb{H} \oplus \mathbb{H};\ \mathcal{M}(2,\mathbb{C}) \mid \mathbb{C}^{\oplus^4}$
$\mathcal{C}\ell_{0,5}$	$\mathbb{C}^4$	$\mathbb{H}^2$	$\mathbb{H} \oplus \mathbb{H};\ \mathcal{M}(2,\mathbb{C}) \mid \mathbb{C}^{\oplus^4}$

There is a suitable way to understand this relation between spinor operators and algebraic spinors or semispinors. First, let us remember that a minimal left ideal of $\mathcal{C}\ell_{p,q}$ is of the form $\mathcal{C}\ell_{p,q}\, f$, where f is given in eqn (4.63), that is, we have

$$S^{A}_{p,q} = \mathcal{C}\ell_{p,q}\, f, \tag{6.41}$$

where f has the form

$$f = \frac{1}{2}(1 + \mathbf{e}_{I_1}) \cdots \frac{1}{2}(1 + \mathbf{e}_{I_k}), \tag{6.42}$$

and $\{\mathbf{e}_{I_1}, \ldots, \mathbf{e}_{I_k}\}$ is a set of $\mathcal{C}\ell_{p,q}$ elements which commute with each other and such that $(\mathbf{e}_{I_\alpha})^2 = 1$ $(\alpha = 1, \ldots, k)$. Moreover, $k = q - r_{q-p}$, where r_j are the Radon–Hurwitz numbers. However, note that

$$f = \mathbf{e}_{I_\alpha} f, \quad \alpha = 1, \ldots, k. \tag{6.43}$$

Since $\mathbf{e}_{I_1}, \ldots, \mathbf{e}_{I_k}$ commute with each other, and $(\mathbf{e}_{I_\alpha})^2 = 1$, we can choose $\kappa = \epsilon$, and ϵ, ϵ', and so on in the definition of a graded representation as the elements $\mathbf{e}_{I_1}, \ldots, \mathbf{e}_{I_k}$. Let us take, for example, $\kappa = \mathbf{e}_{I_1}$. Then, $\mathcal{C}\ell_{p,q} = \mathcal{C}\ell^0_{p,q} \oplus \mathcal{C}\ell^1_{p,q}$, and $\kappa = \mathbf{e}_{I_1} \in \mathcal{C}\ell^1_{p,q}$. But then

$$\mathcal{C}\ell_{p,q}\, f = (\mathcal{C}\ell^0_{p,q} \oplus \mathcal{C}\ell^1_{p,q}) f = \mathcal{C}\ell^0_{p,q}\, f \oplus \mathcal{C}\ell^1_{p,q}\, f = \mathcal{C}\ell^0_{p,q}\, f \oplus \mathcal{C}\ell^1_{p,q}\, \mathbf{e}_{I_1} f, \tag{6.44}$$

and since

$$\mathcal{C}\ell^1_{p,q}\, \mathbf{e}_{I_1} \in \mathcal{C}\ell^0_{p,q}, \tag{6.45}$$

we have

$$\mathcal{C}\ell_{p,q}\, f \subset \mathcal{C}\ell^0_{p,q}\, f. \tag{6.46}$$

Next, we choose $\epsilon = \mathbf{e}_{I_2}$, and a grading such that $\mathcal{C}\ell^0_{p,q} = {}^0\mathcal{C}\ell^0_{p,q} \oplus {}^1\mathcal{C}\ell^0_{p,q}$, with $\mathbf{e}_{I_2} \in {}^1\mathcal{C}\ell^0_{p,q}$. Since ${}^1\mathcal{C}\ell^0_{p,q}\, \mathbf{e}_{I_2} \in {}^0\mathcal{C}\ell^0_{p,q}$, we have

$$\mathcal{C}\ell_{p,q}\, f \subset {}^0\mathcal{C}\ell^0_{p,q}\, f. \tag{6.47}$$

We can continue with this procedure up to the last element $\mathbf{e}_{I_k}$, to obtain

$$\mathcal{C}\ell_{p,q}\, f \subset {}^{0\cdots 0}\mathcal{C}\ell^0_{p,q}\, f = {}^{(0)^{k-1}}\mathcal{C}\ell^0_{p,q}\, f. \tag{6.48}$$

Nevertheless, $\dim \mathcal{C}\ell^0_{p,q} = (\dim \mathcal{C}\ell_{p,q})/2$; $\dim {}^0\mathcal{C}\ell^0_{p,q} = (\dim \mathcal{C}\ell_{p,q})/(2 \cdot 2)$; and so on; in addition,

$$\dim {}^{(0)^{k-1}}\mathcal{C}\ell^0_{p,q} = \frac{\dim \mathcal{C}\ell_{p,q}}{2^k}. \tag{6.49}$$

Since $\dim \mathcal{C}\ell_{p,q} = 2^n$, and the dimension of $\mathcal{C}\ell_{p,q}\, f$ is 2^{n-k}, we can conclude that

$$\mathcal{C}\ell_{p,q}\, f = {}^{(0)^{k-1}}\mathcal{C}\ell^0_{p,q}\, f. \tag{6.50}$$

Thus, from the point of view of vector spaces, we have

$$\mathcal{C}\ell_{p,q}\, f \underset{V}{\simeq} {}^{(0)^{k-1}}\mathcal{C}\ell^0_{p,q}\,. \tag{6.51}$$

Let Λ be the pseudoscalar of $\mathcal{C}\ell_{p,q}$. We leave it as an exercise to show that $\Lambda^2 = 1$ when $p - q = 0, 1 \mod 4$. The semispinors $\phi^\pm \in S^\pm_{p,q}$ are eigenvalues of Λ when $\Lambda^2 = 1$, that is, $\Lambda\phi^{\pm} = \pm\phi^{\pm}$ for $\psi^{\pm} \in S^{\pm}_{p,q}$. But, when $p-q=1 \mod 4$, the volume element $\Lambda \in \mathrm{Cen}(\mathcal{C}\ell_{p,q})$ (Λ is an odd element), and then $\Lambda\phi^{\pm} = \phi^{\pm}\Lambda$. Consequently, we can write the semispinors as

$$\phi^{\pm} = \frac{1}{2}\phi(1 \pm \Lambda), \tag{6.52}$$

where $\phi \in S_{p,q}$, and

$$S^{\pm}_{p,q} = S_{p,q}\frac{1}{2}(1 \pm \Lambda) = ({}^{(0)^{k-1}}\mathcal{C}\ell^{0}_{p,q})f\frac{1}{2}(1 \pm \Lambda). \tag{6.53}$$

Now we consider a grading such that

$$({}^{(0)^{k-1}}\mathcal{C}\ell^{0}_{p,q})^{1}\Lambda \subset ({}^{(0)^{k-1}}\mathcal{C}\ell^{0}_{p,q})^{0} = {}^{(0)^{k}}\mathcal{C}\ell^{0}_{p,q}, \tag{6.54}$$

and then

$$(\mathcal{C}\ell_{p,q}\, f)^{\pm} \underset{V}{\simeq} {}^{(0)^{k}}\mathcal{C}\ell^{0}_{p,q}\,. \tag{6.55}$$

As we see, this situation happens when $p-q=1 \mod 4$, or $p-q=1,5 \mod 8$, which was the condition we started with.

The case of complex Clifford algebras is much simpler than that of the reals. The comparison of algebraic spinors, classical spinors, and spinor operators for complex Clifford algebras is given in table 6.7. Note that, when n is even, all spaces have the same dimension but, when n is odd, the classical spinor and the spinor operator spaces have the same dimension as the space of the algebraic semispinor.

When comparing the different spinor definitions, one may be tempted to ask whether any of the definitions is better than the others. This kind of question is meaningful only if we clearly specify the aspects of the concept we want to examine or emphasise. In this sense, when we are interested in calculations with spinors, the concept of spinor operators proves to be very useful, since the representation space is an algebra. This fact is particularly clear in those cases where the spinor algebra is a Clifford algebra.

There are two cases where the spinor algebra is a Clifford algebra, and which have very important physical applications: (i) the Clifford algebra of the three-dimensional Euclidean space $\mathcal{C}\ell_{3,0}$, and (ii) the Clifford algebra of the four-dimensional Minkowski spacetime $\mathcal{C}\ell_{3,1}$. The concept of the spinor operator proves to be very useful in these cases.

Example 6.11 In order to discuss the case of the three-dimensional Euclidean space, let us establish some notation, that is, $\Psi \in \mathcal{C}\ell_{3,0}\, f$ (where f is a primitive idempotent); $|\Psi\rangle \in \mathbb{C}^2$; $\Psi_+ \in \mathcal{C}\ell^{+}_{3,0}$; $\Psi_0 \in \mathcal{C}\ell^{0}_{3,0}$;

Table 6.7 Algebraic Spinors (S^{A}_{n}), Classical Spinors (S^{C}_{n}), and Spinor Operators (S^{O}_{n}) for Complex Clifford Algebras $\mathcal{C}\ell_{n}(\mathbb{C})$

	S^{A}_{n}	S^{C}_{n}	S^{O}_{n}
$\mathcal{C}\ell_{2k}(\mathbb{C})$	$\mathbb{C}^{2^k}$	$\mathbb{C}^{2^{k-1}} \oplus \mathbb{C}^{2^{k-1}}$	$\mathbb{C}^{\oplus 2^k}$
$\mathcal{C}\ell_{2k+1}(\mathbb{C})$	$\mathbb{C}^{2^k} \oplus \mathbb{C}^{2^k}$	$\mathbb{C}^{2^k}$	$\mathbb{C}^{\oplus 2^k}$

and $\boldsymbol{\Psi} \in \mathbb{H}$. The Clifford algebra $\mathcal{C}\ell_{3,0}$ is generated by $\{\mathbf{e}_1, \mathbf{e}_2, \mathbf{e}_3\}$, and $\sigma_i = \rho(\mathbf{e}_i)$ $(i = 1, 2, 3)$ are the Pauli matrices, as seen in example 3.2. The pseudoscalar of $\mathcal{C}\ell_{3,0}$ is $\mathbf{I} = \mathbf{e}_1\mathbf{e}_2\mathbf{e}_3$.

Let $\psi \in \mathcal{C}\ell_{3,0}$, where

$$\psi = s + v_1\mathbf{e}_1 + v_2\mathbf{e}_2 + v_3\mathbf{e}_3 + b_{12}\mathbf{e}_{12} + b_{13}\mathbf{e}_{13} + b_{23}\mathbf{e}_{23} + t_{123}\mathbf{e}_{123},$$

and where $\mathbf{e}_{12} = \mathbf{e}_1\mathbf{e}_2$, and so on, as usual. The matrix representation of ψ, obtained by using the idempotent $f = \frac{1}{2}(1 + \mathbf{e}_3)$, is

$$\rho(\psi) = \begin{pmatrix} (s + v_3) + i(t_{123} + b_{12}) & (v_1 - b_{13}) + i(b_{23} - v_2) \\ (v_1 + t_{13}) + i(b_{23} + v_2) & (s - v_3) + i(t_{123} - b_{12}) \end{pmatrix}.$$

The algebraic spinor $\Psi = \psi f$ reads

$$\begin{aligned} \Psi &= [(s + v_3) + \mathbf{I}(b_{12} + t_{123})]f + [(v_1 + b_{13}) + \mathbf{I}(v_2 + b_{23})]\mathbf{e}_1 f \\ &= (w_1 + \mathbf{I}w_2)f + (w_3 + \mathbf{I}w_4)\mathbf{e}_1 f. \end{aligned}$$

Its matrix representation is provided by

$$\rho(\Psi) = \begin{pmatrix} (s + v_3) + i(t_{123} + b_{12}) & 0 \\ (v_1 + t_{13}) + i(b_{23} + v_2) & 0 \end{pmatrix} = \begin{pmatrix} w_1 + iw_2 & 0 \\ w_3 + iw_4 & 0 \end{pmatrix}.$$

The space $S^A_{3,0}$ is clearly isomorphic as a vector space to $\mathbb{C}^2$, with the identification

$$\Psi = (w_1 + \mathbf{I}w_2)f + (w_3 + \mathbf{I}w_4)\mathbf{e}_1 f \leftrightarrow \begin{pmatrix} w_1 + iw_2 \\ w_3 + iw_4 \end{pmatrix} = |\Psi\rangle.$$

The spinor operator $\Psi_+ \in \mathcal{C}\ell^+_{3,0}$ has the form

$$\Psi_+ = s + b_{12}\mathbf{e}_{12} + b_{13}\mathbf{e}_{13} + b_{23}\mathbf{e}_{23}.$$

Let us conveniently rewrite the coefficients of Ψ_+ as

$$\Psi_+ = w_1 + w_2\mathbf{e}_{12} + w_3\mathbf{e}_{13} + w_4\mathbf{e}_{23}.$$

The matrix representation of Ψ_+ is given by

$$\rho(\Psi_+) = \begin{pmatrix} w_1 + iw_2 & -w_3 + iw_4 \\ w_3 + iw_4 & w_1 - iw_2 \end{pmatrix},$$

which shows the vector space isomorphism $\mathcal{C}\ell^+_{3,0} \simeq \mathcal{C}\ell_{3,0} f \simeq \mathbb{C}^2$. We also know that $\mathcal{C}\ell^+_{3,0} \simeq \mathcal{C}\ell_{0,2} \simeq \mathbb{H}$, which can be made explicitly by the identification

$$\mathbf{e}_{12} \leftrightarrow \boldsymbol{i}, \quad \mathbf{e}_{23} \leftrightarrow \boldsymbol{j}, \quad \mathbf{e}_{13} \leftrightarrow \boldsymbol{k},$$

that is,

$$\boldsymbol{\Psi} = w_1 + w_2\boldsymbol{i} + w_4\boldsymbol{j} + w_3\boldsymbol{k}.$$

For the spinor algebra $\mathcal{C}\ell^0_{3,0}$, we define the α grading by $\alpha(\psi) = \mathbf{v}\psi\mathbf{v}$, where $\mathbf{v}$ is a unit vector and we choose $\kappa = \mathbf{u}$, where $\mathbf{u}$ is a unit vector orthogonal to $\mathbf{v}$. Specifically, we choose $\mathbf{v} = \mathbf{e}_1$, and $\mathbf{u} = \mathbf{e}_3$. Then, the spinor operator $\Psi_0 \in \mathcal{C}\ell^0_{3,0}$ has the form

$$\Psi_0 = s + v_1\mathbf{e}_1 + b_{23}\mathbf{e}_{23} + t_{123}\mathbf{e}_{123}.$$

Let us also conveniently rewrite the coefficients of Ψ_0 as

$$\Psi_0 = w_1 + w_2\mathbf{e}_{123} + w_3\mathbf{e}_1 + w_4\mathbf{e}_{23}.$$

It is represented by

$$\rho(\Psi_0) = \begin{pmatrix} w_1 + iw_2 & w_3 + iw_4 \\ w_3 + iw_4 & w_1 + iw_2 \end{pmatrix},$$

revealing the vector space isomorphism of $\mathcal{C}\ell^0_{3,0}$ with these spinor spaces. Note that, as an algebra, $\mathcal{C}\ell^0_{3,0} \simeq \mathbb{C} \oplus \mathbb{C}$, with the identification $i \leftrightarrow \mathbf{e}_1\mathbf{e}_2\mathbf{e}_3$ and $(1, 0) \leftrightarrow 1$, $(0, 1) \leftrightarrow \mathbf{e}_1$.

It is important for physical applications to consider bilinear quantities constructed from spinors. In terms of $|\Psi\rangle \in \mathbb{C}^2$, these quantities are given by

$$\sigma = \langle\Psi|\Psi\rangle, \quad J_i = \langle\Psi|\sigma_i|\Psi\rangle, \quad S_{ij} = \langle\Psi|\frac{1}{2}[\sigma_i, \sigma_j]|\Psi\rangle, \quad \omega = \langle\Psi|\sigma_1\sigma_2\sigma_3|\Psi\rangle,$$

with $i, j = 1, 2, 3$. It follows that

$$\begin{aligned} \sigma &= w_1^2 + w_2^2 + w_3^2 + w_4^2, \\ J_1 &= 2(w_1 w_4 + w_2 w_3), \\ J_2 &= 2(w_1 w_3 - w_2 w_4), \\ J_3 &= w_1^2 + w_2^2 - w_3^2 - w_4^2. \end{aligned}$$

The transformation $(w_1, w_2, w_3, w_4) \mapsto (J_1, J_2, J_3)$ is known as the Kustaanheimo–Stiefel transformation.

Now, let us see how to write the bilinear quantities using the other spinor definitions. Clearly, it is enough to consider J_i $(i = 1, 2, 3)$. In terms of $\Psi \in \mathcal{C}\ell_{3,0} f$, we first note that $\widetilde{\Psi} \in f\mathcal{C}\ell_{3,0}$ (a right minimal ideal). The matrix representation of $\widetilde{\Psi}$ is

$$\rho(\widetilde{\Psi}) = \begin{pmatrix} (s + v_3) - i(t_{123} + b_{12}) & (v_1 + b_{13}) - i(b_{23} + v_2) \\ 0 & 0 \end{pmatrix} = \begin{pmatrix} w_1 - iw_2 & w_3 - iw_4 \\ 0 & 0 \end{pmatrix}.$$

Then, we can identify

$$\widetilde{\Psi} = f(w_1 - \mathbf{I}w_2) + f\mathbf{e}_1(w_3 - \mathbf{I}w_4) \leftrightarrow \begin{pmatrix} w_1 - iw_2 & w_3 - iw_4 \end{pmatrix} = \langle \Psi |.$$

Moreover, since $\mathrm{Tr}[\rho(f)] = 2\langle f \rangle_0 = 1$, we have that

$$J_i = 2\langle \widetilde{\Psi} \mathbf{e}_i \Psi \rangle_0 = 2\langle \Psi \widetilde{\Psi} \mathbf{e}_i \rangle_0,$$

where we used $\langle AB \rangle_0 = \langle BA \rangle_0$. But $\Psi = \Psi f$, with $f = (1/2)(1 + \mathbf{e}_3)$, and then

$$J_i = \langle \Psi(1 + \mathbf{e}_3)\widetilde{\Psi}\mathbf{e}_i \rangle_0.$$

Moreover, for spinor operators, we also have

$$J_i = \langle \Psi_+(1 + \mathbf{e}_3)\widetilde{\Psi}_+\mathbf{e}_i \rangle_0, \qquad J_i = \langle \Psi_0(1 + \mathbf{e}_3)\widetilde{\Psi}_0\mathbf{e}_i \rangle_0.$$

It is worth noting that the expression for J_i takes a very interesting and simple form using Ψ_+. In fact, we have

$$J_i = \langle \Psi_+\widetilde{\Psi}_+\mathbf{e}_i \rangle_0 + \langle \Psi_+\mathbf{e}_3\widetilde{\Psi}_+\mathbf{e}_i \rangle_0.$$

However, $\mathbf{e}_i \in \mathcal{C}\ell_{3,0}^-$, and then $\Psi_+\widetilde{\Psi}_+\mathbf{e}_i \in \mathcal{C}\ell_{3,0}^-$, which gives $\langle \Psi_+\widetilde{\Psi}_+\mathbf{e}_i \rangle_0 = 0$, since scalar elements belong to $\mathcal{C}\ell_{p,q}^+$. Hence,

$$J_i = \langle \Psi_+\mathbf{e}_3\widetilde{\Psi}_+\mathbf{e}_i \rangle_0 = \langle \Psi_+\mathbf{e}_3\widetilde{\Psi}_+ \rangle_1 \cdot \mathbf{e}_i.$$

Let us denote

$$\mathbf{J} = \Psi_+\mathbf{e}_3\widetilde{\Psi}_+.$$

Note that $\widetilde{\mathbf{J}} = \mathbf{J}$; $\widehat{\mathbf{J}} = -\mathbf{J}$; and $\bar{\mathbf{J}} = -\mathbf{J}$. The only $\mathcal{C}\ell_{3,0}$ elements which that satisfy these properties are vectors, that is, $\mathbf{J} = \langle \mathbf{J} \rangle_1$. Then,

$$J_i = \mathbf{J} \cdot \mathbf{e}_i.$$

The spinor operator Ψ_+ itself has a simple and interesting interpretation. First, we note that

$$\Psi_+\widetilde{\Psi}_+ = w_1^2 + w_2^2 + w_3^2 + w_4^2 = \sigma \geq 0.$$

Then, we can write

$$\Psi_+ = \sqrt{\sigma}R,$$

where $R \in \mathcal{C}\ell_{3,0}^+$ and such that $R\widetilde{R} = 1$. But this means that $R \in \mathrm{Spin}(3)$ (see eqn (5.86)). The operation $R\mathbf{e}_3\tilde{R}$ therefore represents a rotation of the vector $\mathbf{e}_3$. Consequently,

$$\mathbf{J} = \Psi_+\mathbf{e}_3\widetilde{\Psi}_+ = \sigma R\mathbf{e}_3\tilde{R}$$

can be interpreted as giving the vector $\mathbf{J}$ as the result of the composition of a rotation of the vector $\mathbf{e}_3$ and a dilation by a factor σ (Vaz, 2013). This is an amazing interpretation for a spinor in the three-dimensional Euclidean space. For an example of an application of this result, see the article by Vaz Jr (2013).

Example 6.12 Let us consider now the four-dimensional Minkowski spacetime with signature $(1, 3)$. This is a case where the dimension of the space of classical spinors is half of the dimensions of the space of algebraic spinors and spinor operators. Moreover, this case discussed in section 6.1; in it, there are two non-equivalent representations of the spacetime rotations in terms of 2×2 complex matrices, an idea which takes us to the concept of Weyl spinors. Since this is a very important subject, its discussion from the Clifford algebra point of view will be addressed later in sections 6.10 and 6.11. Let us now focus our attention on algebraic spinors and spinor operators.

Let us use a notation similar to that used in example 6.11, that is, $\Psi \in C\ell_{1,3} f$ (where f is a primitive idempotent); $|\Psi\rangle \in \mathbb{H}^2$; $\Psi_+ \in C\ell^+_{1,3}$; $\Psi_0 \in C\ell^0_{1,3}$; and $\boldsymbol{\Psi} \in \mathbb{C}^2$. The Clifford algebra $C\ell_{1,3}$ is generated by $\{\mathbf{e}_0, \mathbf{e}_1, \mathbf{e}_2, \mathbf{e}_3\}$ such that $(\mathbf{e}_0)^2 = -(\mathbf{e}_1)^2 = -(\mathbf{e}_2)^2 = -(\mathbf{e}_3)^2 = 1$. Let $\psi \in C\ell_{1,3}$ be given as

$$\psi = s + v_0\mathbf{e}_0 + v_1\mathbf{e}_1 + v_2\mathbf{e}_2 + v_3\mathbf{e}_3 + b_{01}\mathbf{e}_{01} + b_{02}\mathbf{e}_{02} + b_{03}\mathbf{e}_{03} + b_{12}\mathbf{e}_{12} + b_{13}\mathbf{e}_{13} + b_{23}\mathbf{e}_{23}$$
$$+ t_{012}\mathbf{e}_{012} + t_{013}\mathbf{e}_{013} + t_{023}\mathbf{e}_{023} + t_{123}\mathbf{e}_{123} + q_{0123}\mathbf{e}_{0123},$$

whose matrix representation using $f = (1/2)(1 + \mathbf{e}_0)$ is (see chapter 4, exercise 1 of chapter 4)

$$\rho(\psi) = \begin{pmatrix} \boldsymbol{p} & \boldsymbol{r} \\ \boldsymbol{q} & \boldsymbol{s} \end{pmatrix},$$

where

$$\begin{aligned} \boldsymbol{p} &= (s + v_0) + \boldsymbol{i}(b_{23} + t_{023}) + \boldsymbol{j}(-b_{13} - b_{013}) + \boldsymbol{k}(b_{12} + t_{012}), \\ \boldsymbol{q} &= (q_{0123} - t_{123}) + \boldsymbol{i}(v_1 - b_{01}) + \boldsymbol{j}(v_2 - b_{02}) + \boldsymbol{k}(v_3 - b_{03}), \\ \boldsymbol{r} &= (-q_{0123} - t_{123}) + \boldsymbol{i}(v_1 + b_{01}) + \boldsymbol{j}(v_2 + b_{02}) + \boldsymbol{k}(v_3 + b_{03}), \\ \boldsymbol{s} &= (s - v_0) + \boldsymbol{i}(b_{23} - t_{023}) + \boldsymbol{j}(t_{013} - b_{13}) + \boldsymbol{k}(b_{12} - t_{012}). \end{aligned}$$

The algebraic spinor $\Psi \in C\ell_{1,3} f$ is

$$\Psi = [(s + v_0) + (b_{12} + t_{012})\mathbf{e}_{12} + (b_{13} + b_{013})\mathbf{e}_{13} + (b_{23} + t_{023})\mathbf{e}_{23}]f$$
$$+ [(q_{0123} - t_{123}) + (v_3 - b_{03})\mathbf{e}_{12} + (-v_2 + b_{02})\mathbf{e}_{13} + (v_1 - b_{01})\mathbf{e}_{23}]\mathbf{e}_{0123}f,$$

and its matrix representation is

$$\rho(\Psi) = \begin{pmatrix} \boldsymbol{p} & 0 \\ \boldsymbol{q} & 0 \end{pmatrix} = \begin{pmatrix} p_0 + p_1\boldsymbol{i} + p_2\boldsymbol{j} + p_3\boldsymbol{k} & 0 \\ q_0 + q_1\boldsymbol{i} + q_2\boldsymbol{j} + q_3\boldsymbol{k} & 0 \end{pmatrix}.$$

The vector space isomorphism between $S^A_{1,3}$ and $\mathbb{H}^2$ is provided by identifying

$$\boldsymbol{i} \leftrightarrow \mathbf{e}_{23}, \quad \boldsymbol{j} \leftrightarrow \mathbf{e}_{31}, \quad \boldsymbol{k} \leftrightarrow \mathbf{e}_{12}, \quad |1\rangle \leftrightarrow f, \quad |2\rangle \leftrightarrow \mathbf{e}_0\mathbf{e}_1\mathbf{e}_2\mathbf{e}_3 f.$$

For the spinor operator Ψ_+, we have

$$\Psi_+ = s + b_{01}\mathbf{e}_{01} + b_{02}\mathbf{e}_{02} + b_{03}\mathbf{e}_{03} + b_{12}\mathbf{e}_{12} + b_{13}\mathbf{e}_{13} + b_{23}\mathbf{e}_{23} + q_{0123}\mathbf{e}_{0123}.$$

We conveniently rewrite the coefficients of Ψ_+ as

$$\Psi_+ = p_0 - q_1\mathbf{e}_{01} - q_2\mathbf{e}_{02} - q_3\mathbf{e}_{03} + p_3\mathbf{e}_{12} - p_2\mathbf{e}_{13} + p_1\mathbf{e}_{23} + q_0\mathbf{e}_{0123}.$$

Then, its matrix representation is

$$\rho(\Psi_+) = \begin{pmatrix} \boldsymbol{p} & -\boldsymbol{q} \\ \boldsymbol{q} & \boldsymbol{p} \end{pmatrix}.$$

Note that as an algebra we have $C\ell^+_{1,3} \simeq C\ell_{3,0} \simeq \mathcal{M}(2, \mathbb{C})$.

Besides the spinor operator Ψ_+, we can define a spinor operator Ψ_0 using the α grading automorphism defined by $\alpha(\psi) = \mathbf{e}_{123}\psi\mathbf{e}_{123}$. In this case, we use $\kappa = \mathbf{e}_0$. Then, Ψ_0 has the form

$$\Psi_0 = s + v_1\mathbf{e}_1 + v_2\mathbf{e}_2 + v_3\mathbf{e}_3 + b_{12}\mathbf{e}_{12} + b_{13}\mathbf{e}_{13} + b_{23}\mathbf{e}_{23} + t_{123}\mathbf{e}_{123}.$$

Note that, as an algebra, $C\ell^0_{1,3} \simeq C\ell_{0,3} \simeq \mathbb{H} \oplus \mathbb{H}$. If we conveniently rewrite the coefficients of Ψ_0 as

$$\Psi_0 = p_0 + q_1\mathbf{e}_1 + q_2\mathbf{e}_2 + q_3\mathbf{e}_3 + p_1\mathbf{e}_{12} - p_2\mathbf{e}_{13} + p_3\mathbf{e}_{23} - q_0\mathbf{e}_{123},$$

its matrix representation is written as

$$\rho(\Psi_0) = \begin{pmatrix} \boldsymbol{p} & \boldsymbol{q} \\ \boldsymbol{q} & \boldsymbol{p} \end{pmatrix}.$$

As in the three-dimensional Euclidean space, the use of spinor operators here has some significant computational advantages, as well as an amazing interpretation. In order to see this, we first note that

$$\Psi_+\widetilde{\Psi}_+ = (s^2 + b_{12}^2 + b_{13}^2 + b_{23}^2 - b_{01}^2 - b_{02}^2 - b_{03}^2 - q_{0123}^2)$$
$$+ (sq_{0123} - b_{01}b_{23} + b_{02}b_{13} - b_{03}b_{12})\mathbf{e}_{0123}.$$

We usually denote

$$\sigma = \langle \Psi_+ \widetilde{\Psi}_+ \rangle_0, \qquad \omega = \langle \Psi_+ \mathbf{e}_{0123} \widetilde{\Psi}_+ \rangle_0,$$

in such a way that

$$\Psi_+ \widetilde{\Psi}_+ = \sigma - \omega \mathbf{e}_{0123}.$$

If we define

$$\rho \cos \beta = \sigma, \qquad \rho \sin \beta = -\omega,$$

we have that Ψ_+ can be written as

$$\Psi_+ = \sqrt{\rho} e^{(\beta/2)\mathbf{e}_{0123}} R,$$

where $R \in \mathcal{C}\ell^+_{1,3}$, and $R\widetilde{R} = 1$. Thus, $R \in \mathrm{Spin}_+(1,3)$. We see that Ψ_+ has an interpretation very similar to that of the three-dimensional case, except for the term $e^{(\beta/2)\mathbf{e}_{0123}}$, which has no obvious interpretation. This term has no effect in an expression like $\Psi_+ \mathbf{e}_\mu \widetilde{\Psi}_+$ because $\mathbf{e}_{0123}$ anti-commutes with vectors, but it has in $\Psi_+ \mathbf{e}_\mu \mathbf{e}_\nu \widetilde{\Psi}_+$ $(\mu \neq \nu)$. In the case of a bivector, we can interpret the term $e^{\beta \mathbf{e}_{0123}}$ as a duality rotation (Rainich, 1925; Misner and Wheeler, 1957; Vaz Jr and Rodrigues Jr., 1993).

6.6 The Inner Product in the Space of Algebraic Spinors

Let us consider in this section the definition of the inner product in the space of spinors according to the algebraic definition of a spinor. This case deserves particular attention, since a classical spinor in a quadratic space that is $\mathbb{R}^{p,q}$ or $\mathbb{R}^{q,p}$ is an algebraic spinor (or an algebraic semispinor) in a quadratic space that is either $\mathbb{R}^{q,p-1}$ or $\mathbb{R}^{p,q-1}$, respectively. Furthermore, the space of spinor operators is isomorphic as vector space to the space of algebraic spinors or semispinors.

The Spinor Structure Map

An $\mathbb{R}$-linear mapping ς in the space of algebraic spinors S is said to be a *spinor structure mapping* if $\varsigma^2 = \pm 1$ and if $\hat{a} = \varsigma a \varsigma^{-1}$, for all $a \in \mathcal{C}\ell_{p,q} \subset \mathrm{End}_{\mathbb{K}}(S)$.

If $n = p + q$ is even, the n-vector η is such that $\mathbf{v}\eta = -\eta\mathbf{v}$, for all $\mathbf{v} \in \bigwedge_1(\mathbb{R}^{p,q})$, in such a way that η commutes with the even elements and anti-commutes with the odd elements. Thus, for n even, this spinor structure map exists and is given by the multiplication by η, in such a way that $\eta a \eta^{-1} = \hat{a}$.

If $n = p + q$ is odd, then the n-vector η commutes with all elements of $\mathcal{C}\ell_{p,q}$. Any odd element of $\mathcal{C}\ell_{p,q}$ can be written in the form of a product of an even element and η, namely, $\mathcal{C}\ell^-_{p,q} = \mathcal{C}\ell^+_{p,q}\,\eta$, and we can write $\mathcal{C}\ell_{p,q} = \mathcal{C}\ell^+_{p,q} \oplus \mathcal{C}\ell^+_{p,q}\,\eta$. Let us now distinguish two cases. In the first case, $p - q = 3, 7 \mod 8$. In this case, we have $\eta^2 = -1$, and η defines a complex structure. We also have $\eta S = iS$, where S denotes the space of algebraic spinors. The spinor structure map corresponds to the complex conjugation in S. In the second case, $p - q = 1, 5 \mod 8$. In this case, $\eta^2 = 1$, and $\eta S = IS$, where $I \in \mathbb{D}$, and $I^2 = 1$. The conjugation in $\mathbb{D}$ (like the conjugation in $\mathbb{C}$) is defined by $I^* = -I$. The application of spinor structure corresponds to the conjugation in S accordingly.

The Two Types of Inner Products in Spinor Space

The spinor inner product is an inner product in the – algebraic – spinor space, with the property that the adjoint with respect to this inner product corresponds to an anti-automorphism in the corresponding Clifford algebra. Since we have two types of anti-automorphisms, namely, the reversion and the conjugation, we have consequently two types of spinor inner products. Let S be the space of algebraic spinors. We can define the spinor inner products $\tilde{h} : S \times S \to \mathbb{K}$, and $\bar{h} : S \times S \to \mathbb{K}$. Given $x, y \in S$, and $a \in \mathcal{C}\ell_{p,q}$, in such a way that $\mathcal{C}\ell_{p,q} \simeq \mathrm{End}_{\mathbb{K}}(S)$, define $\tilde{h}$ as the spinor inner product with the property that

$$\tilde{h}(ax, y) = \tilde{h}(x, \tilde{a}y). \tag{6.56}$$

In this case, the adjoint corresponds to the reversion. The spinor inner product $\bar{h}$ is defined in such a way that

$$\bar{h}(ax, y) = \bar{h}(x, \bar{a}y). \tag{6.57}$$

In this case, the adjoint corresponds to the conjugation.

The conjugation corresponds to the composition between the reversion and the grade involution. As shown in the previous subsection, there always exists a spinor structure map in S – an $\mathbb{R}$-linear mapping ς in the space of algebraic spinors – such that the grade involution reads $\hat{a} = \varsigma a \varsigma^{-1}$, $\forall a \in \mathcal{C}\ell_{p,q}$. Consequently, one of these inner products determines the other, and vice versa, by the expression

$$\bar{h}(x, y) = \tilde{h}(\varsigma x, y), \tag{6.58}$$

where ς denotes the application of spinor structure. Indeed, $\bar{h}(ax, y) = \tilde{h}(\varsigma ax, y) = \tilde{h}(\hat{a}\varsigma x, y) = \tilde{h}(\varsigma x, \bar{a}y) = \bar{h}(x, \bar{a}y)$. Let us denote by $^\circ$ any one of the anti-automorphisms $\tilde{}$ or $\bar{}$, that is, let ψ° denote $\tilde{\psi}$ or $\bar{\psi}$ for $\psi \in \mathcal{C}\ell_{p,q}$. As seen in chapter 4, $\mathcal{C}\ell_{p,q}$ is a simple algebra; therefore, $f\,\mathcal{C}\ell_{p,q}\,f \simeq \mathbb{K}$, where $\mathbb{K} = \mathbb{R}, \mathbb{C}, \mathbb{H}$; and, if $\mathcal{C}\ell_{p,q}$ is a semisimple algebra, then $f\,\mathcal{C}\ell_{p,q}\,f \simeq \mathbb{K} \oplus \mathbb{K}$. For $\psi, \phi \in S$, we have $\psi f = \psi$, and $\phi f = \phi$; and for $\psi^\circ \phi$, it follows that

$$\psi^\circ \phi = (\psi f)^\circ \phi f = f^\circ \psi^\circ \phi f. \tag{6.59}$$

Now, notice that the quantity $h(\psi, \phi) = \psi^\circ \phi$ satisfies $h(a\psi, \phi) = h(\psi, a^\circ \phi)$. Therefore, if $f^\circ = f$, then $\psi^\circ \phi$ is a spinor scalar product. If $f^\circ \neq f$, we can take an element $s \in \mathcal{C}\ell^*_{p,q}$ such that $s f^\circ s^{-1} = f$ and then $s\psi^\circ \phi$ is a spinor scalar product.

In this way, the applications $\tilde{h}$ and $\bar{h}$, defined as

$$\begin{aligned} \tilde{h}(\psi, \phi) &= s\tilde{\psi}\phi, \\ \bar{h}(\psi, \phi) &= s\bar{\psi}\phi, \end{aligned} \tag{6.60}$$

with $s \in \mathcal{C}\ell^*_{p,q}$, are spinor scalar products. The existence of either an element $s \in \mathcal{C}\ell_{p,q}$ satisfying $s\tilde{f}fs^{-1} = f$ or $s \in \mathcal{C}\ell_{p,q}$ satisfying $s\bar{f}fs^{-1} = f$ can be explicitly verified for low dimensions and induced for high dimensions via the periodicity theorem.

Example 6.13 Let us consider as an example $\mathcal{C}\ell_{1,3}$, with $(\mathrm{e}_0)^2 = 1$; $(\mathrm{e}_i)^2 = -1$ $(i = 1, 2, 3)$; and $\mathrm{e}_\mu \mathrm{e}_\nu = -\mathrm{e}_\nu \mathrm{e}_\mu$ $(\mu \neq \nu)$. Let us take algebraic spinors as elements of the minimal left ideal $S = \mathcal{C}\ell_{1,3}\,f$, where $f = (1/2)(1 + \mathrm{e}_0)$. Let us denote $\xi_1 = f$, and $\xi_2 = \mathrm{e}_5 f$, where $\mathrm{e}_5 = \mathrm{e}_0\mathrm{e}_1\mathrm{e}_2\mathrm{e}_3$. The left ideal S has a basis

$$\{\xi_1, \mathrm{e}_2\mathrm{e}_3\xi_1, \mathrm{e}_3\mathrm{e}_1\xi_1, \mathrm{e}_1\mathrm{e}_2\xi_1, \xi_2, \mathrm{e}_2\mathrm{e}_3\xi_2, \mathrm{e}_3\mathrm{e}_1\xi_2, \mathrm{e}_1\mathrm{e}_2\xi_2\}.$$

The set $f\,\mathcal{C}\ell_{1,3}\,f$ is isomorphic to $\mathbb{H}$ and has a basis $\{\xi_1, \mathrm{e}_2\mathrm{e}_3\xi_1, \mathrm{e}_3\mathrm{e}_1\xi_1, \mathrm{e}_1\mathrm{e}_2\xi_1\}$. The ideal S is thus a right $\mathbb{H}$-module with the basis $\{\xi_1, \xi_2\}$. Now we can see that

$$\begin{aligned} \tilde{\xi}_1\xi_1 &= \xi_1, & \tilde{\xi}_1\xi_2 &= 0, \\ \tilde{\xi}_2\xi_1 &= 0, & \tilde{\xi}_2\xi_2 &= \xi_1, \end{aligned} \quad \text{and} \quad \begin{aligned} \bar{\xi}_1\xi_1 &= 0, & \bar{\xi}_1\xi_2 &= \xi_2, \\ \bar{\xi}_2\xi_1 &= \xi_2, & \bar{\xi}_2\xi_2 &= 0. \end{aligned}$$

The spinor scalar product $\tilde{h}$ can be defined by

$$\tilde{h}(\psi, \phi) = \tilde{\psi}\phi,$$

and the spinor scalar product $\bar{h}$ can be defined as

$$\bar{h}(\psi, \phi) = \mathbf{e}_5\bar{\psi}\phi.$$

The Spinor Inner Product, and Charge Conjugation

As the two types of spinor inner products have now been examined, let us now focus on the complex Clifford algebras in this context. Given a spinor ψ in $\mathbb{C}\otimes\mathcal{C}\ell_{p,q}$, the associated adjoint spinor ψ° is represented by

$$\psi^\circ(\phi) = h^\circ(\psi, \phi), \tag{6.61}$$

where ψ is an arbitrary spinor. For all $a \in \mathbb{C}\otimes\mathcal{C}\ell_{p,q}$, we can verify that $h^\circ(a\psi, \phi) = h^\circ(\psi, a^\circ\phi)$, by the equations in (6.60). Except in the cases where p is odd, and q is even, the mapping *, which is the composition of the Clifford conjugation and the complex conjugation – seen as an involutive automorphism which induces a non-trivial automorphism in the centre of the algebra – is an adjoint involution associated with some Hermitian spinor product. When p is even, and q is odd, the Clifford conjugation is substituted by the reversion, in the previous case.

Let us first consider the case where p is odd, and q is even. For a choice of basis for the Clifford algebra representations, consider the Hermitian conjugation †, which is related to * by

$$\bar{a}^* = s^{-1}a^\dagger s, \qquad \forall a \in \mathbb{C}\otimes\mathcal{C}\ell_{p,q}, \tag{6.62}$$

where $\bar{s}^* = s \in \mathbb{C}\otimes\mathcal{C}\ell_{p,q}$ – or, equivalently, $s^\dagger = s$. As a particular case, given spinors ψ and ϕ, we define the spin-invariant product

$$h_*(\phi, \psi) = s\phi^*\psi,$$

which is invariant by elements of the Clifford–Lipschitz group and which takes values in the complex numbers algebra, whose identity is $f\mathcal{C}\ell_{p,q}f \simeq \mathbb{C}$. It follows that we can obtain a product which is a complex number and is given by $\langle\phi, \psi\rangle = \operatorname{Tr} h_*(\phi, \psi)$. The adjoint of ψ with respect to the product in eqn (6.62) is given by

$$\psi^\Delta = s\psi^* = \psi^\dagger s\,. \tag{6.63}$$

If a basis for the representation of $\mathbb{C}\otimes\mathcal{C}\ell_{p,1}$ is chosen such that $(\mathbf{e}^0)^\dagger = -\mathbf{e}^0$, and $(\mathbf{e}^i)^\dagger = \mathbf{e}^i$, $i = 1, \ldots, p$, then eqn (6.62) is equivalent to $\mathbf{e}^{a\dagger} = -s\mathbf{e}^a s^{-1}$. Hence, we can choose $s = i\mathbf{e}^0$, which implies the well-known relation

$$\psi^\Delta = i\psi^\dagger\mathbf{e}^0\,.^2 \tag{6.64}$$

For the case $\mathbb{C}\otimes\mathcal{C}\ell_{1,q}$, the factor i is absent from eqn (6.64).

[2]Here, we denote the adjoint spinor by ψ^Δ, unlike the case in quantum mechanics textbooks where, in general, the authors adopt the notation $\bar{\psi}$ for the case where ψ is a Dirac spinor. We use this notation in order to reduce the possibility of confusing the adjoint spinor with the Clifford conjugation.

When p is even, and q is odd, the Clifford conjugation is substituted by the reversion, and, therefore

$$\tilde{a}^* = s^{-1}a^\dagger s, \qquad \forall a \in \mathbb{C}\otimes \mathcal{C}\ell_{p,q}, \tag{6.65}$$

where $\tilde{s}^* = s^\dagger = s$. The adjoint of ψ with respect to the product in eqn (6.65) is defined by

$$\psi^\Delta = s\tilde{\psi}^* = \psi^\dagger s\,. \tag{6.66}$$

The adjoint spinor in (6.61) is associated with a pseudo-Hermitian product for which $\bar{\psi}^*$ or $\tilde{\psi}^*$ are the adjoint spinors. There also exists spin-invariant products for which the involutions $\bar{\psi}$ or $\tilde{\psi}$ are the adjoint spinors. Except in the case when $n = 1$ mod 8 or $n = 5$ mod 8, the involution $(\bar{\cdot})$ induces an involution in the reducible Clifford algebra simple components. If $(\cdot)^\mathsf{T}$ represents the transposition in some matrix basis, then, except in those cases, it follows that

$$\bar{a} = b^{-1}a^\mathsf{T} b, \qquad \forall a \in \mathbb{C}\otimes \mathcal{C}\ell_{p,q}\,, \tag{6.67}$$

where $\bar{b} = b^\mathsf{T} = \pm b \in \mathbb{C}\otimes \mathcal{C}\ell_{p,q}$. The symmetry of b determines the symmetry of a complex bilinear product defined by

$$\bar{h}(\phi, \psi) = b\bar{\phi}\psi. \tag{6.68}$$

Equation (6.67) is thus equivalent to

$$\mathbf{e}^{a\,\mathsf{T}} = -b\mathbf{e}^a b^{-1}\,, \tag{6.69}$$

whose matrix entries are usually taken as the charge conjugation definition. If ψ^Δ is the adjoint of ψ with respect to the product (6.68), then

$$\psi^\Delta = b\bar{\psi} = \psi^\mathsf{T} b\,. \tag{6.70}$$

This kind of adjoint spinor is called a Majorana spinor. Except when $n = 3$ or 7 mod 8, the reversion induces an involution in the reducible Clifford algebra simple components. We can define

$$\tilde{a} = b^{-1}a^\mathsf{T} b, \qquad \forall\, a \in \mathbb{C}\otimes \mathcal{C}\ell_{p,q}\,, \tag{6.71}$$

where $\tilde{b} = b^\mathsf{T} = \pm b$, and therefore $\mathbf{e}^{a\,\mathsf{T}} = b\mathbf{e}^a b^{-1}$.

The automorphism * regarding the complex conjugation leaves invariant the real subalgebra generated by the real orthogonal space of signature (p, q). Hence, the automorphism * is defined as the complex conjugate of the matrix entries in some basis. The automorphism * depends on the choice of basis. Except when the real subalgebra is isomorphic to the complex matrix algebra for some $m \in \mathbb{C}\otimes \mathcal{C}\ell_{p,q}$, these two automorphisms are related by

$$a^* = m a^\star m^{-1}, \qquad \forall a \in \mathbb{C}\otimes \mathcal{C}\ell_{p,q}\,. \tag{6.72}$$

When the real subalgebra is represented by the algebra of real matrices, we can choose m such that $mm^* = 1$. When the real subalgebra is either the tensor product between a matrix algebra and the quaternions, or the sum of two of these algebras, we can choose $mm^* = -1$. The real subalgebra is isomorphic to the complex matrix algebra when $p - q = 3$ or $7 \mod 8$. In this case, the complex conjugation transposes the simple components and, except in this case, we define the charge conjugate spinor

$$\psi^{\mathsf{c}} = \psi^* m\,. \tag{6.73}$$

This expression can be written by using the Dirac adjoint and the charge conjugation matrix as well. Except in the case when $n = 1$ or $5 \mod 8$, or when p is odd, and q is even, we can use eqns (6.65) and (6.67):

$$\bar{a}^* = \bar{s}b^{-1}a^{\dagger\mathsf{T}}b\bar{s}^{-1}\,. \tag{6.74}$$

Now, the complex conjugation operator commutes with the Clifford conjugation operator, and ${}^{\dagger\mathsf{T}} = {}^{\mathsf{T}\dagger} = {}^\star$, yielding

$$a^* = ma^\star m^{-1}, \qquad\qquad m = s^{-1\,*}b^{-1}\,, \tag{6.75}$$

where we used the fact that $\bar{s}^* = s$. We can choose m in such a way that $mm^* = \pm 1$, which is immediately obtainable from a suitable choice of C. Using eqn (6.73), eqn (6.75) can be expressed as

$$\psi^{\mathsf{c}} = b^{-1}\bar{\psi}^{\mathsf{T}}\,. \tag{6.76}$$

In the same way, for the case where $n = 3$ or $7 \mod 8$, or when p is odd, and q is odd, eqn (6.73) reads

$$\psi^{\mathsf{c}} = b^{-1}\bar{\psi}^{\mathsf{T}}, \tag{6.77}$$

where $\bar{\psi}$ is given by eqn (6.66).

In the particular case concerning Dirac spinors, since they carry a reducible representation of the real Clifford subalgebra, elements of the spaces that carry the irreducible representation are called Majorana spinors. This case is manifested either when the real subalgebra is a real matrix algebra or the direct sum of two such algebras, namely, when $p - q = 0, 1, 2 \mod 8$. In this signature, Dirac spinors can be split into eigenspaces of the charge conjugation operator ${}^{\mathsf{c}}$. Hence, a Majorana spinor is an eigenspinor of the charge conjugation operator:

$$\psi^{\mathsf{c}} = \pm\psi\,. \tag{6.78}$$

It is worth emphasising that, for the complex case, eqn (6.78) reads $\psi^{\mathsf{c}} = e^{i\theta}\psi$, where $\theta \in \mathbb{R}$. When the dimension of the underlying vector space is even, the irreducible representations of the complex Clifford algebra induce a reducible representation of the even subalgebra, where the spinor representation is split into two inequivalent semispinor representations of the even subalgebra. The central idempotents that cause the even subalgebra to be split into simple components are given by $P_\pm = \frac{1}{2}(1 + \mathring{\eta})$, where either $\mathring{\eta} = \eta$, or $\mathring{\eta} = i\eta$, ensuring that $\mathring{\eta}^2 = 1$, where η denotes the n-volume element. A Dirac spinor can thus be split into subspaces which transform as irreducible representations under the even subalgebra:

$$\psi = \psi_+ + \psi_-, \qquad\qquad \text{where} \quad \psi_\pm = P_\pm\psi\,, \tag{6.79}$$

where $\psi_\pm$ are called chiral spinors, or Weyl spinors.

When the dimension of the vector space is odd, the irreducible representation of the complexified Clifford algebra induces irreducible representations of the even subalgebra and can also induce a reducible representation of the real even subalgebra. This is the case where $p - q = 1 \mod 8$, where Dirac spinors carry a reducible representation of the real subalgebra. For the case where $p - q = 7 \mod 8$, the Dirac spinors carry irreducible representations of the real subalgebra and the even subalgebra, but such spinors carry a reducible representation of the real even subalgebra. For $p-q = 7 \mod 8$, we know that

$$\mathcal{C}\ell_{p,q} \simeq \mathbb{C} \otimes \mathcal{M}(2^{(n-1)/2}, \mathbb{R}), \qquad \mathcal{C}\ell^+_{p,q} \simeq \mathcal{M}(2^{(n-1)/2}, \mathbb{R})\,. \tag{6.80}$$

Hence, we can choose a basis for the Clifford algebra in such a way that the grade involution plays the role of complex conjugation of the components. The complexified Clifford algebra $\mathcal{C}\ell_{p,q}(\mathbb{C})$ is reducible and, in this case, the grade involution interchanges the simple components. The complex conjugation also interchanges the components of the algebras, and the automorphism $\widehat{(\,\cdot\,)}^*$ preserves the simple components.

6.7 The Triality Principle in the Clifford Algebraic Context

In this section, the triality principle is introduced in the Clifford algebraic context; the geometric point of view is further explored in the work by Benn and Tucker (1987) and Knus (1998). For other geometric and topological approaches, see, for instance, the work by Porteous (1969) and Harvey (1990).

Let us consider a field $\mathbb{K}$ of characteristic that does not equal 2. Suppose that the spinor space S can be expressed as the direct sum $S = S^+ \oplus S^-$, where $S^\pm$ are semispinor spaces – carrying non-equivalent irreducible representations of the even subalgebra $\mathcal{C}\ell^+(V, g)$. The aim here is to search for a vector space V such that the associated semispinor spaces $S^\pm$ have the same dimension of V. When $\mathbb{K} = \mathbb{R}$ or $\mathbb{C}$, it must happen when dim V = dim S^+ = dim S^- = n and also when the index of g equals the index of h, where h denotes the spinor metric associated with the reversion – when $\mathbb{K} = \mathbb{C}$. The first condition holds when $2^{n/2-1} = n$, namely, when $n = 8$. When $\mathbb{K} = \mathbb{R}$, the spinor metric h associated with the reversion exists only in one of the following three cases: $V \simeq \mathbb{R}^{8,0}$; $\mathbb{R}^{0,8}$; or $\mathbb{R}^{4,4}$, although this last case demands a careful approach, since the metric index is not maximal in this case. For more details, see the work by Benn and Tucker (1987) and de Andrade, Rojas, and Toppan (2001). Hence, in what follows, we consider $V \simeq \mathbb{C}^8$. Moreover, in the real case we have $V \simeq \mathbb{R}^{8,0}$; $\mathbb{R}^{0,8}$; or $\mathbb{R}^{4,4}$.

Let us define the space $\mathrm{E} = V \oplus S^+ \oplus S^-$, which is 24-dimensional. Its elements read $\phi = \mathbf{x} + \mathfrak{u} + \mathfrak{v}$, where in this section we fix the notation, $\mathbf{x} \in V$; $\mathfrak{u} \in S^+$; and $\mathfrak{v} \in S^-$. We first introduce a result that will be used throughout this section.

Lemma 6.1: *Let us consider $x \in V$. Then for all $\mathfrak{v} \in S^-$, there exists $\mathfrak{u} \in S^+$ such that $\mathfrak{v} = \mathbf{x}\mathfrak{u}$.*

Proof: Indeed, consider the volume element $\eta \in \mathcal{C}\ell^+(V, g)$. Since S^+ and S^- are spaces that carry the irreducible representations of $\mathcal{C}\ell^+(V, g)$, and dim $V = 8$, then $S = S^+ \oplus S^-$ carries an irreducible representation of $\mathcal{C}\ell(V, g)$. In this case, it follows that $S^\pm = \frac{1}{2}(1\pm\eta)S^\pm$ since, in the matrix algebra, every element of the even subalgebra $\mathcal{C}\ell^+(V,g)$ can be represented by a block diagonal matrix $\left(\begin{smallmatrix} * & \mathbf{0} \\ \mathbf{0} & * \end{smallmatrix}\right)$ such that η can be represented by $\eta = \left(\begin{smallmatrix} \mathbf{I} & \mathbf{0} \\ \mathbf{0} & -\mathbf{I} \end{smallmatrix}\right)$. For any vector $\mathbf{x} \in V$, it follows that $\eta\mathbf{x} = -\mathbf{x}\eta$. Hence, since $\mathfrak{u} \in S^+$, therefore $\mathbf{x}\mathfrak{u} = \mathbf{x}\frac{1}{2}(1+\eta)\mathfrak{u} = \frac{1}{2}(1-\eta)\mathbf{x}\mathfrak{u}$, and we conclude that $\mathbf{x}\mathfrak{u} \in S^-$. Similarly, given $\mathfrak{v} \in S^-$, we can prove that $\mathbf{x}\mathfrak{v} \in S^+$. ✓

The spinor metric $h : S^\pm \times S^\pm \to \mathbb{K}$ is defined in each one of these spaces $S^\pm$ as

$$h(\mathfrak{u}, \mathbf{x}\mathfrak{v}) = h(\tilde{\mathbf{x}}\mathfrak{u}, \mathfrak{v}) = \widehat{\mathbf{x}\mathfrak{u}}\mathfrak{v}\,, \tag{6.81}$$

where the dual adjoint is given by

$$\begin{aligned} (\widehat{\cdot}) : S^\pm &\longrightarrow (S^\pm)^* = \mathrm{Hom}(S^\pm, \mathbb{K})\,, \\ \psi &\mapsto \widehat{\psi}, \qquad \text{where } \widehat{\psi}(\phi) = h(\phi, \psi),\ \forall \phi \in S^\pm. \end{aligned}$$

From this definition, a symmetric bilinear form $\mathcal{B}$ endowing the vector space $\mathbf{E}$ can be introduced, from the spinor metric h and also from the metric g that endows the space V. Let $\mathbf{x}_i \in V$; $\mathfrak{u}_i \in S^+$; $\mathfrak{v}_i \in S^-$, $i = 1, 2, 3$. Then, we can define (Cartan, 1937; Chevalley, 1954; Benn and Tucker, 1987)

$$\mathcal{B}(\phi_1, \phi_2) = g(\mathbf{x}_1, \mathbf{x}_2) + h(\mathfrak{u}_1, \mathfrak{u}_2) + h(\mathfrak{v}_1, \mathfrak{v}_2)\,. \tag{6.82}$$

Now, a totally symmetric trilinear tensor $T : \mathrm{E} \times \mathrm{E} \times \mathrm{E} \to \mathbb{K}$ can be defined as

$$\begin{aligned} T(\phi_1, \phi_2, \phi_3) = {} & h(\mathfrak{u}_1, \mathbf{x}_2\mathfrak{v}_3) + h(\mathfrak{u}_1, \mathbf{x}_3\mathfrak{v}_2) + h(\mathfrak{u}_2, \mathbf{x}_3\mathfrak{v}_1) \\ & + h(\mathfrak{u}_2, \mathbf{x}_1\mathfrak{v}_3) + h(\mathfrak{u}_3, \mathbf{x}_2\mathfrak{v}_1) + h(\mathfrak{u}_3, \mathbf{x}_1\mathfrak{v}_2). \end{aligned} \tag{6.83}$$

The Chevalley Product

It is possible to endow the space E with a commutative and non-associative product $\circ : \mathrm{E} \times \mathrm{E} \to \mathrm{E}$, called the Chevalley product and implicitly defined as

$$T(\phi_1, \phi_2, \phi_3) = \mathcal{B}(\phi_1 \circ \phi_2, \phi_3)\,. \tag{6.84}$$

From eqn (6.83) – defining the total symmetry property of T – we conclude that the product $\circ$ is indeed commutative ($\phi_1 \circ \phi_2 = \phi_2 \circ \phi_1$). The pair $(\mathrm{E}, \circ)$ is a commutative and non-alternative algebra – in particular, the algebra $(\mathrm{E}, \circ)$ is not associative. In order to prove this last property, we shall show that $\mathbf{x} \circ (\mathbf{x} \circ \mathfrak{u}) \neq (\mathbf{x} \circ \mathbf{x}) \circ \mathfrak{u}$, for all non-isotropic vectors $\mathbf{x} \in V$ and for all $\mathfrak{u} \in S^+$. In fact, $\mathbf{x} \circ (\mathbf{x} \circ \mathfrak{u}) = \mathbf{x} \circ (\mathbf{x}\mathfrak{u}) = \mathbf{x}(\mathbf{x}\mathfrak{u})$, since $\mathbf{x}\mathfrak{u} \in S^-$. Then it yields $\mathbf{x} \circ (\mathbf{x} \circ \mathfrak{u}) = \mathbf{x}^2\mathfrak{u} = g(\mathbf{x}, \mathbf{x})\mathfrak{u}$, whereas $(\mathbf{x} \circ \mathbf{x}) \circ \mathfrak{u} = 0$.

It is useful to observe that, if ϕ_1 and ϕ_2 are elements in the same subspace, then $T(\phi_1, \phi_2, \phi_3) = 0$ and thus $\phi_1 \circ \phi_2 = 0$. Moreover, for all $\mathbf{x} \in V$, for all $\mathfrak{u} \in S^+$, and for all $\mathfrak{v} \in S^-$ we can assert that

$$\mathbf{x} \circ \mathfrak{u} = \mathbf{x}\mathfrak{u}, \qquad \mathbf{x} \circ \mathfrak{v} = \mathbf{x}\mathfrak{v}. \tag{6.85}$$

Indeed, it forthwith follows for the definitions (6.82) and (6.83) that

$$\begin{aligned} \mathcal{B}(\mathbf{x} \circ \mathfrak{u}, \mathfrak{v}) &= T(\mathbf{x}, \mathfrak{u}, \mathfrak{v}) = h(\mathfrak{u}, \mathbf{x}\mathfrak{v}) = h(\mathbf{x}\mathfrak{u}, \mathfrak{v}) = \mathcal{B}(\mathbf{x}\mathfrak{u}, \mathfrak{v})\,, \\ \mathcal{B}(\mathbf{x} \circ \mathfrak{v}, \mathfrak{u}) &= T(\mathbf{x}, \mathfrak{v}, \mathfrak{u}) = h(\mathfrak{v}, \mathbf{x}\mathfrak{u}) = h(\mathbf{x}\mathfrak{v}, \mathfrak{u}) = \mathcal{B}(\mathbf{x}\mathfrak{v}, \mathfrak{u})\,. \end{aligned}$$

The spinor norm $\mathbf{x} \circ \mathfrak{u}$ can be derived from the norms of $\mathbf{x}$ and $\mathfrak{u}$, since $h(\mathbf{x} \circ \mathfrak{u}, \mathbf{x} \circ \mathfrak{u}) = h(\mathbf{x}\mathfrak{u}, \mathbf{x}\mathfrak{u}) = g(\mathbf{x}, \mathbf{x})h(\mathfrak{u}, \mathfrak{u})$. Since the Chevalley product between vectors and semispinors has been defined from the action of the regular representation of $S^\pm$ the equations in (6.85), it is useful to express the Chevalley product between semispinors from the Clifford product between spinors, by the relations

$$\begin{aligned}\mathcal{B}(\mathfrak{u}\circ\mathfrak{v},\mathbf{x}) &= T(\mathfrak{u},\mathfrak{v},\mathbf{x}) = h(\mathbf{x}\mathfrak{u},\mathfrak{v}) = (\widehat{\mathbf{x}\mathfrak{u}})\mathfrak{v} = \langle\widehat{\mathfrak{u}}\mathbf{x}\mathfrak{v}\rangle_0 \\ &= \langle\mathbf{x}\mathfrak{v}\widehat{\mathfrak{u}}\rangle_0 = \langle(\mathbf{x}\langle\mathfrak{v}\widehat{\mathfrak{u}}\rangle_1)\rangle_0 = \mathcal{B}(\mathbf{x},\langle\mathfrak{v}\widehat{\mathfrak{u}}\rangle_1)\,,\end{aligned} \tag{6.86}$$

which implies that

$$\mathfrak{u}\circ\mathfrak{v} = \langle\mathfrak{v}\widehat{\mathfrak{u}}\rangle_1\,. \tag{6.87}$$

A prominent result that introduces the triality principle is that the inclusions

$$V\circ S^+\subseteq S^-,\qquad S^+\circ S^-\subseteq V,\qquad S^-\circ V\subseteq S^+ \tag{6.88}$$

hold. Indeed, if $\phi_1\in V$, $\phi_2\in S^+$, and $\phi_3\in V\oplus S^+$, we denote $\phi_1=\mathbf{x}_1$; $\phi_2=\mathfrak{u}_2$; and $\phi_3=\mathbf{x}_3+\mathfrak{u}_3$. Hence, $T(\phi_1,\phi_2,\phi_3)=T(\mathbf{x}_1,\mathfrak{u}_2,\phi_3)=0=\mathcal{B}(\phi_1\circ\phi_2,\phi_3)$, namely, $\phi_1\circ\phi_2$ is an element in the space orthogonal to $V\oplus S^+$ with respect to the metric $\mathcal{B}$, since $\mathrm{E}=S^-\oplus(S^-)^\perp$ – here we suppose that ker $h=\{0\}$. A similar reasoning holds for the other cases.

A spinor representation σ of the Clifford-Lipschitz group in $S^\pm$, and the vector representation χ of the Clifford-Lipschitz group in V, induce a representation Y in E. Given a unitary element of the Clifford-Lipschitz group $s\in\Gamma^+_{p,q}$, the action of such representation in $Y(s):\mathrm{E}\to\mathrm{E}$ is defined as:

$$Y(s)(\mathbf{x}+\mathfrak{u}+\mathfrak{v}) = \chi(s)\mathbf{x}+\sigma(s)\mathfrak{u}+\sigma(s)\mathfrak{v} = s\mathbf{x}s^{-1}+s\mathfrak{u}+s\mathfrak{v}, \tag{6.89}$$

where such a mapping is orthogonal with respect to the bilinear form $\mathcal{B}$, since

$$\begin{aligned}&\mathcal{B}(Y(s)\phi_1,Y(s)\phi_2)\\ &\quad= \mathcal{B}(\chi(s)\mathbf{x}_1+\sigma(s)\mathfrak{u}_1+\sigma(s)\mathfrak{v}_1,\chi(s)\mathbf{x}_2+\sigma(s)\mathfrak{u}_2+\sigma(s)\mathfrak{v}_2)\\ &\quad= g(\chi(s)\mathbf{x}_1,\chi(s)\mathbf{x}_2)+h(\sigma(s)\mathfrak{u}_1,\sigma(s)\mathfrak{u}_2)+h(\sigma(s)\mathfrak{v}_1,\sigma(s)\mathfrak{v}_2)\\ &\quad= g(s\mathbf{x}_1s^{-1},s\mathbf{x}_2s^{-1})+h(s\mathfrak{u}_1,s\mathfrak{u}_2)+h(s\mathfrak{v}_1,s\mathfrak{v}_2)\\ &\quad= g(\mathbf{x}_1,\mathbf{x}_2)+h(\mathfrak{u}_1,\mathfrak{u}_2)+h(\mathfrak{v}_1,\mathfrak{v}_2)\\ &\quad= \mathcal{B}(\phi_1,\phi_2)\,.\end{aligned}$$

In addition, if $\mathbf{x}_0\in V$ is such that $\mathbf{x}_0^2=1$, then $Y^2(\mathbf{x}_0)=I$. Obviously, $\mathbf{x}_0\in\Gamma_{p,q}$, in such a way that the notation $Y(\mathbf{x}_0)$ does indeed make sense. Since $\mathbf{x}_0^2=1$, therefore $\mathbf{x}_0^{-1}=\mathbf{x}_0$, yielding

$$\begin{aligned}Y^2(\mathbf{x}_0)(\mathbf{x}+\mathfrak{u}+\mathfrak{v}) &= Y(\mathbf{x}_0)(\mathbf{x}_0\mathbf{x}\mathbf{x}_0+\mathbf{x}_0\mathfrak{u}+\mathbf{x}_0\mathfrak{v})\\ &= \mathbf{x}_0^2\mathbf{x}\mathbf{x}_0^2+\mathbf{x}_0^2\mathfrak{u}+\mathbf{x}_0^2\mathfrak{v} = \mathbf{x}+\mathfrak{u}+\mathfrak{v}.\end{aligned}$$

For the tensor T to be invariant under $Y(s)$, it is necessary that $Y(s)$ be an automorphism with respect to the Chevalley product. In fact,

$$\begin{aligned} T(Y(s)\phi_1, Y(s)\phi_2, Y(s)\phi_3) &= \mathcal{B}([Y(s)\phi_1] \circ [Y(s)\phi_2], Y(s)\phi_3) \\ &= \mathcal{B}(Y(s)[\phi_1 \circ \phi_2], Y(s)\phi_3) = \mathcal{B}(\phi_1 \circ \phi_2, \phi_3) \\ &= T(\phi_1, \phi_2, \phi_3). \end{aligned} \tag{6.90}$$

The Triality Principle

In this section, the triality principle is introduced, in order that the algebraic character of this approach can be made explicit. Consider a spinor $\mathfrak{u}_0 \in S^+$ of unitary norm with respect to the spinor metric, namely, $h(\mathfrak{u}_0, \mathfrak{u}_0) = 1$. Define the linear mapping

$$\begin{aligned} \zeta : S^+ &\to \mathcal{L}(V, S^-) \\ \mathfrak{u}_0 &\mapsto \zeta(\mathfrak{u}_0) : V \to S^- \\ &\qquad \mathbf{x} \mapsto \zeta(\mathfrak{u}_0)(\mathbf{x}) := \mathbf{x} \circ \mathfrak{u}_0 \end{aligned} \tag{6.91}$$

where $\mathcal{L}(V, W)$ is the space of linear functions from the space V to the space W. with the consequent property that $\zeta(\mathfrak{u}_0)$ is orthogonal with respect to $\mathcal{B}$, which can be straightforwardly verified:

$$\begin{aligned} h(\zeta(\mathfrak{u}_0)\mathbf{x}_1, \zeta(\mathfrak{u}_0)\mathbf{x}_2) &= h(\mathbf{x}_1 \circ \mathfrak{u}_0, \mathbf{x}_2 \circ \mathfrak{u}_0) = h(\mathbf{x}_1\mathfrak{u}_0, \mathbf{x}_2\mathfrak{u}_0) \\ &= \widetilde{(\mathbf{x}_1\mathfrak{u}_0)}\mathbf{x}_2\mathfrak{u}_0 = \tilde{\mathfrak{u}}_0\tilde{\mathbf{x}}_1\mathbf{x}_2\mathfrak{u}_0 = g(\mathbf{x}_1, \mathbf{x}_2)h(\mathfrak{u}_0, \mathfrak{u}_0) = g(\mathbf{x}_1, \mathbf{x}_2). \end{aligned}$$

This result relates the spinor $\mathbf{x} \circ \mathfrak{u}_0 \in S^-$ norm to the norms of $\mathbf{x} \in V$, and $\mathfrak{v} \in S^-$.

The mapping $\zeta(\mathfrak{u}_0)$ is uniquely extended to an involutive automorphism in $V \oplus S^-$. If $\mathfrak{v} \in S^-$ is such that $\mathfrak{v} = \zeta(\mathfrak{u}_0)\mathbf{x}$ for an unique $\mathbf{x} \in V$, then it is possible to define

$$\zeta(\mathfrak{u}_0)(\mathfrak{v}) = \mathbf{x}. \tag{6.92}$$

Furthermore, the mapping $\zeta(\mathfrak{u}_0)$ is defined in S^+ as a reflection with respect to the spinor $\mathfrak{u}_0 \in S^+$:

$$\zeta(\mathfrak{u}_0)(\mathfrak{u}) = 2h(\mathfrak{u}, \mathfrak{u}_0)\mathfrak{u}_0 - \mathfrak{u} \tag{6.93}$$

One more important result necessary to the existence of triality is the following:

Lemma 6.2: *Let $\mathfrak{u}_0 \in S^+$ and let $\mathbf{x}_0 \in V$ such that $\mathbf{x}_0^2 = 1$, and $h(\mathfrak{u}_0, \mathfrak{u}_0) = 1$. Then, $\zeta(\mathfrak{u}_0)Y(\mathbf{x}_0)\zeta(\mathfrak{u}_0) = Y(\mathbf{x}_0)\zeta(\mathfrak{u}_0)Y(\mathbf{x}_0)$.*

Proof: Consider $\mathfrak{v} \in S^-$. Then,

$$\begin{aligned}
&\zeta(\mathfrak{u}_0)Y(\mathbf{x}_0)\zeta(\mathfrak{u}_0)(\mathfrak{v})\\
&\quad= \zeta(\mathfrak{u}_0)Y(\mathbf{x}_0)(\zeta(\mathfrak{u}_0)(\mathfrak{v})) = \zeta(\mathfrak{u}_0)(\mathbf{x}_0(\zeta(\mathfrak{u}_0)(\mathfrak{v}))\mathbf{x}_0)\\
&\quad= (\mathbf{x}_0(\zeta(\mathfrak{u}_0)(\mathfrak{v}))\mathbf{x}_0)\circ\mathfrak{u}_0, \quad \text{since } (\mathbf{x}_0(\zeta(\mathfrak{u}_0)(\mathfrak{v}))\mathbf{x}_0)\in V\\
&\quad= (\mathbf{x}_0(\zeta(\mathfrak{u}_0)(\mathfrak{v}))\mathbf{x}_0)\mathfrak{u}_0,\\
&\quad= 2g(\mathbf{x}_0,\zeta(\mathfrak{u}_0)\mathfrak{v})\mathbf{x}_0\mathfrak{u}_0 - ((\zeta(\mathfrak{u}_0)\mathfrak{v})\mathbf{x}_0)\mathbf{x}_0\mathfrak{u}_0,\\
&\qquad\qquad \text{since } (\zeta(\mathfrak{u}_0)\mathfrak{v})\mathbf{x}_0 + \mathbf{x}_0(\zeta(\mathfrak{u}_0)\mathfrak{v}) = 2g(\mathbf{x}_0,\zeta(\mathfrak{u}_0)\mathfrak{v})\\
&\quad= 2h(\zeta(\mathfrak{u}_0)\mathbf{x}_0,\zeta(\mathfrak{u}_0)\zeta(\mathfrak{u}_0)\mathfrak{v})\mathbf{x}_0\mathfrak{u}_0 - (\zeta(\mathfrak{u}_0)\mathfrak{v}))\mathfrak{u}_0\\
&\quad= 2h(\mathbf{x}_0\circ\mathfrak{u}_0,\mathfrak{v})\mathbf{x}_0\mathfrak{u}_0 - \zeta(\mathfrak{u}_0)(\mathfrak{v}\mathfrak{u}_0)\\
&\quad= 2h(\mathbf{x}_0\mathfrak{u}_0,\mathfrak{v})\mathbf{x}_0\mathfrak{u}_0 - \zeta(\mathfrak{u}_0)(\mathfrak{v}\mathfrak{u}_0)\\
&\quad= 2h(\mathfrak{u}_0,\mathbf{x}_0\mathfrak{v})\mathbf{x}_0\mathfrak{u}_0 - (\mathfrak{v}\mathfrak{u}_0)\circ\mathfrak{u}_0, \quad \text{since } \mathfrak{v}\mathfrak{u}_0\in V\\
&\quad= 2h(\mathfrak{u}_0,\mathbf{x}_0\mathfrak{v})\mathbf{x}_0\mathfrak{u}_0 - \mathfrak{v}\mathfrak{u}_0\mathfrak{u}_0\\
&\quad= 2h(\mathfrak{u}_0,\mathbf{x}_0\mathfrak{v})\mathbf{x}_0\mathfrak{u}_0 - \mathfrak{v}.
\end{aligned}$$

On the other hand,

$$\begin{aligned}
&Y(\mathbf{x}_0)\zeta(\mathfrak{u}_0)Y(\mathbf{x}_0)(\mathfrak{v}) = Y(\mathbf{x}_0)\zeta(\mathfrak{u}_0)(\mathbf{x}_0\mathfrak{v})\\
&= Y(\mathbf{x}_0)[2h(\mathbf{x}_0\mathfrak{v},\mathfrak{u}_0)\mathfrak{u}_0 - \mathbf{x}_0\mathfrak{v}]\\
&= 2h(\mathbf{x}_0\mathfrak{v},\mathfrak{u}_0)\mathbf{x}_0\mathfrak{u}_0 - \mathbf{x}_0\mathbf{x}_0\mathfrak{v}.
\end{aligned} \tag{6.94}$$

In this case, the lemma is demonstrated for S^-. Now, by taking $\mathbf{x}\in V$, we obtain:

$$\begin{aligned}
&\zeta(\mathfrak{u}_0)Y(\mathbf{x}_0)\zeta(\mathfrak{u}_0)(\mathbf{x})\\
&\quad= \zeta(\mathfrak{u}_0)Y(\mathbf{x}_0)(\mathbf{x}\mathfrak{u}_0) = \zeta(\mathfrak{u}_0)(\mathbf{x}_0\mathbf{x}\mathfrak{u}_0)\\
&\quad= 2h(\mathbf{x}_0\mathbf{x}\mathfrak{u}_0,\mathfrak{u}_0) - \mathbf{x}_0\mathbf{x}\mathfrak{u}_0 = 2h(\mathbf{x}\mathfrak{u}_0,\mathbf{x}_0\mathfrak{u}_0)\mathfrak{u}_0 - \mathbf{x}_0\mathbf{x}\mathfrak{u}_0\\
&\quad= 2h(\zeta(\mathfrak{u}_0)\mathbf{x},\zeta(\mathfrak{u}_0)\mathbf{x}_0)\mathfrak{u}_0 - \mathbf{x}_0\mathbf{x}\mathfrak{u}_0 = 2g(\mathbf{x},\mathbf{x}_0)\mathfrak{u}_0 - \mathbf{x}_0\mathbf{x}\mathfrak{u}_0\\
&\quad= \mathbf{x}\mathbf{x}_0\mathfrak{u}_0.
\end{aligned}$$

In addition,

$$\begin{aligned}
&Y(\mathbf{x}_0)\zeta(\mathfrak{u}_0)Y(\mathbf{x}_0)(\mathbf{x}) = Y(\mathbf{x}_0)\zeta(\mathfrak{u}_0)(\mathbf{x}_0\mathbf{x}\mathbf{x}_0)\\
&\quad= Y(\mathbf{x}_0)(\mathbf{x}_0\mathbf{x}\mathbf{x}_0)\mathfrak{u}_0 = \mathbf{x}_0^2\mathbf{x}\mathbf{x}_0\mathfrak{u}_0 = \mathbf{x}\mathbf{x}_0\mathfrak{u}_0.
\end{aligned}$$

Finally,

$$\zeta(\mathfrak{u}_0)(\mathbf{x}_0\mathfrak{u}_0) = \zeta(\mathfrak{u}_0)(\mathbf{x}_0\circ\mathfrak{u}_0) = \zeta(\mathfrak{u}_0)(\zeta(\mathfrak{u}_0)\mathbf{x}_0) = \mathbf{x}_0\,. \tag{6.95}$$

Since $\zeta(\mathfrak{u}_0)$ is an involutive automorphism, given $\mathfrak{u}\in S^+$, it follows that

$$\begin{aligned}
&\zeta(\mathfrak{u}_0)Y(\mathbf{x}_0)\zeta(\mathfrak{u}_0)(\mathfrak{u})\\
&\quad= \zeta(\mathfrak{u}_0)Y(\mathbf{x}_0)(2h(\mathfrak{u},\mathfrak{u}_0)\mathfrak{u}_0 - \mathfrak{u}) = \zeta(\mathfrak{u}_0)(2h(\mathfrak{u},\mathfrak{u}_0)\mathbf{x}_0\mathfrak{u}_0 - \mathbf{x}_0\mathfrak{u})\\
&\quad= 2h(\mathfrak{u},\mathfrak{u}_0)\zeta(\mathfrak{u}_0)\mathbf{x}_0\mathfrak{u}_0 - \zeta(\mathfrak{u}_0)(\mathbf{x}_0\mathfrak{u})\\
&\quad= 2h(\mathfrak{u},\mathfrak{u}_0)\mathbf{x}_0 - \zeta(\mathfrak{u}_0)(\mathbf{x}_0\mathfrak{u}),
\end{aligned}$$

from eqn (6.95). On the other hand,

$$\begin{aligned}
&Y(\mathbf{x}_0)\zeta(\mathfrak{u}_0)Y(\mathbf{x}_0)(\mathfrak{u}) \\
&\quad = Y(\mathbf{x}_0)\zeta(\mathfrak{u}_0)(\mathbf{x}_0\mathfrak{u}) = \mathbf{x}_0\zeta(\mathfrak{u}_0)(\mathbf{x}_0\mathfrak{u})\mathbf{x}_0 \\
&\quad = -\zeta(\mathfrak{u}_0)(\mathbf{x}_0\mathfrak{u})\mathbf{x}_0\mathbf{x}_0 + 2g(\zeta(\mathfrak{u}_0)(\mathbf{x}_0\mathfrak{u}), \mathbf{x}_0)\mathbf{x}_0 \\
&\quad = 2h(\mathbf{x}_0\mathfrak{u}, \zeta(\mathfrak{u}_0)\mathbf{x}_0)\mathbf{x}_0 - \zeta(\mathfrak{u}_0)(\mathbf{x}_0\mathfrak{u}) \\
&\quad = 2h(\mathbf{x}_0 \circ \mathfrak{u}, \mathbf{x}_0 \circ \mathfrak{u}_0)\mathbf{x}_0 - \zeta(\mathfrak{u}_0)(\mathbf{x}_0\mathfrak{u}) \\
&\quad = 2g(\mathbf{x}_0, \mathbf{x}_0)h(\mathfrak{u}, \mathfrak{u}_0)\mathbf{x}_0 - \zeta(\mathfrak{u}_0)(\mathbf{x}_0\mathfrak{u}) \\
&\quad = 2h(\mathfrak{u}, \mathfrak{u}_0)\mathbf{x}_0 - \zeta(\mathfrak{u}_0)(\mathbf{x}_0\mathfrak{u}). \qquad \checkmark
\end{aligned}$$

Using those results, define now the operator $\Theta : \mathrm{E} \to \mathrm{E}$ as follows:

$$\Theta(\mathbf{x}_0, \mathfrak{u}_0) = Y(\mathbf{x}_0)\zeta(\mathfrak{u}_0)\,. \tag{6.96}$$

Theorem 6.1 ▶ *The operator* $\Theta(\mathbf{x}_0, \mathfrak{u}_0)$ *is an order 3 automorphism.*

Proof: Indeed, using lemma 6.2, we have

$$\begin{aligned}
\Theta^3(\mathbf{x}_0, \mathfrak{u}_0) &= Y(\mathbf{x}_0)\zeta(\mathfrak{u}_0)Y(\mathbf{x}_0)\zeta(\mathfrak{u}_0)Y(\mathbf{x}_0)\zeta(\mathfrak{u}_0) \\
&= Y(\mathbf{x}_0)\zeta(\mathfrak{u}_0)Y(\mathbf{x}_0)Y(\mathbf{x}_0)\zeta(\mathfrak{u}_0)Y(\mathbf{x}_0) = 1. \qquad \checkmark
\end{aligned}$$

Moreover, we can prove that $\Theta(\mathbf{x}_0, \mathfrak{u}_0)$ is orthogonal with respect to the bilinear form $\mathcal{B}$

$$\mathcal{B}(\Theta(\mathbf{x}_0, \mathfrak{u}_0)\phi_1, \Theta(\mathbf{x}_0, \mathfrak{u}_0)\phi_2) = \mathcal{B}(\phi_1, \phi_2) \tag{6.97}$$

and that

$$T(\Theta(\mathbf{x}_0, \mathfrak{u}_0)\phi_1, \Theta(\mathbf{x}_0, \mathfrak{u}_0)\phi_2, \Theta(\mathbf{x}_0, \mathfrak{u}_0)\phi_3) = T(\phi_1, \phi_2, \phi_3)\,, \tag{6.98}$$

since both mappings $Y(\mathbf{x}_0)$ and $\zeta(\mathfrak{u}_0)$ satisfy each of the relations in eqns (6.97) and (6.98).

In addition, the subspaces V, S^+, and S^- in E are cyclically permuted by $\Theta(\mathbf{x}_0, \mathfrak{u}_0)$:

$$\Theta(\mathbf{x}_0, \mathfrak{u}_0)V \subset S^+, \quad \Theta(\mathbf{x}_0, \mathfrak{u}_0)S^+ \subset S^-, \quad \Theta(\mathbf{x}_0, \mathfrak{u}_0)S^- \subset V\,. \tag{6.99}$$

Indeed, given $\mathbf{x} \in V$, it follows that

$$\begin{aligned}
\Theta(\mathbf{x}_0, \mathfrak{u}_0)(\mathbf{x}) &= Y(\mathbf{x}_0)\zeta(\mathfrak{u}_0)\mathbf{x} = Y(\mathbf{x}_0)(\mathbf{x}\mathfrak{u}_0) = \mathbf{x}_0\mathbf{x}\mathfrak{u}_0 \in S^+, \\
\Theta(\mathbf{x}_0, \mathfrak{u}_0)(\mathfrak{u}) &= Y(\mathbf{x}_0)\zeta(\mathfrak{u}_0)\mathfrak{u} = Y(\mathbf{x}_0)[2h(\mathfrak{u}, \mathfrak{u}_0)\mathfrak{u}_0 - \mathfrak{u}] \\
&= 2h(\mathfrak{u}, \mathfrak{u}_0)\mathbf{x}_0\mathfrak{u}_0 - \mathbf{x}_0\mathfrak{u} \in S^-. \\
\Theta(\mathbf{x}_0, \mathfrak{u}_0)(\mathfrak{v}) &= Y(\mathbf{x}_0)\zeta(\mathfrak{u}_0)\mathfrak{v} = Y(\mathbf{x}_0)\mathbf{x} = \mathbf{x}_0\mathbf{x}\mathbf{x}_0 \in V.
\end{aligned}$$

The space of spinors associated with V is written as $S^+ \oplus S^-$. From these results, we can assert that, if the space $S^\pm$ is taken as a vector space, the spinor space associated with $S^\pm$ is $V \oplus S^\mp$. In this sense, semispinors in $S^\pm$ – if we consider the underlying vector space structure of $S^\pm$ – are vectors in V. We can moreover prove the isomorphisms of Clifford algebras:

$$\mathcal{C}\ell(V,g) \simeq \mathcal{C}\ell(S^+,h) \simeq \mathcal{C}\ell(S^-,h). \tag{6.100}$$

Indeed, consider the automorphism obtained from $\Theta(\mathbf{x}_0,\mathfrak{u}_0)$ on the algebras $\mathcal{C}\ell(V,g)$, $\mathcal{C}\ell(S^+,h)$, and $\mathcal{C}\ell(S^-,h)$, already defined. Consider $S^-\oplus V$ as the spinor space of $\mathcal{C}\ell(S^+,h)$ and, using the notation in the book by Benn and Tucker (1987), let us denote by $\diamond$ the Clifford product in the algebra $\mathcal{C}\ell(S^+,h)$. Given $\mathbf{x}\in V$, and $\psi\in S=S^+\oplus S^-$,

$$[\mathbf{x}\psi]=[\mathbf{x}]\diamond[\psi].$$

For more details see the book by Benn and Tucker (1987).

Triality, and Octonionic Realisations

One of the most prominent aspects regarding the triality principle is its underlying geometric content. The realisation of triality in the Clifford algebraic context can be accomplished in the octonionic algebra $\mathbb{O}$, defined as the space $\mathbb{R}\oplus\mathbb{R}^{0,7}$ endowed with the product $\circ\colon (\mathbb{R}\oplus\mathbb{R}^{0,7})\times(\mathbb{R}\oplus\mathbb{R}^{0,7})\to\mathbb{R}\oplus\mathbb{R}^{0,7}$, the so-called octonionic standard product (Lounesto, 2001*b*; da Rocha and Vaz Jr, 2006). Moreover, trialities can be composed upon different octonion products.

Concerning the vector space $\mathbb{R}\oplus\mathbb{R}^{0,7}$ with the basis $\{\mathbf{e}_0=1,\mathbf{e}_a\}_{a=1}^7$, where it is usual to identify $\mathbf{e}_0=1$ with the basis of $\mathbb{R}$, the octonionic product reads (Harvey, 1990; Ivanova, 1993; Baez, 2002; da Rocha and Vaz Jr, 2006).

$$\mathbf{e}_a\circ\mathbf{e}_b=\epsilon_{ab}^{c}\mathbf{e}_c-\delta_{ab}\quad (a,b,c=1,\ldots,7), \tag{6.101}$$

where $\epsilon_{ab}^{c}=1$ for the cyclic permutations

$$(abc)=(126),(237),(341),(452),(563),(674),(715).$$

Explicitly, the multiplication is given by table 6.8, wherein all the relations can be expressed as $\mathbf{e}_a\circ\mathbf{e}_{a+1}=\mathbf{e}_{a+5 \mod 7}$.

Table 6.8 The Octonionic Product between Units in the $\mathbb{O}_{+5}$ Convention

1	$\mathbf{e}_1$	$\mathbf{e}_2$	$\mathbf{e}_3$	$\mathbf{e}_4$	$\mathbf{e}_5$	$\mathbf{e}_6$	$\mathbf{e}_7$
$\mathbf{e}_1$	-1	$\mathbf{e}_6$	$\mathbf{e}_4$	$-\mathbf{e}_3$	$\mathbf{e}_7$	$-\mathbf{e}_2$	$-\mathbf{e}_5$
$\mathbf{e}_2$	$-\mathbf{e}_6$	-1	$\mathbf{e}_7$	$\mathbf{e}_5$	$-\mathbf{e}_4$	$\mathbf{e}_1$	$-\mathbf{e}_3$
$\mathbf{e}_3$	$-\mathbf{e}_4$	$-\mathbf{e}_7$	-1	$\mathbf{e}_1$	$\mathbf{e}_6$	$-\mathbf{e}_5$	$\mathbf{e}_2$
$\mathbf{e}_4$	$\mathbf{e}_3$	$-\mathbf{e}_5$	$-\mathbf{e}_1$	-1	$\mathbf{e}_2$	$\mathbf{e}_7$	$-\mathbf{e}_6$
$\mathbf{e}_5$	$-\mathbf{e}_7$	$\mathbf{e}_4$	$-\mathbf{e}_6$	$-\mathbf{e}_2$	-1	$\mathbf{e}_3$	$\mathbf{e}_1$
$\mathbf{e}_6$	$\mathbf{e}_2$	$-\mathbf{e}_1$	$\mathbf{e}_5$	$-\mathbf{e}_7$	$-\mathbf{e}_3$	-1	$\mathbf{e}_4$
$\mathbf{e}_7$	$\mathbf{e}_5$	$\mathbf{e}_3$	$-\mathbf{e}_2$	$\mathbf{e}_6$	$-\mathbf{e}_1$	$-\mathbf{e}_4$	-1

The octonionic product can be constructed using the Clifford algebra $\mathcal{C}\ell_{0,7}$ as

$$\mathbf{u} \circ \mathbf{v} = \langle \mathbf{u}\mathbf{v}(1 - \psi)\rangle_{0\oplus 1}, \quad \mathbf{u}, \mathbf{v} \in \mathbb{R} \oplus \mathbb{R}^{0,7}, \tag{6.102}$$

where the 3-vector ψ is given by

$$\psi = \mathbf{e}_1\mathbf{e}_2\mathbf{e}_6 + \mathbf{e}_2\mathbf{e}_3\mathbf{e}_7 + \mathbf{e}_3\mathbf{e}_4\mathbf{e}_1 + \mathbf{e}_4\mathbf{e}_5\mathbf{e}_2 + \mathbf{e}_5\mathbf{e}_6\mathbf{e}_3 + \mathbf{e}_6\mathbf{e}_7\mathbf{e}_4 + \mathbf{e}_7\mathbf{e}_1\mathbf{e}_5 \,. \tag{6.103}$$

In a close analogy, the octonionic product can be also expressed with respect to the Clifford algebra on the Euclidean space $\mathbb{R}^{8,0}$, according to Lounesto (2001*a*), in terms of a basis $\{\mathbf{e}_1, \ldots, \mathbf{e}_8\}$ of $\mathbb{R}^{8,0}$. The octonionic product is given in this case by

$$\mathbf{u} \circ \mathbf{v} = \langle \mathbf{u}\,\mathbf{e}_8\,\mathbf{v}(1 + \psi)(1 - \mathbf{e}_{12\ldots 8})\rangle_1, \quad \mathbf{u}, \mathbf{v} \in \mathbb{R}^{8,0}, \tag{6.104}$$

where $\frac{1}{8}(1 + \star\psi)\frac{1}{2}(1 - \mathbf{e}_{12\ldots 8})$ is an idempotent. Both the approaches are equivalent: bivectors in $\mathcal{C}\ell_{8,0}$ correspond to the elements in $\mathbb{R}\oplus\mathbb{R}^{0,7} \subset \mathcal{C}\ell_{0,7}$, when the isomorphism $\mathbf{e}_\sigma\mathbf{e}_8 \mapsto \mathbf{e}_\sigma$, $\sigma = 1, 2, \ldots, 7$ is considered and $\mathbf{e}_8\mathbf{e}_8 = 1 = \mathbf{e}_0$ denotes the octonionic unit in $\mathbb{R} \hookrightarrow \mathbb{R} \oplus \mathbb{R}^{0,7}$. In fact, $\mathbf{e}_\sigma^2 = (\mathbf{e}_\sigma\mathbf{e}_8)^2 = -\mathbf{e}_\sigma\mathbf{e}_\sigma\mathbf{e}_8\mathbf{e}_8 = -1$. More details can be seen, for example, in the work by Lounesto (2001*a*); da Rocha and Traesel (2012), and da Rocha, Traesel, and Vaz Jr (2012).

Table 6.8 can be obtained by the octonionic product defined either by eqn (6.102) or by eqn (6.104). The definition in eqn (6.102) is regarded from hereon, where, in this case $\mathbb{R} \oplus \mathbb{R}^7$ is considered instead of the usual $\mathbb{R}^8$ underlying vector space, concerning the definition in eqn (6.104).

Some useful identities follow from eqn (6.101):

$$\epsilon_{abc}\epsilon_{dcf} + \epsilon_{dbc}\epsilon_{acf} = \delta_{ab}\delta_{df} + \delta_{af}\delta_{db} - 2\delta_{ad}\delta_{bf} \,.$$

Moreover, an analogue of the Jacobi formula in this context reads

$$[\mathbf{e}_i, [\mathbf{e}_j, \mathbf{e}_k]] + [\mathbf{e}_k, [\mathbf{e}_i, \mathbf{e}_j]] + [\mathbf{e}_j, [\mathbf{e}_k, \mathbf{e}_i]] = 3\epsilon_{ijkl}\mathbf{e}_l \,, \tag{6.105}$$

where $\epsilon_{ijkl} := -\epsilon_{mij}\epsilon_{mkl} - \delta_{il}\delta_{jk} + \delta_{ik}\delta_{jl}$ (Gunaydin and Ketov, 1996). Since the underlying vector space of $\mathbb{O}$ can be considered as being $\mathbb{R} \oplus \mathbb{R}^{0,7} \hookrightarrow \mathcal{C}\ell_{0,7}$, the Clifford conjugation of $\mathbf{v} = v^0 + v^a\mathbf{e}_a \in \mathbb{O}$ is given by $\bar{\mathbf{v}} = v^0 - v^a\mathbf{e}_a$, where v^0 and v^a are real coefficients. The underlying structure of the vector space is unable to assert whether the $\mathbb{O}$-conjugation is equivalent to the grade involution, since the octonionic conjugation $\bar{\mathbf{v}}$ can be written either as $\hat{\mathbf{v}}$ or $\bar{\mathbf{v}}$, with respect to Clifford algebra morphisms. However, the octonionic conjugation $\bar{\mathbf{v}}$ is involutive and thus an anti-automorphism, which immediately excludes the graded involution.

To invoke an explicit realisation of the triality principle, we observe that all automorphisms of $\mathrm{SO}(n)$ are of the form $A \mapsto SAS^{-1}$, where $S \in \mathrm{O}(n)$. Moreover, the automorphisms of $\mathrm{Spin}(n)$ can be written as $a \mapsto sas^{-1}$, for $s \in \mathrm{Pin}(n)$, for $n \neq 8$. However, the group Spin(8) has exceptional automorphisms, which cyclically permute the elements of the set $\{-1, \pm\mathbf{e}_{12\ldots 8}\}$, which lies in the centre of Spin(8). Such an order 3 automorphism of Spin(8) is the triality automorphism. Moreover, an order 2 automorphism of Spin(8) interchanges the element -1 with either of $\pm\mathbf{e}_{12\ldots 8}$. Such an automorphism of Spin(8) is the so-called swap automorphism, denoted by swap(a), for $a \in \text{Spin}(8)$ (Lounesto, 2001*b*).

From the representation point of view, triality can be viewed as permuting the vector space $\mathbb{R}^8$ and the two even spinor spaces, namely, the minimal left ideals $\mathcal{C}\ell_8^+ \frac{1}{8}(1+\star\psi)(1\pm\mathbf{e}_{12\ldots8})$, where ψ is given by eqn (6.103). Moreover, it corresponds to a 120° rotation of the Coxeter–Dynkin diagram of the Lie algebra $D_4 \simeq \mathfrak{so}(8)$ (Knus, 1998; Lounesto, 2001*b*).

Since the octonionic framework is helpful in this presentation, it is worth regarding triality in the context of the Clifford algebra $\mathcal{C}\ell_{0,7} \simeq \mathcal{M}(8,\mathbb{R}) \oplus \mathcal{M}(8,\mathbb{R})$ and the \$pin group

$$\$\text{pin}(8) = \{a \in \mathcal{C}\ell_{0,7}^+ \mid |\langle \bar{a}a\rangle_0| = 1, \text{ and } a\mathbf{v}\hat{a}^{-1} \in \mathbb{R}\oplus\mathbb{R}^{0,7}, \forall \mathbf{v} \in \mathbb{R}\oplus\mathbb{R}^{0,7}\}.$$

For $a \in \$\text{pin}(8)$, and $\mathbf{v} \in \mathbb{R}\oplus\mathbb{R}^{0,7}$, the two linear transformations A_1 and A_2 of $\text{O}(\mathbb{R}\oplus\mathbb{R}^{0,7})$ can be defined, respectively, by

$$A_1(\mathbf{v}) = 16\langle a\mathbf{v}f\rangle_{0\oplus1}, \qquad A_2(\mathbf{v}) = 16\langle a\mathbf{v}\hat{f}\rangle_{0\oplus1}, \tag{6.106}$$

where $f = \frac{1}{8}(1+\star\psi)(1\pm\mathbf{e}_{12\ldots8})$. The action of a on the left ideal $\mathcal{C}\ell_{0,7}\frac{1}{8}(1+\star\psi)$ has the matrix representation

$$[a] = \begin{pmatrix} A_1 & 0 \\ 0 & A_2 \end{pmatrix} \in \$\text{pin}(8). \tag{6.107}$$

Now, given $\mathbf{u} \in \mathbb{R}\oplus\mathbb{R}^{0,7}$, the linear transformation $U \in \text{End}(\mathbb{R}\oplus\mathbb{R}^{0,7})$ can be defined by

$$U(\mathbf{v}) = 16\langle \mathbf{u}\mathbf{v}f\rangle_{0\oplus1}. \tag{6.108}$$

Thus,

$$A(\mathbf{v}) = a \circ \mathbf{v}.$$

Therefore, the space $\mathbb{R}\oplus\mathbb{R}^{0,7}$ is the underlying vector space for the octonion algebra $\mathbb{O}$.

Now, using a notation similar to the one used in the article by Lounesto (2001*b*), let us define the so-called companion matrix $\check{U}$ that he associated with $U \in \text{SO}(\mathbb{R}\oplus\mathbb{R}^{0,7})$ in the same article:

$$\check{U}(\mathbf{v}) := \widehat{U(\hat{\mathbf{v}})}, \qquad \mathbf{v} \in \mathbb{R}\oplus\mathbb{R}^{0,7}. \tag{6.109}$$

Since $\check{U}^\intercal(\mathbf{v}) = 16\langle \mathbf{u}\mathbf{v}\hat{f}\rangle_{0\oplus1}$, therefore,

$$[u] = \begin{pmatrix} U & 0 \\ 0 & \check{U}^\intercal \end{pmatrix} \in \$\text{pin}(8), \tag{6.110}$$

which from now on, in this section, we shall denote by $\mathbf{u} \sim U$. Now, by computing the matrix product

$$A(\mathbf{u}) = a\mathbf{u}\hat{a}^{-1} \sim \begin{pmatrix} A_1 & 0 \\ 0 & A_2 \end{pmatrix}\begin{pmatrix} U & 0 \\ 0 & \check{U}^\intercal \end{pmatrix}\begin{pmatrix} \check{A}_2^{-1} & 0 \\ 0 & \check{A}_1^{-1} \end{pmatrix}, \tag{6.111}$$

we find the correspondence $A(\mathbf{u}) \sim A_1 U \check{A}_2^{-1}$. By denoting $A_0 = \check{A}$, for all $\mathbf{u}, \mathbf{v} \in \mathbb{R} \oplus \mathbb{R}^{0,7}$ we obtain (Lounesto, 2001*b*)

$$\check{A}_0(\mathbf{u}) \circ \mathbf{v} = A_1 U \check{A}_2^{-1}(\mathbf{v}) = A_1(\mathbf{u} \circ \check{A}_2^{-1}(\mathbf{v})). \tag{6.112}$$

The ordered triple (A_0, A_1, A_2) in SO(8) is called a triality triplet with respect to the octonion product of $\mathbb{O}$. Lounesto (2001*b*) observed that, if (A_0, A_1, A_2) is a triality triplet, then (A_2, A_0, A_1) and $(\check{A}_2, \check{A}_1, \check{A}_0)$ are also triality triplets. Hence, Cartan's triality principle reads

$$\check{A}_0(\mathbf{u} \circ \mathbf{v}) = A_1(\mathbf{u}) \circ A_2(\mathbf{v}), \qquad \mathbf{u}, \mathbf{v} \in \mathbb{R} \oplus \mathbb{R}^{0,7}. \tag{6.113}$$

6.8 Pure Spinors

In this section, we introduce the main concepts involving pure spinors. For more details and further developments see the work by Benn and Tucker (1987), Cartan (1937), Chevalley (1954), Budinich and Trautmann (1989), Crumeyrolle (1990), and Budinich (2002). It is also worth pointing, out that, in four and in six dimensions, pure spinors accidentally coincide with Weyl spinors (Lounesto, 2001*a*; da Rocha and da Silva, 2010), while there are quadratic constraints that pure spinors obey in higher dimensions (Budinich, 2002). In particular, the constraints in ten dimensions play an important role in Berkovits's approach to superstrings (Berkovits and Howe, 2002; Berkovits, 2004).

Given a complex vector space $\mathbb{C}^{2r}$ and its associated Clifford algebra $C\ell(2r, \mathbb{C})$, we have already seen from the classical definition of spinors that a spinor $\mathfrak{u}$ is a vector of the $2r$-dimensional representation space of the Spin group, associated with the Clifford algebra $C\ell(2r, \mathbb{C}) = \text{End } S$. A pure spinor is defined by the Cartan equation $\mathbf{v}\mathfrak{u} = 0$ (Cartan, 1937). For $\mathfrak{u} \neq 0$, $\mathbf{v} \in \mathbb{C}^{2r}$ is isotropic. Before introducing this concept, let us remember some prerequisites in what follows.

Given a bilinear form $B : V \times V \to \mathbb{K}$, the set $\{\mathbf{v} \in V \,|\, B(\mathbf{u}, \mathbf{v}) = 0, \forall \mathbf{u} \in V\}$ is a subspace of the space V – denominated radical of V (rad V). A vector space V endowed with a bilinear form B – either symmetric or antisymmetric – is said to be a direct sum of orthogonal vector subspaces V_i $(i = 1, \ldots, r)$ – with respect to B – if $V = \bigoplus_{i=1}^r V_i$, and $B(\mathbf{v}_i, \mathbf{v}_j) = 0$ for all $\mathbf{v}_i \in V_i$, and $\mathbf{v}_j \in V_j$. A subspace $U \subseteq V$ is said to be isotropic (totally isotropic) if the restriction of B to the subspace U is degenerate (null). A vector $\mathbf{v} \in V$ is said to be isotropic if $B(\mathbf{v}, \mathbf{v}) = 0$, and a subspace $U \subseteq V$ is said to be isotropic if it contains a non-trivial isotropic vector. A two-dimensional quadratic space that has a null radical is called a hyperbolic plan (Crumeyrolle, 1990), and the orthogonal direct sum of r hyperbolic plans P_i is a $2r$-dimensional hyperbolic space over a field $\mathbb{K}$.

Now we exhibit the Witt decomposition, relating hyperbolic and isotropic subspaces of V. For details about proofs and correlated topics see, for example, the work by Chevalley (1954), Lam (1980), Crumeyrolle (1990), and Abłamowicz (1995). Considering (V, B) a finite-dimensional vector space endowed by a non-degenerate quadratic form B over a field $\mathbb{K}$, and $W \subset V$ a maximal isotropic subspace, the Witt decomposition shows that there exists a maximal isotropic subspace $U \subset V$ such that

(a) $\dim \mathrm{U} = \dim \mathrm{W}, \quad U \cap W = \{0\}$.

(b) $V = W \oplus U \oplus (W \oplus U)^{\perp}$.
(c) For all $\mathbf{v} \in (W \oplus U)^{\perp} \backslash \{0\}$, we have $B(\mathbf{v}, \mathbf{v}) \neq 0$.

In addition, among all bases $\{\mathbf{w}_j\} \subset W$, there exists a basis $\{\mathbf{u}_i\} \subset U$ satisfying $B(\mathbf{u}_i, \mathbf{w}_j) = \delta_{ij}$, $1 \leq i, j \leq k$, called the Witt basis.

Another result is based upon the straightforward definition of a vector space V whose correlation is non-degenerate – namely, the mapping $\mathbf{v} \in V \mapsto B(\mathbf{v}, \cdot)$ is one-to-one – and is endowed with the quadratic form $Q : V \to \mathbb{K}$. Then,

(a) Every r-dimensional totally isotropic subspace $U \subset V$ is an element of a hyperbolic subspace H_{2r} of V.
(b) V is isotropic if and only if V contains a hyperbolic plan.
(c) If V is isotropic, the quadratic form Q takes values in all non-null elements in $\mathbb{K}$, namely, for all $a \in \mathbb{K}^*$, there exists $\mathbf{v} \in V \backslash \{0\}$ such that $Q(\mathbf{v}) = B(\mathbf{v}, \mathbf{v}) = a$.

A hyperbolic plan P has a basis $(\mathbf{u}, \mathbf{v})$ consisting of a pair of isotropic vectors ($\mathbf{u}^2 = 0 = \mathbf{v}^2$) satisfying $B(\mathbf{u}, \mathbf{v}) = 1$ and called a hyperbolic pair. Any hyperbolic space H_{2r} has a basis of hyperbolic vectors $(\mathbf{u}_i, \mathbf{v}_i)$ such that $B(\mathbf{u}_i, \mathbf{u}_j) = 0 = B(\mathbf{v}_i, \mathbf{v}_j)$, and $B(\mathbf{u}_i, \mathbf{v}_j) = \delta_{ij}$, $i, j = 1, \ldots, r$. Moreover, any quadratic space can be split in an orthogonal direct sum, consisting of a totally isotropic subspace – rad V – an anisotropic subspace, and a hyperbolic subspace. For more details see the work by Chevalley (1954), Benn and Tucker (1987), and Crumeyrolle (1990). Consequently, given any totally isotropic maximal subspace F of V, we can find another maximal totally isotropic subspace $F' \subset V$ such that $F \oplus F' = V$. In particular, if B is anti-symmetric and non-degenerate, then V is the hyperbolic space – $H_{2r} = V$ – and the hyperbolic pairs $(\mathbf{u}_i, \mathbf{v}_j)$ form the so-called symplectic basis of H_{2r}.

Given a non-degenerate bilinear form B, any isometry $\sigma : F \to F'$ can be extended to an isometry of V. In particular, this result implies that, when F and F' are maximal totally isotropic subspaces, they have the same dimension r, which is called the Witt index of V. Spaces where the associated Witt index is maximal are called neutral.[3] According to the Sylvester theorem, the index of V is given by the maximum between p and q, where $p - q$ is the metric signature.

Example 6.14 Given an orthonormal basis $\{\mathbf{e}_1, \ldots, \mathbf{e}_n\}$ of $\mathbb{R}^{p,q}$, where $p + q = n$, and denoting by r the index of Q, the Witt decomposition for $\mathbb{R}^{p,q}$ is given by $(F \oplus F') \perp G$, where F and F' are maximal totally isotropic subspaces, dim F = dim $F' = r$, and the subspace $F \oplus F'$ is generated by the hyperbolic pairs $(\mathbf{u}_i, \mathbf{v}_j)$ defined by

$$\begin{aligned} \mathbf{u}_1 &= \frac{1}{2}(\mathbf{e}_1 + \mathbf{e}_n), \ \mathbf{u}_2 = \frac{1}{2}(\mathbf{e}_2 + \mathbf{e}_{n-1}), \ldots, \ \mathbf{u}_r = \frac{1}{2}(\mathbf{e}_r + \mathbf{e}_{n-r+1}), \\ \mathbf{v}_1 &= \frac{1}{2}(\mathbf{e}_1 - \mathbf{e}_n), \ \mathbf{v}_2 = \frac{1}{2}(\mathbf{e}_2 - \mathbf{e}_{n-1}), \ldots, \ \mathbf{v}_r = \frac{1}{2}(\mathbf{e}_r - \mathbf{e}_{n-r+1}). \end{aligned} \tag{6.114}$$

These elements satisfy the relations

$$\mathbf{u}_i\mathbf{u}_j + \mathbf{u}_j\mathbf{u}_i = 0 = \mathbf{v}_i\mathbf{v}_j + \mathbf{v}_j\mathbf{v}_i, \qquad (\Rightarrow \mathbf{v}_i^2 = 0 = \mathbf{u}_i^2), \tag{6.115}$$

$$\mathbf{u}_i\mathbf{v}_j + \mathbf{v}_j\mathbf{u}_i = \delta_{ij}, \tag{6.116}$$

where $i, j = 1, \ldots, r$, and G is an anisotropic subspace generated by $\{\mathbf{e}_{r+1}, \mathbf{e}_{r+2}, \ldots, \mathbf{e}_{n-r}\}$. The sets $\{\mathbf{u}_i\}$ and $\{\mathbf{v}_j\}$ generate, respectively, the maximal totally isotropic subspaces F and F'. In particular, when $p = q = r$, the vector space $\mathbb{R}^{r,r}$ is neutral and equal to H_{2r}.

[3] Quadratic spaces over algebraically closed fields are always neutral.

In the complex case, the space is $V_{\mathbb{C}} = \mathbb{C}^{2r}$, and we can consider a basis $\{\mathbf{e}_1, \mathbf{e}_2, \ldots, \mathbf{e}_{2r}\}$ for $V = \mathbb{R}^{p,q}$, with $p + q = 2r$, and $p \leq q$. Hence, the quadratic form $Q_{\mathbb{C}}$ has maximal Witt index r, and $V_{\mathbb{C}} = F \oplus F'$, where the sets constituted by the vectors

$$\begin{aligned}
\mathbf{u}_1 &= \frac{1}{2}(\mathbf{e}_1 + \mathbf{e}_{2r}), \ldots, \mathbf{u}_k = \frac{1}{2}(\mathbf{e}_k + \mathbf{e}_{2r-k+1}),\\
\mathbf{u}_{k+1} &= \frac{1}{2}(i\mathbf{e}_{k+1} - \mathbf{e}_{2r-k}), \ldots, \mathbf{u}_r = \frac{1}{2}(i\mathbf{e}_r + \mathbf{e}_{r+1}),\\
\mathbf{v}_1 &= \frac{1}{2}(\mathbf{e}_1 - \mathbf{e}_{2r}), \ldots, \mathbf{v}_k = \frac{1}{2}(\mathbf{e}_k - \mathbf{e}_{2r-k+1}),\\
\mathbf{v}_{k+1} &= \frac{1}{2}(i\mathbf{e}_{k+1} - \mathbf{e}_{2r-k}), \ldots, \mathbf{v}_r = \frac{1}{2}(i\mathbf{e}_r - \mathbf{e}_{r+1})
\end{aligned}$$

generate, respectively, the maximal totally isotropic subspaces F and F' for $V_{\mathbb{C}}$, and $(\mathbf{u}_i, \mathbf{v}_j)$ are then hyperbolic pairs with respect to $Q_{\mathbb{C}}$.

In the particular case of the Minkowski spacetime $\mathbb{R}^{1,3}$ with orthonormal basis $\{\mathbf{e}_\mu\}$, the quadratic form Q has index 1, and the Witt decomposition of $\mathbb{R}^{1,3}$ is given by $(F \oplus F')\perp G$, where $\mathbf{u}_1 = \frac{1}{2}(\mathbf{e}_0 + \mathbf{e}_1)$, $\mathbf{v}_1 = \frac{1}{2}(-\mathbf{e}_0 + \mathbf{e}_1)$. The space G is generated by $\mathbf{e}_2$ and $\mathbf{e}_3$ – this case is commonly defined in the bosonic strings formalism. For a gentle introduction see, for example, the work by Zwiebach (2001) – whereas, for the case $\mathbb{C}^{1+3} = F \oplus F'$, the maximal totally isotropic subspaces F and F' are, respectively, generated by (Zwiebach, 2001)

$$\begin{aligned}
\mathbf{u}_1 &= \frac{1}{2}(\mathbf{e}_0 + \mathbf{e}_1), \quad \mathbf{u}_2 = \frac{1}{2}(i\mathbf{e}_2 + \mathbf{e}_3),\\
\mathbf{v}_1 &= \frac{1}{2}(\mathbf{e}_1 - \mathbf{e}_0), \quad \mathbf{v}_2 = \frac{1}{2}(i\mathbf{e}_2 - \mathbf{e}_3).
\end{aligned}$$

The spinor space S for $\mathcal{C}\ell_{r,r}$ can be constructed from the Witt decomposition $\mathbb{R}^{r,r} = F \oplus F'$, where the maximal totally isotropic subspaces F and F' – both of dimension r – are generated by

$$\begin{aligned}
\mathbf{u}_1 &= \frac{1}{2}(\mathbf{e}_1 + \mathbf{e}_{2r}),\ \mathbf{u}_2 = \frac{1}{2}(\mathbf{e}_2 + \mathbf{e}_{2r-1}), \ldots,\ \mathbf{u}_r = \frac{1}{2}(\mathbf{e}_r + \mathbf{e}_{r+1}),\\
\mathbf{v}_1 &= \frac{1}{2}(\mathbf{e}_1 - \mathbf{e}_{2r}),\ \mathbf{v}_2 = \frac{1}{2}(\mathbf{e}_2 - \mathbf{e}_{2r-1}), \ldots,\ \mathbf{v}_r = \frac{1}{2}(\mathbf{e}_r - \mathbf{e}_{r+1}).
\end{aligned} \tag{6.117}$$

Define now the volume element associated with the maximal totally isotropic subspaces F' as $\Omega_{F'} = \mathbf{v}_1\mathbf{v}_2 \ldots \mathbf{v}_r$ so that $\Omega^2_{F'} = 0$. A basis for S is given by $\mathbf{u}_{i_1 i_2 \ldots i_k}\Omega_{F'}$, $1 \leq i_1 \leq \cdots \leq i_r \leq r$. The elements $\{\mathbf{u}_{i_1 i_2 \ldots i_k}, \mathbf{v}_{i_1 i_2 \ldots i_k}\}$ form a basis of the algebra $\mathcal{C}\ell_{r,r}$. It then follows that dim $\mathcal{C}\ell_{r,r}\Omega_{F'} = 2^r$ and that $S \simeq \mathcal{C}\ell_{r,r}\Omega_{F'}$. Hence, S is a minimal left ideal of $\mathcal{C}\ell_{r,r}$. Since $B|_F \equiv 0 \equiv B|_{F'}$, in particular, the elements $\{\mathbf{u}_i\}$ generate the exterior algebra $\bigwedge(F) \hookrightarrow \mathcal{C}\ell_{r,r}$.

Furthermore, using the relation given in eqn (6.116), the elements of F can be positioned on the right side of any term in $\mathcal{C}\ell(V, g)$. Hence, we obtain

$$\mathcal{C}\ell(V, g)\Omega_F = \mathcal{C}\ell(F', g)\Omega_F = \bigwedge(F')\Omega_F,$$

where the left ideal $\mathcal{C}\ell(V, g)\Omega_F$ has the dimension of the exterior algebra associated to F' – equal to 2^r – which thus is a minimal ideal (Abłamowicz, 1995). Hence $\mathcal{C}\ell(F', g)\Omega_F$ is the spinor space.

Although the r-vector Ω_F defined in this way is isotropic, $\mathbf{e}_1\mathbf{e}_2 \ldots \mathbf{e}_r\,\Omega_F$ is a primitive idempotent, and it can be written as the sum of mutually annihilating idempotents.

Example 6.15 Consider the subspaces F and F' given by (6.114), and define new elements $\mathring{\mathbf{v}}_i = \mathbf{e}_i\mathbf{v}_i$, which satisfy $\mathring{\mathbf{v}}_i^2 = \mathring{\mathbf{v}}_i$, and $\mathring{\mathbf{v}}_i\mathring{\mathbf{v}}_j = \mathring{\mathbf{v}}_j\mathring{\mathbf{v}}_i$. The product $\epsilon = \mathring{\mathbf{v}}_1\mathring{\mathbf{v}}_2\ldots\mathring{\mathbf{v}}_r = (-1)^{r(r-1)/2}\mathbf{e}_1\mathbf{e}_2\ldots\mathbf{e}_r\,\Omega_{F'}$ is a primitive idempotent, and a complete set of primitive idempotents $\{\epsilon_1,\epsilon_2,\ldots,\epsilon_r\}$ can be obtained from ϵ by applying the reversion in $\mathcal{C}\ell_{r,r}$ to one or more factors $\mathbf{v}_i$ of ϵ, with the properties

$$\epsilon_1+\epsilon_2+\cdots+\epsilon_r = 1,\quad \epsilon_i\epsilon_j = 0\ (i\neq j),\ \epsilon_i^2=\epsilon_i,\ i,j=1,2,\ldots,r. \tag{6.118}$$

From this decomposition, we can write

$$\mathcal{C}\ell_{r,r} = \mathcal{C}\ell_{r,r}\epsilon_1\oplus\mathcal{C}\ell_{r,r}\epsilon_2\oplus\cdots\oplus\mathcal{C}\ell_{r,r}\epsilon_r\,, \tag{6.119}$$

where the direct sum corresponds to the sum of non-decomposable ideals. Therefore, $\epsilon\mathcal{C}\ell_{r,r}\epsilon\simeq\mathbb{R}$, and $\mathcal{C}\ell_{r,r}\epsilon$ is a minimal ideal.

Since $\mathcal{C}\ell_{r,r}$ is a simple algebra, then a representation $\rho:\mathcal{C}\ell_{r,r}\to\mathrm{End}\,S$ is faithful and induces a representation ρ^+ of $\mathcal{C}\ell_{r,r}^+$ in End S. The spinor space S decomposes in two subspaces $S^\pm = \mathcal{C}\ell_{r,r}^\pm\Omega_{F'}$ of dimension 2^{r-1} and which are irreducible and invariant with respect to ρ^+. Moreover, the elements of $S^\pm$ are called semispinors, where the spinor space is written as the direct sum of two semispinor spaces $S=S^+\oplus S^-$. The representation ρ^+ is an isomorphism between $\mathcal{C}\ell_{r,r}^+$ and End $S^+\times$ End S^-.

Considering again the subspaces F and F' given in (6.114), the Witt decomposition of $\mathbb{R}^{r,r+1}$ is given by $(F\oplus F')\perp\mathbf{e}_{2r+1}$. The algebra $\mathcal{C}\ell_{r,r+1}\simeq\mathcal{M}(2^r,\mathbb{C})$ is simple, and its centre is generated by $\{1,\mathbf{e}_{123\ldots 2r+1}\}\simeq\mathbb{C}$. The spinor space $S=\mathcal{C}\ell_{r,r+1}\Omega_{F'}$ has dimension 2^{r+1} over $\mathbb{R}$. The algebra $\mathcal{C}\ell_{r+1,r}\simeq\mathcal{M}(2^r,\mathbb{R})\oplus\mathcal{M}(2^r,\mathbb{R})$ is semisimple, and can be decomposed as the direct sum of minimal ideals $\mathcal{C}\ell_{r,r+1}\frac{1}{2}(1\pm\mathbf{e}_{123\ldots 2r+1})$ of dimension 2^r, from the central primitive idempotents.

Definition 6.4 ▶ *If F and T are maximal totally isotropic subspaces in V, an element of $\Omega_F\mathcal{C}\ell(V,g)\Omega_T$ is a representative spinor for T – with respect to F. A spinor that represents some subspace T is called a pure spinor.*

Since we used a maximal totally isotropic subspace F to define the spinor space, any other maximal totally isotropic subspace T can be used to define a right minimal ideal and consequently to define a 1-dimensional subspace $\Omega_F\mathcal{C}\ell(V,g)\Omega_T$ in the spinor space.

In general, a mapping $\rho:\mathcal{C}\ell(V,g)\to\mathrm{End}(S)$ defined by $\rho(u):\mathbf{v}\Omega_F\mapsto u\mathbf{v}\Omega_F$ is used to provide a representation of $\mathcal{C}\ell(V,g)$ in $S=\mathcal{C}\ell(V,g)\Omega_F$. Such a representation also induces equivalent representations in its subgroups, that is, the Clifford–Lipschitz group, the Pin group, and the Spin group as well. Since $\psi^{-1}\mathbf{v}\psi\in V$, for all $\psi\in\Gamma_{p,q}$ and for all $\mathbf{v}\in V$, a Witt basis $\{\mathbf{u}_i,\mathbf{v}_i\}$ can be led to another Witt basis $\{\mathbf{u}_i',\mathbf{v}_i'\}$, where $\mathbf{u}_i'=\psi^{-1}\mathbf{u}_i\psi$, and $\mathbf{v}_i'=\psi^{-1}\mathbf{v}_i\psi$. Therefore, another basis for S is given by $\mathbf{u}'_{i_1i_2\ldots i_k}\Omega_{F'}$, where $\Omega_{F'}=\psi^{-1}\Omega_F\psi$, and such bases are said to be geometrically equivalent. The choice of basis is unique, and the geometric equivalence between any maximal totally isotropic subspaces leads to the concept of pure spinors. Indeed, every maximal totally isotropic subspace F defining a spinor space S is an equivalence class of pure spinors, characterised as the non-trivial elements in the intersection $(\mathcal{C}\ell(V,g)\Omega_F)\cap(\Omega_{F'}\mathcal{C}\ell(V,g))$ between minimal ideals. The pure spinor space is 1-dimensional in $\mathbb{K}$. Theorem 6.2 is used in some texts as the definition of a pure spinor.

Theorem 6.2 ▶ *If the space of the representative spinors for F is generated by Ω_F, and if ψ is a representative spinor of F, then $\mathbf{u}\psi=0$, $\forall\mathbf{u}\in F$.*[4]

[4]Note that $\mathbf{u}\psi$ is the regular representation of the group $\mathrm{Spin}(p,q)$.

Proof: If ψ is any element in $\mathcal{C}\ell(V,g)\Omega_F$, then $\psi = \phi\Omega_F$, for some $\phi \in \bigwedge(F')$. For any vector $\mathbf{v}_i \in F'$, we can split $\phi = \mathbf{v}_i\phi_1 + \phi_2$, where $\phi_1, \phi_2 \in \bigwedge(F')\setminus\langle\mathbf{v}_i\rangle$. Hence, $\mathbf{u}_i\psi = \mathbf{u}_i\phi\Omega_F = \kappa\mathbf{u}_i\Omega_F$, for some $\kappa \in \bigwedge(F')$, where in the last equality we used the relations in (6.115) and (6.116). It follows that $\mathbf{u}_i\psi = 0$, since $\mathbf{u}_i\Omega_F = \mathbf{u}_i\,\mathbf{u}_1\ldots\mathbf{u}_r = (-1)^{i-1}\mathbf{u}_1\ldots\mathbf{u}_i^2\ldots\mathbf{u}_r = 0$. Since this result holds for any $i = 1,\ldots,r$, therefore $\mathbf{u}\psi = 0$, for all $\mathbf{u} \in F$. ✓

Given a maximal totally isotropic subspace F, there is no unique space F' such that $V = F \oplus F'$. If T is another maximal totally isotropic subspace where dim $(T \cap F) = h$, then it is possible to choose a Witt basis such that $\{\mathbf{u}_i\}$ is a basis of F and $\{\mathbf{u}_1,\ldots,\mathbf{u}_h,\mathbf{v}_{h+1},\ldots,\mathbf{v}_t\}$ is a basis of T. From some basis $\{\mathbf{u}_1,\ldots,\mathbf{u}_h\}$ for $T\cap F$, the Witt basis can be completed by the Gram–Schmidt procedure. A representative for T is therefore given by $\mathbf{u} = \mathbf{v}_{h+1}\ldots\mathbf{v}_r\Omega_F$, which is known as the canonical form for a pure spinor.

Now, given $s \in \Gamma_{p,q}$, if $T = \sigma(s)(F) = sFs^{-1}$, a representative spinor for T can be always written as $\psi = s\Omega_F$. Indeed, any isomorphism between vector (sub)spaces F and T can be extended to an isometry $\sigma(s)$ in $\mathrm{O}(V)$ by the Witt theorem. Hence, $\Omega_T = s\Omega_F s^{-1} \neq 0$, and Ω_T is a product of elements of a basis of T. Since $s\Omega_F \in \mathcal{C}\ell(V,g)\Omega_F$, and $s\Omega_F = \Omega_T\, s \in \Omega_T\mathcal{C}\ell(V,g)$, it follows that $s\Omega_F$ generates the intersection $(\mathcal{C}\ell(V,g)\Omega_F) \cap (\Omega_T\mathcal{C}\ell(V,g))$.

It then follows that pure spinors have definite parity. Indeed, when $n = p+q$ is even, all elements of the twisted Clifford–Lipschitz group $\hat{\Gamma}_{p,q}$ have defined parity since, given $s \in \hat{\Gamma}_{p,q}$, we have $\hat{s}\mathbf{v}s^{-1} \in V$, $\forall\mathbf{v} \in V$. On the other hand, $\widetilde{\hat{s}\mathbf{v}s^{-1}} = \hat{s}\mathbf{v}s^{-1} \Rightarrow \tilde{s}^{-1}\mathbf{v}\bar{s} = \hat{s}\mathbf{v}s^{-1} \Rightarrow \tilde{s}\hat{s}\mathbf{v} = \mathbf{v}\bar{s}s \Rightarrow \widehat{(\bar{s}s)}\mathbf{v}(\bar{s}s)^{-1} = \mathbf{v} \Rightarrow \bar{s}s \in \ker\hat{\sigma} \Rightarrow \bar{s}s \in \mathbb{R}^*$. Hence, $s^{-1} = \lambda\bar{s}$ for some $\lambda \in \mathbb{R}^*$. However, $\overline{\hat{s}\mathbf{v}s^{-1}} = -\hat{s}\mathbf{v}s^{-1} \Rightarrow \overline{s^{-1}}\mathbf{v}\tilde{s} = \hat{s}\mathbf{v}s^{-1} \Rightarrow \bar{s}\hat{s}\mathbf{v} = \mathbf{v}\tilde{s}s \Rightarrow \widehat{(\tilde{s}s)}\mathbf{v}(\tilde{s}s)^{-1} = \mathbf{v} \Rightarrow \tilde{s}s \in \ker\hat{\sigma} \Rightarrow \tilde{s}s \in \mathbb{R}^*$, and therefore $s^{-1} = k\tilde{s}$, for some $k \in \mathbb{R}^*$. It follows that $\bar{s} = c\tilde{s}$ for some $c \in \mathbb{R}^*$. Now, by calculating the reversion of both members in the previous equation, we obtain $\hat{s} = cs$, which implies that $c = \pm 1$. Hence, $\hat{s} = \pm s$, and it follows that a representative spinor for T – written as $\psi = s\Omega_F$ – is respectively even or odd.

In addition, $\hat{\sigma}(s) \in \mathrm{O}(p,q)$, for all $s \in \hat{\Gamma}_{p,q}$. In fact, given $\mathbf{v} \in V$, we have $\|\hat{\sigma}(\mathbf{v})\|^2 = \hat{s}\mathbf{v}s^{-1}\hat{s}\mathbf{v}s^{-1} = s\mathbf{v}s^{-1}s\mathbf{v}s^{-1}$, which can be derived from the previous paragraph. Hence, it yields $\|\hat{\sigma}(\mathbf{v})\|^2 = s\mathbf{v}\mathbf{v}s^{-1} = \|\mathbf{v}\|^2 ss^{-1} = \|\mathbf{v}\|^2$. Moreover, the twisted Clifford–Lipschitz group can also be characterised as

$$\hat{\Gamma}_{p,q} = \{\mathbf{v}_1\ldots\mathbf{v}_k \in \mathcal{C}\ell_{p,q} \mid \mathbf{v}_i \text{ are non-isotropic vectors}, i = 1,\ldots,k\},$$

since from the previous result we have $\hat{\sigma}(s) \in \mathrm{O}(p,q)$ and, from the Cartan–Dieudonné theorem, there exist non-isotropic vectors $\mathbf{v}_1,\ldots,\mathbf{v}_k$ such that $\hat{\sigma}(s) = S_{\mathbf{v}_1} \circ \cdots \circ S_{\mathbf{v}_k}$, where $S_{\mathbf{v}}$ denotes a reflexion with respect to the hyperplane orthogonal to $\mathbf{v}$. Since $\hat{\sigma}(\mathbf{v}) = S_{\mathbf{v}}$, it follows that $\hat{\sigma}(s) = \hat{\sigma}(\mathbf{v}_1)\ldots\hat{\sigma}(\mathbf{v}_k) = \hat{\sigma}(\mathbf{v}_1\ldots\mathbf{v}_k)$. In addition, since ker $\hat{\sigma} = \mathbb{R}^*$, we arrive at $s = \lambda\mathbf{v}_1\ldots\mathbf{v}_k$, for some $\lambda \in \mathbb{R}^*$. Finally, if we redefine $\mathbf{v}_1 \mapsto \lambda\mathbf{v}_1$, then we obtain the aforementioned result.

Let us suppose now that $\mathfrak{u}$ is a pure spinor.[5] Supposing that $\mathfrak{u}$ is an even semispinor, then $\mathfrak{u}_C$ is an odd semispinor, where the notation $(\ )_C$ indicates charge conjugation.

[5] Considering $n = \dim_{\mathbb{R}} V$, if $n = 4$ or 6, all Weyl spinors are pure, namely, $\mathbf{x}\mathfrak{u} = 0, \forall\mathbf{x} \in F \subset V_{\mathbb{C}}$.

As the spinor $\mathfrak{u}$ is pure, it represents some maximal totally isotropic subspace, let us say M_1. Since $\mathfrak{u}_C$ is also a pure spinor, it represents some other maximal totally isotropic subspace, M_2. If M_1 and M_2 are two maximal totally isotropic subspaces, then M_1 and M_2 have the same parity if and only if $\dim_{\mathbb{K}}(M_1 \cap M_2) = r \mod 2$. Indeed, by the Witt theorem, there exists $\sigma \in \mathrm{O}(p,q)$ – equivalently, $s \in \hat{\Gamma}_{r,r}$ – which leads M_2 to F via the application $\hat{\sigma}(s)$. If ψ_1 and ψ_2 are representatives of M_1 and M_2, respectively, then $\Omega_F = s\psi_2$ (since $\psi_2 = s'\Omega_F$ for some $s' \in \Gamma$) and, if $\psi = s\psi_1$, then ψ is a representative for $M = \hat{\sigma}(s)(M_1)$. However, since $s \in \hat{\Gamma}_{r,r}$ has defined parity, then ψ and Ω_F have the same parity with respect to the grade involution if and only if ψ_1 and ψ_2 have the same parity. In addition, $M \cap F = \hat{\sigma}(s)(M_1 \cap M_2)$, and it is sufficient to prove that the representative spinors for M and F are either both odd or both even if $\dim(M \cap F) = r \mod 2$. When a Witt basis is adopted for M and F, then a representative ψ for M has the canonical form $\psi = \mathbf{v}_{h+1} \dots \mathbf{v}_r\Omega_F$, where $\dim(M \cap F) = h$. Therefore, ψ and Ω_F are either both even or both odd if $r - h = 0 \mod 2$, namely, $h = r \mod 2$.

Another property asserts the conditions under which a linear combination of pure spinors comprises a pure spinor, the demonstration of which can be seen in the work by Chevalley (1954), Benn and Tucker (1987), and Crumeyrolle (1990). Given spinors ψ_1 and ψ_2 which respectively represent the maximal totally isotropic subspaces T_1 and T_2 of V, a necessary and sufficient condition for $\psi_1 + \psi_2$ be pure is that $\dim(T_1 \cap T_2) = r$ or $r - 2$.

Observation ☞ In general, it is always possible to choose a set of pure spinors as a basis for the spinor space, and any semispinor is a linear combination of pure spinors with definite parity. From the result $\dim(T_1 \cap T_2) = r$, or $r - 2$, another result previously obtained asserts that $\dim(T_1 \cap T_2) = r \mod 2$. This result then implies that a linear combination of pure spinors of same parity is a pure spinor if $r \leq 3$, and therefore all semispinors are pure for $r \leq 3$.

Now, every $(r-1)$-dimensional maximal totally singular subspace M_1 is contained in exactly one maximal totally isotropic subspace, which is either even or odd, of V. To see this fact, remember that it is always possible to transform M_1 in a subspace of F from the action of an element of the orthogonal group. Hence, it suffices to consider this assertion for a subspace $M_1 \subset F$. Let Z be a maximal totally isotropic subspace containing M_1. If Z has the same parity as F, then we have $\dim\,(Z \cap F) = r \mod 2$. However, since $\dim\,(Z \cap F) \geq r - 1$, and $\dim Z = r$, and taking into account that F contains at least a subspace M_1 of dimension $r - 1$, then $Z = F$. If Z does not have the same parity as F, then for the same reasoning it follows that $Z \cap F = M_1$. Let $\mathfrak{u}$ be a representative spinor for Z, written as

$$\mathfrak{u} = a\mathbf{v}_1 \dots \mathbf{v}_{r-1} + b\mathbf{v}_1 \dots \mathbf{v}_r, \quad a, b \in \mathbb{K}, \quad \mathbf{v}_r \in F, \quad \mathbf{v}_r \notin M_1\,. \tag{6.120}$$

Since $\mathfrak{u}$ is an element of defined parity, then $\mathfrak{u} = a\mathbf{v}_1 \dots \mathbf{v}_r$, and therefore Z is uniquely determined by the representative spinor.

A very interesting case for its applications for the use of the Clifford algebras $C\ell_{1,3}(\mathbb{C}) \simeq C\ell_{3,1}(\mathbb{C})$ is to consider the space $V = \mathbb{R}^{3,1}$. If F is a maximal totally isotropic subspace of $V_{\mathbb{C}}$, then, supposing that M_2 is another maximal totally isotropic subspace, F and M_2 have the same dimension, $\dim_{\mathbb{C}} M_2 = 2$, like all maximal totally isotropic subspaces of $V_{\mathbb{C}}$. Let M_1 and M_2 have, respectively, semispinors $\mathfrak{u}_1$ – even – and $\mathfrak{u}_2$ – odd – which represent them and which are pure, from the previous observation. We already demonstrated that $\dim_{\mathbb{C}}(M_1 \cap M_2) = r \mod 2$ if and only if M_1 and M_2 have the same parity. Now, since in this case we have $r = 2$, then necessarily $\dim_{\mathbb{C}}(M_1 \cap M_2)$ must be odd. Indeed, $\dim_{\mathbb{C}}(M_1 \cap M_2) = 1$, since M_1 and M_2 have different parities. Hence, $M_1 \cap M_2 = \text{span}\{\mathbf{y}\}$, where $\mathbf{y} \in V_{\mathbb{C}}$, and it follows from a former result that $\mathbf{y}^2 = 0$.

Defining $\mathring{\Omega}$ as a $2r$-form such that $\mathring{\Omega}^2 = 1$, the idempotents $P_\pm = \frac{1}{2}(1 \pm \mathring{\Omega})$ reduce $\mathcal{C}\ell^+_{3,1}(\mathbb{C})$ to simple ideals. Then, $\mathring{\Omega} = i e_{0123} = i\Omega$. Denoting $\mathfrak{u}_C$ as the charge conjugation of $\mathfrak{u}$, then both $\mathfrak{u}$ and $\mathfrak{u}_C$ have different parity.

If $\mathfrak{u}$ is a pure spinor, then

$$\mathbf{y}\mathfrak{u} = 0, \qquad \forall \mathbf{y} \in M_1 \cap M_2 . \tag{6.121}$$

In the same way, $\mathfrak{u}_C$ is a pure spinor, and eqn (6.121) implies that

$$\mathbf{y}^* \mathfrak{u}_C = 0 . \tag{6.122}$$

In addition, $\mathbf{y} \in M_1 \cap M_2$ implies that $\mathbf{y}^* \in M_1 \cap M_2$, where $\mathbf{y}^*$ denotes the $\mathbb{C}$-conjugate of $\mathbf{y}$, since $M_1 \cap M_2 = \text{span}\{\mathbf{y}\}$ has complex dimension equal to 1. Hence, $\mathbf{y}^* = \lambda \mathbf{y}$, where $\lambda \in \mathbb{C}$; since $(\mathbf{y}^*)^* = \mathbf{y}$, it follows that

$$\mathbf{y} = (\mathbf{y}^*)^* = (\lambda \mathbf{y})^* = \lambda^* \mathbf{y}^* = \lambda^* \lambda \mathbf{y} \Rightarrow \lambda^* \lambda = 1,$$

and we can conclude that $\lambda \in S^1$. There exists $\mu \in S^1$ such that $\lambda = \mu^2$ and there also exists an element $\mathbf{p} \in V_{\mathbb{C}}$ such that $\mathbf{p} = \mu \mathbf{y}$. This result yields

$$\begin{aligned} \mathbf{p}^* &= (\mu \mathbf{y})^* = \mu^* \mathbf{y}^* = \mu^* \lambda \mathbf{y} \\ &= \mu^* (\mu^2) \mathbf{y} = (\mu^* \mu) \mu \mathbf{y} = \mu \mathbf{y} = \mathbf{p}. \end{aligned} \tag{6.123}$$

Therefore, $\mathbf{p}$ is a real isotropic vector lying in the intersection of two maximal totally isotropic subspaces. Moreover, the sum of a pure spinor with its charged conjugated spinor is annihilated by $\mathbf{p}$. In fact, we have

$$\mathbf{p}(\mathfrak{u} + \mathfrak{u}_C) = \mathbf{p}\mathfrak{u} + \mathbf{p}\mathfrak{u}_C = 0, \tag{6.124}$$

where eqns (6.121) and (6.122) are taken into account. The vector $\mathbf{p}$ is determined up to the multiplication by a real constant, defining thus a projective space.

Let us suppose now that $\mathfrak{u}$ represents T – we know that $\dim_{\mathbb{C}} M_1 = 2$ – and let us consider a basis $\{\mathbf{p}, \mathbf{x}\}$ of T. Then, $\mathbf{p}(\mathbf{x}\mathfrak{u}_C) = -\mathbf{x}(\mathbf{p}\mathfrak{u}_C) = 0$, from the properties $\mathbf{p}\mathfrak{u}_C = 0$, and $\mathbf{p}\mathbf{x} = -\mathbf{x}\mathbf{p}$. Moreover, obviously the equation

$$\mathbf{x}(\mathbf{x}\mathfrak{u}_C) = \mathbf{x}^2 \mathfrak{u}_C = 0 \tag{6.125}$$

holds, since $\mathbf{x}$ is isotropic. Therefore, $\mathbf{p}(\mathbf{x}\mathfrak{u}_C) = 0$; $\mathbf{p}\mathfrak{u} = 0$; and $\mathbf{x}\mathfrak{u}_C$ and $\mathfrak{u}$ are pure spinors that represent T.

Since the space of the spinors that represent T is one-dimensional over the complex field, there exists $\eta \in \mathbb{C}^*$ – that the possibility that $\eta = 0$ does not exist since $\mathbf{x}$ is not in the subspace represented by $\mathfrak{u}_C$ – such that $\mathbf{x}\mathfrak{u}_C = \eta\mathfrak{u}$. Defining $\boldsymbol{\omega} = \eta^{-1}\mathbf{x} \in M_1$, we then obtain $\boldsymbol{\omega}\mathfrak{u}_C = \eta^{-1}\mathbf{x}\mathfrak{u}_C = \eta^{-1}\eta\mathfrak{u} = \mathfrak{u}$. It follows that

$$\boldsymbol{\omega}\mathfrak{u}_C = \mathfrak{u}\,. \tag{6.126}$$

By taking the charge conjugation in eqn (6.126), we obtain

$$(\boldsymbol{\omega}\mathfrak{u}_C)_C = \mathfrak{u}_C \Rightarrow \boldsymbol{\omega}^*\mathfrak{u} = \mathfrak{u}_C. \tag{6.127}$$

Moreover, $\boldsymbol{\omega}\boldsymbol{\omega}^*\mathfrak{u} = \boldsymbol{\omega}\mathfrak{u}_C = \mathfrak{u}$; since $\boldsymbol{\omega} \in T$, then $\boldsymbol{\omega}\mathfrak{u} = 0$, yielding $\boldsymbol{\omega}^*\boldsymbol{\omega}\mathfrak{u} = 0$. In addition, $(\boldsymbol{\omega}\boldsymbol{\omega}^* + \boldsymbol{\omega}^*\boldsymbol{\omega})\mathfrak{u} = \mathfrak{u}$, implying that

$$(\boldsymbol{\omega}\boldsymbol{\omega}^* + \boldsymbol{\omega}^*\boldsymbol{\omega}) = 1. \tag{6.128}$$

Define a vector $\mathbf{a} \in V_{\mathbb{C}}$ as $\mathbf{a} = \boldsymbol{\omega} + \boldsymbol{\omega}^*$. We assert that $\mathbf{a}$ is a unit vector. Indeed,

$$\begin{aligned}\mathbf{a}^2 &= (\boldsymbol{\omega} + \boldsymbol{\omega}^*)(\boldsymbol{\omega} + \boldsymbol{\omega}^*) = \boldsymbol{\omega}^2 + \boldsymbol{\omega}\boldsymbol{\omega}^* + \boldsymbol{\omega}^*\boldsymbol{\omega} + (\boldsymbol{\omega}^*)^2 \\ &= \boldsymbol{\omega}\boldsymbol{\omega}^* + \boldsymbol{\omega}^*\boldsymbol{\omega} = 1.\end{aligned}$$

On the other hand,

$$\begin{aligned}\mathbf{a}(\mathfrak{u} + \mathfrak{u}_C) &= \mathbf{a}\mathfrak{u} + \mathbf{a}\mathfrak{u}_C = (\boldsymbol{\omega} + \boldsymbol{\omega}^*)\mathfrak{u} + (\boldsymbol{\omega} + \boldsymbol{\omega}^*)\mathfrak{u}_C \\ &= \boldsymbol{\omega}\mathfrak{u} + \boldsymbol{\omega}^*\mathfrak{u} + \boldsymbol{\omega}\mathfrak{u}_C + \boldsymbol{\omega}^*\mathfrak{u}_C = \mathfrak{u} + \mathfrak{u}_C.\end{aligned}$$

The equivalence class $[\mathbf{a}]$ is defined from the relation $\mathbf{a} \sim \mathbf{a} + \sigma\mathbf{p}$, where $\sigma \in \mathbb{C}$. Indeed, the unit vector $\mathbf{a}$ is defined up to the sum of a scalar multiple of $\mathbf{y}$, since

$$(\mathbf{a} + \sigma\mathbf{p})(\mathfrak{u} + \mathfrak{u}_C) = \mathbf{a}(\mathfrak{u} + \mathfrak{u}_C) + \sigma\mathbf{p}(\mathfrak{u} + \mathfrak{u}_C) = \mathbf{a}(\mathfrak{u} + \mathfrak{u}_C). \tag{6.129}$$

The sum $\mathfrak{u} + \mathfrak{u}_C$ is a Majorana spinor, as in fact this spinor is an eigenspinor of the charge conjugation operator C:

$$C[\psi] \equiv \psi_C = (\mathfrak{u} + \mathfrak{u}_C)_C = \mathfrak{u}_C + \mathfrak{u} = \psi. \tag{6.130}$$

6.9 Dual Rotations, and the Penrose Flagpole

Penrose flagpoles can be characterised immediately from the formalism introduced in section 6.8. For more details, see, for example, the work by Penrose and Rindler (1984) and Benn and Tucker (1987). First, let $\mathfrak{u}_1$ and $\mathfrak{u}_2$ be pure spinors representing M_1 and M_2, where $\dim_{\mathbb{C}}(M_1 \cap M_2) = k$; then, $\langle \mathfrak{u}_2\widehat{\mathfrak{u}_1}\rangle_k = \Omega_{M_1 \cap M_2}$. Here, $\Omega_{M_1 \cap M_2}$ denotes the volume element $M_1 \cap M_2$, and $\widehat{\psi}$ denotes the adjoint spinor associated with ψ with respect to the spinor inner product $\tilde{h}(\psi, \phi)$ in eqn (6.56). This result is a theorem whose proof can be seen in the work by Benn and Tucker (1987). Hence, since $\mathrm{span}\{\mathbf{y}\} = M_1 \cap M_2$, and $\dim_{\mathbb{C}}(M_1 \cap M_2) = 1$, therefore $\langle i\mathfrak{u}\mathfrak{u}_C\rangle_1$ is a scalar multiple of $\mathbf{p} = \Omega_{M_1 \cap M_2}$. The factor $i = \sqrt{-1}$ is introduced in order to turn $\mathbf{p} = \langle i\mathfrak{u}\mathfrak{u}_C\rangle_1$ into a real vector. In the Lorentzian case, we can always choose a basis for the spinor space such that the charge conjugation is equivalent to the spinor component conjugation (Chevalley, 1954). In this way,

$$(\mathfrak{u}_1, \mathfrak{u}_2)^* = ((\mathfrak{u}_1)_C, (\mathfrak{u}_2)_C), \tag{6.131}$$

where (,) is the spin product with adjoint induced by the reversion, namely, $(\psi, \phi) = \widehat{\psi}(\phi) = \tilde{h}(\psi, \phi)$. By taking a basis $\{\mathbf{e}_j\}$ of $V_\mathbb{C}$, it follows that

$$\langle i\mathfrak{u}\widehat{\mathfrak{u}}_C\rangle_1 = \langle i\mathfrak{u}\widehat{\mathfrak{u}}_C\mathbf{e}_j\rangle_0\mathbf{e}^j = \langle \mathfrak{u}_C i\mathbf{e}_j\mathfrak{u}\rangle_0\mathbf{e}^j = (\mathfrak{u}_C, i\mathbf{e}_j\mathfrak{u})\mathbf{e}^j. \tag{6.132}$$

Then,

$$\langle i\mathfrak{u}\widehat{\mathfrak{u}}_C\rangle_1^* = -(\mathfrak{u}, i\mathbf{e}_j\mathfrak{u}_C)\mathbf{e}^j = -(i\mathbf{e}_j\mathfrak{u}, \mathfrak{u}_C)\mathbf{e}^j = (\mathfrak{u}_C, i\mathbf{e}_j\mathfrak{u})\mathbf{e}^j = \langle i\mathfrak{u}\widehat{\mathfrak{u}}_C\rangle_1. \tag{6.133}$$

Therefore, $\mathbf{p}^* = \mathbf{p}$, where $\mathbf{p} \in V$ is determined up to a real scalar. The vector $\mathbf{p}$ is identified modulo a real scalar to a family of coplanar vectors, which determine the Penrose flagpole.

Let now $\{\mathbf{p}, \boldsymbol{\omega}\}$ be another basis of T, where $\boldsymbol{\omega} \in \bigwedge_1(\mathbb{C}^{3,1})$ satisfies $\boldsymbol{\omega}\mathfrak{u}_C = \mathfrak{u}$ and is well defined up to a scalar multiple of $\mathbf{x}$, since $\mathbf{x}\mathfrak{u}_C = 0$. From the previously obtained results, we know that $i\mathfrak{u}\widehat{\mathfrak{u}}$ is a complex multiple of $\mathbf{x}\boldsymbol{\omega}$. Since $\mathbf{p}\mathfrak{u}_C = 0$, therefore $[\omega] = \boldsymbol{\omega} + \zeta\mathbf{p}$, $\zeta \in \mathbb{C}$. Hence, $\boldsymbol{\omega}$ is determined up to the sum with a scalar multiple of $\mathbf{p}$.

Likewise, another possible characterisation for a pure spinor is given by the assertion that a spinor $\mathfrak{u}$ is pure if and only if $\langle \mathfrak{u}\widehat{\mathfrak{u}}\rangle_k = 0$, for all $k \neq r$, where $r = \dim_\mathbb{C} M_1$ (Benn and Tucker, 1987). Therefore, in this case, if $M_1 = M_2$, then $M_1 \cap M_2 = M_1$, and $\dim_\mathbb{C}(M_1 \cap M_2) = \dim_\mathbb{C} M_1 = 2$, where one concludes that $\langle \mathfrak{u}\widehat{\mathfrak{u}}\rangle_2$, which is the volume element of $M_1 \cap M_2 = M_1$, is a scalar multiple of $\mathbf{p}\boldsymbol{\omega}$. Suppose that this scalar is $2e^{i\theta}$. It follows that

$$\langle i\mathfrak{u}\widehat{\mathfrak{u}}\rangle_2 = i\mathfrak{u}\widehat{\mathfrak{u}} = 2\exp(i\theta)\mathbf{p}\boldsymbol{\omega}. \tag{6.134}$$

When the *flagpole* is defined as

$$G = \frac{1}{2}(i\mathfrak{u}\widehat{\mathfrak{u}} - i\mathfrak{u}_C\widehat{\mathfrak{u}}_C), \tag{6.135}$$

it is possible to write it as

$$G = \exp(i\theta)\mathbf{p}\boldsymbol{\omega} + \exp(-i\theta)\mathbf{p}\boldsymbol{\omega}^*, \tag{6.136}$$

and, therefore,

$$\begin{aligned}
G(\omega + \omega^*) &= (\exp(i\theta)\mathbf{p}\boldsymbol{\omega} + \exp(-i\theta)\mathbf{p}\boldsymbol{\omega}^*)(\boldsymbol{\omega} + \boldsymbol{\omega}^*)\\
&= \exp(i\theta)\mathbf{p}\boldsymbol{\omega}^2 + \exp(i\theta)\mathbf{p}\boldsymbol{\omega}\boldsymbol{\omega}^* + \exp(-i\theta)\mathbf{p}\boldsymbol{\omega}^*\boldsymbol{\omega} + \exp(-i\theta)\mathbf{p}(\boldsymbol{\omega}^*)^2\\
&= \cos\theta\mathbf{p}\boldsymbol{\omega}\boldsymbol{\omega}^* + i\sin\theta\mathbf{p}\boldsymbol{\omega}\boldsymbol{\omega}^* + \cos\theta\mathbf{p}\boldsymbol{\omega}^*\boldsymbol{\omega} - i\sin\theta\mathbf{p}\boldsymbol{\omega}^*\boldsymbol{\omega}\\
&= \cos\theta\mathbf{p}(\boldsymbol{\omega}\boldsymbol{\omega}^* + \boldsymbol{\omega}^*\boldsymbol{\omega}) + i\sin\theta\mathbf{p}(\boldsymbol{\omega}\boldsymbol{\omega}^* - \boldsymbol{\omega}^*\boldsymbol{\omega})\\
&= \cos\theta\mathbf{p} + 2i\sin\theta\mathbf{p}\boldsymbol{\omega}\wedge\boldsymbol{\omega}^*\\
&= \cos\theta\mathbf{p} + 2i\sin\theta\mathbf{p}\wedge\boldsymbol{\omega}\wedge\boldsymbol{\omega}^*.
\end{aligned} \tag{6.137}$$

The last equality comes from the fact that M_1 is a maximal totally isotropic subspace, and $\mathbf{p}\cdot\omega = 0$. From the expression (6.136), when $\theta = 0$ we obtain

$$G\big|_{\theta=0} = \mathbf{p}(\omega+\omega^*) = \mathbf{pa} = \mathbf{p}\wedge\mathbf{a}, \tag{6.138}$$

since $\mathbf{p}$ and $\mathbf{a}$ are elements of M_1, which is isotropic. Hence,

$$G\big|_{\theta=0} = F = \mathbf{p}\wedge\mathbf{a}. \tag{6.139}$$

From the definition given in eqn (6.135), we can assert that

$$\begin{aligned} 2G(\omega+\omega^*) &= i\mathrm{u}\widehat{\mathrm{u}}\,(\omega+\omega^*) - i\mathrm{u}_C\widehat{\mathrm{u}}_C(\omega+\omega^*) \\ &= i\mathrm{u}[(\omega+\omega^*)\mathrm{u}]^{\frown} - i\mathrm{u}_C[(\omega+\omega^*)\mathrm{u}_C]^{\frown} \\ &= i\mathrm{u}(\widehat{\omega\mathrm{u}} + \widehat{\omega^*\mathrm{u}}) - i\mathrm{u}_C(\widehat{\omega\mathrm{u}_C} + \widehat{\omega^*\mathrm{u}_C}) \\ &= i\mathrm{u}\widehat{\mathrm{u}_C} - i\mathrm{u}_C\widehat{\mathrm{u}}, \end{aligned} \tag{6.140}$$

since $\omega^*\mathrm{u} = \mathrm{u}_C$; $\omega\mathrm{u}_C = \mathrm{u}$; and $\omega\mathrm{u} = 0$.

Equation (6.139) can be written as

$$F = \mathrm{Re}(i\mathrm{u}\widehat{\mathrm{u}}). \tag{6.141}$$

Indeed, considering φ, ξ, ϕ, and ψ as arbitrary spinors, we know that

$$\begin{aligned} (\varphi, \widetilde{(\phi\widehat{\psi})}\,\xi) &= ((\phi\widehat{\psi})\varphi, \xi) = (\psi,\varphi)(\phi,\xi) = -(\varphi,\psi)(\phi,\xi) \\ &= -(\varphi, \widetilde{(\psi\widehat{\phi})}\,\xi), \end{aligned} \tag{6.142}$$

and then $\widetilde{(\phi\widehat{\psi})} = -\widetilde{(\psi\widehat{\phi})}$. It follows from eqn (6.140) that

$$\begin{aligned} G(\omega+\omega^*) &= i\mathrm{u}\widehat{\mathrm{u}_C} - i\mathrm{u}_C\widehat{\mathrm{u}} \\ &= i\mathrm{u}\widehat{\mathrm{u}_C} + \widetilde{(i\mathrm{u}\widehat{\mathrm{u}_C})}, \end{aligned} \tag{6.143}$$

and, since $i\Omega\psi = \psi$, therefore $\widetilde{\mathrm{u}\widehat{\mathrm{u}_C}} = -\Omega\mathrm{u}\widehat{\mathrm{u}_C}\Omega = -\Omega\mathrm{u}(\widetilde{\Omega\widehat{\mathrm{u}_C}}) = -i\Omega\mathrm{u}(\widetilde{i\Omega\widehat{\mathrm{u}_C}}) = -\mathrm{u}\widehat{\mathrm{u}_C}$, where we conclude that $\mathrm{u}\widehat{\mathrm{u}_C} \in \bigwedge_1(\mathbb{C}^{3+1})\oplus\bigwedge_3(\mathbb{C}^{3+1})$. Since $\langle\mathrm{u}\widehat{\mathrm{u}_C}\rangle_3$ changes sign under reversion, in order for eqn (6.143) to give a non-trivial solution for $G(\omega+\omega^*)$, it follows that $G(\omega+\omega^*) = \langle\mathrm{u}\widehat{\mathrm{u}_C}\rangle_1 = \mathbf{p}$, from eqn (6.133). For the equation $G\big|_{\theta=0} = F = \mathbf{p}\wedge\mathbf{a}$ to hold, the required result must follow.

Now, both the real vector $\mathbf{p}$ and the bivector F can be written from a Majorana spinor and the volume element of a maximal totally isotropic subspace $\mathbb{C}^{3,1}$ as $\mathbf{p} = \frac{1}{2}\langle\psi(\widehat{\Omega\psi})\rangle_1$, and respectively $F = \frac{1}{2}\langle\psi(\widehat{\Omega\psi})\rangle_2$. Here, ψ is a Majorana spinor, and

$\Omega_F = \mathbf{u}_{0123}$ is the volume element of $F \subset V_{\mathbb{C}}$. In fact, a Majorana spinor can be written as $\psi = \mathfrak{u} + \mathfrak{u}_C$; since $\mathfrak{u} = i\Omega\mathfrak{u}$, therefore

$$\begin{aligned}
\psi(\widehat{\Omega\psi}) &= ((\mathfrak{u}+\mathfrak{u}_C)(z\widehat{(\mathfrak{u}+\mathfrak{u}_C)})) = (\mathfrak{u}+\mathfrak{u}_C)(\widehat{(\mathfrak{u}+\mathfrak{u}_C)}\widehat{z}) \\
&= (\mathfrak{u}+\mathfrak{u}_C)((\widehat{\mathfrak{u}}+\widehat{\mathfrak{u}_C})\widehat{z}) = (\mathfrak{u}+\mathfrak{u}_C)(\widehat{\mathfrak{u}}\widehat{z}+\widehat{\mathfrak{u}_C}\widehat{z}) \\
&= (\mathfrak{u}+\mathfrak{u}_C)(-i\widehat{\mathfrak{u}}+i\widehat{\mathfrak{u}_C}) \\
&= -i\mathfrak{u}\widehat{\mathfrak{u}}+i\mathfrak{u}_C\widehat{\mathfrak{u}_C}+i\mathfrak{u}\widehat{\mathfrak{u}_C}-i\mathfrak{u}_C\widehat{\mathfrak{u}} \\
&= (-i\mathfrak{u}\widehat{\mathfrak{u}}+i\mathfrak{u}_C\widehat{\mathfrak{u}_C})+(i\mathfrak{u}\widehat{\mathfrak{u}_C}+i\widetilde{(\mathfrak{u}\widehat{\mathfrak{u}_C})}),
\end{aligned} \tag{6.144}$$

since $\mathfrak{u} = iz\mathfrak{u}$ implies that $\mathfrak{u}_C = -iz\mathfrak{u}_C$. Now, since $\widetilde{(\psi\widehat{\phi})} = -\phi\widehat{\psi}$, the term $-i\mathfrak{u}\widehat{\mathfrak{u}} + i\mathfrak{u}_C\widehat{\mathfrak{u}_C}$ is odd under a reversion. However, the term $i\mathfrak{u}\widehat{\mathfrak{u}_C} + i\widetilde{(\mathfrak{u}\widehat{\mathfrak{u}_C})}$ is even; this result implies from eqn (6.133) that $\mathbf{p} = \frac{1}{2}\langle\psi(\widehat{z\psi})\rangle_1$, and $F = \frac{1}{2}\langle\psi(\widehat{z\psi})\rangle_2$.

Here, we stated that $i\Omega\mathfrak{u} = \mathfrak{u}$, where $\Omega = \mathbf{u}_{0123}$ is the volume element associated with the maximal totally isotropic subspace $F \subset V_{\mathbb{C}}$; this statement can be forthwith proved. The object $\mathring{\Omega}$ was defined as a $2r$-form such that $\mathring{\Omega}^2 = 1$, so that the idempotents $P_\pm = \frac{1}{2}(1+\mathring{\Omega})$ reduce $\mathcal{C}\ell_{3,1}^+(\mathbb{C})$ to simple ideals; consequently, $\mathring{\Omega} = i\mathbf{e}_{0123} = i\Omega$. Since $\mathfrak{u}$ was defined as an even spinor, therefore $\mathring{\Omega}\mathfrak{u} = \mathring{\Omega}s\Omega_F = s\mathring{\Omega}\Omega_F = s\Omega_F = \mathfrak{u}$, since $\mathfrak{u} = s\Omega_F$ is the form of $\mathfrak{u}$ that represents F. Finally, in order to see that $\Omega\Omega_F = \Omega_F$, it suffices to see that, in general, given a Witt basis $\{\mathbf{u}_i, \mathbf{v}_j\}$ for $V_{\mathbb{C}}$, we have that $\Omega = \mathbf{u}_1 \wedge \mathbf{v}_1 \wedge \mathbf{u}_2 \wedge \mathbf{v}_2 \wedge \cdots \wedge \mathbf{u}_r \wedge \mathbf{v}_r = [\mathbf{u}_1, \mathbf{v}_1][\mathbf{u}_2, \mathbf{v}_2]\ldots[\mathbf{u}_r, \mathbf{v}_r]$ and, therefore,

$$\begin{aligned}
\Omega\Omega_F &= [\mathbf{u}_1, \mathbf{v}_1][\mathbf{u}_2, \mathbf{v}_2]\ldots[\mathbf{u}_r, \mathbf{v}_r]\mathbf{u}_1\mathbf{u}_2\ldots\mathbf{u}_r \\
&= [\mathbf{u}_1, \mathbf{v}_1]\mathbf{u}_1[\mathbf{u}_2, \mathbf{v}_2]\mathbf{u}_2\ldots[\mathbf{u}_r, \mathbf{v}_r]\mathbf{u}_r \\
&= \mathbf{u}_1\mathbf{u}_2\ldots\mathbf{u}_r = \mathbf{u}_1 \wedge \mathbf{u}_2 \wedge \cdots \wedge \mathbf{u}_r = \Omega_F,
\end{aligned} \tag{6.145}$$

since $[\mathbf{u}_i, \mathbf{v}_i]\mathbf{u}_i = \mathbf{u}_i\mathbf{v}_i\mathbf{u}_i = (1 - \mathbf{v}_i\mathbf{u}_i)\mathbf{u}_i = \mathbf{u}_i$. Here, the obvious notation for the commutator $[a, b] = ab - ba$ is used, for $a, b \in \mathcal{C}\ell_{p,q}$.

If $\mathfrak{v}$ denotes a pure spinor related to $\mathfrak{u}$ by

$$\mathfrak{v} = \exp(i\theta)\mathfrak{u}, \tag{6.146}$$

then

$$\langle i\mathfrak{v}\widehat{\mathfrak{v}_C}\rangle_1 = \langle i\exp(i\theta)\mathfrak{u}\exp(-i\theta)\widehat{\mathfrak{u}_C}\rangle_1 = \langle i\mathfrak{u}\widehat{\mathfrak{u}_C}\rangle_1 = \mathbf{p}, \tag{6.147}$$

and therefore the spinor $\mathfrak{v}$ determines the same null direction that $\mathfrak{u}$ does.

If $\mathfrak{v}$ determines the 2-form F', then

$$\begin{aligned}
F' &= \mathrm{Re}(i\mathfrak{v}\widehat{\mathfrak{v}}) = \mathrm{Re}(i\exp(i\theta)\mathfrak{u}\exp(i\theta)\widehat{\mathfrak{u}}) \\
&= \mathrm{Re}(i\cos(2\theta)\mathfrak{u}\widehat{\mathfrak{u}} - \sin(2\theta)\mathfrak{u}\widehat{\mathfrak{u}}) \\
&= \cos(2\theta)\mathrm{Re}(i\mathfrak{u}\widehat{\mathfrak{u}}) - \sin(2\theta)\mathrm{Re}(\Omega i\mathfrak{u}\widehat{\mathfrak{u}}) \\
&= \cos(2\theta)F - \sin(2\theta)\Omega F.
\end{aligned} \tag{6.148}$$

Consequently,

$$F' = \exp(-2\theta\Omega)F. \tag{6.149}$$

Therefore, F' is associated with F by a dual rotation, since

$$\star F = \tilde{F}\Omega = \Omega\tilde{F} = -\Omega F, \tag{6.150}$$

and it follows that

$$\Omega F = -\star F. \tag{6.151}$$

It finally reads

$$F' = \cos(2\theta)F + \sin(2\theta)\star F\,, \tag{6.152}$$

and we thus identify F with the Penrose flagpole structure. In addition, the flagpole F rotates by the angle 2θ, and the spinor $\mathfrak{u}$ associated with F rotates by θ. This flagpole is a generalisation of the Penrose flagpole, which is a particular case when $\theta = 0$.

6.10 Weyl Spinors in $\mathcal{C}\ell_{3,0}$

This section aims to uniquely introduce the Weyl spinors from the Clifford algebra $\mathcal{C}\ell_{3,0}$, which is exactly the Penrose formalism in an algebraic language which is more general and accessible than that used in the Penrose formalism (Penrose and Rindler, 1984). Moreover, this formulation emulates the van der Waerden framework (van der Waerden, 1928; Veblen, 1933), which is revisited in the appendix. First, let us define the idempotents $f_\pm = \frac{1}{2}(1 \pm \mathbf{e}_3)$ – which satisfy the relations $f_+f_- = f_-f_+ = 0$, and $f_\pm^2 = f_\pm$. Using the isomorphism $\mathcal{C}\ell_{3,0} \simeq \mathcal{M}(2,\mathbb{C})$ given by $\mathbf{e}_i \mapsto \sigma_i$, where σ_i denotes the Pauli matrices, it is immediately obvious that

$$\begin{aligned} f_+ &\mapsto \begin{pmatrix} 1 & 0 \\ 0 & 0 \end{pmatrix}, \quad \mathbf{e}_1 f_+ \mapsto \begin{pmatrix} 0 & 0 \\ 1 & 0 \end{pmatrix}, \\ f_- &\mapsto \begin{pmatrix} 0 & 0 \\ 0 & 1 \end{pmatrix}, \quad \mathbf{e}_1 f_- \mapsto \begin{pmatrix} 0 & 1 \\ 0 & 0 \end{pmatrix}. \end{aligned} \tag{6.153}$$

The isomorphism $\mathcal{C}\ell_{3,0}f_+ \simeq \mathcal{C}\ell_{3,0}^+ f_+$ is straightforward to realise

$$\begin{aligned} \mathcal{C}\ell_{3,0}^+ f_+ \ni \phi_+ f_+ &= \begin{pmatrix} w_1 & -w_2^* \\ w_2 & w_1^* \end{pmatrix} f_+ = \begin{pmatrix} w_1 & 0 \\ w_2 & 0 \end{pmatrix} \\ &\simeq \begin{pmatrix} w_1 & w_3 \\ w_2 & w_4 \end{pmatrix} \begin{pmatrix} 1 & 0 \\ 0 & 0 \end{pmatrix} \in \mathcal{C}\ell_{3,0}f_+. \end{aligned}$$

An algebraic spinor can be written as $\mathcal{K} = \psi f_+$, where $\psi = s + b^{12}\mathbf{e}_{12} + b^{13}\mathbf{e}_{13} + b^{23}\mathbf{e}_{23} \in \mathcal{C}\ell_{3,0}^+$, and the coefficients are real numbers. We can define the

- *undotted contravariant spinor:*

$$\begin{aligned} \mathcal{K} = \psi f_+ &= (s + b^{12}\mathbf{e}_{123})(f_+) + (b^{13} + b^{23}\mathbf{e}_{123})(\mathbf{e}_1 f_+) \\ &= k^1(f_+) + k^2(\mathbf{e}_1 f_+), \end{aligned} \tag{6.154}$$

where $k^1 = s + b^{12}\mathbf{e}_{123}$, and $k^2 = b^{13} + b^{23}\mathbf{e}_{123}$. Spinors are expressed in such a way that their components commute with the basis $\{f_+, \mathbf{e}_1 f_+\}$ of the space of the algebraic spinors. Hence, all components are written as elements of the centre of $\mathcal{C}\ell_{3,0}$.

From the spinor $\mathcal{K}$, three other types of spinors can be defined. The first is the

- *undotted covariant spinor:*

$$\begin{aligned} \mathcal{K}^* = \mathbf{e}_1\overline{\mathcal{K}} = \mathbf{e}_1\overline{(k_1 f_+ + k_2\mathbf{e}_1 f_+)} &= \mathbf{e}_1(f_- k^1 + f_-(-\mathbf{e}_1)k^2) \\ f &= (-k^2)f_+ + (k^1)(f_+\mathbf{e}_1). \end{aligned} \tag{6.155}$$

Since $\mathcal{K}^* \in f_+\mathcal{C}\ell_{3,0}$, we can write $\mathcal{K}^* = k_1(f_+) + k_2(f_+\mathbf{e}_1)$. The relationships between the upper and lower components read

$$k_1 = -k^2, \qquad\qquad k_2 = k^1\,,$$

and are similar to those in the standard van der Waerden formalism. For more details, see the work by van der Waerden (1928), Veblen (1933), Penrose (1967), and Penrose and Rindler (1984).

Given an algebraic spinor $\mathcal{K}^* \in f_+\mathcal{C}\ell_{3,0}$ and $\mathcal{L} = \eta^1 f_+ + \eta^2\mathbf{e}_1 f_+ \in \mathcal{C}\ell_{3,0}f_+$ – where η^1 and η^2 are the spinor $\mathcal{L}$ components that are elements of the centre $\mathcal{C}\ell_{3,0}$ – the spinor metric associated with the idempotent f_+ can be defined as follows:

$$\begin{aligned} G_{f_+} : f_+\mathcal{C}\ell_{3,0} \times \mathcal{C}\ell_{3,0}f_+ &\to f_+\mathcal{C}\ell_{3,0}f_+ \simeq \mathbb{C}f_+ \\ (\mathcal{K},\mathcal{L}) &\mapsto G_{f_+}(\mathcal{K},\mathcal{L}) = \mathcal{K}^*\mathcal{L} \\ &\qquad = (-k^2 f_+ + k^1 f_+\mathbf{e}_1)(\eta^1 f_+ + \eta^2\mathbf{e}_1 f_+)\,, \end{aligned} \tag{6.156}$$

which leads to

$$G_{f_+}(\mathcal{K},\mathcal{L}) = \mathcal{K}^*\mathcal{L} = (-k^2\eta^1 + k^1\eta^2)f_+. \tag{6.157}$$

This definition coincides with the classical spinor definition, where the scalar product has mixed and antisymmetric components, and f_+ plays the role of the unit in the algebra $f_+\mathcal{C}\ell_{3,0}f_+ \simeq \mathbb{C}$.

Moreover, from $\mathcal{K}$ we define the

- *dotted contravariant spinor:*

$$\begin{aligned} \underline{\mathcal{K}} = \mathbf{e}_1\tilde{\mathcal{K}} = \mathbf{e}_1(k^1 f_+ + k^2\mathbf{e}_1 f_+)^{\sim} &= \mathbf{e}_1(f_+\tilde{k^1} + f_+\mathbf{e}_1\tilde{k^2}) \\ = \tilde{k^1}(\mathbf{e}_1 f_+) + \tilde{k^2}f_- &= \tilde{k^1}(f_-\mathbf{e}_1) + \tilde{k^2}(f_-). \end{aligned} \tag{6.158}$$

Since $\underline{\mathcal{K}} \in f_-\mathcal{C}\ell_{3,0}$, we can write $\underline{\mathcal{K}} = \overline{k}^{\dot{1}}(f_-\mathbf{e}_1) + \overline{k}^{\dot{2}}(f_-)$, obtaining the equivalences

$$\overline{k}^{\dot{1}} = \tilde{k^1}, \qquad\qquad \overline{k}^{\dot{2}} = \tilde{k^2}.$$

In addition,[6] $\tilde{k^\alpha} = (a + b\mathbf{e}_{123})^{\sim} = (a + b\mathbf{e}_{321}) = a - b\mathbf{e}_{123}$, which suggests the notation $\tilde{k^\alpha} = \overline{k^\alpha}$.[7]

Finally, we construct the

[6] $\alpha = 1, 2$.

[7] By denoting $\mathcal{C}\ell_s$ the scalars and $\mathcal{C}\ell_p$ the pseudoscalars, we obtain the isomorphism $\mathcal{C}\ell_s \oplus \mathcal{C}\ell_p \simeq \mathbb{C}$, since $(\mathbf{e}_{123}^2 = -1)$. The notation $\tilde{k^\alpha} = \overline{k^\alpha}$ is therefore evident, since $\mathbf{e}_{123}$ plays the role of the imaginary unit of $\mathbb{C}$, and the reversion in $\mathcal{C}\ell_p$ is equivalent to the complex conjugation.

- *dotted covariant spinor:*

$$\begin{aligned}\underline{\mathcal{K}}^* &= \overline{(\mathbf{e}_1\underline{\mathcal{K}})} = -\overline{(\underline{\mathcal{K}})}\mathbf{e}_1 = -\overline{\bar{k}^{\dot{1}}(f_-\mathbf{e}_1) + \bar{k}^{\dot{2}}(f_-)}\mathbf{e}_1 \\ &= -(-\mathbf{e}_1 f_+ \bar{k}^{\dot{1}} + f_+ \bar{k}^{\dot{2}})\mathbf{e}_1 = \bar{k}^{\dot{1}} f_- - \bar{k}^{\dot{2}} f_+ \mathbf{e}_1 \\ &= (-\bar{k}^{\dot{2}})(\mathbf{e}_1 f_-) + (\bar{k}^{\dot{1}})(f_-).\end{aligned} \tag{6.159}$$

Since $\underline{\mathcal{K}}^* \in \mathcal{C}\ell_{3,0}f_-$, we can write $\underline{\mathcal{K}}^* = (\bar{k}_{1'})(\mathbf{e}_1 f_-) + (\bar{k}_{\dot{2}})(f_-)$, and the expressions

$$\bar{k}_{\dot{1}} = -\bar{k}^{\dot{2}}, \qquad \bar{k}_{\dot{2}} = \bar{k}^{\dot{1}} \tag{6.160}$$

hold.

Given $\underline{\mathcal{K}} \in f_-\mathcal{C}\ell_{3,0}$, and $\underline{\mathcal{L}}^* = \bar{\eta}^{\dot{2}}\mathbf{e}_1 f_+ + \bar{\eta}^{\dot{1}} f_- \in \mathcal{C}\ell_{3,0}$, the spinor metric associated with the idempotent f_- is defined as follows:

$$\begin{aligned} G_{f_-} &: f_-\mathcal{C}\ell_{3,0} \times \mathcal{C}\ell_{3,0}f_- \to f_-\mathcal{C}\ell_{3,0}f_- \simeq \mathbb{C}f_- \\ &(\underline{\mathcal{K}}, \underline{\mathcal{L}}^*) \mapsto \underline{\mathcal{K}}\underline{\mathcal{L}}^* = (\bar{k}^{\dot{1}} f_-\mathbf{e}_1 + \bar{k}^{\dot{2}} f_-)(-\bar{\eta}^{\dot{2}}\mathbf{e}_1 f_- + \bar{\eta}^{\dot{1}} f_-),\end{aligned}$$

and, therefore,

$$G_{f_-} = (\bar{\eta}^{\dot{1}}\bar{k}^{\dot{2}} - \bar{\eta}^{\dot{2}}\bar{k}^{\dot{1}})f_- \,. \tag{6.161}$$

The expressions for the four types of Weyl spinors, as elements of an ideal of $\mathcal{C}\ell_{3,0}$, are as follows:

- undotted contravariant spinor:

$$\boxed{\mathcal{K} = k^1(f_+) + k^2(\mathbf{e}_1 f_+) \in \mathcal{C}\ell_{3,0}f_+}$$

- undotted covariant spinor:

$$\boxed{\mathcal{K}^* = \mathbf{e}_1\widetilde{\mathcal{K}} = k_1(f_+) + k_2(f_+\mathbf{e}_1) \in f_+\mathcal{C}\ell_{3,0}}$$

- dotted contravariant spinor:

$$\boxed{\underline{\mathcal{K}} = \mathbf{e}_1\bar{\mathcal{K}} = \bar{k}^{\dot{1}}(f_-\mathbf{e}_1) + \bar{k}^{\dot{2}}(f_-) \in f_-\mathcal{C}\ell_{3,0}}$$

- dotted covariant spinor:

$$\boxed{\underline{\mathcal{K}}^* = -(\overline{\underline{\mathcal{K}}})\mathbf{e}_1 = (\overline{\mathbf{e}_1\underline{\mathcal{K}}}) = \bar{k}_{\dot{1}}(\mathbf{e}_1 f_-) + \bar{k}_{\dot{2}}(f_-) \in \mathcal{C}\ell_{3,0}f_-}$$

Taking now into account the operations already defined, the diagram in figure 6.1 illustrates how to pass from one ideal to another in $\mathcal{C}\ell_{3,0}$.

$\mathcal{K}$	$\underset{*}{\rightarrow}$	$\mathcal{K}^*$	$\underset{-(\hat{\ })}{\rightarrow}$	$\underline{\mathcal{K}}$	$\underset{*}{\rightarrow}$	$\underline{\mathcal{K}}^* = \widehat{\mathcal{K}}$
↑		↑		↑		↑
contravariant		covariant		contravariant		covariant
undotted		undotted		dotted		dotted
$\mathcal{C}\ell_{3,0}f_+$		$f_+\mathcal{C}\ell_{3,0}$		$f_-\mathcal{C}\ell_{3,0}$		$\mathcal{C}\ell_{3,0}f_-$

Figure 6.1 Passing from One Ideal to Another in $\mathcal{C}\ell_{3,0}$

6.11 Weyl Spinors in the Clifford Algebra $\mathcal{C}\ell_{0,3} \simeq \mathbb{H} \oplus \mathbb{H}$

Consider the Euclidean space $\mathbb{R}^3$, and an orthonormal basis $\{\mathfrak{e}_1, \mathfrak{e}_2, \mathfrak{e}_3\}$. The Clifford algebra $\mathcal{C}\ell_{0,3}$ is defined by the relations $g(\mathfrak{e}_i, \mathfrak{e}_j) = -\delta_{ij} = \frac{1}{2}(\mathfrak{e}_i\mathfrak{e}_j + \mathfrak{e}_j\mathfrak{e}_i)$, where $i, j = 1, 2, 3$, and, in particular, $\mathfrak{e}_i^2 = -1$.

By using the very same procedure as we did for $\mathcal{C}\ell_{3,0}$, we first reduce the redundant dimensions, proving that $\mathcal{C}\ell_{0,3}\mathfrak{f}_+ = \mathcal{C}\ell_{0,3}^+\mathfrak{f}_+$, where $\mathfrak{f}_\pm = \frac{1}{2}(1 \pm \mathbf{e}_{123})$. Indeed, the left ideal $\mathcal{C}\ell_{0,3}\mathfrak{f}_+$ is algebraically isomorphic to $\mathbb{H}$. In addition, writing an arbitrary element of $\mathcal{C}\ell_{0,3}$ as

$$\begin{aligned} A &= a^0 + a^k\mathfrak{e}_k + b^1\mathfrak{e}_{23} + b^2\mathfrak{e}_{31} + b^3\mathfrak{e}_{12} + b^0\mathfrak{e}_{123} \\ &= a^0 + a^k\mathfrak{e}_k - b^k\mathfrak{e}_k\mathfrak{e}_{123} + b^0\mathfrak{e}_{123} \in \mathcal{C}\ell_{0,3}, \end{aligned}$$

we can see that

$$A\mathfrak{f}_+ = [(a^0 + b^0) + (a^k - b^k)\mathfrak{e}_k]\mathfrak{f}_+ = [(a^0 + b^0) + (a^k - b^k)\mathfrak{e}_k\mathfrak{e}_{123}]\mathfrak{f}_+.$$

Hence, given $A\mathfrak{f}_+ \in \mathcal{C}\ell_{0,3}\mathfrak{f}_+$, by writing $A' = (a^0 + b^0) + (a^k - b^k)\mathfrak{e}_k\mathfrak{e}_{123}$, we can see that $A' \in \mathcal{C}\ell_{0,3}^+$, and $A\mathfrak{f}_+ = A'\mathfrak{f}_+$. These results show that $\mathcal{C}\ell_{0,3}\mathfrak{f}_+ \subset \mathcal{C}\ell_{0,3}^+\mathfrak{f}_+$. The other inclusion immediately follows.

Consider an even element $Q \in \mathcal{C}\ell_{0,3}^+$:

$$\begin{aligned} Q &= a + b\mathfrak{e}_{12} + c\mathfrak{e}_{13} + d\mathfrak{e}_{23} = (a + b\mathfrak{e}_{12}) + \mathfrak{e}_{13}(c - d\mathfrak{e}_{12}) \\ &= k^1 + \mathfrak{e}_{13}k^2. \end{aligned}$$

Another way to describe Weyl spinors is to consider the algebra $\mathcal{C}\ell_{0,3} \simeq \mathbb{H} \oplus \mathbb{H}$, where a spinor $\mathcal{K} = Q\mathfrak{f}_+ \in \mathcal{C}\ell_{0,3}\mathfrak{f}_+$ can be expressed as an *undotted contravariant spinor*: $\mathcal{K} = (k^1\mathfrak{f}_+ + \mathfrak{e}_{13}k^2\mathfrak{f}_+)$.

Likewise, we can also define a *dotted contravariant spinor* $\underline{\mathcal{K}} = (\mathfrak{f}_+\bar{k}^1 - \mathfrak{f}_+\bar{k}^2\mathfrak{e}_{13})$.

Multiplying the conjugate of $\mathcal{K}$ to the left by $\mathfrak{e}_{13}$, we obtain

$$\mathfrak{e}_{13}\overline{\mathcal{K}} = \mathfrak{e}_{13}(\mathfrak{f}_+\overline{k^1} + \mathfrak{f}_+\overline{k^2}\mathfrak{e}_{31}) = \mathfrak{f}_+k^1\mathfrak{e}_{13} + \mathfrak{f}_+k^2,$$

and by performing the product with another spinor $\mathcal{L} \in \mathcal{C}\ell_{0,3}f_+$, we obtain

$$\mathfrak{e}_{13}\mathcal{K}\mathcal{L} = k^1\overline{\eta^1}\mathfrak{e}_{13}\mathfrak{f}_+ - k^1\eta^2\mathfrak{f}_+ + k^2\eta^1\mathfrak{f}_+ + k^2\overline{\eta^2}\mathfrak{e}_{13}\mathfrak{f}_+ .$$

Hence, the spinor metric in $\mathcal{C}\ell_{0,3}$ can be obtained:

$$\mathcal{G}(\mathcal{K}, \mathcal{L}) = 2\langle(\mathbf{e}_{13}\overline{\mathcal{K}})\mathcal{L}\rangle_0 = (k^2\eta^1 - k^1\eta^2)\mathfrak{f}_+$$

Consider now a mapping $\sigma : \mathcal{C}\ell_{0,3}^+ \to \mathcal{C}\ell_{0,3}^+$ defined by the expression

$$\sigma(Q) = \mathfrak{e}_{32}\overline{Q}\mathfrak{e}_{23}.$$

The mapping σ turns a right module into a right module. Indeed,

$$\begin{aligned}\sigma(k^1 + \mathfrak{e}_{13}k^2) &= \sigma(a + b\mathfrak{e}_{12} + c\mathfrak{e}_{13} + d\mathfrak{e}_{23}) \\ &= \mathfrak{e}_{32}(a + b\mathfrak{e}_{12} + c\mathfrak{e}_{13} + d\mathfrak{e}_{23})\mathfrak{e}_{23} \\ &= (a + b\mathfrak{e}_{12}) + (c - d\mathfrak{e}_{12})\mathfrak{e}_{13} = k^1 + k^2\mathfrak{e}_{13}.\end{aligned}$$

For a spinor $\mathcal{K} = Q\mathfrak{f}_+$, it follows that

$$\begin{aligned}\sigma(\psi) = \sigma(Q\mathfrak{f}_+) &= \mathfrak{e}_{32}(\overline{Q\mathfrak{f}_+})\mathfrak{e}_{23} = \mathfrak{e}_{32}(\mathfrak{f}_+\overline{Q}\mathfrak{e}_{23}) \\ &= \mathfrak{f}_+\mathfrak{e}_{32}\overline{Q}\mathfrak{e}_{23} = \mathfrak{f}_+\sigma(Q) \\ &= \mathfrak{f}_+(k^1 + k^2\mathfrak{e}_{13}).\end{aligned}$$

In this way,

$$\sigma(\mathcal{K})\mathfrak{e}_{13} = \mathfrak{f}_+(k^1\mathfrak{e}_{13} - k^2) = \mathfrak{f}_+(-k^2 + k^1\mathfrak{e}_{13}) = \mathcal{K}^*.$$

The spinor metric can also be defined as follows:

$$G(\psi, \phi) = \langle\sigma(\psi)\mathfrak{e}_{13}\phi\rangle_{\mathbb{C}} = \frac{1}{2}[\sigma(\psi)\mathfrak{e}_{13}\phi + \mathfrak{e}_{21}\sigma(\psi)\mathfrak{e}_{13}\phi\mathfrak{e}_{12}].$$

The algebra $\mathcal{C}\ell_{0,3}$ is not as natural for the spacetime metric description as $\mathcal{C}\ell_{3,0}$ is. In fact, it is suitable for the Euclidean space $\mathbb{R}^4$ since, for an element $\mathbf{u} \in \mathbb{R}^4$, we have $\mathbf{u}\bar{\mathbf{u}} = u_0^2 + \vec{u}^2$, where $u_0 \in \mathbb{R}$, $\vec{u} \in \mathbb{R}^3$. In addition, $\mathcal{C}\ell_{0,3} \simeq \mathbb{H} \oplus \mathbb{H}$ is a semisimple algebra, and the ring $\mathbb{H}$ is not commutative. Thus, we must distinguish the left and right products in $\mathbb{H}$. Furthermore, it has been proven that there exists a mapping σ that leads one module to the other.

6.12 Spinor Transformations

Consider now an arbitrary element

$$R = s + v^i\mathbf{e}_i + b^{ij}\mathbf{e}_{ij} + p\mathbf{e}_{123} = \alpha + \beta\mathbf{e}_{12} + \gamma\mathbf{e}_{13} + \delta\mathbf{e}_{23} \in \mathcal{C}\ell_{3,0},$$

where

$$\alpha = s + p\mathbf{e}_{123}, \quad \beta = b^{12} - v^3\mathbf{e}_{123}, \quad \gamma = b^{13} + v^2\mathbf{e}_{123}, \quad \text{and} \quad \delta = b^{23} - v^1\mathbf{e}_{123}$$

are elements of the centre of $\mathcal{C}\ell_{3,0}$. Under the action of R, an undotted contravariant spinor $\mathcal{K}$ behaves as $R\mathcal{K} = R(\psi f_+) = k^1(Rf_+) + k^2(R\mathbf{e}_1 f_+)$, where

$$\begin{aligned}Rf_+ &= (\alpha + \beta\mathbf{e}_{123})f_+ + (\gamma + \delta\mathbf{e}_{123})(\mathbf{e}_1 f_+), \\ R\mathbf{e}_1 f_+ &= (\alpha - \beta\mathbf{e}_{123})f_+ + (-\gamma + \delta\mathbf{e}_{123})(\mathbf{e}_1 f_+).\end{aligned}$$

A matrix representation $\rho : \mathcal{C}\ell_{3,0} \to \mathcal{M}(2, \mathbb{C})$ of R is given by

$$\rho(R) = \begin{pmatrix} \alpha + \beta i & -\gamma + \delta i \\ \gamma + \delta i & \alpha - \beta i \end{pmatrix}.$$

It follows that $\det \rho(R) = \alpha^2+\beta^2+\gamma^2+\delta^2$. Under the morphisms in $\mathcal{C}\ell_{3,0}$, the element $R \in \mathcal{C}\ell_{3,0}$ transforms as

$$\begin{aligned}\hat{R} &= \overline{\alpha} + \overline{\beta}\mathbf{e}_{12} + \overline{\gamma}\mathbf{e}_{13} + \overline{\delta}\mathbf{e}_{23},\\ \tilde{R} &= \overline{\alpha} - \overline{\beta}\mathbf{e}_{12} - \overline{\gamma}\mathbf{e}_{13} - \overline{\delta}\mathbf{e}_{23},\\ \overline{R} &= \alpha - \beta\mathbf{e}_{12} - \gamma\mathbf{e}_{13} - \delta\mathbf{e}_{23}.\end{aligned}$$

Hence, the relation

$$R\overline{R} = \alpha^2 + \beta^2 + \gamma^2 + \delta^2 = \det \rho(R)$$

holds. In this way, given $R \in \mathcal{C}\ell_{3,0}$ and demanding that $R \in \$\text{pin}_+(1,3)$, namely, $R\overline{R} = 1$, we find that $\det \rho(R) = 1$, and hence $R \in \mathrm{SL}(2,\mathbb{C})$. Consequently, the isomorphism

$$\$\text{pin}_+(1,3) \simeq \mathrm{SL}(2,\mathbb{C})$$

has been established. From the condition $R\overline{R} = 1$, it follows that $\overline{R} = R^{-1}$. Therefore, the transformation rules

$$\mathcal{K} \longmapsto R\mathcal{K}, \tag{6.162}$$

$$\mathcal{K}^* \longmapsto \mathbf{e}_1(\overline{R\mathcal{K}}) = \mathbf{e}_1\overline{\mathcal{K}}\overline{R} = \mathcal{K}^* R^{-1}, \tag{6.163}$$

$$\underline{\mathcal{K}} \longmapsto \mathbf{e}_1(\widetilde{R\mathcal{K}}) = \mathbf{e}_1\tilde{\mathcal{K}}\tilde{R} = \underline{\mathcal{K}}(\overline{\hat{R}}) = \underline{\mathcal{K}}(\hat{R})^{-1}, \tag{6.164}$$

$$\underline{\mathcal{K}}^* \longmapsto (\widehat{R\mathcal{K}}) = \hat{R}\underline{\mathcal{K}}^*, \tag{6.165}$$

hold for Weyl spinors. Such transformations on algebraic Weyl spinors are well known from the point of view of classical spinors. Indeed,

$$\rho(\hat{R}) = [\rho(R)^\dagger)]^{-1}.$$

When the representation of each one of the four spinors is taken into account, the correspondences

$$\mathcal{K} \longleftrightarrow \begin{pmatrix} k^1 & 0 \\ k^2 & 0 \end{pmatrix}, \quad \underline{\mathcal{K}} \longleftrightarrow \begin{pmatrix} 0 & 0 \\ \overline{k}^{\dot{1}} & \overline{k}^{\dot{2}} \end{pmatrix} = \begin{pmatrix} 0 & 0 \\ -\overline{k}_{\dot{2}} & \overline{k}_{1'} \end{pmatrix},$$

$$\mathcal{K}^* \longleftrightarrow \begin{pmatrix} -k^2 & k^1 \\ 0 & 0 \end{pmatrix}, \quad \underline{\mathcal{K}}^* \longleftrightarrow \begin{pmatrix} 0 & \overline{k}_{1'} \\ 0 & \overline{k}_{\dot{2}} \end{pmatrix}$$

hold. To summarise, in order to formulate the four types of Weyl spinors – and subsequently the Dirac spinor (Hladik, 1999) – it suffices to consider the ideal $\mathcal{C}\ell_{3,0}f_+$. Thus, the right and the left Clifford multiplication by the element $\mathbf{e}_1$ yield the other ideals – $f_+\mathcal{C}\ell_{3,0}$, $\mathcal{C}\ell_{3,0}f_-$, $f_-\mathcal{C}\ell_{3,0}$, whose elements are the other three types of Weyl spinors. The results obtained in this section can be compared to those in the appendix, which uses the dotted/undotted van der Waerden notation.

6.13 Spacetime Vectors as Paravectors of $C\ell_{3,0}$ from Weyl Spinors

Paravectors

One of the main benefits of the formalism presented in this chapter is the Penrose interpretation of a spinor as being a structure that is more basic than a point in spacetime is (Penrose and Rindler, 1984). Thus, it is natural to construct spacetime vectors from Weyl spinors.

Let us discuss first the concept of the paramultivector (Baylis, 1996), in particular, the paravector, which is the sum of a scalar and a vectors. One of the main purposes for the introduction of this concept is to reduce the dimension of the algebra, in order to formulate a theory. The main advantage of a minimalist formulation is that redundancies are precluded and only those elements that are strictly needed are used. A paravector is defined as an element of $\mathbb{R} \oplus \mathbb{R}^{p,q} \subset C\ell_{p,q}$. We are here interested in the case $n = 3$, where $C\ell_{3,0}$ is the Pauli algebra. We shall now show how an arbitrary element ψ of $C\ell_{3,0}$ can be written as a complex paravector.

When the orthonormal basis for $C\ell_{3,0}$ is $\{1, \mathbf{e}_i, \mathbf{e}_i\mathbf{e}_j, \mathbf{e}_{123}\}$, the pseudoscalar $\mathbf{e}_1\mathbf{e}_2\mathbf{e}_3 \equiv \mathbf{I}$ satisfies $\mathbf{I}^2 = -1$ and is an element in the centre of $C\ell_{3,0}$. Hence, the algebra $C\ell_{3,0}$ is isomorphic to an algebra with half of the number of elements – when considered over $\mathbb{C}$ – where $\mathbf{I}$ plays the role of the imaginary unit.

With respect to a $\mathbb{C}$-structure, every bivector in $C\ell_{3,0}$ can be written as an imaginary vector, for instance, $\mathbf{e}_1\mathbf{e}_2 = \mathbf{e}_1\mathbf{e}_2\mathbf{e}_3\mathbf{e}_3 = \mathbf{I}\mathbf{e}_3 = \mathbf{e}_3\mathbf{I}$. This vector is precisely the dual Hodge operator acting on basis vectors, and a basis for $C\ell_{3,0}$ in this context is given by $\{\mathbf{e}_2\mathbf{e}_3, \mathbf{e}_3\mathbf{e}_1, \mathbf{e}_1\mathbf{e}_2\}$, where we use the isomorphism $\rho : C\ell_{3,0} \to C\ell_{1,3}^+$, $\mathbf{e}_i \mapsto \rho(\mathbf{e}_i) = \gamma_i\gamma_0$. In addition, we see that

$$\mathbf{e}_i\mathbf{e}_j = -\gamma_i\gamma_j, \quad \mathbf{e}_j\mathbf{e}_k\mathbf{e}_l = \gamma_0\gamma_j\gamma_k\gamma_l. \tag{6.166}$$

Remember that $\{\gamma_\mu\}_{\mu=0}^3$ denotes a basis for $C\ell_{1,3}$.

Parabivectors and paratrivectors are also introduced in $C\ell_{3,0}$ and defined respectively as elements of $\bigwedge_1(\mathbb{R}^3) \oplus \bigwedge_2(\mathbb{R}^3)$, and $\bigwedge_2(\mathbb{R}^3) \oplus \bigwedge_3(\mathbb{R}^3)$. From (6.166), bivectors of $C\ell_{1,3}$ correspond to parabivectors of $C\ell_{3,0}$. Here, we will focus on paravectors; for further details, see, for instance, the work by Baylis (1996).

An element $\psi \in C\ell_{3,0}$ is the sum of a scalar p_0 and a vector $\mathbf{p}$, both of which are, in general, complex structures, where $p_0 = \mathrm{Re}(p_0) + \mathrm{Im}(p_0)$, and $\mathbf{p} = \mathrm{Re}(\mathbf{p}) + \mathrm{Im}(\mathbf{p})$. In addition, $\mathrm{Re}(p_0) = \langle\psi\rangle_0$; $\mathrm{Im}(\mathbf{p}) = \langle\psi\rangle_2$; $\mathrm{Re}(\mathbf{p}) = \langle\psi\rangle_1$; and $\mathrm{Im}(p_0) = \langle\psi\rangle_3$. The object ψ is said to be a $\mathbb{C}$-paravector.

Paravector Automorphisms

We shall describe now how the grade involution, the reversion, and the conjugation act on paravectors. The conjugation reverses the vector part of ψ:

$$\psi = p_0 + \mathbf{p} \mapsto \bar{\psi} = p_0 - \mathbf{p}. \tag{6.167}$$

The conjugation is then called spatial reversion in this context. Any element of $C\ell_{3,0}$ can be written as $\psi = \frac{1}{2}(\psi + \bar{\psi}) + \frac{1}{2}(\psi - \bar{\psi}) = \langle\psi\rangle_0 + \langle\psi\rangle_1$, where $\langle\psi\rangle_0$ and $\langle\psi\rangle_1$

are, respectively, the scalar and the vector components of ψ. They can be regarded, respectively, as the scalar and the vector products

$$\langle\psi\phi\rangle_0 = \frac{1}{2}(\psi\phi + \overline{\psi\phi}), \qquad \langle\psi\phi\rangle_1 = \frac{1}{2}(\psi\phi - \overline{\psi\phi}). \tag{6.168}$$

Thus, a multivector $\psi \in \mathcal{C}\ell_{3,0}$ is a scalar if and only if $\psi = \bar{\psi}$.

Likewise, the complex conjugation is given by the reversion in $\mathcal{C}\ell_{3,0}$. Indeed, if $\psi \in \mathcal{C}\ell_{3,0}$ is written in the paravector basis $\{\mathbf{e}_0 = 1, \mathbf{e}_1, \mathbf{e}_2, \mathbf{e}_3\}$, the complex conjugation is obtained when the complex conjugate is taken with respect to each coefficient:

$$\psi = \psi^\mu \mathbf{e}_\mu \mapsto \tilde{\psi} = \psi^{\mu *}\mathbf{e}_\mu, \quad \mu = 0, 1, 2, 3. \tag{6.169}$$

The complex conjugation is used to split multivectors into real and imaginary parts:

$$\psi = \frac{1}{2}(\psi + \tilde{\psi}) + \frac{1}{2}(\psi - \tilde{\psi}) = \ \operatorname{Re}(\psi)\ +\ \operatorname{Im}(\psi). \tag{6.170}$$

The composition between the complex conjugation and the Clifford conjugation results in the grade involution, which decomposes a multivector into even and odd parts:

$$\psi = \frac{1}{2}(\psi + \widehat{\psi}) + \frac{1}{2}(\psi - \widehat{\psi}) = \langle\psi\rangle_+ + \langle\psi\rangle_- \,. \tag{6.171}$$

The norm of a paravector is given by the quadratic form

$$\psi\bar{\psi} = (p_0 + \mathbf{p})(p_0 - \mathbf{p}) = p_0^2 - \mathbf{p}^2. \tag{6.172}$$

The Lorentzian metric tensor components $\eta_{\mu\nu}$ associated with the Minkowski spacetime defines the norm of a vector in the same way. We conclude that vectors of $\mathbb{R}^{1,3}$ in special relativity can be represented by real paravectors of $\mathcal{C}\ell_{3,0}$. The metric is given by

$$\eta_{\mu\nu} = \langle\mathbf{e}_\mu \bar{\mathbf{e}}_\nu\rangle_0 = \begin{cases} 1, & \text{if } \mu = \nu = 0, \\ -1, & \text{if } \mu = \nu = 1, 2, 3, \\ 0, & \text{if } \mu \neq \nu. \end{cases} \tag{6.173}$$

Using Weyl Spinors

Weyl spinors can generate arbitrary paravectors as

$$2\mathcal{K}\mathbf{e}_1\underline{\mathcal{K}} = 2\mathcal{K}\mathbf{e}_1\mathbf{e}_1\tilde{\mathcal{K}} = 2\mathcal{K}\tilde{\mathcal{K}} \in \mathbb{R} \oplus \mathbb{R}^3 \,.$$

Given Weyl spinors $\mathcal{K} = k^1 f_+ + k^2\mathbf{e}_1 f_+ \in \mathcal{C}\ell_{3,0}f_+$, and $\overline{\mathcal{K}} = \overline{k}^{\dot{1}}(f_-\mathbf{e}_1) + \overline{k}^{\dot{2}}(f_-)$, it follows that

$$\mathcal{K}\tilde{\mathcal{K}} = \mathcal{K}\mathbf{e}_1\underline{\mathcal{K}} = k^1\overline{k}^{\dot{1}} f_+ + k^1\overline{k}^{\dot{2}} f_+\mathbf{e}_1 + k^2\overline{k}^{\dot{1}} f_-\mathbf{e}_1 + k^2\overline{k}^{\dot{2}} f_-, \tag{6.174}$$

which leads, from eqns (6.153) and (6.154), to the expression

$$\mathcal{K}\tilde{\mathcal{K}} = \begin{pmatrix} k^1\overline{k}^{\dot{1}} & k^1\overline{k}^{\dot{2}} \\ k^2\overline{k}^{\dot{1}} & k^2\overline{k}^{\dot{2}} \end{pmatrix}. \tag{6.175}$$

Indeed, a paravector $\mathfrak{a} \in \mathbb{R} \oplus \mathbb{R}^3 \in \mathcal{C}\ell_{3,0}$ can be written as

$$\mathfrak{a} = 2\mathcal{K}\mathbf{e}_1\underline{\mathcal{K}} = 2\mathcal{K}\tilde{\mathcal{K}}, \tag{6.176}$$

since, given a spinor operator $\psi \in \mathcal{C}\ell_{3,0}^+$, it follows that

$$\begin{aligned} \mathfrak{a} = 2\mathcal{K}\tilde{\mathcal{K}} = 2\psi f_+ f_+ \tilde{\psi} = 2\psi f_+ \tilde{\psi} = \psi(1+\mathbf{e}_3)\tilde{\psi} \\ = \psi\tilde{\psi} + \psi\mathbf{e}_3\tilde{\psi} = \mathfrak{a}^0 + \mathfrak{a}^i\mathbf{e}_i. \end{aligned} \tag{6.177}$$

The paravector $\mathfrak{a}$ is future pointed, since

$$\begin{aligned} \mathbb{R} \ni \psi\tilde{\psi} = \mathfrak{a}^0 &= (a + b\mathbf{e}_{12} + c\mathbf{e}_{13} + d\mathbf{e}_{23})(a - b\mathbf{e}_{12} - c\mathbf{e}_{13} - d\mathbf{e}_{23}) \\ &= a^2 + b^2 + c^2 + d^2 > 0. \end{aligned}$$

In addition, since $\mathfrak{a}^i\mathfrak{a}_i = (\psi\tilde{\psi})^2 = (\mathfrak{a}^0)^2$, the paravector $\mathfrak{a}$ is null: $\mathfrak{a}^2 = (\mathfrak{a}^0)^2 - \mathfrak{a}^i\mathfrak{a}_i = 0$. The last equality in eqn (6.177) arises from the fact that any vector in $\mathbb{R}^3$ can be written as $\mathbf{x} = x^i\mathbf{e}_i = \psi\mathbf{e}_3\tilde{\psi}$, which is derived from the reference vector $\mathbf{e}_3$ via a rotation and a dilatation. This expression is the spin density multiplied by the factor $\hbar/2$.

Now we will obtain the spacetime metric from Weyl spinors. According to eqn (6.174), let the two paravectors $\mathfrak{a}$ and $\mathfrak{b}$ read

$$\begin{aligned} \mathfrak{a} &= k^1\overline{k}^{\dot{1}} f_+ + k^1\overline{k}^{\dot{2}} f_+\mathbf{e}_1 + k^2\overline{k}^{\dot{1}} f_-\mathbf{e}_1 + k^2\overline{k}^{\dot{2}} f_- = \mathfrak{a}_0 + \mathfrak{a}^i\mathbf{e}_i, \\ \mathfrak{b} &= r^1\overline{r}^{\dot{1}} f_+ + r^1\overline{r}^{\dot{2}} f_+\mathbf{e}_1 + r^2\overline{r}^{\dot{1}} f_-\mathbf{e}_1 + r^2\overline{r}^{\dot{2}} f_- = \mathfrak{b}_0 + \mathfrak{b}^i\mathbf{e}_i, \end{aligned} \tag{6.178}$$

and their respective conjugations be given by

$$\begin{aligned} \mathfrak{a} &= \widehat{k^1\overline{k}^{\dot{1}}} f_- - \widehat{k^1\overline{k}^{\dot{2}}}\mathbf{e}_1 f_- - \widehat{k^2\overline{k}^{\dot{1}}}\mathbf{e}_1 f_+ + \widehat{k^2\overline{k}^{\dot{2}}} f_+ = \mathfrak{a}_0 - \mathfrak{a}^i\mathbf{e}_i, \\ \mathfrak{b} &= \widehat{r^1\overline{r}^{\dot{1}}} f_- - \widehat{r^1\overline{r}^{\dot{2}}}\mathbf{e}_1 f_- - \widehat{r^2\overline{r}^{\dot{1}}}\mathbf{e}_1 f_+ + \widehat{r^2\overline{r}^{\dot{2}}} f_+ = \mathfrak{b}_0 - \mathfrak{b}^i\mathbf{e}_i. \end{aligned} \tag{6.179}$$

The Clifford relation is then obtained for paravectors:

$$\begin{aligned} \mathfrak{a}\hat{\mathfrak{b}} + \mathfrak{b}\hat{\mathfrak{a}} &= (k^1\overline{k}^{\dot{1}} r^2\overline{r}^{\dot{2}} + k^1\overline{k}^{\dot{2}} r^2\overline{r}^{\dot{1}} + k^2\overline{k}^{\dot{1}} r^1\overline{r}^{\dot{2}} + k^2\overline{k}^{\dot{2}} r^1\overline{r}^{\dot{1}}) \\ &= 2(\mathfrak{a}^0\mathfrak{b}^0 - \mathfrak{a}^i\mathfrak{b}_i) = 2g(\mathfrak{a}, \mathfrak{b}). \end{aligned} \tag{6.180}$$

From light-like paravectors obtained from the generators $\{f_+, f_-, \mathbf{e}_1 f_+, \mathbf{e}_1 f_-\}$ of the four ideals of $\mathcal{C}\ell_{3,0}$, it is possible to obtain the tetrad $\{\mathbf{e}_0 = 1, \mathbf{e}_1, \mathbf{e}_2, \mathbf{e}_3\}$ in the paravector space. To start, let us define a basis of light-like paravectors $\{\pi^\mu\}$:

$$\begin{aligned} \pi^0 &= f_+\widetilde{f_+} = f_+, & \pi^1 &= f_-\widetilde{f_-} = f_-, \\ \pi^2 &= \mathbf{e}_1 f_+\widetilde{f_+} = \mathbf{e}_1 f_+, & \pi^3 &= f_+\widetilde{\mathbf{e}_1 f_+} = \mathbf{e}_1 f_-. \end{aligned} \tag{6.181}$$

The orthonormal tetrad is obtained from the basis of paravectors $\{\pi^\mu\}$:

$$\begin{aligned} \mathbf{e}_0 &= \pi^0 + \pi^1, & \mathbf{e}_1 &= \pi^2 - \pi^3, \\ \mathbf{e}_2 &= \pi^3 - \pi^2, & \mathbf{e}_3 &= \pi^0 - \pi^1, \end{aligned} \tag{6.182}$$

since

$$\begin{aligned} \mathbf{e}_0 &= f_+ + f_-, & \mathbf{e}_1 &= \mathbf{e}_1 f_+ + \mathbf{e}_1 f_-, \\ \mathbf{e}_2 &= -\mathbf{e}_{123}(\mathbf{e}_1 f_- - \mathbf{e}_1 f_+), & \mathbf{e}_3 &= f_+ - f_-. \end{aligned} \tag{6.183}$$

Hence, both the Minkowski space tetrad and the paravector space $\mathbb{R} \oplus \mathbb{R}^3$ tetrad are obtained.

6.14 Paravectors of $\mathcal{C}\ell_{4,1}$ in $\mathcal{C}\ell_{3,0}$ via the Periodicity Theorem

Consider the basis $\{\varepsilon_{\mathring{A}}\}_{\mathring{A}=0}^{5}$ of $\mathbb{R}^{2,4}$; it obviously satisfies the relations

$$\varepsilon_{\mathring{0}}^2 = \varepsilon_{\mathring{5}}^2 = 1, \qquad \varepsilon_{\mathring{1}}^2 = \varepsilon_{\mathring{2}}^2 = \varepsilon_{\mathring{3}}^2 = \varepsilon_{\mathring{4}}^2 = -1, \qquad \varepsilon_{\mathring{A}} \cdot \varepsilon_{\mathring{B}} = 0 \quad (\mathring{A} \neq \mathring{B}).$$

Moreover, consider also $\mathbb{R}^{4,1}$, with the basis $\{E_A\}_{A=0}^{4}$, where

$$E_0^2 = -1, \qquad E_1^2 = E_2^2 = E_3^2 = E_4^2 = 1, \qquad E_A \cdot E_B = 0 \quad (A \neq B).$$

The basis $\{E_A\}$ can be obtained from the basis $\{\varepsilon_{\mathring{A}}\}$ if we define the isomorphism

$$\begin{aligned} \xi : \mathcal{C}\ell_{4,1} &\to \textstyle\bigwedge_2(\mathbb{R}^{2,4}), \\ E_A &\mapsto \xi(E_A) = \varepsilon_A \varepsilon_5. \end{aligned} \tag{6.184}$$

The basis $\{E_A\}$ defined by eqn (6.184) obviously satisfies eqn (6.14).

Given a vector $\alpha = \alpha^{\mathring{A}} \varepsilon_{\mathring{A}} \in \mathbb{R}^{2,4}$, we obtain a paravector $\mathfrak{b} \in \mathbb{R} \oplus \mathbb{R}^{4,1} \hookrightarrow \mathcal{C}\ell_{4,1}$ if the element ε_5 is left multiplied by $\mathfrak{b}$:

$$\mathfrak{b} = \alpha \varepsilon_5 = \alpha^A E_A + \alpha^5. \tag{6.185}$$

From the periodicity theorem, the isomorphism $\mathcal{C}\ell_{4,1} \simeq \mathcal{C}\ell_{1,1} \otimes \mathcal{C}\ell_{3,0}$ follows and thus it is possible to express an element of $\mathcal{C}\ell_{4,1}$ as a 2×2 matrix with entries in $\mathcal{C}\ell_{3,0}$.

A homomorphism $\vartheta : \mathcal{C}\ell_{4,1} \to \mathcal{C}\ell_{3,0}$ can be defined by

$$E_i \mapsto \vartheta(E_i) = E_i E_0 E_4 \equiv \mathbf{e}_i. \tag{6.186}$$

It follows that $\mathbf{e}_i^2 = 1$, $E_i = \mathbf{e}_i E_4 E_0$ and that $E_4 = E_+ + E_-$, $E_0 = E_+ - E_-$, where $E_\pm := \frac{1}{2}(E_4 \pm E_0)$. Hence, the paravector $\mathfrak{b}$ can be split into

$$\mathfrak{b} = \alpha^5 + (\alpha^0 + \alpha^4)E_+ + (\alpha^4 - \alpha^0)E_- + \alpha^i \mathbf{e}_i E_4 E_0. \tag{6.187}$$

If we choose E_4 and E_0 to be, respectively, represented by $E_4 = \begin{pmatrix} 0 & 1 \\ 1 & 0 \end{pmatrix}$, and $E_0 = \begin{pmatrix} 0 & -1 \\ 1 & 0 \end{pmatrix}$, then

$$E_+ = \begin{pmatrix} 0 & 0 \\ 1 & 0 \end{pmatrix}, \quad E_- = \begin{pmatrix} 0 & 1 \\ 0 & 0 \end{pmatrix}, \quad E_4E_0 = \begin{pmatrix} 1 & 0 \\ 0 & -1 \end{pmatrix}, \tag{6.188}$$

and hence the paravector $\mathfrak{b} \in \mathbb{R} \oplus \mathbb{R}^{4,1} \hookrightarrow C\ell_{4,1}$ in eqn (6.187) is represented by

$$\mathfrak{b} = \begin{pmatrix} \alpha^5 + \alpha^i \mathbf{e}_i & \alpha^4 - \alpha^0 \\ \alpha^0 + \alpha^4 & \alpha^5 - \alpha^i \mathbf{e}_i \end{pmatrix}. \tag{6.189}$$

The vector $\alpha \in \mathbb{R}^{2,4}$ is an element of the Klein absolute, that is, $\alpha^2 = 0$. In addition, this condition implies that $\alpha^2 = 0 \Leftrightarrow \mathfrak{b}\bar{\mathfrak{b}} = 0$, since $\alpha^2 = \alpha\alpha = \alpha 1 \alpha = \alpha\varepsilon_5^2\alpha = \alpha\varepsilon_5\varepsilon_5\alpha = \mathfrak{b}\bar{\mathfrak{b}}$. We denote

$$\lambda = \alpha^4 - \alpha^0, \qquad \mu = \alpha^4 + \alpha^0. \tag{6.190}$$

Using the matrix representation of $\mathfrak{b}\bar{\mathfrak{b}}$, the entry $(\mathfrak{b}\bar{\mathfrak{b}})_{11}$ of the matrix is given by

$$(\mathfrak{b}\bar{\mathfrak{b}})_{11} = x\bar{x} - \lambda\mu = 0, \tag{6.191}$$

where

$$x := (\alpha^5 + \alpha^i \mathbf{e}_i) \in \mathbb{R} \oplus \mathbb{R}^3 \hookrightarrow C\ell_{3,0}. \tag{6.192}$$

If we fix $\mu = 1$, then $\lambda = x\bar{x} \in \mathbb{R}$. This choice corresponds to a projective description. Then, the paravector $\mathfrak{b} \in \mathbb{R} \oplus \mathbb{R}^{4,1} \hookrightarrow C\ell_{4,1}$ can be represented as

$$\mathfrak{b} = \begin{pmatrix} x & \lambda \\ \mu & \bar{x} \end{pmatrix} = \begin{pmatrix} x & x\bar{x} \\ 1 & \bar{x} \end{pmatrix}. \tag{6.193}$$

From eqn (6.191), we obtain $(\alpha^5 + \alpha^i \mathbf{e}_i)(\alpha^5 - \alpha^i \mathbf{e}_i) = (\alpha^4 - \alpha^0)(\alpha^4 + \alpha^0)$, which implies that

$$(\alpha^5)^2 - (\alpha^i \mathbf{e}_i)(\alpha^j \mathbf{e}_j) = (\alpha^4)^2 - (\alpha^0)^2, \tag{6.194}$$

yielding

$$(\alpha^5)^2 + (\alpha^0)^2 - (\alpha^1)^2 - (\alpha^2)^2 - (\alpha^3)^2 - (\alpha^4)^2 = 0\,, \tag{6.195}$$

which is the Klein absolute (eqn (5.110)).

6.15 Twistors as Geometric Multivectors

In this section, we present and discuss the Keller approach and also introduce our definition of twistors, showing how the twistor formulation can be led to the Keller approach and, consequently, to the Penrose classical twistor framework. The twistor defined as a minimal lateral ideal is further examined in the book by Crumeyrolle (1990). Robinson congruences and the incidence relation, which determines a point in spacetime as a secondary concept obtained from the intersection between two twistors, are also investigated here.

The Keller Approach

The twistor approach by Keller (1997) uses the projectors $\mathbb{P}_{\mathcal{R},L} := \frac{1}{2}(1 \pm i\gamma_5)$ (where R and L, respectively, denote the right the left projections, as they are usually called) and the element $T_{\mathbf{x}} = 1 + \gamma_5\mathbf{x}$, where $\mathbf{x} = x^\mu\gamma_\mu \in \mathbb{R}^{1,3}$. Now we introduce some of the results obtained by Keller (1997). Define the *reference twistor* $\eta_{\mathbf{x}}$, which is associated with the vector $\mathbf{x} \in \mathbb{R}^{1,3}$ and with the Weyl covariant dotted spinor (written as the left-handed projection of a Dirac spinor ω) $\Pi = \mathbb{P}_L\omega = \binom{0}{\xi}$ by

$$\eta_{\mathbf{x}} = T_{\mathbf{x}}\mathbb{P}_L\omega = (1 + \gamma_5\mathbf{x})\mathbb{P}_L\omega = (1 + \gamma_5\mathbf{x})\Pi\,. \tag{6.196}$$

In order to show the equivalence of this definition with the Penrose classical twistor formalism, the Weyl representation is used:

$$\eta_{\mathbf{x}} = (1 + \gamma_5\mathbf{x})\Pi = \left[\begin{pmatrix} I & 0 \\ 0 & I \end{pmatrix} + \begin{pmatrix} -i_2 & 0 \\ 0 & i_2 \end{pmatrix}\begin{pmatrix} 0 & \vec{x} \\ \vec{x}^c & 0 \end{pmatrix}\right]\begin{pmatrix} 0 \\ \xi \end{pmatrix}. \tag{6.197}$$

Each entry in these matrices denotes a 2×2 matrix, and the vector representation

$$\vec{x} = \begin{pmatrix} x^0 + x^3 & x^1 + ix^2 \\ x^1 - ix^2 & x^0 - x^3 \end{pmatrix} \tag{6.198}$$

is related to the point $\mathbf{x} \in \mathbb{R}^{1,3}$, where $\vec{x}^c$ is the $\mathbb{H}$-conjugation of $\vec{x} \in \mathbb{R}^{1,3}$, given by eqn (6.198). Hence, the reference twistor reads

$$\eta_{\mathbf{x}} = \begin{pmatrix} -i\vec{x}\xi \\ \xi \end{pmatrix}, \tag{6.199}$$

which is the index-free version of the Penrose classical twistor (Penrose, 1967). The sign in the first component is different, since we use the Weyl representation:

$$\gamma(e_0) = \gamma_0 = \begin{pmatrix} 0 & I \\ I & 0 \end{pmatrix}, \quad \gamma(e_k) = \gamma_k = \begin{pmatrix} 0 & -\sigma_k \\ \sigma_k & 0 \end{pmatrix}. \tag{6.200}$$

In order to get the correct sign, Keller uses a representation which is similar to the Weyl one, but in which the vectors in $\mathbb{R}^3$ are reflected ($\vec{x} \mapsto -\vec{x}$) through the origin:

$$\gamma(e_0) = \gamma_0 = \begin{pmatrix} 0 & I \\ I & 0 \end{pmatrix}, \quad \gamma(e_k) = \gamma_k = \begin{pmatrix} 0 & \sigma_k \\ -\sigma_k & 0 \end{pmatrix}. \tag{6.201}$$

Thus, it is possible to get the Penrose twistor

$$\eta_{\mathbf{x}} = \begin{pmatrix} i\vec{x}\xi \\ \xi \end{pmatrix}. \tag{6.202}$$

Therefore, twistors are completely described by the multivector structure of the Dirac algebra $\mathbb{C} \otimes \mathcal{C}\ell_{1,3} \simeq \mathcal{C}\ell_{4,1} \simeq \mathcal{M}(4, \mathbb{C})$.

An Alternative Approach to Twistors

We now define twistors as a special class of algebraic spinors in $\mathcal{C}\ell_{4,1}$. The isomorphism $\mathcal{C}\ell_{4,1} \simeq \mathbb{C} \otimes \mathcal{C}\ell_{1,3}$, where

$$E_0 = i\gamma_0, \quad E_1 = \gamma_{10}, \quad E_2 = \gamma_{20}, \quad E_3 = \gamma_{30}, \quad E_4 = \gamma_5\gamma_0 = -\gamma_{123}, \tag{6.203}$$

explicitly gives rise to the relations $E_0^2 = -1$, and $E_1^2 = E_2^2 = E_3^2 = E_4^2 = 1$. A paravector $x \in \mathbb{R} \oplus \mathbb{R}^{4,1} \hookrightarrow \mathcal{C}\ell_{4,1}$ is written as

$$x = x^0 + x^A E_A = x^0 + \alpha^0 E_0 + x^1 E_1 + x^2 E_2 + x^3 E_3 + \alpha^4 E_4. \tag{6.204}$$

We also define an element $\chi := xE_4 \in \bigwedge^0(\mathbb{R}^{4,1}) \oplus \bigwedge^1(\mathbb{R}^{4,1}) \oplus \bigwedge^2(\mathbb{R}^{4,1})$ as

$$\chi = xE_4 = x^0 E_4 + \alpha^0 E_0 E_4 + x^1 E_1 E_4 + x^2 E_2 E_4 + x^3 E_3 E_4 + \alpha^4.$$

It can be seen that

$$\chi\frac{1}{2}(1 + i\gamma_5) = T_{\mathbf{x}}\frac{1}{2}(1 + i\gamma_5) = T_{\mathbf{x}}\mathbb{P}_L. \tag{6.205}$$

We define the twistor as the algebraic spinor

$$\chi\mathbb{P}_L U f \in (\mathbb{C} \otimes \mathcal{C}\ell_{1,3})f\,,$$

where f is a primitive idempotent of $\mathbb{C} \otimes \mathcal{C}\ell_{1,3} \simeq \mathcal{C}\ell_{4,1}$, and $U \in \mathcal{C}\ell_{4,1}$ denotes an arbitrary element. Hence, Uf is a Dirac spinor, and $\mathbb{P}_L U f = \binom{0}{\xi} = \Pi \in \frac{1}{2}(1+i\gamma_5)(\mathbb{C} \otimes \mathcal{C}\ell_{1,3})$ is a covariant dotted Weyl spinor. The twistor reads

$$\begin{aligned}\chi\Pi &= xE_4\Pi \\ &= (x^0 E_4 + \alpha^0 E_0 E_4 + x^1 E_1 E_4 + x^2 E_2 E_4 + x^3 E_3 E_4 + \alpha^4)\Pi.\end{aligned}$$

From the relation $E_4\Pi = \gamma_5\gamma_0\Pi = -\gamma_0\gamma_5\Pi = -i\gamma_0\Pi$, it follows that

$$\begin{aligned}\chi\Pi &= (x^0 E_4 + \alpha^0 E_0 E_4 + x^1 E_1 E_4 + x^2 E_2 E_4 + x^3 E_3 E_4 + \alpha^4)\Pi \\ &= x^0(E_4\Pi) + x^k E_k(E_4\Pi) + \alpha^0 E_0(E_4\Pi) + \alpha^4\Pi \\ &= -ix^0\gamma_0\Pi - ix^k\gamma_k\Pi + \alpha^0\Pi + \alpha^4\Pi \\ &= (1 + \gamma_5)\Pi \\ &= \begin{pmatrix} i\vec{x}\xi \\ \xi \end{pmatrix}.\end{aligned} \tag{6.206}$$

Hence, our definition is shown to be equivalent to Keller's and, therefore, to the Penrose classical twistor, by eqn (6.202).

The incidence relation, which determines a point in spacetime from the intersection between two twistors (Penrose, 1967; Penrose and Rindler, 1984), is given by

$$J_{\bar{\chi}\chi} = \overline{xE_4U}xE_4U = -\bar{U}E_4\bar{x}xE_4U = 0,$$

since the paravector $x \in \mathbb{R} \oplus \mathbb{R}^{4,1} \hookrightarrow \mathcal{C}\ell_{4,1}$ is an element in the Klein absolute; consequently, $x\bar{x} = 0$. Finally, the Robinson congruence is defined in our formalism from the product

$$J_{\bar{\chi}\chi'} = \overline{xE_4U}x'E_4U = -\bar{U}E_4\bar{x}x'E_4U. \tag{6.207}$$

This product is null if $x = x'$, and the Robinson congruence is defined when we fix x and let x' vary.

6.16 Spinor Classification According to Bilinear Covariants

There is a spinor classification, due to Lounesto (2001*a*), which is particularly interesting both for mathematicians and physicists. The essence of this classification relies upon the so-called bilinear covariants, which describe physical observables in field theory. Moreover, the so-called Fierz aggregate can bring a robust geometrical interpretation of these quantities (Lounesto, 2001*a*). Within the Lounesto classification, a specific bilinear covariant plays a prominent role. In fact, the 1-covector current density has such interpretation, at least for the case of a regular spinor describing the electron in Dirac theory (Dirac, 1928). The current density reads $\mathbf{J} = J_\mu \mathbf{e}^\mu = \psi^\dagger\gamma_0\gamma_\mu\psi\mathbf{e}^\mu$, where ψ denotes a classical spinor, $\{\gamma_\mu\}$ stands for the gamma matrices, and $\{\mathbf{e}^\mu\}$ is a dual basis in $\mathbb{C}\otimes \mathcal{C}\ell_{1,3}$. Regarding the electron theory, $\mathbf{J}$ is a conserved current. Consequently, the time component $J_0 = \psi^\dagger\psi$ provides the probability density associated with the electron and which should be non-null.[8] Hence, we must have $\mathbf{J} \neq 0$.

The Lounesto spinor classification is derived when a classical spinor ψ is taken into account. Indeed, the well-known Lounesto spinor classification is based on bilinear covariants and the underlying multivector structure (Crawford, 1985; Lounesto, 2001*a*). The physical nature of the classification focuses on the bilinear covariants, which are physical observables which describe physical features of fermionic particles. The observable quantities are given by the following multivector structure:

$$\sigma = \psi^\dagger\gamma_0\psi, \tag{6.208}$$

$$\mathbf{J} = J_\mu\,\mathbf{e}^\mu = \psi^\dagger\gamma_0\gamma_\mu\psi\,\mathbf{e}^\mu, \tag{6.209}$$

$$\mathbf{S} = S_{\mu\nu}\,\mathbf{e}^\mu\wedge\mathbf{e}^\nu = \frac{1}{2}\psi^\dagger\gamma_0 i\gamma_{\mu\nu}\psi\,\mathbf{e}^\mu\wedge\mathbf{e}^\nu, \tag{6.210}$$

$$\mathbf{K} = K_\mu\,\mathbf{e}^\mu = \psi^\dagger\gamma_0 i\gamma_{0123}\gamma_\mu\psi\,\mathbf{e}^\mu, \tag{6.211}$$

$$\omega = -\psi^\dagger\gamma_0\gamma_{0123}\psi, \tag{6.212}$$

where $\gamma_{0123} = i\gamma_5$. The expression for these quantities using algebraic spinors and spinor operators can be found in the book by Lounesto (2001*a*).

The bilinear covariants have a physical interpretation in the Dirac theory, after a suitable multiplication by some physical constants. Indeed, eJ_0 is interpreted as charge density, ecJ_k $(k = 1, 2, 3)$ as electric current density, $(e\hbar/2mc)S^{ij}$ as magnetic moment density, $(e\hbar/2mc)S^{0j}$ as electric moment density, and $(\hbar/2)K_\mu$ as spin density. The interpretation of the scalar σ and pseudoscalar ω bilinear covariants is less clear than this, but when combined in $\rho^2 = \sigma^2 + \omega^2 = |J|^2$ (by the Fierz–Pauli–Kokink (FPK) identities), ρ can be interpreted as probability density. A prominent requirement for the Lounesto spinor classification is that the bilinear covariants satisfy quadratic algebraic relations known as the FPK identities (Holland, 1986; Crawford, 1985; Lounesto, 2001*a*):

[8] It is worth emphasising that the reason for considering $\mathbf{J}$ as the current density is clear when the spinor obeys the usual dynamics ruled by the Dirac equation (Villalobos, da Silva, and da Rocha, 2015). The mass dimension in this case is the same mass dimension 3/2 associated with the usual spin-1/2 fermions in the standard model of elementary particles. When $\mathbf{J} = 0$ is required, the underlying dynamics are not be provided by the Dirac equation. Since the construction is relativistic, the emergent spinors with $\mathbf{J} = 0$ respect the Klein–Gordon equation.

$$\mathbf{J}^2 = \omega^2 + \sigma^2, \quad \mathbf{K}^2 = -\mathbf{J}^2, \quad \mathbf{J} \lrcorner \mathbf{K} = 0, \quad \mathbf{J} \wedge \mathbf{K} = -(\omega + \sigma\gamma_{0123})\mathbf{S}\,. \tag{6.213}$$

By taking a classical spinor ξ which satisfies $\xi^\dagger \gamma_0 \psi \neq 0$, the original spinor ψ can be recovered from its aggregate $\mathbf{Z}$, which is provided by

$$\mathbf{Z} = \sigma + \mathbf{J} + i\mathbf{S} + i\mathbf{K}\gamma_{0123} + \omega\gamma_{0123}\,, \tag{6.214}$$

using the Takahashi algorithm (Takahashi, 1982; Vaz, 1998). In fact, the spinor ψ can be written as the multivector $\mathbf{Z}\xi$ by

$$\psi = \frac{1}{2\sqrt{\xi^\dagger \gamma_0 \mathbf{Z}\xi}}\, e^{-i\theta}\mathbf{Z}\xi, \tag{6.215}$$

where $e^{-i\theta} = 2(\xi^\dagger \gamma_0 \mathbf{Z}\xi)^{-1/2}\xi^\dagger\gamma_0\psi \in \mathrm{U}(1)$. For more details see, for example, the article by Crawford (1985). Moreover, if the bilinear covariants satisfy the Fierz identities, then the complex multivector $\mathbf{Z}$ is called a Fierz aggregate. If $\gamma_0 \mathbf{Z}^\dagger \gamma_0 = \mathbf{Z}$, then $\mathbf{Z}$ is said to be a boomerang (Lounesto, 2001*a*).

Within the Lounesto classification scheme, the condition $\mathbf{J} \neq 0$ is fundamental for defining the so-called boomerang (Lounesto, 2001*a*). With this condition, there exist just six types of spinors, according to the bilinear covariants.[9] Indeed, in the Lounesto scheme, spinors are classified as regular or singular spinors. Regular spinors present either $\sigma \neq 0$, or $\omega \neq 0$ (or even both non-null quantities). On the other hand, singular spinors present $\sigma = 0 = \omega$, in which case the Fierz identities are in general replaced by the following conditions, which are more general than those for regular spinors (Crawford, 1985):

$$\begin{aligned} \mathbf{Z}^2 = 4\sigma\mathbf{Z}, \qquad \mathbf{Z}\gamma_\mu\mathbf{Z} = 4J_\mu\mathbf{Z}, \qquad \mathbf{Z}\gamma_{0123}\mathbf{Z} = -4\omega\mathbf{Z}, \\ i\mathbf{Z}\gamma_{\mu\nu}\mathbf{Z} = 4S_{\mu\nu}\mathbf{Z}, \qquad i\mathbf{Z}\gamma_{0123}\gamma_\mu\mathbf{Z} = 4K_\mu\mathbf{Z}\,. \end{aligned} \tag{6.216}$$

The aggregate plays an essential role within the Lounesto classification, since $\mathbf{Z}$ has to be promoted to a boomerang, satisfying

$$\mathbf{Z}^2 = 4\sigma\mathbf{Z}. \tag{6.217}$$

For regular spinors, eqn (6.217) holds and $\mathbf{Z}$ is a boomerang. However, for singular spinors, we must ensure that the aggregate is a boomerang. In fact, $\mathbf{J}$ must be parallel to $\mathbf{K}$, and both must lie in the plane defined by $\mathbf{S}$. Hence, by using eqn (6.214) and taking into account singular spinors, we can clearly see that the aggregate reads

$$\mathbf{Z} = \mathbf{J}(1 + i\mathbf{s} + ih\gamma_{0123}), \tag{6.218}$$

where $\mathbf{s}$ is a space-like vector orthogonal to $\mathbf{J}$, and h is a real number. The multivector expressed in eqn (6.218) is a boomerang (da Rocha and da Silva, 2010). Equation (6.217) yields, for singular spinors, $\mathbf{Z}^2 = 0$. However, for the FPK identities to hold,

[9] More three classes can be obtained when the condition $\mathbf{J} \neq 0$ is not demanded (da Rocha, Fabbri, da Silva, Cavalcanti, and Silva-Neto, 2013).

both conditions $\mathbf{J}^2 = 0$, and $(\mathbf{s} + h\gamma_{0123})^2 = -1$, must hold in order to constrain the possible spinor classes.[10]

Equation (6.215) implies that different bilinear covariants combinations may lead to different spinors, by the constraints imposed by the FPK identities. Hence, the algebraic constraints reduce the possibilities to six different spinor classes, namely:

(1) $\sigma \neq 0, \quad \omega \neq 0$
(2) $\sigma \neq 0, \quad \omega = 0$
(3) $\sigma = 0, \quad \omega \neq 0$
(4) $\sigma = 0 = \omega, \quad \mathbf{K} \neq 0, \quad \mathbf{S} \neq 0$
(5) $\sigma = 0 = \omega, \quad \mathbf{K} = 0, \quad \mathbf{S} \neq 0$
(6) $\sigma = 0 = \omega, \quad \mathbf{K} \neq 0, \quad \mathbf{S} = 0$

Classes (1), (2), and (3) contain regular spinors. Class (4) spinors are called flag-dipole spinors (da Rocha, Fabbri, da Silva, Cavalcanti, and Silva-Neto, 2013), while class (5) spinors are called flagpole spinors (Benn and Tucker, 1987). Majorana spinor (Majorana, 1932) and Elko dark spinors are elements of the fifth class (da Rocha and Rodrigues, 2006; Ahluwalia and Grumiller, 2005; Ahluwalia, Lee, and Schritt, 2010; da Rocha and da Silva, 2009; da Rocha and da Silva, 2010; Cavalcanti, 2014; da Rocha and da Silva, 2014). Finally, class (6) dipole spinors are exemplified by Weyl spinors. New physical particles have been proposed via the use of this classification in, for example, the work by da Rocha and Rodrigues Jr (2006) and by da Rocha, Fabbri, da Silva, Cavalcanti, and Silva-Neto (2013), and also in exotic structure frameworks (da Rocha, Bernardini, and da Silva, 2011; Bernardini and da Rocha, 2012; Cavalcanti, da Silva, and da Rocha, 2014).

Note that there are only six different spinor fields. In fact, for regular spinors, since $\mathbf{J} \neq 0$, it follows that $\mathbf{S} \neq 0$, and $\mathbf{K} \neq 0$, from the identities in (6.213). On the other hand, for the singular case, the geometry asserts that $\mathbf{J}(\mathbf{s} + h\gamma_{0123}) = \mathbf{S} + \mathbf{K}\gamma_{0213}$. Hence, $\mathbf{J} \neq 0$ provides all the possibilities. The most general form of the respective spinors in each class have been introduced in the article by Cavalcanti (2014). High-dimensional spaces have a similar spinor classification (Bonora, de Brito, and da Rocha, 2015; Bonora and da Rocha, 2016); however, the so-called geometric Fierz identities (Lazaroiu, Babalic, and Coman, 2013) obstruct the proliferation of new spinor classes in high-dimensions (Bonora, de Brito, and da Rocha, 2015).

6.17 Additional Readings

The standard references for spinor theory, from the classical point of view, are (Brauer and Weyl, 1935; Cartan, 1937), and those from the algebraic point of view are the work by Chevalley (1954) and that by Riesz (1993). One of the most important applications of spinor theory is the Dirac theory of the electron, and Lounesto (2001*a*) gives a detailed comparison of the concepts of classical spinors, algebraic spinors, and spinor operators in the realm of the Dirac theory. The concept of spinor operators is further approached in the book by Dorac and Lasenby (2003), who compared that concept

[10] Note that $\mathbf{J}$ must be non-zero in the Lounesto classification.

with the classical and algebraic definitions. Pauli spinors and Dirac spinors are discussed in detail in the book by Hladik (1999). A classical reference for Weyl spinors and field theory is the book by Penrose and Rindler (1984). For more information about spinor theory we also suggest the book by Harvey (1990), mainly for a different approach to triality, as well as the book by Knus (1998). Moreover, additional details on generalisations of the octonionic algebra can be seen in the articles by da Rocha and Vaz Jr (2006*a*), da Rocha and Traesel (2012), and da Rocha, Traesel, and Vaz Jr (2012). Particular applications of triality to physics are further presented by Hasiewicz and Kwasniewski (1985), de Andrade, Rojas, and Toppan (2001), and Baez (2002). Pure spinors have been also used in superstring theory (Berkovits, 2004) and particle physics (Hasiewicz and Kwasniewski, 1985; Benn and Tucker, 1987; Budinich and Trautman, 1989; de Andrade, Rojas, and Toppan, 2001; Budinich, 2002; Ahluwalia, Lee, and Schritt, 2010; da Rocha, Bernardini, and Vaz, 2010). For additional information about the Lounesto spinor classification and its applications in field theory and theories of gravity, see, for example, the work by Rocha and Rodrigues Jr (2006), da Rocha and da Silva (2009), da Rocha and da Silva (2010), da Rocha, Bernardini, and da Silva (2011), Bernardini and da Rocha (2012), da Rocha, Fabbri, da Silva, Cavalcanti, and Silva-Neto (2013), da Rocha and da Silva (2014), Cavalcanti, da Silva, and da Rocha (2014); for their high-dimensional extensions, see the article by Bonora, de Brito, and da Rocha, (2015). In addition, spinor classification in the context of quantum Clifford algebras is presented by Abłamowicz, Gonçalves, and da Rocha (2014). (Abłamowicz, Gonçalves, and da Rocha, 2014)

6.18 Exercises

(1) The Clifford algebra $\mathcal{C}\ell_{2,1}$ is isomorphic to $\mathcal{M}(2, \mathbb{R} \oplus \mathbb{R})$. (a) Show that $f = \frac{1}{2}(1+\mathbf{e}_1)\frac{1}{2}(1+\mathbf{e}_2\mathbf{e}_3)$ is a primitive idempotent in $\mathcal{C}\ell_{2,1}$ and that the subalgebra $f\mathcal{C}\ell_{2,1}f$ is isomorphic to the scalars. (b) Show that the elements

$$f_1 = \frac{1}{4}(1 + \mathbf{e}_1 + \mathbf{e}_2\mathbf{e}_3 + \mathbf{e}_1\mathbf{e}_2\mathbf{e}_3), \qquad f_2 = \frac{1}{4}(\mathbf{e}_2 - \mathbf{e}_1\mathbf{e}_2 + \mathbf{e}_3 - \mathbf{e}_1\mathbf{e}_3)$$

form a basis for the space of algebraic spinors in $\mathcal{C}\ell_{2,1}$ and that (i) $\tilde{f}_1 f_1 = 0 = \tilde{f}_1 f_2 = \tilde{f}_2 f_1 = \tilde{f}_2 f_2$; (ii) $\bar{f}_1 f_1 = 0 = \bar{f}_2 f_2$; (iii) $\bar{f}_2 f_1 = -f_2$; and (iv) $\bar{f}_1 f_2 = f_2$. (c) Given arbitrary algebraic spinors in $\mathcal{C}\ell_{2,1}$, show that the spinor scalar products $\tilde{\psi}\phi$ and $\mathbf{e}_2\bar{\psi}\phi$ take values in $f\mathcal{C}\ell_{2,1}f$ and that $\tilde{\psi}\phi$ is identically null. Show that $\mathbf{e}_2\bar{\psi}\phi$ is antisymmetric.

(2) Given an orthonormal basis $\{\gamma_0, \gamma_1, \gamma_2, \gamma_3\}$ of $\mathcal{C}\ell_{1,3} \simeq \mathcal{M}(2, \mathbb{H})$, show that

$$\begin{aligned}
h_1 &= \frac{1}{2}(1 + \gamma_0), & h_2 &= \frac{1}{2}(-\gamma_1\gamma_2\gamma_3 + \gamma_0\gamma_1\gamma_2\gamma_3),\\
i_1 &= \frac{1}{2}(\gamma_2\gamma_3 + \gamma_0\gamma_2\gamma_3), & i_2 &= \frac{1}{2}(\gamma_1 - \gamma_0\gamma_1),\\
j_1 &= \frac{1}{2}(\gamma_3\gamma_1 + \gamma_0\gamma_3\gamma_1), & j_2 &= \frac{1}{2}(\gamma_2 - \gamma_0\gamma_2),\\
k_1 &= \frac{1}{2}(\gamma_1\gamma_2 + \gamma_0\gamma_1\gamma_2), & k_2 &= \frac{1}{2}(\gamma_3 - \gamma_0\gamma_3)
\end{aligned}$$

form a basis for the real vector space $C\ell_{1,3}\frac{1}{2}(1+\gamma_0)$. Show that the set $\{h_1, i_1, j_1, k_1\}$ is a basis for the real vector space $fC\ell_{1,3}f$ and is, in addition, a ring isomorphic to $\mathbb{H}$. Moreover, show that the right module $fC\ell_{1,3}f$-linear $C\ell_{1,3}f$ is two-dimensional, with the basis $\{h_1, h_2\}$. In this basis, show that the left multiplication by γ_μ is represented by the matrices with quaternionic entries

$$\gamma_0 = \begin{pmatrix} 1 & 0 \\ 0 & -1 \end{pmatrix}, \quad \gamma_1 = \begin{pmatrix} 0 & \boldsymbol{i} \\ \boldsymbol{i} & 0 \end{pmatrix}, \quad \gamma_2 = \begin{pmatrix} 0 & \boldsymbol{j} \\ \boldsymbol{j} & 0 \end{pmatrix}, \quad \gamma_3 = \begin{pmatrix} 0 & \boldsymbol{k} \\ \boldsymbol{k} & 0 \end{pmatrix}.$$

Show also that

$$\begin{aligned} &\tilde{h}_1 h_1 = h_1, \qquad \tilde{h}_1 h_2 = 0 = \tilde{h}_2 h_1, \qquad \tilde{h}_2 h_2 = -h_1, \\ &\bar{h}_1 h_1 = 0 = \bar{h}_2 h_2, \qquad \bar{h}_2 h_1 = h_2, \quad \bar{h}_1 h_2 = h_2 . \end{aligned}$$

(3) Let us introduce in $\mathbb{C}\otimes C\ell_{p,q}$ the complex conjugation such that $u^* = A - iB$ for $u = A + iB$, where $A, B \in C\ell_{p,q}$. Show according to the Clifford algebras in table 6.9 that $f(\mathbb{C}\otimes C\ell_{p,q})f \simeq \mathbb{C}$, where the idempotents f associated with each of the Clifford algebras are given in the second column of the table. Prove in addition that, given $\psi \neq 0$ and ϕ arbitrary algebraic spinors in $(\mathbb{C}\otimes C\ell_{p,q})f$, the properties given in the third column of Table 6.9 hold.

Table 6.9 Clifford Algebras and Associated Idempotents for $u^* = A - iB$ for $u = A + iB$, Where $A, B \in C\ell_{p,q}$

Clifford Algebra	Idempotent (f)	Property
$\mathbb{C}\otimes C\ell_{0,2}$	$\frac{1}{2}(1+i\mathbf{e}_1)$	$\bar{\psi}^*\psi > 0$
$\mathbb{C}\otimes C\ell_{1,1}$	$\frac{1}{2}(1+\mathbf{e}_1)$	
$\mathbb{C}\otimes C\ell_{2,0}$	$\frac{1}{2}(1+\mathbf{e}_1)$	$\tilde{\psi}^*\psi > 0$
$\mathbb{C}\otimes C\ell_{0,3}$	$\frac{1}{2}(1+i\mathbf{e}_1)\frac{1}{2}(1+i\mathbf{e}_2\mathbf{e}_3)$	$\bar{\psi}^*\psi > 0, \quad \tilde{\psi}^*\phi = 0$
$\mathbb{C}\otimes C\ell_{1,2}$	$\frac{1}{2}(1+\mathbf{e}_1)\frac{1}{2}(1+i\mathbf{e}_2\mathbf{e}_3)$	$\bar{\psi}^*\phi = 0$
$\mathbb{C}\otimes C\ell_{2,1}$	$\frac{1}{2}(1+\mathbf{e}_1)\frac{1}{2}(1+i\mathbf{e}_2\mathbf{e}_3)$	$\bar{\psi}^*\phi > 0, \quad \tilde{\psi}^*\psi > 0$
$\mathbb{C}\otimes C\ell_{3,0}$	$\frac{1}{2}(1+\mathbf{e}_1)\frac{1}{2}(1+i\mathbf{e}_2\mathbf{e}_3)$	$\bar{\psi}^*\phi = 0, \quad \tilde{\psi}^*\psi > 0$
$\mathbb{C}\otimes C\ell_{0,4}$	$\frac{1}{2}(1+i\mathbf{e}_1)\frac{1}{2}(1+i\mathbf{e}_2\mathbf{e}_3)$	$\bar{\psi}^*\psi > 0$
$\mathbb{C}\otimes C\ell_{1,3}$	$\frac{1}{2}(1+\mathbf{e}_1)\frac{1}{2}(1+i\mathbf{e}_2\mathbf{e}_3)$	

Define now

$$\mathbf{A} = \begin{cases} \mathbf{e}_1\mathbf{e}_2\ldots\mathbf{e}_p & \text{if} \quad p = 0,1 \mod 4 \\ i\mathbf{e}_1\mathbf{e}_2\ldots\mathbf{e}_p & \text{if} \quad p = 2,3 \mod 4 \end{cases}$$

and show the properties in the second and third columns of table 6.10, for all $\mathbf{v} \in \mathbb{R}^{p,q}$, for $\psi \neq 0$, and for ϕ arbitrary algebraic spinors in $(\mathbb{C}\otimes C\ell_{p,q})f$.

Table 6.10 Clifford Algebras and Associated Idempotents for $\mathbf{A} = \mathbf{e}_1\mathbf{e}_2 \ldots \mathbf{e}_p$ if $p = 0, 1 \mod 4$ and for $\mathbf{A} = i\mathbf{e}_1\mathbf{e}_2 \ldots \mathbf{e}_p$ if $p = 2, 3 \mod 4$

Clifford Algebra	Idempotent (f)	Property
$\mathbb{C} \otimes \mathcal{C}\ell_{0,2}$	$\bar{\psi}^* \mathbf{A}\psi > 0$	$\overline{\mathbf{v}\psi}^* \phi + \bar{\psi}^* \mathbf{v}\phi = 0$
$\mathbb{C} \otimes \mathcal{C}\ell_{1,1}$		$\widetilde{\mathbf{v}\psi}^* \phi - \tilde{\psi}^* \mathbf{v}\phi = 0$
$\mathbb{C} \otimes \mathcal{C}\ell_{2,0}$	$\tilde{\psi}^* \mathbf{A}\psi > 0$	$\overline{\mathbf{v}\psi}^* \phi + \bar{\psi}^* \mathbf{v}\phi = 0$
$\mathbb{C} \otimes \mathcal{C}\ell_{0,3}$	$\bar{\psi}^* \mathbf{A}\psi > 0$	$\overline{\mathbf{v}\psi}^* \phi + \bar{\psi}^* \mathbf{v}\phi = 0$
$\mathbb{C} \otimes \mathcal{C}\ell_{1,2}$	$\bar{\psi}^* \mathbf{A}\phi = 0$	$\widetilde{\mathbf{v}\psi}^* \phi - \bar{\psi}^* \mathbf{v}\phi = 0$
$\mathbb{C} \otimes \mathcal{C}\ell_{2,1}$	$\tilde{\psi}^* \mathbf{A}\phi = 0$	$\overline{\mathbf{v}\psi}^* \phi + \bar{\psi}^* \mathbf{v}\phi = 0$
$\mathbb{C} \otimes \mathcal{C}\ell_{3,0}$	$\bar{\psi}^* \mathbf{A}\phi = 0, \;\; \tilde{\psi}^* \mathbf{A}\psi > 0$	$\widetilde{\mathbf{v}\psi}^* \phi - \tilde{\psi}^* \mathbf{v}\phi = 0$
$\mathbb{C} \otimes \mathcal{C}\ell_{0,4}$	$\bar{\psi}^* \mathbf{A}\psi > 0$	$\overline{\mathbf{v}\psi}^* \phi + \bar{\psi}^* \mathbf{v}\phi = 0$
$\mathbb{C} \otimes \mathcal{C}\ell_{1,3}$		$\widetilde{\mathbf{v}\psi}^* \phi - \tilde{\psi}^* \mathbf{v}\phi = 0$

(4) Given the Clifford algebra $\mathcal{C}\ell_{1,7} \simeq \mathcal{M}(16, \mathbb{R})$ with generators $\{\gamma_0, \gamma_a\}$ such that $\gamma_0^2 = 1$, $\gamma_a^2 = -1$, $a = 1, \ldots, 7$ and the idempotent

$$f = \frac{1}{2}(1 + \gamma_0)\frac{1}{2}(1 + i\gamma_1\gamma_2)\frac{1}{2}(1 + i\gamma_3\gamma_4)\frac{1}{2}(1 + i\gamma_5\gamma_6),$$

an open question is whether there exists a spinor $\psi \in (\mathbb{C}\otimes\mathcal{C}\ell_{1,7})f$ such that $(\langle\psi\tilde{\psi}^*\rangle_0)^2 < 0$. Can the reader answer that? Consider an element

$$\psi = (1 - \mathbf{e}_1\mathbf{e}_2\mathbf{e}_3\mathbf{e}_4\mathbf{e}_5\mathbf{e}_6\mathbf{e}_7\mathbf{e}_8)(1 + \mathbf{w}) \in \mathcal{C}\ell_8,$$

where

$$\mathbf{w} = \mathbf{e}_1\mathbf{e}_2\mathbf{e}_3\mathbf{e}_6 - \mathbf{e}_1\mathbf{e}_2\mathbf{e}_5\mathbf{e}_7 - \mathbf{e}_1\mathbf{e}_3\mathbf{e}_4\mathbf{e}_5 + \mathbf{e}_1\mathbf{e}_4\mathbf{e}_6\mathbf{e}_7 + \mathbf{e}_2\mathbf{e}_3\mathbf{e}_4\mathbf{e}_7 - \mathbf{e}_2\mathbf{e}_4\mathbf{e}_5\mathbf{e}_6 - \mathbf{e}_3\mathbf{e}_5\mathbf{e}_6\mathbf{e}_7.$$

Show that ψ satisfies $\psi\mathbf{v}\tilde{\psi} = 0$, for all $\mathbf{v} \in \mathbb{R}^8$, and that $\psi^2 = 16\psi$. Show that the factor $(1 + \mathbf{w})$ is not invertible (hint: show that $(1 + \mathbf{w})^2 = 8(1 + \mathbf{w})$).

(5) Compute the matrix associated with the multivector $\frac{1}{4}(\mathbf{w} - 3)$, where $\mathbf{w}$ is given as in exercise 4, in the basis $(g_1, g_2, \ldots, g_8)$, where $g_a = \mathbf{e}_a\mathbf{e}_8 f$, $a = 1, 2, \ldots, 8$, and $f = \frac{1}{8}(1 + \mathbf{w})(1 + \mathbf{e}_{12\ldots8})$. In other words, prove that

$$4\langle\tilde{g}_a(\mathbf{w} - 3)g_b\rangle_0 = \text{diag}(-1, -1, -1, -1, -1, -1, -1, 1)\,.$$

(6) Show that the quantities given by eqn (6.208) in terms of a classical spinor ψ can be written in terms of the spinor operator $\Psi = \Psi_+$ (see example 6.11) as

$$\sigma + e_5\omega = \Psi\tilde{\Psi}, \quad J = \Psi e_0\tilde{\Psi}, \quad S = \frac{1}{2}\Psi e_{21}\tilde{\Psi}, \quad e_5 K = \Psi e_3\tilde{\Psi},$$

and, supposing that $\Psi\tilde{\Psi} \neq 0$, prove the FPK identities in (6.213).

Appendix A
The Standard Two-Component Spinor Formalism

In chapter 6, we presented Weyl spinors in $\mathcal{C}\ell_{3,0}$; now, the connection with the ordinary notation in either field theory or supersymmetry books is briefly presented.

A spacetime vector $\mathbf{v} \in \mathbb{R}^{1,3}$ can be expressed as $\mathbf{v} = x^0\mathbf{e}_0 + x^1\mathbf{e}_1 + x^2\mathbf{e}_2 + x^3\mathbf{e}_3$, where (x^0, x^1, x^2, x^3) denotes components of $\mathbf{v}$ with respect to an orthonormal basis $\{\mathbf{e}_0, \mathbf{e}_1, \mathbf{e}_2, \mathbf{e}_3\}$. Null vectors are isotropic vectors and satisfy $(x^0)^2 - (x^1)^2 - (x^2)^2 - (x^3)^2 = 0$. They present null directions in $\mathbb{R}^{1,3}$ with respect to the origin $\mathcal{O}$ of an arbitrary frame in $\mathbb{R}^{1,3}$.

The space of null directions that are future (past) pointed are denoted by $\mathcal{S}^+$ $[\mathcal{S}^-]$, and represented by the intersections $\mathbb{E}^+$ $[\mathbb{E}^-]$ of the future (past) light cones with the hyperplanes $x^0 = 1$ $(x^0 = -1)$.[1] The space $\mathbb{S}^\pm$ is a sphere with the equation $x^2 + y^2 + z^2 = 1$, where (x, y, z) are coordinates in $\mathbb{E}^\pm$ (Penrose and Rindler, 1984).

Generally, the direction of any null vector $\mathbf{v} \in \mathbb{R}^{1,3}$, unless the vector is an element of the plane defined by the equation $x^0 = 0$, can be represented by two points. This description results from the intersection of $\mathbf{v}$ and the hyperplanes $x^0 = \pm 1$. The future-pointed $\mathbf{v}$ is thus represented by $(x^1/\|x^0\|, x^2/\|x^0\|, x^3/\|x^0\|)$. The inner points of $\mathbb{E}^+$ $(\mathbb{E}^-)$ represent the set of future-pointed (past-pointed) light-like directions.

By considering $\mathbb{E}^+$ and by performing a stereographic projection on the Argand–Gauss plane, we obtain a representation of the union between the set of complex numbers and the point at infinity, the latter corresponding to the north pole of $\mathbb{E}^+$.

By defining the complex number

$$\beta = \frac{x + iy}{1 - z}, \tag{A.1}$$

we obtain $\beta\overline{\beta} = \frac{x^2+y^2}{(1-z)^2}$ and, consequently,

$$x = \frac{\beta + \overline{\beta}}{\beta\overline{\beta} + 1}, \quad y = \frac{\beta - \overline{\beta}}{i(\beta\overline{\beta} + 1)}, \quad z = \frac{\beta\overline{\beta} - 1}{\beta\overline{\beta} + 1}. \tag{A.2}$$

The correspondence between points of $\mathbb{E}^+$ and the Argand–Gauss plane is injective if the point $\beta \sim \infty$ is added to the complex plane, making it correspond to the north pole with components $(1, 0, 0, 1)$. However, to avoid this point, it is convenient to associate a point of $\mathbb{E}^+$ not to a complex number β but to a pair of complex numbers [2] (ξ, η),

[1] This space is a Riemann sphere.

[2] With the condition that both numbers are not simultaneously equal to 0.

where

$$\beta = \xi/\eta. \tag{A.3}$$

The pairs (ξ, η) and $(\lambda\xi, \lambda\eta)$, where $\lambda \in \mathbb{C}$, represent the same point in $\mathbb{E}^+$. Such components are called *projective coordinates.*

The point $\beta = \xi/\eta \sim \infty$ corresponds to the point of coordinates $\binom{\xi}{\eta} = \binom{1}{0}$. The equations in (A.2) can then be expressed as

$$x = \frac{\xi\bar{\eta} + \eta\bar{\xi}}{\xi\bar{\xi} + \eta\bar{\eta}}, \quad y = \frac{\xi\bar{\eta} - \eta\bar{\xi}}{i(\xi\bar{\xi} + \eta\bar{\eta})}, \quad z = \frac{\xi\bar{\xi} - \eta\bar{\eta}}{\xi\bar{\xi} + \eta\bar{\eta}}. \tag{A.4}$$

The point $P = (1, x, y, z)$ is an arbitrary point of the light-cone transversal section with constant time and represents a null future-pointed direction, which can be represented by any point of the line $\mathcal{O}P$. In particular, if a point R is taken in the line $\mathcal{O}P$ by multiplying P by the factor $(\xi\bar{\xi} + \eta\bar{\eta})/\sqrt{2}$, then R has coordinates

$$\begin{aligned} x^1 &= \frac{1}{\sqrt{2}}(\xi\bar{\eta} + \eta\bar{\xi}), \quad x^2 = \frac{1}{i\sqrt{2}}(\xi\bar{\eta} - \eta\bar{\xi}), \\ x^3 &= \frac{1}{\sqrt{2}}(\xi\bar{\xi} - \eta\bar{\eta}), \quad x^0 = \frac{1}{\sqrt{2}}(\xi\bar{\xi} + \eta\bar{\eta}). \end{aligned} \tag{A.5}$$

Unlike the point P, the point R is not invariant under $(\xi, \eta) \mapsto (r\xi, r\eta), r \in \mathbb{R}$, although it is independent of phases $(\xi, \eta) \mapsto (e^{i\theta}\xi, e^{i\theta}\eta), \theta \in \mathbb{R}$.

Consider now the following complex linear transformation

$$\begin{aligned} \xi &\mapsto \tilde{\xi} = \alpha\xi + \mu\eta, \\ \eta &\mapsto \tilde{\eta} = \gamma\xi + \delta\eta, \end{aligned} \tag{A.6}$$

where $\alpha, \mu, \gamma, \delta \in \mathbb{C}$ satisfy $\alpha\delta - \mu\gamma \neq 0$, so that the transformation is invertible. It can be rewritten as

$$\beta \mapsto f(\beta) = \frac{\alpha\beta + \mu}{\gamma\beta + \delta}, \tag{A.7}$$

and called a Möbius transformation, from the set $\mathbb{C}\backslash\{-\delta/\gamma\}$ to $\mathbb{C}\backslash\{\alpha/\gamma\}$. Moreover, if $f(-\delta/\gamma) \sim \infty$, and $f(\infty) \sim \alpha/\gamma$, then f is an injective function from the complex plane, compactified by the point at the infinity; this point is denoted by $(\mathbb{C} \cup \{\infty\})$.

Hence, the space of light-like vectors on Minkowski spacetime is naturally a Riemann sphere. The restricted Lorentz group $\mathcal{L}^+$ is, on the other hand, the automorphism group of the Riemann sphere. The equations in (A.6), when taken with the condition

$$\alpha\delta - \mu\gamma = 1$$

are called spinor transformations, where $\beta = \xi/\eta$ is related to the null vectors by the equations in (A.5), implying that

$$\beta = \frac{x^1 + ix^2}{x^0 - x^3} = \frac{x^0 - x^3}{x^1 - ix^2}. \tag{A.8}$$

The spinor matrix $\mathbf{A} \in \mathrm{SL}(2, \mathbb{C})$ is defined as

$$\mathbf{A} = \begin{pmatrix} \alpha & \mu \\ \gamma & \delta \end{pmatrix}, \quad \det \mathbf{A} = 1. \tag{A.9}$$

The equations in (A.6), with respect to $\mathbf{A}$, read

$$\begin{pmatrix} \tilde{\xi} \\ \tilde{\eta} \end{pmatrix} = \mathbf{A} \begin{pmatrix} \xi \\ \eta \end{pmatrix}.$$

The spinor matrices $\{\pm\mathbf{A}\}$ induce the same transformation of $\beta = \xi/\eta$. The equations in (A.5) yield

$$\frac{1}{\sqrt{2}} \begin{pmatrix} x^0 + x^3 & x^1 + ix^2 \\ x^1 - ix^2 & x^0 - x^3 \end{pmatrix} = \begin{pmatrix} \xi\bar{\xi} & \xi\bar{\eta} \\ \eta\bar{\xi} & \eta\bar{\eta} \end{pmatrix} = \begin{pmatrix} \xi \\ \eta \end{pmatrix} (\bar{\xi} \quad \bar{\eta}). \tag{A.10}$$

Hence, up to a factor $1/\sqrt{2}$, it follows that

$$\begin{aligned} \begin{pmatrix} x^0 + x^3 & x^1 + ix^2 \\ x^1 - ix^2 & x^0 - x^3 \end{pmatrix} &\mapsto \begin{pmatrix} \tilde{x^0} + \tilde{x^3} & \tilde{x^1} + i\tilde{x^2} \\ \tilde{x^1} - i\tilde{x^2} & \tilde{x^0} - \tilde{x^3} \end{pmatrix} \\ &= \mathbf{A} \begin{pmatrix} x^0 + x^3 & x^1 + ix^2 \\ x^1 - ix^2 & x^0 - x^3 \end{pmatrix} \mathbf{A}^{\dagger}. \end{aligned} \tag{A.11}$$

The transformation acting on the point $\mathbf{v} = (x^0, x^1, x^2, x^3)$ is real and preserves the light-cone structure $(x^0)^2 - (x^1)^2 - (x^2)^2 - (x^3)^2 = 0$. Thus, this relation defines a restricted Lorentz transformation.

Hence, the group SL(2,$\mathbb{C}$) is the twofold covering of the restricted Lorentz group $\mathrm{SO}_+(1,3) \simeq \mathcal{L}^+$.

A more general case than this is the spin space $\mathbb{G}^\alpha$, which has three basic operations (Penrose and Rindler, 1984):

- *multiplication by scalars*: $\mathbb{C} \times \mathbb{G}^\alpha \to \mathbb{G}^\alpha$ $(\lambda \in \mathbb{C}, k^\alpha \in \mathbb{G}^\alpha \mapsto \lambda k^\alpha \in \mathbb{G}^\alpha)$
- *sum*: $\mathbb{G}^\alpha \times \mathbb{G}^\alpha \to \mathbb{G}^\alpha$ $(k^\alpha, \omega^\alpha \in \mathbb{G}^\alpha \mapsto k^\alpha + \omega^\alpha \in \mathbb{G}^\alpha)$
- *scalar product*: $\mathbb{G}^\alpha \times \mathbb{G}^\alpha \to \mathbb{C}$ $(k^\alpha, \omega^\alpha \in \mathbb{G}^\alpha \mapsto \{k^\alpha, \omega^\alpha\} \in \mathbb{C})$

The dual spin space $\mathbb{G}_\alpha$ is similarly defined: $\mathbb{G}_\alpha \ni \pi_\alpha : \mathbb{G}^\alpha \to \mathbb{C}$. Thus, $k_\alpha \omega^\alpha \equiv \{k^\alpha, \omega^\alpha\} \in \mathbb{C}$. With respect to the null vectors of Minkowski spacetime, Penrose proposed that the algebra of null vectors must be contained in the algebra of spinors. The spin space $\mathbb{G}^{\dot{\alpha}}$ is defined by the application $\varrho : \mathbb{G}^\alpha \to \mathbb{G}^{\dot{\alpha}}$ such that

$$\begin{aligned} &\varrho(k^\alpha + \omega^\alpha) = \varrho(k^\alpha) + \varrho(\omega^\alpha), \\ &\varrho(\lambda k^\alpha) = \bar{\lambda}\varrho(k^\alpha), \quad \forall k^\alpha, \omega^\alpha \in \mathbb{G}^\alpha, \quad \lambda \in \mathbb{C}\ (\bar{\lambda} \text{ is the } \mathbb{C}\text{-conjugate of } \lambda). \end{aligned}$$

Instead of by the notation $\varrho(k^\alpha)$, this transformation is denoted by $\overline{k^\alpha}$, according to the Penrose notation, characterised by the composition between the $\mathbb{C}$-conjugation and the transposition. From now on, the notation

$$\overline{k^\alpha} \equiv \bar{k}^{\dot{\alpha}} \in \mathbb{G}^{\dot{\alpha}} \tag{A.12}$$

shall be used. From (A.10), the space $\mathbb{R}^{1,3}$ endowed with the coordinates (x^0, x^1, x^2, x^3) can be expressed from the components of the spin vector k $(\xi = k^0, \eta = k^1)$. The basic operations in this space are defined by

$$\lambda(k^0, k^1) = (\lambda k^0, \lambda k^1), \tag{A.13}$$
$$(k^0, k^1) + (\omega^0, \omega^1) = (k^0 + \omega^0, k^1 + \omega^1), \tag{A.14}$$
$$\{(k^0, k^1), (\omega^0, \omega^1)\} = k^0\omega^1 - k^1\omega^0. \tag{A.15}$$

With the antisymmetric bilinear form in eqn (A.15), the representation of a spin vector is given by the choice of a pair of normalised spin vectors o^α and ι^α:

$$\{o^\alpha, \iota^\alpha\} = o_\alpha \iota^\alpha = 1 = -\iota_\alpha o^\alpha = -\{\iota^\alpha, o^\alpha\}, \tag{A.16}$$

where (o_α, ι_α) stands for the dual pair of (o^α, ι^α). Moreover, the antisymmetry of (A.15) implies that $o_\alpha o^\alpha = \iota_\alpha \iota^\alpha = 0$. The pair (o^α, ι^α), with the condition in eqn (A.16), is called the spin basis, and the components of k with respect to the spin basis are provided by (Figueiredo, de Oliveira, and Rodrigues, 1990)

$$k^0 = \{k^\alpha, \iota^\alpha\}, \quad k^1 = -\{k^\alpha, o^\alpha\}. \tag{A.17}$$

Hence, $k^\alpha = k^0 o^\alpha + k^1 \iota^\alpha$.

The antisymmetric element

$$\begin{aligned} \mathbb{G}_{\alpha\beta} \ni \epsilon_{\alpha\beta} : \mathbb{G}^\alpha &\to \mathbb{G}_\beta \\ k^\alpha &\mapsto k_\beta = k^\alpha \epsilon_{\alpha\beta}, \end{aligned} \tag{A.18}$$

is responsible for lowering and raising indices, such that

$$\{k^\alpha, \omega^\alpha\} = \epsilon_{\alpha\beta} k^\alpha \omega^\beta = k^0\omega^1 - k^1\omega^0 = -\{\omega^\alpha, k^\alpha\}. \tag{A.19}$$

We can write

$$\epsilon_{\alpha\beta} = o_\alpha \iota_\beta - \iota_\alpha o_\beta \tag{A.20}$$

since, for any spin basis, it follows that

$$\begin{aligned} k^\alpha = k^0 o^\alpha + k^1 \iota^\alpha &= (o^\alpha \iota^\beta - \iota^\alpha o^\beta)(k^0 o_\beta + k^1 \iota_\beta) \\ &= \epsilon^{\alpha\beta} k_\beta. \end{aligned} \tag{A.21}$$

Here, $\mathbb{G}_{\alpha\beta}$ denotes the tensor product $\mathbb{G}_\alpha \otimes \mathbb{G}_\beta$. The dual tensor $\mathbb{G}^{\alpha\beta} \ni \epsilon^{\alpha\beta} : \mathbb{G}_\beta \to \mathbb{G}^\alpha$ satisfies

$$\epsilon_{\alpha\beta}\epsilon^{\gamma\beta} = \delta_\alpha^{\ \gamma}, \tag{A.22}$$

where $\mathbb{G}^\alpha \to \mathbb{G}_\beta \to \mathbb{G}^\gamma$ is defined. Similarly, we can express

$$\epsilon_\alpha^{\ \beta} = o_\alpha \iota^\beta - \iota_\alpha o^\beta, \tag{A.23}$$

and

$$\{k, \omega\} = k_\beta \omega^\beta = k_0\omega^0 + k_1\omega^1 = k^0\omega^1 - \omega^0 k^1 \Rightarrow \left\{ \begin{array}{c} k_0 = -k^1 \\ k_1 = k^0 \end{array} \right\}. \tag{A.24}$$

Now, given arbitrary spinors $k_\alpha, \omega_\alpha \in \mathbb{G}_\alpha$ such that $k_\alpha \omega^\alpha = 1$, we can write

$$\begin{aligned}\epsilon_{\alpha\beta} &= k_\alpha \omega_\beta - \omega_\alpha k_\beta = (k_0 \iota_\alpha + k_1 o_\alpha)(\omega_0 \iota_\beta + \omega_1 o_\beta)\\ &= (k_1\omega_0 - k_0\omega_1) o_\alpha \iota_\beta - (k_1\omega_0 - k_0\omega_1)\iota_\alpha o_\beta\\ &= o_\alpha \iota_\beta - \iota_\alpha o_\beta,\end{aligned} \tag{A.25}$$

where the equivalence to eqn (A.20) is then accomplished.

Now the spacetime metric and the spacetime vectors as well can be constructed from spin vectors. Spacetime vectors of $\mathbb{R}^{1,3}$ present a spinor description via spin vectors. Latin indices are employed here to label the elements x^μ of a real vector space, when the indices α and $\dot\alpha$ are grouped together. Moreover, the notation $\{a, b, \ldots\} = \{\alpha\dot\alpha, \beta\dot\beta, \ldots\}$ will be used. The null tetrad $(l^\mu, n^\mu, m^\mu, \overline{m}^\mu)$

$$l^\mu = o^\alpha o^{\dot\alpha}, \quad n^\mu = \iota^\alpha \iota^{\dot\alpha}, \quad m^\mu = o^\alpha \iota^{\dot\alpha}, \quad \overline{m}^\mu = \iota^\alpha o^{\dot\alpha} \tag{A.26}$$

and the metric $\eta_{\mu\nu} = \epsilon_{\alpha\beta}\epsilon_{\dot\alpha\dot\beta}$ can be defined by verifying that the vectors of the null tetrad are null vectors with respect to η_{ab}:

$$\eta_{\mu\nu} l^\mu l^\nu = l^\mu l_\mu = 0\,. \tag{A.27}$$

Similarly, $n^\mu n_\mu = m^\mu m_\mu = \overline{m}^\mu \overline{m}_\mu = 0$. In addition, the following expressions hold:

$$\begin{aligned} l^\mu n_\mu = 1, \qquad m^\mu \overline{m}_\mu = -1, \qquad l^\mu m_\mu = l^\mu \overline{m}_\mu = n^\mu m_\mu = n^\mu \overline{m}_\mu = 0\,,\\ \eta_\mu^{\ \nu} \equiv \eta_{\mu\rho} g^{\rho\nu} = n_\mu l^\nu + l_\mu n^\nu - \overline{m}_\mu m^\nu - m_\mu \overline{m}^\nu\,. \end{aligned} \tag{A.28}$$

It is sometimes convenient to define another tetrad $(t^\mu, x^\mu, y^\mu, z^\mu)$ as

$$\begin{aligned} t^\mu &= \tfrac{1}{\sqrt{2}}(l^\mu + n^\mu) = \tfrac{1}{\sqrt{2}}(o^\alpha o^{\dot\alpha} + \iota^\alpha \iota^{\dot\alpha}),\\ x^\mu &= \tfrac{1}{\sqrt{2}}(m^\mu + \overline{m}^\mu) = \tfrac{1}{\sqrt{2}}(o^\alpha \iota^{\dot\alpha} + \iota^\alpha o^{\dot\alpha}),\\ y^\mu &= \tfrac{i}{\sqrt{2}}(m^\mu + \overline{m}^\mu) = \tfrac{i}{\sqrt{2}}(o^\alpha \iota^{\dot\alpha} - \iota^\alpha o^{\dot\alpha}),\\ z^\mu &= \tfrac{1}{\sqrt{2}}(l^\mu - n^\mu) = \tfrac{1}{\sqrt{2}}(o^\alpha o^{\dot\alpha} - \iota^\alpha \iota^{\dot\alpha}). \end{aligned} \tag{A.29}$$

We obtain from (A.28) that

$$\begin{aligned} t^\mu x_\mu &= t^\mu y_\mu = t^\mu z_\mu = x^\mu y_\mu = y^\mu z_\mu = z^\mu x_\mu = 0,\\ t^\mu t_\mu &= 1 = -x^\mu x_\mu = -y^\mu y_\mu = -z^\mu z_\mu\,. \end{aligned} \tag{A.30}$$

Consequently, the metric components $\eta_\mu^{\ \nu}$ (A.28) have the form $\eta_\mu^{\ \nu} = t_\mu t^\nu - x_\mu x^\nu - y_\mu y^\nu - z_\mu z^\nu$, thus identifying the tetrad $(t^\mu, x^\mu, y^\mu, z^\mu)$ as the Minkowski tetrad. Hence, we can write

$$K^\mu = K^0 t^\mu + K^1 x^\mu + K^2 y^\mu + K^3 z^\mu. \tag{A.31}$$

By considering the spin basis $\{o^\alpha, \iota^\alpha\}$, we can express the vector K^μ with respect to the basis as

$$K^\mu = K^{0\dot0} l^\mu + K^{0\dot1} m^\mu + K^{1\dot0} \overline{m}^\mu + K^{1\dot1} n^\mu. \tag{A.32}$$

When the two last equations are compared, it follows that

$$K^{\mu} = \frac{1}{\sqrt{2}} \begin{pmatrix} K^0 + K^3 & K^1 + iK^2 \\ K^1 - iK^2 & K^0 - K^3 \end{pmatrix} = \begin{pmatrix} K^{0\dot{0}} & K^{0\dot{1}} \\ K^{1\dot{0}} & K^{1\dot{1}} \end{pmatrix}. \tag{A.33}$$

Thus,

$$K^{\mu} = \pm k^{\alpha} \bar{k}^{\dot{\alpha}}, \tag{A.34}$$

where the sign defines a future (+) (past (−)). The vector K^{μ} is real and null, since $K^{\mu}K_{\mu} = |k_{\alpha}k^{\alpha}|^2 = 0$. If $\xi = k^0$, and $\eta = k^1$, it follows that

$$K^{0\dot{0}} = \xi\bar{\xi}, \quad K^{0\dot{1}} = \xi\bar{\eta}, \quad K^{1\dot{0}} = \eta\bar{\xi}, \quad K^{1\dot{1}} = \eta\bar{\eta}, \tag{A.35}$$

which causes eqn (A.33) to be led to

$$\frac{1}{\sqrt{2}} \begin{pmatrix} x^0 + x^3 & x^1 + ix^2 \\ x^1 - ix^2 & x^0 - x^3 \end{pmatrix} = \begin{pmatrix} \xi\bar{\xi} & \xi\bar{\eta} \\ \eta\bar{\xi} & \eta\bar{\eta} \end{pmatrix} = \begin{pmatrix} \xi \\ \eta \end{pmatrix} (\bar{\xi} \;\; \bar{\eta}), \tag{A.36}$$

where $x^0 = K^0$; $x^1 = K^1$; $x^2 = K^2$; and $x^3 = K^3$.

A.1 Weyl Spinors

Given the formalism presented for spin vectors, which are known as 2-spinors, it is also necessary to study then from the point of view of representations of the Lorentz group $\mathrm{SL}(2, \mathbb{C})$, the 2-fold covering of the restricted Lorentz group $\mathcal{L}^{+} \simeq \mathrm{SO}_{+}(1,3)$. Linear transformations with a unit determinant, with respect to the spin space, determine the group $\mathrm{SL}(2, \mathbb{C})$. We have already shown that there are two non-equivalent representations of $\mathrm{SL}(2, \mathbb{C})$; these are denoted by $D^{(1/2,0)}$ and $D^{(0,1/2)}$, respectively, and the elements of the carrier space associated with them are called Weyl spinors.

Both the left-handed $D^{(1/2,0)}$ and the right-handed $D^{(0,1/2)}$ representations of the Lorentz group determine the rules of transformation obeyed by fermions of spin-1/2. It is well known that the Hermitian conjugation can be used to interchange these two representations. Dirac spinors take into account reducible representations of the form $D^{(1/2,0)} \oplus D^{(0,1/2)}$. From here on, $D^{(1/2,0)}$ spinors will carry undotted indices $\alpha, \beta, \ldots = 1, 2$, and $D^{(0,1/2)}$ spinors will carry dotted indices $\dot{\alpha}, \dot{\beta}, \ldots = 1, 2$.

A.2 Contravariant Undotted Spinors

Contravariant undotted spinors are elements of a complex two-dimensional space endowed with the spinor metric

$$\begin{aligned} G : \mathbb{C}^2 \times \mathbb{C}^2 &\to \mathbb{C}, \\ (\zeta, \chi) &\mapsto G(\zeta, \chi) = \zeta^{\dagger} J \chi, \end{aligned} \tag{A.37}$$

where

$$J = \begin{pmatrix} 0 & 1 \\ -1 & 0 \end{pmatrix}. \tag{A.38}$$

The spinor ζ is represented by the column vector

$$\zeta = \begin{pmatrix} \zeta^1 \\ \zeta^2 \end{pmatrix}. \tag{A.39}$$

This spinor can be identified with its algebraic counterpart in eqn (6.154).

It also carries the $D^{(1/2,0)}$ representation of $\mathrm{SL}(2,\mathbb{C})$ and is transformed under $R \in \mathrm{SL}(2,\mathbb{C})$ as

$$\zeta \mapsto R\zeta .$$

Moreover, this transformation corresponds to the rule established in chapter 6, eqn (6.162).

A.3 Covariant Undotted Spinors

Covariant undotted spinors are elements of a complex two-dimensional dual space $\mathbb{C}^2_\flat$, and are defined by

$$\begin{aligned} \mathbb{C}^2_\flat \ni \zeta_\flat : \mathbb{C}^2 \to \mathbb{C}\,, \\ \chi \mapsto \zeta_\flat(\chi) = \zeta_\flat \chi = G(\zeta,\chi) = \zeta^\dagger J \chi\,, \end{aligned} \tag{A.40}$$

which implies that $\zeta_\flat = \zeta^\dagger J$. Hence, the spinor $\zeta_\flat$ is represented by

$$\zeta_\flat = (\zeta_1, \zeta_2) = (\zeta^2, -\zeta^1)\,. \tag{A.41}$$

This spinor can be identified with its algebraic counterpart in eqn (6.155). For a spinor metric to be invariant under $R \in \mathrm{SL}(2,\mathbb{C})$, it is necessary that

$$\zeta_\flat \mapsto \zeta_\flat R^{-1}, \quad R \in \mathrm{SL}(2,\mathbb{C})\,, \tag{A.42}$$

which corresponds to the transformation in eqn (6.163).

Contravariant undotted spinors and covariant undotted spinors represent respectively elements of $\mathbb{G}^\alpha$ and $\mathbb{G}_\alpha$.

A.4 Contravariant Dotted Spinors

Covariant dotted spinors are elements of a complex two-dimensional space $(\dot{\mathbb{C}}^2) \ni \dot{\zeta}$, $\zeta \in \mathbb{C}^2$, endowed with the spinor metric

$$\begin{aligned} \dot{G} : \dot{\mathbb{C}}^2 \times \dot{\mathbb{C}}^2 \to \mathbb{C}\,, \\ (\dot{\zeta}, \dot{\chi}) \mapsto \dot{G}(\dot{\zeta}, \dot{\chi}) = \dot{\zeta} J \dot{\chi}^\dagger . \end{aligned} \tag{A.43}$$

A covariant dotted spinor is represented by $\dot{\zeta} = (\overline{\zeta}^{\dot{1}}, \overline{\zeta}^{\dot{2}})$ and can be identified with its algebraic counterpart in eqn (6.158).

A.5 Covariant Dotted Spinors

Covariant dotted spinors are elements of a complex two-dimensional dual space $\dot{\mathbb{C}}^2_\flat$, which is defined by

$$\begin{aligned} \dot{\mathbb{C}}^2_\flat \ni \dot{\chi}_\flat : \dot{\mathbb{C}}^2 \to \mathbb{C}\,, \\ \dot{\zeta} \mapsto \dot{\zeta}(\dot{\chi}_\flat) = \dot{\zeta}\dot{\chi}_\flat = \dot{G}(\dot{\zeta},\dot{\chi}) = \dot{\zeta} J \dot{\chi}_\flat\,, \end{aligned} \tag{A.44}$$

which implies that $\dot{\chi}_\flat = J(\dot{\chi})^\dagger$; a covariant dotted spinor is represented by

$$\dot{\chi}_\flat = \begin{pmatrix} \overline{\chi}_{\dot{1}} \\ \overline{\chi}_{\dot{2}} \end{pmatrix} = \begin{pmatrix} \overline{\chi}^{\dot{2}} \\ -\overline{\chi}^{\dot{1}} \end{pmatrix}. \tag{A.45}$$

Hence, we can identify it with its algebraic equivalent counterpart in eqn (6.160).

Clearly, the transformation rule for dotted spinors under the transformation $R \in \mathrm{SL}(2,\mathbb{C})$ is provided by

$$\dot{\zeta} \mapsto \dot{\zeta} R^\dagger\,, \qquad \dot{\zeta}_\flat \mapsto (R^\dagger)^{-1}\dot{\zeta}_\flat\,, \tag{A.46}$$

carrying the $D^{(0,1/2)}$ representation of $\mathrm{SL}(2,\mathbb{C})$. In fact, it corresponds to the transformation in eqn (6.164).

Dotted spinors are elements of $\mathbb{G}^{\dot{\alpha}}$ and $\mathbb{G}_{\dot{\alpha}}$. The action of the Lorentz group on Weyl spinors can be depicted as follows:

$$\zeta \mapsto R\zeta\,, \tag{A.47}$$

$$\zeta_\flat \mapsto \zeta_\flat R^{-1}\,, \tag{A.48}$$

$$\dot{\zeta} \mapsto \dot{\zeta} R^\dagger\,, \tag{A.49}$$

$$\dot{\zeta}_\flat \mapsto (R^\dagger)^{-1}\dot{\zeta}_\flat\,. \tag{A.50}$$

These transformations are emulated in eqns (6.162–6.165), in the algebraic spinor framework.

From 2-spinors, Dirac spinors can defined as elements of $\mathbb{G}^\alpha \oplus \mathbb{G}_{\dot{\alpha}}$. They are classically realised as elements of $\mathbb{C}^4$, equipped with the spinor metric

$$\begin{aligned} \mathcal{G} : \mathbb{C}^4 \times \mathbb{C}^4 \to \mathbb{C} \\ (\psi_1,\psi_2) \mapsto \mathcal{G}(\psi_1,\psi_2) = \psi_1^\dagger J_d \psi_2, \end{aligned} \tag{A.51}$$

where ψ is defined by

$$\mathbb{C}^2 \oplus (\dot{\mathbb{C}}^2)_\flat \ni \psi = \zeta + \dot{\chi}_\flat = \begin{pmatrix} \zeta^1 \\ \zeta^2 \\ \overline{\chi}_{\dot{1}} \\ \overline{\chi}_{\dot{2}} \end{pmatrix}. \tag{A.52}$$

With respect to the standard basis of $\mathbb{C}^4$, the matrix J_d is the representation $J_d = \mathrm{diag}(J,J)$, where J denotes the symplectic matrix defined by eqn (A.38).

Dirac spinors carry the $D^{(1/2,0)} \oplus D^{(0,1/2)}$ representation of $\mathrm{SL}(2,\mathbb{C})$. Under the condition $\mathcal{G}(\psi_1,\psi_2) = \mathcal{G}(\rho(R)\psi_1, \rho(R)\psi_2)$, requiring that the spinor metric $\mathcal{G}$ be invariant under R, the following important representation is obtained:

$$\rho(R) = \begin{pmatrix} R & 0 \\ 0 & (R^\dagger)^{-1} \end{pmatrix}, \qquad R \in \mathrm{SL}(2,\mathbb{C}). \tag{A.53}$$

A.6 Null Flags and Flagpoles

This section describes the classical framework corresponding to the algebraic formulation provided by section 6.9.

We have already associated, via eqn (A.34), a future-pointed null vector, which contains components K^μ, with the spin-vector k^α, which has coordinates (ξ,η). From eqn (A.35), the 2-uple (ξ,η) can be further identified as the coordinates associated with the components K^μ, which are invariant under transformations $(\xi,\eta) \mapsto (e^{i\theta}\xi, e^{i\theta}\eta)$. Moreover, this ambiguity can be reduced up to a sign by introducing a structure that is composed of a null vector $\mathbf{K} = K^\mu \mathbf{e}_\mu$, called a *pole*, and a null half-plane – tangent to the light cone and having $\mathbf{K}$ as the intersection – called a *flagpole*.

Given a contravariant undotted spinor k^α, a geometric object can be constructed, namely, the flagpole. As in eqn (A.36), we shall change the notation by establishing $K^\mu = x^\mu$. The pole is defined by eqn (A.34), namely,

$$x^\mu = \frac{1}{\sqrt{2}} x^{\alpha\dot\alpha} = k^\alpha \bar{k}^{\dot\alpha} = \frac{1}{\sqrt{2}} \begin{pmatrix} x^0 + x^3 & x^1 + ix^2 \\ x^1 - ix^2 & x^0 - x^3 \end{pmatrix} = \begin{pmatrix} k^1\bar{k}^{\dot 1} & k^1\bar{k}^{\dot 2} \\ k^2\bar{k}^{\dot 1} & k^2\bar{k}^{\dot 2} \end{pmatrix}.$$

The vector x^μ is dilated by ρ^2, when k^α is multiplied by $\lambda = \rho e^{i\theta}$. Notwithstanding, the vector x^μ does not change its direction and is independent of the choice of θ. Hence, the null vector x^μ is uniquely determined by the spinor k^α. However, the spinor k^α is not uniquely determined by x^μ, which corresponds to a family of spinors. They form a projective space and differ from each other by a phase $e^{i\theta}$.

The momentum is defined as

$$F^{\mu\nu} = F^{\alpha\beta\dot\alpha\dot\beta} = k^\alpha k^\beta \epsilon^{\dot\alpha\dot\beta} + \epsilon^{\alpha\beta}\bar{k}^{\dot\alpha}\bar{k}^{\dot\beta}. \tag{A.54}$$

The antisymmetric tensor F^{ab} is real and determines a half-plane that is tangent to the light cone along the vector $x^\mu = k^\alpha \bar{k}^{\dot\alpha}$.

By taking a spin basis $\{k^\alpha, \omega^\alpha\}$, where $k_\alpha \omega^\alpha = 1$, we find that $\epsilon^{\alpha\beta} = k^\alpha\omega^\beta - \omega^\beta k^\alpha$. Thus, the quantity $F^{\mu\nu}$ can be characterised as the angular momentum, since

$$\begin{aligned} F^{\mu\nu} &= F^{\alpha\beta\dot\alpha\dot\beta} = k^\alpha k^\beta(\bar{k}^{\dot\alpha}\bar\omega^{\dot\beta} - \bar\omega^{\dot\alpha}\bar{k}^{\dot\beta}) + (k^\alpha\omega^\beta - \omega^\alpha k^\beta)\bar{k}^{\dot\alpha}\bar{k}^{\dot\beta} \\ &= k^\alpha\bar{k}^{\dot\alpha}(k^\beta\bar\omega^{\dot\beta} - \omega^\beta\bar{k}^{\dot\beta}) + (k^\alpha\bar\omega^{\dot\alpha} + \omega^\alpha\bar{k}^{\dot\alpha})k^\beta\bar{k}^{\dot\beta} = X^{\alpha\dot\alpha}Y^{\beta\dot\beta} - Y^{\alpha\dot\alpha}X^{\beta\dot\beta} \\ &= x^\mu y^\nu - y^\mu x^\nu. \end{aligned} \tag{A.55}$$

The tensor F^{ab} hence represents a bivector constituted of two vectors with components x^μ and y^μ in $\mathbb{R}^{1,3}$. The pole x^μ is the null flagpole vector, uniquely determined by the spinor o^α. The second vector, given by

$$y^\mu = Y^{\alpha\dot{\alpha}} = (k^\alpha \bar{\omega}^{\dot{\alpha}} + \omega^\alpha \bar{k}^{\dot{\alpha}}), \tag{A.56}$$

is also determined by k^α, although not uniquely, since the pair $(k^\alpha, \omega^\alpha)$ is not the only way to construct $\epsilon^{\alpha\beta}$. Indeed, any spinor of type

$$\omega_0^\alpha = \omega^\alpha + \lambda k^\alpha, \quad \lambda \in \mathbb{C} \tag{A.57}$$

satisfies $k_\alpha \omega_0^\alpha = 1$. With this freedom, the vector y^μ transforms as $y_0^\mu = y^\mu + (\lambda + \bar{\lambda}) x^\mu$. Each scalar $(\lambda + \bar{\lambda})$ can be thus associated to a family of coplanar vectors y_0^μ. This is the flagpole, as proposed by Penrose. Some prominent properties can be now derived.

The vector y^μ is orthogonal to the null vector x^μ. Indeed,

$$x^\mu \cdot y^\mu = x_\mu y^\mu = -\frac{1}{2} X_{\alpha\dot{\alpha}} Y^{\alpha\dot{\alpha}} = -\frac{1}{2} k_\alpha \bar{k}_{\dot{\alpha}} (k^\alpha \bar{\omega}^{\dot{\alpha}} + \omega^\alpha \bar{k}^{\dot{\alpha}}) = 0\,. \tag{A.58}$$

Moreover, y^μ is a space-like unit vector. In fact,

$$\begin{aligned} y^\mu \cdot y^\mu = y_\mu y^\mu &= \frac{1}{2}(k_\alpha \bar{\omega}_{\dot{\alpha}} + \omega_\alpha \bar{k}_{\dot{\alpha}})(k^\alpha \bar{\omega}^{\dot{\alpha}} + \omega^\alpha \bar{k}^{\dot{\alpha}}) \\ &= \frac{1}{2}(k_\alpha \omega^\alpha)(\bar{\omega}_{\dot{\alpha}} \bar{k}^{\dot{\alpha}}) + \frac{1}{2}(\omega_\alpha k^\alpha)(\bar{k}_{\dot{\alpha}} \bar{\omega}^{\dot{\alpha}}) = -1. \end{aligned} \tag{A.59}$$

By multiplying the spinor k^α by $e^{i\theta}$, the vector y^μ spins around the pole by the angle 2θ. Actually, we have

$$\begin{aligned} y^\mu_{rot} &= Y^{A\dot{\alpha}}_{rot} = e^{2i\theta} k^\alpha \bar{\omega}^{\dot{\alpha}} + e^{-2i\theta} \omega^\alpha \bar{k}^{\dot{\alpha}} \\ &= \cos 2\theta (k^\alpha \bar{\omega}^{\dot{\alpha}} + \omega^\alpha \bar{k}^{\dot{\alpha}}) + \sin 2\theta (i k^\alpha \bar{\omega}^{\dot{\alpha}} - i \omega^\alpha \bar{k}^{\dot{\alpha}}) \\ &= y^\mu \cos 2\theta + z^\mu \sin 2\theta\,. \end{aligned} \tag{A.60}$$

In addition, $z^\mu = i(k^\alpha \bar{\omega}^{\dot{\alpha}} - \omega^\alpha \bar{k}^{\dot{\alpha}})$ is a space-like unit vector, orthogonal to the vectors x^μ and y^μ. Together with y^μ, it constitutes the flagpole.

In order to fix the notation, we know that two-dimensional spinor representations of the Lorentz group can be derived from the property that, under a Lorentz transformation, a contravariant 4-vector x^μ transforms as $x^\mu \mapsto x'^\mu = R^\mu{}_\nu x^\nu$, where $R \in \mathrm{SO}(1,3)$ satisfies $R^\mu{}_\nu \eta_{\mu\rho} R^\rho{}_\lambda = \eta_{\nu\lambda}$. The corresponding covariant 4-vector $x_\mu \mapsto \eta_{\mu\nu} x^\nu$ satisfies $x_\nu = x'_\mu R^\mu{}_\nu$. The most general proper orthochronous Lorentz transformation, corresponding to a rotation by an angle of θ about an axis $\hat{\mathbf{n}}$, where $\vec{\theta} = \theta \hat{\mathbf{n}}$, and a boost vector $\zeta \mapsto \hat{\mathbf{v}} \tanh^{-1} \beta$, where $\hat{\mathbf{v}} = \mathbf{v}/\|\mathbf{v}\|$ and where $\beta = \|\mathbf{v}\|$, is a 4×4 matrix given by

$$R = \exp\left(-\frac{i}{2} \theta^{\rho\sigma} S_{\rho\sigma}\right) = \exp\left(-i\vec{\theta} \cdot \vec{S} - i\zeta \cdot \vec{K}\right), \tag{A.61}$$

where $\theta^i = \frac{1}{2}\epsilon^{ijk}\theta_{jk}$; $\zeta^i = \theta^{i0} = -\theta^{0i}$; $S^i = \frac{1}{2}\epsilon^{ijk} S_{jk}$; $K^i = S^{0i} = -S^{i0}$; and

$$(S_{\rho\sigma})^\mu{}_\nu = i(\eta_\rho{}^\mu \eta_{\sigma\nu} - \eta_\sigma{}^\mu \eta_{\rho\nu})\,. \tag{A.62}$$

Here, the indices $i, j, k = 1, 2, 3$, and $\epsilon^{123} = +1$ (Dreiner, Haber, and Martin, 2010).

It follows from (A.61, A.62) that an infinitesimal orthochronous Lorentz transformation is given by $R^\mu{}_\nu \approx \delta^\mu_\nu + \theta^\mu{}_\nu$. Moreover, the infinitesimal boost parameter reads

$\beta\hat{\mathbf{v}}$, since $\beta \ll 1$ for an infinitesimal boost. Hence, the actions of the infinitesimal boosts and rotations on the spacetime coordinates are respectively given by

$$\vec{x} \mapsto \vec{x}' \approx \vec{x} + (\vec{\theta} \times \vec{x}), \quad (t \mapsto t' \approx t),$$
$$\vec{x} \mapsto \vec{x}' \approx \vec{x} + \vec{\beta} t, \quad (t \mapsto t' \approx t + \vec{\beta} \cdot \vec{x}). \tag{A.63}$$

For contravariant 4-vectors, the reasoning is similar.

With respect to the Lorentz transformation R, a general n-component field Φ transforms according to a representation R of the Lorentz group as $\Phi(x^\mu) \mapsto \Phi'(x'^\mu) = [R]\,\Phi(x^\mu)$, where $[R]$ is the corresponding (finite) d-dimensional matrix representation. Equivalently, the functional form of the transformed field Φ obeys

$$\Phi'(x^\mu) = [R]\Phi([R^{-1}]^\mu{}_\nu x^\nu). \tag{A.64}$$

For proper orthochronous Lorentz transformations,

$$R = \exp\left(-\frac{i}{2}\theta_{\mu\nu} J^{\mu\nu}\right) \approx \mathbb{I}_{d\times d} - i\vec{\theta}\cdot\vec{J} - i\vec{\zeta}\cdot\vec{K}, \tag{A.65}$$

where $\mathbb{I}_{d\times d}$ is the $d \times d$ identity matrix, and $\theta_{\mu\nu}$ parameterises the Lorentz transformation R by (A.61). The six independent components of the matrix-valued antisymmetric tensor $J^{\mu\nu}$ are the d-dimensional generators of the Lorentz group and satisfy the commutation relations

$$[J^{\mu\nu}, J^{\lambda\kappa}] = i(g^{\mu\kappa} J^{\nu\lambda} + g^{\nu\lambda} J^{\mu\kappa} - g^{\mu\lambda} J^{\nu\kappa} - g^{\nu\kappa} J^{\mu\lambda}). \tag{A.66}$$

The vectors $\vec{J}$ and $\vec{K}$ are defined as the generators of rotations parameterised by $\vec{\theta}$ and the boosts parameterised by $\vec{\zeta}$, respectively, where $J^i = \frac{1}{2}\epsilon^{ijk} J_{jk}$, and $K^i = J^{0i}$.

Here, we focus on the inequivalent non-trivial irreducible representations of the Lorentz algebra $D^{(1/2,0)}$ and $D^{(0,1/2)}$. In the $D^{(1/2,0)}$ representation, $\vec{J} = \vec{\sigma}/2$, and $\vec{K} = -i\vec{\sigma}/2$, in eq. (A.65), so

$$R_{(\frac{1}{2},0)} \mapsto R \approx \mathbb{I}_{2\times 2} - i\vec{\theta}\cdot\vec{\sigma}/2 - \vec{\zeta}\cdot\vec{\sigma}/2, \tag{A.67}$$

where $\vec{\sigma} = (\sigma^1, \sigma^2, \sigma^3)$ represents the Pauli matrices. The transformation R carries undotted spinor indices, as indicated by $R_\alpha{}^\beta$. A two-component spinor in the $D^{(1/2,0)}$ representation is already denoted by ψ_α, which transforms as $\psi_\alpha \mapsto R_\alpha{}^\beta \psi_\beta$.

On the other hand, in the $D^{(0,1/2)}$ representation, $\vec{J} = -\vec{\sigma}^*/2$, and $\vec{K} = -i\vec{\sigma}^*/2$, in eqn (A.65). Hence, its representation matrix is R^*, the complex conjugate of eqn (A.67). By definition, the indices carried by R^* are dotted, as indicated by $(R^*)_{\dot\alpha}{}^{\dot\beta}$. It is already known that a two-component $D^{(0,1/2)}$ spinor $\psi^\dagger_{\dot\alpha}$ transforms as $\psi^\dagger_{\dot\alpha} \mapsto (R^*)_{\dot\alpha}{}^{\dot\beta}\psi^\dagger_{\dot\beta}$.

It follows that the $D^{(1/2,0)}$ and $D^{(0,1/2)}$ representations are related by Hermitian conjugation. In fact, if ψ_α denotes a $D^{(1/2,0)}$ spinor, then $(\psi_\alpha)^\dagger$ transforms as a $D^{(0,1/2)}$ spinor. In combining spinors to make Lorentz tensors, it is useful to regard $\psi^\dagger_{\dot\alpha}$ as a row vector, and ψ_α as a column vector, with

$$\psi^\dagger_{\dot\alpha} \mapsto (\psi_\alpha)^\dagger. \tag{A.68}$$

The Lorentz transformation property of $\psi^\dagger_{\dot\alpha}$ then follows from $(\psi_\alpha)^\dagger \mapsto (\psi_\beta)^\dagger (R^\dagger)^{\dot\beta}{}_{\dot\alpha}$, where $(R^\dagger)^{\dot\beta}{}_{\dot\alpha} = (R^*)_{\dot\alpha}{}^{\dot\beta}$.

In the dotted-index notation, the dagger is used to denote Hermitian conjugation, as in (A.68). In fact, the dagger is used to denote the Hermitian conjugation of spinors in most textbooks (Srednicki, 2007). However, it is worth emphasising that many references in supersymmetry e.g. (Sohnius, 1985; Srivastava, 1986; West, 1990; Wess and Bagger, 1992; Bailin and Love, 1994; Mohapatra, 2003) employ the Wess and Bagger (1992) notation, where $\overline{\psi}_{\dot{\alpha}} \equiv \psi^{\dagger}_{\dot{\alpha}} = (\psi_{\alpha})^{\dagger}$.

There are two additional spin-1/2 irreducible representations of the Lorentz group, namely, $(R^{-1})^{\intercal}$ and $(R^{-1})^{\dagger}$, However, they are equivalent to the $D^{(1/2,0)}$ and the $D^{(0,1/2)}$ representations, respectively. The spinors that transform under these representations have the raised spinor indices ψ^{α} and $\psi^{\dagger\dot{\alpha}}$, respectively, with the transformation laws $\psi^{\alpha} \mapsto [(R^{-1})^{\intercal}]^{\alpha}{}_{\beta}\psi^{\beta}$, and $\psi^{\dagger\dot{\alpha}} \mapsto [(R^{-1})^{\dagger}]^{\dot{\alpha}}{}_{\dot{\beta}}\psi^{\dagger\dot{\beta}}$, respectively. Lorentz tensors can be derived from spinors by regarding ψ^{α} as a row vector, and $\psi^{\dagger\,\dot{\alpha}}$ as a column vector, with

$$\psi^{\dagger\,\dot{\alpha}} \mapsto (\psi^{\alpha})^{\dagger}\,. \tag{A.69}$$

The Lorentz transformation property $\psi^{\dagger\,\dot{\alpha}}$ then follows from

$$(\psi^{\alpha})^{\dagger} \mapsto [(R^{-1})^{\dagger}]^{\dot{\alpha}}{}_{\dot{\beta}}(\psi^{\beta})^{\dagger}\,.$$

The spinor indices are raised and lowered with the two-index antisymmetric epsilon symbol with non-zero components $\epsilon^{12} = -\epsilon^{21} = \epsilon_{21} = -\epsilon_{12} = 1$, and similar sign conventions for the dotted spinor indices. In particular, $\epsilon^{\dot{\alpha}\dot{\beta}} = (\epsilon^{\alpha\beta})^{*}$, and $\epsilon_{\dot{\alpha}\dot{\beta}} = (\epsilon_{\alpha\beta})^{*}$, as well.

Moreover, the Kronecker delta symbol reads $\delta^{\dot{\beta}}_{\dot{\alpha}} = (\delta^{\beta}_{\alpha})^{*}$. The epsilon symbols with undotted and with dotted indices, respectively, satisfy

$$\epsilon_{\alpha\beta}\epsilon^{\gamma\delta} = -\delta^{\gamma}_{\alpha}\delta^{\delta}_{\beta} + \delta^{\delta}_{\alpha}\delta^{\gamma}_{\beta}, \qquad \epsilon_{\dot{\alpha}\dot{\beta}}\epsilon^{\dot{\gamma}\dot{\delta}} = -\delta^{\dot{\gamma}}_{\dot{\alpha}}\delta^{\dot{\delta}}_{\dot{\beta}} + \delta^{\dot{\delta}}_{\dot{\alpha}}\delta^{\dot{\gamma}}_{\dot{\beta}}\,, \tag{A.70}$$

yielding the so-called Schouten identities

$$\epsilon_{\alpha\beta}\epsilon^{\beta\gamma} = \epsilon^{\gamma\beta}\epsilon_{\beta\alpha} = \delta^{\gamma}_{\alpha}, \tag{A.71}$$

$$\epsilon_{\alpha\beta}\epsilon_{\gamma\delta} + \epsilon_{\alpha\gamma}\epsilon_{\delta\beta} + \epsilon_{\alpha\delta}\epsilon_{\beta\gamma} = 0\,. \tag{A.72}$$

The same equations hold for dotted indices. To construct Lorentz invariant Lagrangians and observables, in particular, Lorentz vectors are obtained by introducing the sigma matrices $\sigma^{\mu}_{\alpha\dot{\beta}}$ and $\bar{\sigma}^{\mu\,\dot{\alpha}\beta}$ defined by

$$\sigma^0 = \bar{\sigma}^0 = \begin{pmatrix} 1 & 0 \\ 0 & 1 \end{pmatrix}, \qquad \sigma^1 = -\bar{\sigma}^1 = \begin{pmatrix} 0 & 1 \\ 1 & 0 \end{pmatrix},$$

$$\sigma^2 = -\bar{\sigma}^2 = \begin{pmatrix} 0 & -i \\ i & 0 \end{pmatrix}, \qquad \sigma^3 = -\bar{\sigma}^3 = \begin{pmatrix} 1 & 0 \\ 0 & -1 \end{pmatrix}. \tag{A.73}$$

Hence, eqn (A.73) is equivalent to $\sigma^{\mu} = (\mathbb{I}_{2\times 2}\,, \vec{\sigma})$, and $\bar{\sigma}^{\mu} = (\mathbb{I}_{2\times 2}\,, -\vec{\sigma})$, which can be related by

$$\sigma^{\mu}_{\alpha\dot{\alpha}} = \epsilon_{\alpha\beta}\epsilon_{\dot{\alpha}\dot{\beta}}\bar{\sigma}^{\mu\,\dot{\beta}\beta}\,, \qquad \bar{\sigma}^{\mu\,\dot{\alpha}\alpha} = \epsilon^{\alpha\beta}\epsilon^{\dot{\alpha}\dot{\beta}}\sigma^{\mu}_{\beta\dot{\beta}}\,,$$
$$\epsilon^{\alpha\beta}\sigma^{\mu}_{\beta\dot{\alpha}} = \epsilon_{\dot{\alpha}\dot{\beta}}\bar{\sigma}^{\mu\dot{\beta}\alpha}\,, \qquad \epsilon^{\dot{\alpha}\dot{\beta}}\sigma^{\mu}_{\alpha\dot{\beta}} = \epsilon_{\alpha\beta}\bar{\sigma}^{\mu\dot{\alpha}\beta}\,. \tag{A.74}$$

There is a one-to-one correspondence between each 2-spinor construction $V_{\alpha\dot{\beta}}$ and the associated Lorentz 4-vector V^{μ}, provided by the Infeld–van der Waerden symbols

$$V^{\mu} = \frac{1}{2}\bar{\sigma}^{\mu\dot{\beta}\alpha}V_{\alpha\dot{\beta}}\,, \qquad V_{\alpha\dot{\beta}} = V^{\mu}\sigma_{\mu\alpha\dot{\beta}}\,. \tag{A.75}$$

In particular, if V^{μ} is a real 4-vector, then $V_{\alpha\dot{\beta}}$ is Hermitian. Moreover, it is often useful to further simplify the notation by defining $V^{\dot{\alpha}\beta} = (V_{\alpha\dot{\beta}})^{*}$. In this notation, an Hermitian 2-spinor satisfies $V_{\alpha\dot{\beta}} = V^{\dot{\alpha}\beta}$. Then,

$$(V^{\intercal})_{\alpha\dot{\beta}} = V_{\beta\dot{\alpha}}\,, \qquad (V^{*})^{\dot{\alpha}\beta} = (V_{\alpha\dot{\beta}})^{*}\,, \qquad (V^{\dagger})_{\alpha\dot{\beta}} = (V^{\beta\dot{\alpha}})^{*} = (V^{*})_{\dot{\beta}\alpha}\,.$$

A Hermitian 2-spinor satisfies $V = V^{\dagger}$ or, equivalently, $V_{\alpha\dot{\beta}} = (V^{*})_{\dot{\beta}\alpha}$.

In addition, 2-spinors can be interpreted as 2×2 matrices. It is indeed convenient to define the following:

$$(V^{\intercal})_{\alpha}{}^{\beta} = V^{\beta}{}_{\alpha}\,, \quad (V^{*})_{\dot{\alpha}}{}^{\dot{\beta}} = (V_{\alpha}{}^{\beta})^{*}\,, \quad (V^{\dagger})^{\dot{\beta}}{}_{\dot{\alpha}} = (V_{\alpha}{}^{\beta})^{*} = (V^{*})_{\dot{\alpha}}{}^{\dot{\beta}}\,.$$

Note that the matrix transposition of $V_{\alpha}{}^{\beta}$ interchanges the rows and columns of W without modifying the relative heights of the α and β indices. Similar results hold for $V_{\alpha\beta}$ and $V^{\alpha\beta}$ by either lowering or raising the spinor indices.

For an anti-commuting two-component spinor ψ, the product $\psi^{\alpha}\psi^{\beta}$ is antisymmetric with respect to the interchange of the spinor indices α and β. Hence, it must be proportional to $\epsilon^{\alpha\beta}$. Similar conclusions hold for the corresponding spinor products with raised undotted indices and with lowered and raised dotted indices, respectively. Thus,

$$\psi^{\alpha}\psi^{\beta} = -\frac{1}{2}\epsilon^{\alpha\beta}\psi\psi\,, \qquad \psi_{\alpha}\psi_{\beta} = \frac{1}{2}\epsilon_{\alpha\beta}\psi\psi\,,$$
$$\psi^{\dagger\,\dot{\alpha}}\psi^{\dagger\,\dot{\beta}} = \frac{1}{2}\epsilon^{\dot{\alpha}\dot{\beta}}\psi^{\dagger}\psi^{\dagger}\,, \qquad \psi^{\dagger}_{\dot{\alpha}}\psi^{\dagger}_{\dot{\beta}} = -\frac{1}{2}\epsilon_{\dot{\alpha}\dot{\beta}}\psi^{\dagger}\psi^{\dagger}\,, \tag{A.76}$$

where $\psi\psi = \psi^{\alpha}\psi_{\alpha}$ and $\psi^{\dagger}\psi^{\dagger} = \psi^{\dagger}_{\dot{\alpha}}\psi^{\dagger\,\dot{\alpha}}$.

The van der Waerden symbols in eqns (A.75) provide antisymmetrised products, from the sigma matrices (Dreiner, Haber, and Martin, 2010):

$$(\sigma^{\mu\nu})_{\alpha}{}^{\beta} = \frac{i}{4}\left(\sigma^{\mu}_{\alpha\dot{\rho}}\bar{\sigma}^{\nu\dot{\rho}\beta} - \sigma^{\nu}_{\alpha\dot{\rho}}\bar{\sigma}^{\mu\dot{\rho}\beta}\right)\,, \tag{A.77}$$

$$(\bar{\sigma}^{\mu\nu})^{\dot{\alpha}}{}_{\dot{\beta}} = \frac{i}{4}\left(\bar{\sigma}^{\mu\dot{\alpha}\rho}\sigma^{\nu}_{\rho\dot{\beta}} - \bar{\sigma}^{\nu\dot{\alpha}\rho}\sigma^{\mu}_{\rho\dot{\beta}}\right)\,. \tag{A.78}$$

Now we can introduce the infinitesimal forms for the 4×4 Lorentz transformation matrix; the corresponding matrices R and $(R^{-1})^{\dagger}$ which transform the $D^{(\frac{1}{2},0)}$ and $D^{(0,\frac{1}{2})}$ spinors, respectively, are given by

$$R^{\mu}{}_{\nu} \approx \delta^{\mu}_{\nu} + \frac{1}{2}\left(\theta_{\alpha\nu}g^{\alpha\mu} - \theta_{\nu\beta}g^{\beta\mu}\right), \tag{A.79}$$

$$R \approx \mathbb{I}_{2\times 2} - \frac{1}{2}i\theta_{\mu\nu}\sigma^{\mu\nu}, \tag{A.80}$$

$$(R^{-1})^{\dagger} \approx \mathbb{I}_{2\times 2} - \frac{1}{2}i\theta_{\mu\nu}\bar{\sigma}^{\mu\nu}. \tag{A.81}$$

The inverses of these quantities are obtained up to the first order in θ by replacing $\theta \mapsto -\theta$ in the formulæ. Equations (A.80) and (A.81) yield

$$(R^{-1})_{\rho}{}^{\sigma} = \epsilon^{\sigma\alpha}R_{\alpha}{}^{\beta}\epsilon^{\beta\rho}, \qquad (R^{-1\dagger})^{\dot\rho}{}_{\dot\sigma} = \epsilon_{\dot\sigma\dot\alpha}\,(R^{\dagger})^{\dot\alpha}{}_{\dot\beta}\,\epsilon^{\dot\beta\dot\rho}. \tag{A.82}$$

These results prove the covariance of the spinor index raising and lowering properties of the epsilon symbols. The infinitesimal forms given by (A.79) and (A.81) imply that

$$R^{\dagger}\bar{\sigma}^{\mu}R = R^{\mu}{}_{\nu}\,\bar{\sigma}^{\nu}, \qquad R^{-1}\sigma^{\mu}(R^{-1})^{\dagger} = R^{\mu}{}_{\nu}\,\sigma^{\nu}. \tag{A.83}$$

Using the Lorentz transformation properties of the undotted and dotted two-component spinor fields, eqn (A.83) yields the proof that the spinor products $\xi^{\dagger}\bar{\sigma}^{\mu}\eta$ and $\xi\sigma^{\mu}\eta^{\dagger}$ transform as Lorentz 4-vectors.

The usual framework use in field theory regards a pure boost from the rest frame to a frame where $p^{\mu} = (E_{\mathbf{p}}\,,\,\mathbf{p})$, which corresponds to $\theta_{ij} = 0$, and $\zeta^{i} = \theta^{i0} = -\theta^{0i}$. The so-called mass-shell condition is satisfied: $p^{0} = E_{\mathbf{p}} = (\|\mathbf{p}\|^{2} + m^{2})^{1/2}$. The matrices $R_{\alpha}{}^{\beta}$ and $[(R^{-1})^{\dagger}]^{\dot\alpha}{}_{\dot\beta}$, which describe Lorentz transformations of spinor fields, are given, respectively, for the $D^{(1/2,0)}$ and $D^{(0,1/2)}$ representations by

$$\exp\left(-\frac{i}{2}\theta_{\mu\nu}J^{\mu\nu}\right) = \begin{cases} R = \exp\left(-\frac{1}{2}\vec{\zeta}\cdot\vec{\sigma}\right) = \sqrt{\dfrac{p\cdot\sigma}{m}}, \\ (R^{-1})^{\dagger} = \exp\left(\frac{1}{2}\vec{\zeta}\cdot\vec{\sigma}\right) = \sqrt{\dfrac{p\cdot\bar{\sigma}}{m}}, \end{cases} \tag{A.84}$$

where

$$\sqrt{p\cdot\sigma} = \frac{(E_{\mathbf{p}} + m)\,\mathbb{I}_{2\times 2} - \vec{\sigma}\cdot\mathbf{p}}{\sqrt{2(E_{\mathbf{p}} + m)}},$$
$$\sqrt{p\cdot\bar{\sigma}} = \frac{(E_{\mathbf{p}} + m)\,\mathbb{I}_{2\times 2} + \vec{\sigma}\cdot\mathbf{p}}{\sqrt{2(E_{\mathbf{p}} + m)}}. \tag{A.85}$$

According to (A.84), the spinor index structure of $\sqrt{p\cdot\sigma}$ and $\sqrt{p\cdot\bar{\sigma}}$ corresponds to that of $R_{\alpha}{}^{\beta}$ and $[(R^{-1})^{\dagger}]^{\dot\alpha}{}_{\dot\beta}$, respectively. Hence, the equations in (A.85) yield

$$\left[\sqrt{p\cdot\sigma}\,\right]_{\alpha}{}^{\beta} = \left[\sqrt{p\cdot\sigma\,\bar{\sigma}^{0}}\,\right]_{\alpha}{}^{\beta} = \frac{(p\cdot\sigma_{\alpha\dot\alpha})\bar{\sigma}^{0\,\dot\alpha\beta} + m\delta^{\beta}_{\alpha}}{\sqrt{2(E_{\mathbf{p}} + m)}}, \tag{A.86}$$

$$\left[\sqrt{p\cdot\bar{\sigma}}\,\right]^{\dot\alpha}{}_{\dot\beta} = \left[\sqrt{p\cdot\bar{\sigma}\,\sigma^{0}}\,\right]^{\dot\alpha}{}_{\dot\beta} = \frac{(p\cdot\bar{\sigma}^{\dot\alpha\alpha})\sigma^{0}_{\alpha\dot\beta} + m\delta^{\dot\alpha}_{\dot\beta}}{\sqrt{2(E_{\mathbf{p}} + m)}}, \tag{A.87}$$

since $\sigma^{0} = \bar{\sigma}^{0} = \mathbb{I}_{2\times 2}$.

A.7 The Supersymmetry Algebra

The two operators ϕ_a and ϕ_b in a graded Lie algebra satisfy

$$\phi_a\phi_b - (-1)^{|a||b|}\phi_b\phi_a = C^d_{ab}\phi_d\,,$$

where either $|a| = 0$ for even (bosonic) ϕ_a or $|a| = 1$ for odd (fermionic) operators and the C^d_{ab} denote the algebra structure constants. The Poincaré generators $J^{\mu\nu}$ in eqn (A.66), together with the P^μ, are bosonic generators. Nevertheless, in supersymmetry, fermionic generators Q^A_α and $\bar{Q}^\beta_{\dot{\alpha}}$ are introduced, respectively denoting elements of the $D^{(\frac{1}{2},0)}$ and $D^{(0,\frac{1}{2})}$ representations of the Lorentz group, and $A, B = 1,\ldots,\mathcal{N}$ label the number of supercharges (West, 1990; Wess and Bagger, 1992).

For $\mathcal{N} = 1$, the supersymmetry algebra reads

$$[M^{\mu\nu}, M^{\sigma\rho}] = i(M^{\mu\nu}\eta^{\nu\rho} + M^{\nu\rho}\eta^{\mu\sigma} - M^{\mu\rho}\eta^{\nu\sigma} - M^{\nu\sigma}\eta^{\mu\rho})\,, \tag{A.88}$$

$$[P^\mu, P^\nu] = 0 = [Q_\alpha, P_\mu] = \{Q_\alpha, Q^\beta\}\,, \tag{A.89}$$

$$[M^{\mu\nu}, P^\sigma] = i(P^\mu\eta^{\nu\sigma} - P^\nu\eta^{\mu\sigma})\,, \tag{A.90}$$

$$[Q_\alpha, M^{\mu\nu}] = (\sigma^{\mu\nu})_\alpha{}^\beta Q_\beta\,, \tag{A.91}$$

$$\{Q_\alpha, \bar{Q}_{\dot{\beta}}\} = 2(\sigma^\mu)_{\alpha\dot{\beta}}P_\mu\,. \tag{A.92}$$

The first three equations describe the usual Poincaré algebra.

A spinor Q_α transforms under an infinitesimal Lorentz transformation as

$$Q_\alpha \mapsto Q'_\alpha = (e^{-\frac{i}{2}\omega_{\mu\nu}\sigma^{\mu\nu}})_\alpha{}^\beta Q_\beta \approx \left(I - \frac{i}{2}\omega_{\mu\nu}\sigma^{\mu\nu}\right)_\alpha^{\ \beta} Q_\beta\,. \tag{A.93}$$

From the operator point of view, it transforms, by denoting $U = e^{-\frac{i}{2}\omega_{\mu\nu}M^{\mu\nu}}$, as

$$Q_\alpha \mapsto Q'_\alpha = U^\dagger Q_\alpha U \approx \left(I + \frac{i}{2}\omega_{\mu\nu}M^{\mu\nu}\right) Q_\alpha \left(I - \frac{i}{2}\omega_{\mu\nu}M^{\mu\nu}\right)\,. \tag{A.94}$$

When eqn (A.94) is compared to eqn (A.93), eqn (A.91) can be derived. Indeed,

$$Q_\alpha - \frac{i}{2}\omega_{\mu\nu}(\sigma^{\mu\nu})_\alpha{}^\beta Q_\beta = Q_\alpha - \frac{i}{2}\omega_{\mu\nu}(Q_\alpha M^{\mu\nu} - M^{\mu\nu}Q_\alpha) + \mathcal{O}(\omega^2), \tag{A.95}$$

which implies that

$$[Q_\alpha, M^{\mu\nu}] = (\sigma^{\mu\nu})_\alpha{}^\beta Q_\beta\,. \tag{A.96}$$

The commutator for the right-handed representation reads

$$[\bar{Q}^{\dot{\alpha}}, M^{\mu\nu}] = (\bar{\sigma}^{\mu\nu})^{\dot{\alpha}}{}_{\dot{\beta}}\bar{Q}^{\dot{\beta}}\,. \tag{A.97}$$

For more detail, the reader can see, for example, the references by West (1990), Wess and Bagger (1992), Bailin and Love (1994), and Dreiner, Haber, and Martin (2010). Moreover, the Clifford–Hopf algebra associated with the super-Poincaré algebra was formulated in the article by da Rocha, Bernardini, and Vaz Jr (2010).

Appendix B
List of Symbols

[1]The Greek letters and some special symbols (e. g. tensor product $\otimes$, etc.) can be found at the end of this list.

References

Abłamowicz, R. (1995). Construction of spinors via Witt decomposition and primitive idempotents: A review. In *Clifford Algebras and Spinor Structures* (ed. R. Abłamowicz and P. Lounesto), pp. 113–123. Kluwer, Amsterdam.

Abłamowicz, R., Gonçalves, I., and da Rocha, R. (2014). Bilinear covariants and spinor fields duality in quantum Clifford algebras. *Journal of Mathematical Physics*, **55**, 103501.

Abłamowicz, R., and Sobczyk, G. (eds) (2004). *Lectures on Clifford (Geometric) Algebras and Applications.* Birkhäuser, Berlin.

Ahlford, L. V. (1986). Möbius transformations in $\mathbb{R}^n$ expressed through 2×2 matrices of Clifford numbers. *Complex Variables, Theory and Application*, **5**(2–4), 215–224.

Ahluwalia, D. V., and Grumiller, D. (2005). Spin-half fermions with mass dimension one: Theory, phenomenology, and dark matter. *Journal of Cosmology and Astroparticle Physics*, **2005**(07), 012.

Ahluwalia, D. V., Lee, C.-Y., and Schritt, D. (2010). Elko as self-interacting fermionic dark matter with axis of locality. *Physics Letters B*, **687**(2–3), 248–252.

Anglès, P. (2008). *Conformal Groups in Geometry and Spin Structures.* Birkhäuser, Providence, RI.

Baez, J. (2002). The octonions. *Bulletin of the American Mathematical Society*, **39**(2), 145–205.

Bailin, D., and Love, A. (1994). *Supersymmetric Gauge Field Theory and String Theory.* Institute of Physics Publishing, Bristol.

Baylis, W. (1996). *Clifford (Geometric) Algebras with Applications in Physics, Mathematics and Engineering.* Birkhäuser, Boston, MA.

Baylis, W. (1998). *Electrodynamics: a Modern Geometric Approach.* Birkhäuser, Boston, MA.

Benn, I., and Tucker, R. (1987). *An Introduction to Spinors and Geometry with Applications in Physics.* Adam Hilger, Bristol.

Berkovits, N. (2004). Multiloop amplitudes and vanishing theorems using the pure spinor formalism for the superstring. *Journal of High Energy Physics*, **2004**(09), 047.

Berkovits, N., and Howe, P. S. (2002). Ten-dimensional supergravity constraints from the pure spinor formalism for the superstring. *Nuclear Physics B*, **635**(1–2), 75–105.

Bernardini, A. E., and da Rocha, R. (2012). Dynamical dispersion relation for ELKO dark spinor fields. *Physics Letters B*, **717**(13), 238–241.

Bethe, H. A., and Salpeter, E. E. (1957). *Quantum Mechanics of One- and Two-Electron Atoms.* Springer, Berlin.

Binz, E., de Gosson, M. A., and Hiley, B. J. (2013). Clifford algebras in symplectic geometry and quantum mechanics. *Foundations of Physics*, **43**(4), 424–439.

Birkhoff, G., and MacLane, S. (1997). *A Survey of Modern Algebras* (4th edn). Macmillan Publishing Company, New York, NY.

Bonora, L., and da Rocha, R. (2016). New spinor fields on lorentzian 7-manifolds. *Journal of High Energy Physics*, **2016**(39), 1–17.

Bonora, L., de Brito, K. P. S., and da Rocha, R. (2015). Spinor fields classification in arbitrary dimensions and new classes of spinor fields on 7-manifolds. *Journal of High Energy Physics*, **2015**(2).

Bourbaki, N. (1989). *Elements of Mathematics: Algebra I, Chapters 1–3*. Number 1. Springer, Berlin.

Brauer, R., and Weyl, H. (1935). Spinors in n dimensions. *American Journal of Mathematics*, **35**(1), 425–449.

Budinich, P. (2002). From the geometry of pure spinors with their division algebra to fermion physics. *Foundations of Physics*, **32**(9), 1347–1398.

Budinich, P., and Trautman, A. (1989). *The Spinorial Chessboard*. Springer, New York, NY.

Cartan, É. (1908). Nombres complexes. In *Encyclopédie des sciences mathématiques* (ed. J. Molk), *Vol. 1*. Gauthiers-Villars, Paris.

Cartan, É. (1937). *Leçons sur la theorie des spineurs*. Hermann, Paris. Reprinted as: *The Theory of Spinors*, Dover Publications, New York, NY, 1981.

Cavalcanti, R. T. (2014). Classification of singular spinor fields and other mass dimension one fermions. *International Journal of Modern Physics D*, **23**(14), 1444002.

Cavalcanti, R. T., da Silva, J. M. H., and da Rocha, R. (2014). VSR symmetries in the DKP algebra: The interplay between Dirac and ELKO spinor fields. *The European Physical Journal Plus*, **129**(246), 1–10.

Charlier, A., Bérard, A., Charlier, M.-F., and Fristot, D. (1992). *Tensors and the Clifford Algebra; Application to the Physics of Bosons and Fermions*. Marcek Dekker, Inc, New York.

Chevalley, C. (1954). *The Algebraic Theory of Spinors*. Columbia University Press, New York, NY. Reprinted as *The Algebraic Theory of Spinors and Clifford Algebras, Collected Works*, Vol. 4, Springer, Berlin, 1997.

Clifford, W. K. (1878). Applications of Grassmann's extensive algebra. *American Journal of Mathematics*, **1**(4), 350–358.

Clifford, W. K. (1882). On the classification of geometric algebras, pp. 397–401. Macmillan, London. In *Mathematical Papers* (ed. R. Tucker), Reprinted by Chelsea, New York, NY 1985.

Crawford, J. P. (1985). On the algebra of Dirac bispinors densities: Factorization and inversion theorems. *Journal of Mathematical Physics*, **26**(7), 1439–1441.

Crowe, M. J. (1994). *A History of Vector Analysis*. Dover Publications, New York, NY.

Crumeyrolle, A. (1990). *Orthogonal and Symplectic Clifford Algebras: Spinor Structures*. Kluwer, Dordrecht.

da Rocha, R., Bernardini, A. E., and da Silva, J. M. H. (2011). Exotic dark spinor fields. *Journal of High Energy Physics*, **2011**(4), 1–26.

da Rocha, R., Bernardini, A. E., and Vaz, J. Jr. (2010). κ-deformed Poincare algebras and quantum Clifford–Hopf algebras. *International Journal of Geometric Methods*

in *Modern Physics*, **7**(5), 821–836.
da Rocha, R., and da Silva, J. M. H. (2009). Elko spinor fields: Lagrangians for gravity derived from supergravity. *International Journal of Geometric Methods in Modern Physics*, **6**(3), 461–477.
da Rocha, R., and da Silva, J. M. H. (2010). Elko, flagpole and flag-dipole spinor fields, and the instanton Hopf fibration. *Advances in Applied Clifford Algebras*, **20**(1), 847–870.
da Rocha, R., and da Silva, J. M. H. (2014). Hawking radiation from Elko particles tunnelling across black-strings horizon. *EPL (Europhysics Letters)*, **107**(5), 50001.
da Rocha, R., Fabbri, L., da Silva, J. M. H., Cavalcanti, R. T., and Silva-Neto, J. A. (2013). Flag-dipole spinor fields in ESK gravities. *Journal of Mathematical Physics*, **54**(10), 102505.
da Rocha, R., and Rodrigues, W. A. Jr. (2006). Where are ELKO spinor fields in lounesto spinor field classification? *Modern Physics Letters A*, **21**(1), 65–74.
da Rocha, R., and Rodrigues, W. A. Jr. (2010). Pair and impair, even and odd form fields, and electromagnetism. *Annalen der Physik*, **19**(1–2), 6–34.
da Rocha, R., and Traesel, M. A. (2012). Generalized non-associative structures on the 7-sphere. *Journal of Physics: Conference Series*, **343**(1), 012026.
da Rocha, R., Traesel, M. A., and Vaz, J. Jr. (2012). Non-associativity in the Clifford bundle on the parallelizable torsion 7-sphere. *Advances in Applied Clifford Algebras*, **22**(3), 595–610.
da Rocha, R., and Vaz, J. Jr. (2006). Extended Grassmann and Clifford algebras. *Advances in Applied Clifford Algebras*, **16**(1), 103–125.
da Rocha, R., and Vaz, J. Jr. (2007). Isotopic liftings of Clifford algebras and applications in elementary particle mass matrices. *International Journal of Theoretical Physics*, **46**, 2464–2487.
da Rocha, R., and Vaz Jr, J. (2006). Clifford algebra-parametrized octonions and generalizations. *Journal of Algebra*, **301**(2).
da Rocha, R., and Vaz Jr, J. (2007). Conformal structures and twistors in the paravector model of spacetime. *International Journal of Geometric Methods in Modern Physics*, **4**(4), 547–576.
Darling, R. W. R. (1994). *Differential Forms and Connections*. Cambridge University Press, Cambridge.
de Andrade, M. A., Rojas, M., and Toppan, F. (2001). The signature triality of Majorana–Weyl space-times. *International Journal of Modern Physics A*, **16**(27), 4453–4479.
de Andrade, M. A., and Toppan, F. (1999). Real structures in Clifford algebras and Majorana conditions in any space-time. *Modern Physics Letters A*, **14**(26), 1797–1814.
Dirac, P. A. M. (1928). The quantum theory of the electron. *Proceedings of the Royal Society of London A: Mathematical, Physical and Engineering Sciences*, **117**(778), 610–624.
do Carmo, M. P. (1994). *Differential Forms and Applications*. Springer, Berlin.
Doran, C., and Lasenby, A. (2003). *Geometric Algebra for Physicists*. Cambridge University Press, Cambridge.

Dorst, L., Fontijne, D., and Mann, S. (2007). *Geometric Algebra for Computer Science.* Morgan Kaufmann, Burlington, MA.

Dorst, L., and Lasenby, J. (eds.) (2011). *Guide to Geometric Algebra in Practice.* Springer, Berlin.

Dreiner, H. K., Haber, H. E., and Martin, S. P. "Two-component spinor techniques and Feynman rules for quantum field theory and supersymmetry," Phys. Rept. **494** (2010) 1 [arXiv:0812.1594 [hep-ph]].

Edelen, D. G. B. (1985). *Applied Exterior Calculus.* John Wiley & Sons, New York, NY.

Fearnley-Sandre, D., and Stokes, T. (1997). Area in Grassmann geometry. *Lecture Notes in Computer Science*, **1360**(1), 141–170.

Figueiredo, V. L., de Oliveira, E. C., and Rodrigues, W. A. Jr. (1990). Covariant, algebraic and operator spinors. *International Journal of Theoretical Physics*, **29**(2), 371–395.

Fillmore, J. P., and Springer, A. (1990). möbius group over general fields using Clifford algebras associated with spheres. *International Journal of Theoretical Physics*, **29**(3), 225–246.

Fjelstad, P. (1986). Extending special relativity via the perplex numbers. *American Journal of Physics*, **54**(5), 416–422.

Flanders, H. (1963). *Differential Forms with Applications to the Physical Sciences.* Academic Press, New York, NY.

Frankel, T. (2012). *The Geometry of Physics: An Introduction* (3rd edn). Cambridge University Press, Cambridge.

Gallier, J. (1997). Clifford algebras, Clifford groups, and a generalization of the quaternions. arXiv:0805.0311 [math.GM].

Garling, D. J. H. (2011). *Clifford Algebras. An Introduction.* Cambridge University Press, Cambridge.

Gilbert, J. E., and Murray, M. A. M. (1991). *Clifford Algebras and Dirac Operators in Harmonic Analysis.* Cambridge University Press, Cambridge.

Goldstein, H., Poole, C. P., and Safko, J. L. (2001). *Classical Mechanics* (3rd edn). Addison-Wesley, Harlow.

Grassmann, H. (1844). *Die lineale Ausdehnungslehre.* Wiegand, Leipzig.

Grassmann, H. (1847). *Geometrische Analyse: Geknüpft an die von Leibniz erfundene geometrische Charakteristik.* Weidmannsche Buchhandlung, Leipzig.

Grassmann, H. (1862). *Die Ausdehnungslehre: Vollstandig und in strenger Form bearbeitet.* Enslin, Berlin.

Grassmann, H. (1894). *Hermann Grassmanns gesammelte Mathematische und physikalische Werke.* Druck und Verlag von B. G. Teubner, Leipzig. (3 volumes 1894–1911).

Greub, W. H. (1978). *Multilinear Algebra* (2nd edn). Springer, Berlin.

Gsponer, A., and Hurni, J.-P. (1994). Lanczos's equation to replace Dirac's equation?, pp. 509–512. SIAM, Philadelphia, PA. In Proceedings of the Cornelius Lanczos International Centenary Conference (ed. J. D. Brown, M. T. Chu, D. C. Ellison, and R. J. Plemmons).

Gunaydin, M., and Ketov, S. V. (1996). Seven-sphere and the exceptional $n = 7$ and $n = 8$ superconformal algebras. *Nuclear Physics B*, **467**(1–2), 215–246.

Gürsey, F. (1956). Correspondence between quaternions and four-spinors. *Review of the Faculty of Science, University of Istanbul, Series A*, **21**(1), 33–54.

Gürsey, F. (1958). Relation of charge independence and baryon conservation to Pauli's transformation. *Il Nuovo Cimento (1955-1965)*, **7**(3), 411–415.

Hamilton, W. R. (1866). *Elements of Quaternions.* Longmans Green, London.

Harvey, F. R. (1990). *Spinors and Calibrations.* Academic Press, New York, NY.

Hasiewicz, Z., and Kwasniewski, A. K. (1985). Triality principle and g2 group in spinor language. *Journal of Mathematical Physics*, **26**(1), 6–11.

Hehl, F. W., and Obukhov, Y. N. (2003). *Foundations of Classical Electrodynamics: Charge, Flux and Metric.* Birkhäuser, Berlin.

Helmstetter, J., and Micali, A. (2008). *Quadratic Mappings and Clifford Algebras.* Birkhäuser, Berlin.

Hestenes, D. (1966). *Spacetime Algebra.* Gordon and Breach, New York.

Hestenes, D. (1967). Real spinor fields. *Journal of Mathematical Physics*, **8**(4), 798 – 808.

Hestenes, D. (1991). The design of linear algebra and geometry. *Acta Applicandae Mathematica*, **23**(1), 65–93.

Hiley, B. J. (2011). Process, distinction, groupoids and Clifford algebras: An alternative view of the quantum formalism. *Lecture Notes in Physics*, **813**(1), 705 – 750.

Hladik, J. (1999). *Spinors in Physics.* Springer, Berlin.

Hoffman, K., and Kunze, R. (1971). *Linear Algebra* (2nd edn). Pearson, London.

Holland, P. R. (1986). Relativistic algebraic spinors and quantum motions in phase space. *Foundations of Physics*, **16**(2), 701–719.

Ivanova, T. (1993). Octonions, self-duality and strings. *Physics Letters B*, **315**(3), 277–282.

Keller, J. (1997). Spinors, twistors, mexors and the massive spinning electron. *Advances in Applied Clifford Algebras*, **7**, 439 – 455.

Klotz, F. S. (1974). Twistors and the conformal group. *Journal of Mathematical Physics*, **15**, 2242–2247.

Knus, M.-A. (1998). *The Book of Involutions.* Vol. 44, American Mathematical Society Colloquium Publications.

Kostrykin, A. I., and Manin, Yu. I. (1997). *Linear Algebra and Geometry.* Gordon and Breach, New York, NY.

Lam, T. Y. (1980). *The Algebraic Theory of Quadratic Forms.* Benjamin, Reading, MA.

Laufer, A. (1997). *The Exponential Map, Clifford Algebras, Spin Representations and Spin Gauge Theory of $U(1)\times T^4$.* Phd thesis, Universität Konstanz, Konstanz.

Lawson, H. B., and Michelson, M.-L. (1990). *Spin Geometry.* Princeton University Press, Princeton, NJ.

Lazaroiu, C. I., Babalic, E. M., and Coman, I. A. (2013). The geometric algebra of Fierz identities in arbitrary dimensions and signatures. *Journal of High Energy Physics*, **2013**(9), 1–74.

Lounesto, P. (1996). Counter-examples in Clifford algebras. *Advances in Applied Clifford Algebras*, **6**(1), 69–104.

Lounesto, P. (2001*a*). *Clifford Algebras and Spinors* (2nd edn). Cambridge University Press, Cambridge.

Lounesto, P. (2001*b*). Octonions and triality. *Advances in Applied Clifford Algebras*, **11**(2), 191–213.

Lounesto, P., and Wene, G. P. (1987). Idempotent structure of Clifford algebras. *Acta Applicandae Mathematica*, **9**(3), 165–173.

Majorana, E. (1932). Oriented atoms in a variable magnetic field. *Il Nuovo Cimento*, **9**(1), 43–50.

Maks, J. (1989). *Modulo (1,1) Periodicity of Clifford Algebras and Generalized (anti)Möbius Transformations*. Phd thesis, Technische Universiteit Delft, Delft.

Meinrenken, E. (2013). *Clifford Algebras and Lie Theory*. Springer, Berlin.

Misner, C. W., and Wheeler, J. A. (1957). Classical physics as geometry. *Annals of Physics*, **2**(1), 525–603.

Mohapatra, R. N. (2003). *Unification and Supersymmetry: The Frontiers of Quark-Lepton Physics* (3rd edn). Springer, Berlin.

Mosna, R. A., Miralles, D., and Vaz Jr, J. (2003). z_2-gradings of Clifford algebras and multivector structures. *Journal of Physics A: Mathematical and General*, **36**(15), 4395–4405.

O'Neill, B. (2006). *Elementary Differential Geometry* (revised 2nd edn). Academic Press, New York, NY.

Pauli, W. (1927). Zur Quantenmechanik des magnetischen elektrons. *Zeitschrift für Physik*, **43**(1), 601–623.

Penrose, R. (1967). Twistor algebra. *Journal of Mathematical Physics*, **8**(2), 345–366.

Penrose, R., and Rindler, W. (1984). *Spinors and Space-Time: Volume 1, Two-Spinor Calculus and Relativistic Fields*. Cambridge University Press, Cambridge.

Perwass, C. (2008). *Geometric Algebra with Applications in Engineering*. Springer, Berlin.

Plymen, R. J., and Robinson, P. L. (1990). *Spinors in Hilbert Spaces*. Cambridge University Press, Cambridge.

Porteous, I. R. (1969). *Topological Geometry*. Cambridge University Press, Cambridge.

Porteous, I. R. (1995). *Clifford Algebras and the Classical Groups*. Cambridge University Press, Cambridge.

Rainich, G. Y. (1925). Electrodynamics in the general relativity theory. *Transactions of the American Mathematical Society*, **27**(1), 106–136.

Riesz, M. (1993). *Clifford Numbers and Spinors*. Kluwer, Dordrecht.

Rodrigues, W. A. Jr, da Rocha, R., Bernardini, A. E., and Vaz, J. Jr. (2005). Hidden consequence of active local Lorentz invariance. *International Journal of Geometric Methods in Modern Physics*, **2**(22), 305–357.

Rodrigues, W. A. Jr, and de Oliveira, E. C. (2007). *The Many Faces of Maxwell, Dirac and Einstein Equations: A Clifford Bundle Approach*. Volume 722, Lecture Notes in Physics. Springer, Berlin.

Rogers, A. (2007). *Supermanifolds: Theory and Applications*. World Scientific, Singapore.

Rotman, J. J. (2000). *Advanced Modern Algebra*. American Mathematical Society, Providence, RI.

Schafer, R. D. (1954). On the algebras formed by the Cayley-Dickson process. *American Journal of Mathematics*, **76**(2), 435–446.

Snygg, J. (1997). *Clifford Algebra: A Computational Tool for Physicists.* Oxford University Press, Oxford.

Snygg, J. (2010). *A New Approach to Differential Geometry Using Clifford's Geometric Algebra.* Oxford University Press, Oxford.

Sohnius, M. F. (1985). Introducing supersymmetry. *Physics Reports*, **128**(2–3), 39–204.

Srednicki, M. (2007). *Quantum Field Theory.* Cambridge University Press, Cambridge.

Srivastava, P. P. (1986). *Supersymmetry, Superfields and Supergravity: An Introduction.* Adam Hilger, Bristol.

Takahashi, Y. (1982). Reconstruction of a spinor via Fierz identities. *Physical Review D*, **26**(8), 2169–2171.

Todorov, I. (2011). Clifford algebras and spinors. *Bulgarian Journal of Physics*, **38**(1), 3–28.

Vahlen, K. T. (1902). Über Bewegungen und complexe Zahlen. *Mathematische Annalen*, **55**(1), 585–593.

van der Waerden, B. L. (1928). Spinoranalyse. *Nachrichten von der Gesellschaft der Wissenschaften zu Göttingen, Mathematisch-Physikalische Klasse*, **1929**(1), 100–109.

Vaz, J. Jr. (1998). Construction of monopoles and instantons by using spinors and the inversion theorem. In *Clifford Algebras and Their Application in Math. Physics* (ed. V. Dietrich, K. Habetha, and G. Jank), Fundamental Theories of Physics **94**, pp. 401–421. Kluwer, Amsterdam.

Vaz, J. Jr. (2013). A Clifford algebra approach to the classical problem of a charge in a magnetic monopole field. *International Journal of Theoretical Physics*, **52**(5), 1440–1454.

Vaz Jr, J., and Rodrigues Jr., W. A. (1993). Equivalence of Dirac and Maxwell equations and quantum mechanics. *International Journal of Theoretical Physics*, **32**(6), 945–959.

Veblen, O. (1933). Geometry of two-component spinors. *Proceedings of the National Academy of Sciences*, **19**(4), 462–474.

Villalobos, C. H., da Silva, J. M. H., and da Rocha, R. (2015). Questing mass dimension 1 spinor fields. *The European Physical Journal C*, **75**(6), 1–7.

Wess, J., and Bagger, J. (1992). *Supersymmetry and Supergravity.* Princeton University Press, Princeton.

West, P. (1990). *Introduction to Supersymmetry and Supergravity* (2nd edn). World Scientific, Singapore.

Winitzki, S. (2010). Linear algebra via exterior products. https://sites.google.com/site/winitzki/linalg, accessed 2 December, 2015.

Zwiebach, B. (2001). *A First Course in String Theory.* Cambridge University Press, Cambridge.

Index

The manufacturer's authorised representative in the EU for product safety is Oxford University Press España S.A. of El Parque Empresarial San Fernando de Henares, Avenida de Castilla, 2 - 28830 Madrid (www.oup.es/en or product.safety@oup.com). OUP España S.A. also acts as importer into Spain of products made by the manufacturer.

Printed and bound by CPI Group (UK) Ltd, Croydon, CR0 4YY

06/07/2026

02157632-0012